𝕮𝖆𝖒𝖇𝖗𝖎𝖉𝖌𝖊 𝕲𝖊𝖔𝖌𝖗𝖆𝖕𝖍𝖎𝖈𝖆𝖑 𝕾𝖊𝖗𝖎𝖊𝖘.

GENERAL EDITOR: F. H. H. GUILLEMARD, M.D.,

FORMERLY LECTURER IN GEOGRAPHY IN THE UNIVERSITY OF CAMBRIDGE.

W0042102

THE

GEOGRAPHY OF DISEASE

THE

GEOGRAPHY OF DISEASE

BY

FRANK G. CLEMOW, M.D. EDIN., D.P.H. CAMB.,

BRITISH DELEGATE TO THE OTTOMAN BOARD OF HEALTH,
PHYSICIAN TO H. M. EMBASSY AT CONSTANTINOPLE,
SECRETARY FOR RUSSIA TO THE EPIDEMIOLOGICAL SOCIETY OF LONDON,
HON. MEMBER OF THE RUSSIAN NATIONAL HEALTH SOCIETY, ETC.

CAMBRIDGE:
AT THE UNIVERSITY PRESS.

1903

CAMBRIDGE UNIVERSITY PRESS
Cambridge, New York, Melbourne, Madrid, Cape Town,
Singapore, São Paulo, Delhi, Tokyo, Mexico City

Cambridge University Press
The Edinburgh Building, Cambridge CB2 8RU, UK

Published in the United States of America by
Cambridge University Press, New York

www.cambridge.org
Information on this title: www.cambridge.org/9781107600300

First published 1903
First paperback edition 2011

A catalogue record for this publication is available from the British Library

ISBN 978-1-107-60030-0 Paperback

Additional resources for this publication at www.cambridge.org/9781107600300

TO MY MOTHER

PREFACE.

THE study of the manner in which disease is spread over the earth's surface, or in other words the geography of disease, is a science of comparatively new growth, and it may be regarded as still in its infancy. It is only, indeed, within quite recent times that treatment of the subject in at all a thorough manner has become possible. Authentic materials for a study of the kind were formerly not available, and it was not until the latter half of the nineteenth century—marked as it has been by an enormous increase in the means and rapidity of travel, by a vastly greater intercommunication of nation with nation, by the opening up of great tracts of the earth's surface hitherto unexplored, and finally by an immense increase in our knowledge of the causation and mode of spread of the maladies affecting mankind—that an attempt to consider the distribution of disease the world over became feasible.

Still more recently the demonstration of the origin and manner of diffusion of malaria—giving rise, as it has, to the hope of removing from mankind one of the worst of disease-scourges—has given an immense impetus to the study of geographical pathology. The brilliant work of British and Italian observers in this connection has in no country met with more practical recognition than in the British Isles, whose example in founding schools of tropical medicine and journals dealing with the same subject has been followed by many other European nations.

The revival—if it be not rather a new birth—of interest in the distribution of diseases, and the recognition of the extreme importance of the study, owe not a little to the spirit of colonial expansion which has marked the history of more than one nation in the latter years of the nineteenth century. In no nation has this spirit been so active or led to such eminently practical results as in our own country. England has slowly awakened to a sense of her splendid position as the greatest colonising force, and as possessing the greatest colonial empire the world has ever seen ; and with a recognition of her greatness has come a recognition of her responsibility. In the interests of the colonies in all parts of the globe, and of the navy which connects and guards the whole, a knowledge of the diseases met with in various climes and in various countries, and of the best methods of dealing with them, is of the first importance ; and the necessity of a special training in this subject for the large number of medical graduates who annually leave the universities and schools at home to practise under the most widely differing conditions in the colonies abroad is no less obvious. Thanks not only to private initiative but also to the wise statesmanship of the present Secretary of State for the Colonies, provision is now made for this special training, and means afforded for a practical study of the subject, to one aspect of which this present book is devoted.

Nor have other nations been behindhand in this work. Hirsch's monumental volumes on Geographical and Historical Pathology contributed greatly to elevate this study as a science, and to bring it on a level with the other medical sciences. Many French, German, and Italian authors have written books dealing with the subject which have taken their places as classics. A number of journals in those languages are now published, and in the columns of such papers as the *Archives de Médecine Navale*, the *Archiv für Schiffs- und Tropen-Hygiene*, the *Annali di Medicina Navale*,

Janus, and others, an abundance of material for study of disease-distribution is available.

The principal sources of information upon which the present work is based are enumerated in a list which will be found at the end of the book. For the colonies I have consulted the latest returns in the Official Health Reports of each colony; and I am indebted to Dr Manson and the Colonial Office for facilities to study those reports in the library of the Office. I must also acknowledge my indebtedness to the libraries of the Royal Statistical Society, the Royal College of Physicians, and the Royal College of Surgeons. Of the many books from which I have borrowed information the most valuable have been those dealing with the special subject of Geographical Pathology, such as the writings of Hirsch, Davidson, Manson, Lombard, and others; but much additional information has been obtained from recent books of travel, and from many official and unofficial publications dealing with special countries, in which the facts as to disease-prevalence have formed but a chapter, or even but a paragraph. Of the many medical journals from which much useful information has been extracted it will suffice to name, in addition to those mentioned above, the *Journal of Tropical Medicine*, the various Indian journals for India, the *Sei-i-Kwai* for Japan, the *Medical Reports of the Chinese Customs Service* for China, and many American and colonial journals for the United States and the colonies. It has been impossible in the text to insert references for every fact stated. Where references are not given it may be accepted that the statement is based on one of the authorities named in the list at the end of the book.

A few words must be said as to the classification of diseases adopted in the following pages. In a work on systematic Medicine some attempt to group diseases on a scientific basis is of real importance. In a work such as the present, which deals primarily with one aspect only of

disease—its relation, that is to say, to the earth's surface—
a purely scientific basis of classification is less necessary.
The book will, it is hoped, prove useful as a work of
reference; and in such works some convenient arrangement
of the material, to the end that any fact looked for may be
rapidly and easily found, is of the first importance. "To
say the truth" (as Sir Thomas Watson wrote in his
classical Lectures) "I shall consider convenience and use-
fulness in framing my plan, rather than an appearance of
scientific precision." The only attempt made in these
pages to classify on an etiological or other basis will
be found in the separation of certain diseases of the
skin and those due to animal parasites from the large
group of other diseases. The last, forming the greater
part of the book, are grouped alphabetically. When
a malady is known by more than one name, the simplest
name or the one most frequently used has been chosen to
determine its alphabetical position. While this arrangement has certain obvious disadvantages—as, for example,
the apparently greater prominence given to ainhum over
plague or malaria, and the grouping together of diseases so
wholly unlike as urinary calculus and scarlet-fever, typhus-
fever and gout, cancer and pneumonia—it has the great
compensating advantage of facilitating reference.

In the following chapters the greatest prominence is
of course given to the facts of the geographical distribution of each disease over our globe. In most instances
only a very brief account of the general characters and
etiology of the disease has been thought necessary, except
in the case of certain rare or little-known affections. The
history of each malady also, in most of the chapters, is little
more than a very brief summary. Only in the case of those
whose distribution has varied widely at different times (as
for example cholera, plague, and influenza), and where such
knowledge is essential to a complete understanding of the
relation of the disease to the earth's surface, is the history

given in any detail. Finally, after a statement of the known facts as to the distribution of the particular disease in recent times, the factors which have seemed to determine that distribution are discussed as fully as space will allow. The systematic arrangement of the material followed throughout the book has seemed to render an index unnecessary.

This book was begun in London, within easy access of splendid medical and general libraries, and of every authority upon its subject-matter that it was desirable to consult. It has been completed in a foreign country where no such facilities exist. In correcting the proofs I have consequently been unable in many instances to compare the facts and figures contained in them with the original sources from which they are quoted, but solely with my MS. notes and abstracts from those sources. Should any errors thus have crept in I can only trust that they are not of great magnitude, or such as to affect any conclusions that I may have based upon them.

F. G. C.

CONSTANTINOPLE,
   March, 1903.

TABLE OF CONTENTS.

BOOK II.

DISEASES OF THE SKIN.

BOOK III.

ANIMAL PARASITES AND THE DISEASES ASSOCIATED WITH THEM.

MAPS AND CHARTS.

BOOK I.

GENERAL MEDICAL AND SURGICAL DISEASES.

INTRODUCTION

THE GEOGRAPHY OF DISEASE, AND THE FACTORS WHICH DETERMINE IT.

THE study which forms the subject of this volume differs in many respects from a study of the geographical distribution of more concrete objects, such as certain groups of animals, plants, or minerals. Though constantly spoken of as if it were a material, tangible entity, disease is, in fact, no such thing. It is only a morbid phenomenon, or rather a group of morbid processes, in the tissues of a particular animal organism. In the language of logic, it is not even a phenomenon, but an epiphenomenon.

It is only in that class of diseases known or believed to be of parasitic origin that there exists, in addition to the group of intangible signs and symptoms which in ordinary language constitute the disease, a tangible, palpable something, the distribution of which over the earth's surface may be justly compared with the distribution of mammals or insects, herbs or trees. But, on the one hand, it has not yet been shown that all, or nearly all, diseases are of this character; and, on the other hand, even where the parasitic origin of a human disease has been proved, it is well to bear in mind that the parasite and the disease are not one and the same thing, nor is their geographical distribution always or of necessity identical. There is good reason to believe, for example, that the parasites which are the known or suspected causes of such diseases as cholera, blackwater fever, malaria, guinea-worm disease, hydatids, and perhaps enteric and other fevers, may and do exist for long periods together outside the human body, and that there are uninhabited or sparsely inhabited tracts of the earth's surface where these parasites remain in the soil, or in water, or in the bodies of the lower animals, and where the human disease associated with them is only set up when man visits those tracts. In other words, the area of distribution of some disease parasites may be wider than the area of distribution of the human disease caused by them. In these instances the disease has a wider distribution potentially than actually. In others the reverse may be the case, and the area of distribution of the disease may at any given moment be actually wider than that of the parasite which gave rise to it. This, however, is exceptional, and can only occur in the case of certain affections of long duration, such, for example, as elephantiasis arabum, where the symptoms of the disease remain long after the filaria or other parasite which first caused them has disappeared from the tissues.

Any deviation from the normal, physiological processes of health comes into the category of disease. Such deviations are as numerous in kind and causation as they are in degree. A large proportion of them are due to local, temporary, individual causes. Of the great group of diseases usually classed as surgical, perhaps the majority would come under this definition. Fractures and dislocations, burns and scalds, sprains and bruises, and generally the class of injuries due to "accidents" are all of this character, and a discussion of their geographical distribution would serve no useful end. On the other hand there is a large group of diseases, often classed as surgical, which are due, not to causes of the local or transient character just alluded to, but to some more general cause. Such diseases, for example, as "cancer," certain forms of ulcer, urinary calculus, mycetoma, ainhum, trachoma and other eye diseases, though frequently needing the aid of the surgeon in their treatment, are truly on the border-land of medicine and surgery. They show more or less definite relations to external influences, they occur in more or less definite geographical areas, and hence a study of their geographical distribution not only offers many points of interest, but is essential to the solution of problems which the often obscure etiology of such diseases presents.

While on the one hand, therefore, only a few of the diseases commonly classed as "surgical" will come within the scope of the present study, a very large number, on the other hand, of the great group of "medical" diseases will require consideration. Even of these, however, not a few are due to other than general causes, or they present problems of comparatively little interest to the student of geographical pathology, and will consequently not be discussed here.

In the convenient, though not wholly satisfactory classification of diseases usually adopted, it is customary to group certain disorders together, such as the infective fevers, diseases due to animal parasites, constitutional diseases and the like, and to place the remainder in groups of diseases of the various "systems," such as the alimentary, the respiratory, or the circulatory system. As knowledge advances the tendency is for these latter groups to become smaller and smaller. Many diseases which were formerly classified as affections of the respiratory system (as, for example, diphtheria, pneumonia, and pulmonary consumption), of the alimentary system (as cholera and dysentery), or of the nervous system (as cerebro-spinal meningitis) are now known or suspected to be due to a specific infection, and are removed from the general class to some more definite class, as that of the infective fevers or infective granulomata. As such, their geographical distribution presents points of the greatest interest, and will be discussed in these pages. But there still remains a considerable number of affections which must, for various reasons, be passed over here. Some—for example, many diseases of the genito-urinary and nervous systems—must be omitted because their distribution, although presenting points of interest, is too imperfectly known to be of value. Others must be passed over, because their distribution, although more or less known, offers no special points of interest, or because the conclusions to be drawn from its study are quite disproportionate to the large amount of space needed to fully set it forth.

A study of the geography of disease would serve but little purpose without constantly bearing in mind the causation of the various maladies dealt with. In no branch of Medicine has knowledge advanced more rapidly in modern times than in etiology. With improved powers of the microscope; with the birth and development of the science of bacteriology; with an ever-growing list of discoveries of parasitic infection in various diseases; with increased knowledge of the relation of disease to surrounding conditions, and its possible carriage by such media as air, water, milk, dust, fomites; and finally with a wealth of new observations upon the part played by insects and by certain of the lower animals in disseminating infection—much that was formerly obscure and uncertain in the causation and mode of spread of disease has been made clear.

A large number of maladies have now been shown, beyond the reach of doubt, to be associated with the presence in the body of certain parasites, belonging either to the animal world, or to the still lower world of microscopic fungi. A still larger and ever-increasing group of diseases are believed to be produced in the same manner, though the specific organisms associated with them have not as yet been identified. To these groups of diseases the term parasitic, whether macro-parasitic, or micro-parasitic, may be conveniently applied. Until within the last few decades only a small number of diseases—such as scabies, and the various affections brought about by the intestinal worms, by the Guinea-worm and other comparatively large animal organisms—were regarded as in any sense parasitic. The great majority of maladies were believed to be due to such causes as excess or diminution of some chemical constituent of the blood or tissues, or to some change in the chemico-vital processes of the body, brought about by errors in diet, by chill, or some other similar, non-living, agent. It is still permissible-to suppose that some diseases, as, for example, diabetes, myxœdema, and perhaps scurvy, rickets, and gout, are of this character. But modern research has not only shown that the large group of infective or zymotic fevers are as truly of parasitic origin as the itch or the fluke-disease, but it has also shown that such superficially different diseases as elephantiasis and tetanus, leprosy and pneumonia, chyluria and malaria, endemic hæmaturia and pulmonary consumption, are no less truly caused by parasites.

This process of gradually transferring diseases from the group of affections of chemical or chemico-vital causation to the group of parasitic diseases is constantly going on. A few years ago no one would have ventured to regard pneumonia or rheumatic fever as other than simple inflammatory processes, the result of chill. But the one has been practically proved, and the other is suspected, to be due to the agency of micro-parasites. It is, further, believed by many that such diseases as "cancer," scurvy, rickets, gout, urinary calculus, and a host of others, will ultimately prove to be of microbial origin. Some observers have even hazarded the suggestion that practically all diseases which are not due to such obvious external causes as poison, fire and frost, lightning-stroke or traumatic injury will be found to be due to parasites.

This view is an extreme one, yet not a little might be said in its favour. Disease is essentially a vital process—a deranged action, it is true, of living tissues, leading sometimes to their death, but, even then, still a living, vital process. The proof, then, if that were possible, that it is in all or most cases set up by the action of a living thing, would only harmonise with what we already know of its nature and results. Moreover there are very few diseases that are certainly known to be brought about by the sole, unaided action of non-living matter on the human frame. Apart from the mineral poisons, and certain harmful gases and irritating dusts, the inanimate surroundings of man seem to have little power of producing disease in him, however potent some of them may be for traumatic injuries. Even the irritating dusts (such as those which set up certain well-known trade diseases) can scarcely be regarded as the sole agent in causing the disease, but rather as preparing the tissues for the action of harmful micro-organisms. But the case is different with the living surroundings of man. The human animal passes his existence, as it were, in a medium of minute living organisms. The dust in the air he breathes is loaded with them ; the water he drinks often swarms with them ; the earth he walks on, the food he eats, the clothes he wears, all contain and bring into immediate contact with his body countless numbers of such organisms. Fortunately they are imperceptible to the unaided senses, and still more fortunately the vast majority are harmless. Some are, perhaps, beneficial, and may aid in the natural functions of the human body. Others are, and perhaps always have been, quite innocuous. Others again, there is reason to believe, were originally injurious and capable of disturbing the normal processes of healthy life in man, thus producing disease, but, as generation after generation has gone by, man has acquired an immunity to their attacks. In regard to many such micro-organisms the immunity acquired is, it may be supposed, not merely an individual immunity, but a true racial immunity, transmissible from generation to generation, and truly permanent so long as man shall continue to live in an atmosphere of these particular organisms.

There are other micro-parasites which appear to be harmless to man so long as he remains in health, but attack him and set up disease as soon as his powers of resistance are in any way weakened. Whether by original endowment, or by gradual adaptation to the needs of his existence, certain protective forces are now known to exist in the blood and tissues of the human body, which combat and destroy any harmful micro-parasites that may enter it. These forces are believed to be partly of the nature of cells (phagocytes), partly of the nature of chemical substances. Thousands of harmful micro-parasites may enter the body with each breath or with each mouthful, yet, provided the protective forces are present in sufficient number and activity, the invading parasites are destroyed or rendered harmless. But should these forces be diminished in number or activity, as the result of depressing influences, general or local disease is set up. Some of these parasites appear to be widely diffused all around us; others are found only in the more or less near neighbourhood of persons or animals actually suffering from the disease they produce. As examples of the former class may be named the various pyogenic organisms. These exist, there is every reason to believe, in considerable numbers, in the air or soil or dust, in most parts of the world, and must constantly gain access to the healthy human body; yet they do not set up suppuration unless and until they reach a tissue the vitality of which has been already lowered by chill or injury. In this group also may be placed some other pathogenic organisms, as those of tetanus and, perhaps, pneumonia, nasal catarrh, bronchitis, pleurisy, and others.

Finally there is a group of nocuous micro-organisms—including the larger number of pathogenic organisms—which are fortunately not found in abundance all around us, but only in the neighbourhood of a patient actually suffering from the disease they produce. Some of these parasites, as, for example, those of certain of the infective fevers, and the organism (not yet finally identified) of syphilis, seem to be nocuous to most people, no matter what their previous state of health, provided only that they are actually introduced into the tissues, and that an immunity has not been acquired by natural or artificial means. Others, as those of measles and scarlet-fever, are specially nocuous to young children. Some affect particular races. Some set up an acute febrile attack of short duration, while others are the cause of prolonged, chronic suffering and deformity.

Between the macro-parasites and the micro-parasites found

in the human body, there is no true dividing line. The local irritation and general reaction set up by the presence in the body of a number of trichinæ spirales, or of the echinococci of hydatid disease, find their analogues in the ulcers of the intestine and the febrile attack set up by the bacillus of typhoid, and in the bubo and the acute fever and delirium of bubonic plague. It would seem that a large proportion of the lower animal creation is capable of inflicting injury on man. The mode of inflicting the injury differs widely. Some animals have teeth and claws and hoofs, with which they bite and scratch and kick. Others, such as many snakes, the scorpion, tarantula, mosquito, flea, and bed-bug can not only puncture or tear the skin but can at the same time inject a poisonous substance, which may merely cause local irritation or may be absorbed and produce general symptoms and death. Others actually penetrate and take up their abode in man's tissues, in which they spend the whole or a portion of their existence.

The term parasite may be held to cover not only the last-named group, but also such organisms as lice, fleas, and bugs, which do not themselves enter the tissues, but live on the surface of the body and obtain nutriment from it. Closely analogous to these are the intestinal parasites, such as the tape-worms, the round, whip, and thread worms, which live in the intestinal canal, and, as they do not penetrate to the tissues, may in a sense be regarded as external parasites. A more serious form of parasitism is that in which the parasite actually enters the tissues. Internal parasites of this kind belong to very different zoological groups. The highest are those of the *Insecta* class, such as the chigo, or pulex penetrans. Slightly lower are the parasites of the *Arachnida* class, such as the acarus of itch, and the demodex folliculorum. Others belong to the sub-kingdom *Annuloida*, including the parasite of hydatid disease, the Guinea-worm, the blood filariæ, and the many other, nematode and trematode, worms which will be discussed in later chapters. Still lower in the scale of creation are the amœbæ found in some cases of dysentery and in the contents of some liver abscesses; the protozoal organism of malaria; and the large group of schizomycetes, or bacterial parasites, the true zoological position of which is still uncertain.

It would be beyond the scope of this volume to discuss in detail the exact manner in which the micro-parasite produces the tissue changes which we call the disease. It is certain that in many, it is possible that in all diseases of the acute infective class, the multiplication of the micro-organism is associated with the production of poisonous substances, or toxins, the absorption of which leads to the febrile reaction and the other general symptoms of the disease. The ultimate explanation of all disease, whether acute or chronic, infective or non-infective, is still to be worked out by the joint labours of the physicist, the chemist, and the pathologist. The molecular repulsions and attractions of different cells or varieties of protoplasmic substances, the chemical combinations and decompositions associated with morbid cell-changes, the altered functional activity of cells or tissues or organs are all concerned in the production of the simplest disease. *A priori*, nothing is perhaps more remarkable than that some of the deadliest of human diseases should be due to the action of the smallest and simplest of living creatures—organisms consisting of a single cell of protoplasm, apparently differing in no way from the protoplasm which is the ultimate constituent of the cells of the human body itself. Yet this is no more remarkable than that the union of a single atom of hydrogen, a single atom of carbon and a single atom of nitrogen—that is to say, of three out of the four main chemical constituents of the human body—should result in the formation of prussic acid, one of the most deadly of known poisons, the absorption of which into the human body causes instant death. That there is any true analogy between the two phenomena it would perhaps be premature to suggest, but the resemblance is at least a striking one.

It is unnecessary to dwell at greater length upon the subject of parasitology, and the closely allied subjects of immunity and the development of pathogenic from harmless organisms. For a fuller consideration of these questions the reader is referred to other text-books devoted to such subjects. No discussion on the general causation of disease would have been complete without some account, however brief, of what this comparatively new science has taught us. It has discovered with certainty the cause of many diseases, and it has shown in what direction the cause of many others must be sought. Whether it is destined to explain the causation of all disease and to reign supreme over the whole domain of pathology, the future alone can show. For the present it would seem permissible to retain the belief that at least some diseases are due to causes other than parasitic in their nature. Some—as for example myxœdema, and, as many believe, exophthalmic goitre—are almost certainly due to a diminution or excess in the tissues of certain substances produced by the action of a particular gland or set of glands ; and there is nothing at present to show that the changes in the activity of these glands, leading to such diminution or excess, is in any way due to parasitic agency. Other diseases, such as epilepsy, some forms of rheumatism, migraine, neuralgia, atheroma of the arteries, paralysis agitans, aneurisms of non-syphilitic origin, locomotor ataxia, chronic Bright's disease, waxy kidney, lymphadenoma, Raynaud's disease and many others, seem to be due to disturbances in the chemico-vital processes of the body or of special tissues, in the causation of which there is no reason at present to suppose that living parasites have any share.

While the recognition of the parasitic origin of a large number of diseases has thrown a flood of light, not only upon their etiology, but upon their mode of spread and upon their relation to surrounding conditions, it has not yet explained all that is obscure in the natural history of disease. Even where—as, for example, in the case of cholera or plague—the parasite has been found, has been shown to be constantly associated with the disease, and may, within certain limits, be spoken of as its cause, there are still a number of phenomena in the behaviour of the disease which require explanation. These phenomena are particularly found in connection with the class of epidemic infectious fevers.

The sudden and marked changes in the behaviour of the two diseases just named, the great variations in prevalence and intensity of a score of other disorders, the appearance of a disease in one country and its disappearance from another, all open up a number of interesting questions to which at present no quite satisfactory answer has been found. It has been suggested that all these phenomena find analogies in the insect world. The idea is no new one. The close resemblances between the behaviour of epidemic diseases and the flights and disappearances of insects was clearly recognised and discussed by many writers in the eighteenth century. Linnæus included in his published works several memoirs by others on the subject. One of the most interesting of the late Sir Henry Holland's essays discusses

the question at length. The knowledge gained since then by
the actual proof of the existence of micro-parasites as the
cause of disease—then only suspected—has scarcely altered the
character of the problem, though it has altered its terms. In
Linnæus's time, when an author hazarded the suggestion that
small-pox, measles, the plague and other diseases might be caused
by minute animals, he spoke of them as insects—"acari" of
different species. In Sir Henry Holland's time the microscope
had already shown the occurrence in nature of animals still
lower than insects—of infusoria, and similar minute, lowly
organisms, and it was to these that the causation of disease was
tentatively ascribed ; while already the probability that other
still lowlier organisms existed and would ultimately be found,
was fully recognised.

The still lowlier organisms have now been found, in regard at
least to many diseases. And with their discovery, the terms of
the problem, as just stated, have changed. It is now known that
these diseases are not due to insects, but to forms of life infinitely
lower—almost, it would seem, at the lowest end of the scale of
creation. It is now known that if the general behaviour of these
diseases presents an analogy with that of insects, it is only an
analogy. The resemblance, it is true, is often very striking. Some
insects, for example, like some diseases, have permanent breeding-
grounds in special and relatively limited areas. They are numerous
here in some years, rare in others. They may change their
breeding-grounds. They migrate in certain years to other areas ;
they fly in swarms or in small numbers ; they take long flights
over thousands of miles and appear in countries where they were
before unknown ; they disappear almost as suddenly as they
came, or they break up into patches, remain more or less permanently in the newly-invaded territory, or gradually die out.
In all these respects, their behaviour shows complete analogy
with the behaviour of such diseases as cholera or plague, influenza
or dengue.

But if the attempt be made to push the analogy too far, it
breaks down. Insects are highly organised creatures ; they are
adapted to a highly developed form of social existence ; they are
sentient; many of them have extraordinary powers of intelligence;
and finally they are endowed with powers of locomotion, enabling
them to pass over distances which in some instances may be

measured by hundreds of miles. But the lowly micro-organisms believed to be the causes of plague or cholera, of yellow-fever or diphtheria, have none of these characters. It is more especially in their powers of locomotion that they differ from insects. They have, indeed, alone and unaided, almost no powers of locomotion. The most that an individual micro-parasite can do in this respect is to wriggle over short distances in a liquid medium, and the most that the mass of micro-parasites can do is to spread in a saprophytic state by multiplication at the edge, as a patch of ringworm spreads. To pass from one region of the earth's surface to another they are dependent upon the movements of other things, animate or inanimate. They must be carried in the water of a running stream; in the tissues of a man, a rat, or an insect; in a bundle of "infected" goods; perhaps by the wind. Their locomotion is involuntary. They cannot, as insects appear to do, choose when and whither they will migrate. They cannot, when their breeding-grounds become unsuited to them, or when their supply of food seems failing, take wing in a body and fly, as insects do, with almost conscious, intelligent, ordered movement in a definite line to other and more hospitable lands. They seem to be left—so far as locomotion is concerned—to the chapter of accident.

Yet, in the natural history of many epidemic diseases there are phenomena which it seems difficult to explain on the theory of a mere accidental carriage of a micro-parasite. Much has been written upon the "cyclical" character of many diseases. Not merely such classical examples as influenza, cholera, and plague, but other diseases, as diphtheria, scarlet-fever, small-pox, and measles, have all shown a tendency to years of special prevalence, with intervals of comparative inactivity. These "cycles" are of no definite length; the average for any given disease is often widely departed from by that disease. But they exist, and to ascribe them to a mere chapter of accidents is an unsatisfactory explanation. Still less satisfactory is it in regard to such diseases as cholera and plague and influenza, which remain for long years together confined to certain limited areas, and at irregular intervals spread, without apparent reason, to a large portion or the whole of the inhabited world. No doubt in such instances the accidental carriage of the micro-parasite plays some part; probably of more importance are the wide variations which may be supposed to take place from year to year in the vitality and powers of multiplication of the micro-parasite; possibly there is some additional factor, still to be discovered, to explain fully all that is obscure in the behaviour of these diseases.

What that factor may be it is difficult even to surmise. The micro-parasites of disease are an integral part of organic creation. Like all other living creatures they are subject to the laws of evolution, of the struggle for existence, and the survival of the fittest. They are perhaps preyed upon by other micro-organisms or by creatures less lowly in the scale of creation. The antagonism between different disease-germs has often been suspected, though it has never yet been proved. From early historical times it has been repeatedly noted that an epidemic of one disease has been accompanied by a lessening or a disappearance of some other disease. But the experience of one epidemic has been contradicted by that of another of the same disease elsewhere or at a later date; and no proof has as yet been brought forward of a constant, essential antagonism between any two diseases, or the parasites that cause them.

But if there is no proved antagonism between any two disease-germs, there may, for aught that is known to the contrary, be a keen antagonism between one disease-germ and some other living creature—animal, vegetable or fungoid. The wide fluctuations from year to year in the numbers and activity of a given disease-germ may perhaps depend not only upon fluctuations in its own vitality and powers of multiplication, but also upon fluctuations in the particular creature of which it is the normal prey. Should this prove to be the case, then it may be supposed that when the latter are numerous and active the former will be destroyed or kept within normal limits, and that when the reverse is the case the disease-germ will flourish uncontrolled, and the disease associated with it will become epidemic.

It may be that this, or something analogous to this is the explanation of the "cyclical" variations of many epidemic diseases. It presupposes, however, that a large portion, or even a principal portion, of the life of a disease-germ is spent in a saprophytic state, outside the human body, where alone it could fall a prey to the other supposed living organism. It can scarcely be applied in explanation of a large number of the infective fevers, the micro-parasite of which appears to pass the whole, or almost the whole, of its existence in the tissues of persons actually suffering from the disease. In regard to such diseases, if their wide fluctuations from year to year are to be explained by fluctuations in the number and activity of external forces, favourable or inimical to the parasites of those diseases, it seems almost necessary to suppose that these forces exist within the human body. In other words, while it is necessary to suppose that the virulence and powers of multiplication of disease-parasites differ widely in different years, it is equally necessary to admit the possibility that the receptivity of their human host, or his powers to withstand their attacks, may also differ widely.

In olden days great epidemics were attributed either to such highly improbable causes as eclipses of the sun or moon, combinations of the stars or planets, the appearance of a comet or some other celestial event; or to such possible causes as great and unusual incidents upon this globe itself—earthquakes, openings of cracks and chasms in the earth's crust, inundations and overflows of great rivers, the pestilential emanations from the unburied bodies of persons overwhelmed by famine, by war, or by some great cataclysm. There is nothing to show positively that one of these particular events has been the cause of any of the great historical epidemics of disease. Famine and war bring many diseases in their train, it is true, and the emanations from masses of unburied bodies may lead to the local prevalence of certain diseases; but none of the historical pandemics of cholera, of influenza, of plague, of small-pox, or of any other disease has been traced to such causes. It is impossible now to accept so simple an explanation of a plague pandemic as the opening of a hole or gulf in the earth, in Chaldea, whence issued a pestilential vapour that struck all it met with the plague—an explanation which satisfied the historians of the time of Marcus Aurelius.

Yet in regard to these earth-changes or terrestrial spasms and their relation to disease, it may be well perhaps to guard against over-scepticism, and to admit the possibility that such explanations of epidemics, though never absolutely proved in any individual instance, may contain a germ of truth. Some twelve or thirteen years ago one disease—influenza—swept round the entire globe, and more than round it. This behaviour of the malady was attributed by some to great inundations in China;

while others pointed out the close resemblance between the
sweep of the disease round the earth and that of the dust raised
some years before by the great cataclysm of Krakatau, near Java.
That dust was believed to have been carried in the course of a
few months not once, but many times, round the world. While
the Chinese inundation theory was certainly wrong, as the
last pandemic began, not in China, but on the borders of
European and Asiatic Russia, the other suggestion is deserving
of some consideration. The analogy between the behaviour
of influenza and that of the dust of the Krakatau eruption is
at least as striking as the analogy, already discussed at length,
between the behaviour of many epidemic diseases and that of
insect-swarms. Attention, it will be remembered, was drawn
to this latter analogy long before the true causation of such
diseases was known; and, although the suggestion that these
diseases were actually caused by insects was later shown to be
wrong, yet it went strikingly near the truth in that it presupposed
an animate instead of an inanimate cause for them. May not
the suggestion that earth-changes or terrestrial shocks share in
the causation of epidemics go equally near to some other, yet
undiscovered, truth—the sought-for, but yet unknown something
which seems to be needed to explain all the phenomena of
epidemic disease?

Attempts have been repeatedly made to classify diseases
on a geographical basis, grouping together those that have
a universal distribution, those that are found mostly in cold
or temperate climates, and those that occur mainly or solely
in tropical or sub-tropical countries. Such classifications are
always interesting, though they can never, from the nature of
things, be of a final character. With the exception of those disorders which are practically universal, almost no disease can
be said to have a permanent geographical distribution. Apart
from such marked examples as cholera, influenza, and the
other typical "epidemic" diseases, whose distribution varies
enormously from year to year, or from century to century, there
are many diseases whose limits of diffusion are not yet reached—
whose history is, so to say, "in the making." Small-pox, for
example, has not, as yet, been introduced into New Guinea,
Tasmania, New Zealand, and many of the Pacific Islands. The
chigo has in the past few years spread for the first time in

history over the continent of Africa and appeared in Madagascar and India. Sleeping sickness is gradually extending over wider and wider areas in Africa. Syphilis has not yet reached the possible limits of its diffusion, and Greenland, parts of America, and of South Africa have not as yet seen the disease—though it can scarcely be doubted that they will sooner or later, as their traffic with the rest of the world increases. Plague has in the past few years appeared in the western hemisphere and in many places south of the equator; yet but a few years ago it was held—and apparently with justice—that this malady could not extend to the New World or to southern latitudes. Many other examples might be quoted in illustration of the same fact, and of the caution needed in any attempt to classify diseases on a geographical basis.

Without aiming at any strict classification of this kind it will suffice here to group together certain diseases whose distribution over the earth's surface has been more or less alike.

Some diseases have an almost world-wide distribution. Cancer, diarrhœal disorders, mumps, measles, pneumonia, leprosy, rheumatism, typhoid fever, tubercle, whooping-cough, and many others have occurred at some time or other in almost all parts of the inhabited world. They have shown themselves quite independent of lines of latitude or isotherms.

Some diseases, as scarlet fever and diphtheria, are found mainly in temperate or cool climates, and are rare or unknown in and near the tropics. Others, as whooping-cough, cerebro-spinal fever, rheumatic fever, scurvy, and typhus fever, though occurring to some extent in all latitudes are more common or more severe, or both, in the cooler parts of the earth.

In a few instances, of which malaria and dysentery are the most striking, the reverse is the case, and the disease, though found in almost all latitudes, becomes more common and more severe towards the equator.

Many diseases are practically confined to warm latitudes; among the most typical examples of this class are dengue, mycetoma, liver abscess, phagedenic tropical ulcers, oriental sore, yaws, the bilharzia disease, ankylostomiasis, the various filarial disorders (elephantiasis, chyluria, etc.), the guinea-worm disease, and many others.

Some have a still more restricted range and are found only

in comparatively small portions of particular continents, as, for example, verruga peruviana, ainhum and goundou.

Another important group includes certain diseases which remain endemic in comparatively limited areas for long periods together, and only at irregular intervals spread more or less widely over the earth's surface. To this class belong cholera, plague, and yellow fever. The first of these three diseases has several more or less permanent endemic centres in the Far East, all in or near the tropics; yet as an epidemic it has spread at times to almost all parts of the world. The second has also many endemic centres, but of a less permanent character, and not all in or even near the tropics; its epidemic diffusion has in the past been quite as wide as that of cholera. The third is endemic only on the tropical coasts, eastern and western, of the Atlantic Ocean, and its epidemic diffusion has been much more restricted than that of either plague or cholera.

Ignoring for the moment the disturbing influence of season and elevation above sea-level, the earth may be regarded as becoming progressively warmer from the poles to the equator; and this variation in temperature has clearly a powerful influence upon the distribution of many diseases. But it is not the only factor in determining their distribution, nor is it in all instances the most important one. Even in regard to diseases caused by micro-parasites the direct effect of temperature is by no means uniform. The favouring action of warmth upon all living organisms and the greater wealth and variety of the larger flora and fauna of tropical and sub-tropical regions over those of the temperate and cooler zones of the earth, might well warrant the expectation that all such diseases would be more rife and more severe in the former than in the latter. Yet this can scarcely be said unreservedly to be the case. It would seem that the influence of climate upon disease is shown rather in regard to the variety of disorders met with than to the frequency or intensity of each disorder. As might have been anticipated, the disease-list is a longer one in warm countries than in cold. In most tropical and sub-tropical lands all, or nearly all of the ordinary diseases of temperate climes are found to exist, while there are in addition others that are unknown in the latter. The disease-flora is, in brief, richer in the warmer regions of the earth. But when each disorder of temperate climes is considered separately, and traced from the pole to the equator, it is not always found to become progressively more common or progressively more severe. A gradation of this kind was, it is true, mentioned above in regard to malaria and dysentery, and to these may perhaps be added cholera and plague in their epidemic manifestations. But a still longer list has already been given of diseases which are commonest in temperate or hot climates, and actually become rarer or less severe as the equator is approached. It will, further, be shown presently that many—almost the majority—of the ordinary infective fevers are most prevalent in the cool and not the warm season of the year. That this seasonal variation in disease is due not solely to changes of temperature but also to other factors—such as the closer aggregation of people in houses, lessened ventilation, diminished sunlight in the cool season of the year—there can be little doubt; and the same or similar conditions may to some extent be held to account for the distribution of these diseases, on a large scale, over the earth's surface.

The seasonal variation of diseases has an obvious, if indirect, bearing upon their geographical distribution. In general terms the diseases which have shown a tendency to prevail mostly in the cooler season of the year, both in warm and cool latitudes, have been small-pox, influenza, measles, diphtheria, croup, whooping-cough, pneumonia, inflammatory affections of the respiratory passages, typhus fever, erysipelas, cerebro-spinal meningitis, puerperal fever, and rheumatism. On the other hand, many affections of the digestive organs, as diarrhœa, dysentery, and cholera, have always been most active in the warmer months. Plague has been for the most part a warm-season disease in temperate climates and a cool-season disease in the tropics. Scarlet fever, relapsing fever, and many others have shown no very definite or constant seasonal relations. In regard to a large number of diseases, particularly those occurring in the tropics, the seasonal variations have been less dependent upon temperature than upon rainfall; in other words, they have been less a question of the difference between cold and warm seasons than of the difference between wet and dry seasons. In many warm countries the whole course of social life is largely determined by the annual recurrence, at a more or less constant date, of "the rains"; and each period of the year—whether it be the commencement of the rains, the rainy season proper, the decline of the rains, or the dry season—has its own character in regard to the prevalence or scarcity of some particular disease or diseases.

The influence of altitude above sea-level upon the distribution of disease is far from uniform and is not always easy to determine. The lowered atmospheric pressure and the lessened temperature found at considerable heights are usually combined with a greater freedom of the air from organic impurities, and with the presence of a scantier population, living a more or less isolated existence, and holding rare communication with the rest of the world. Such conditions not only lessen the chances of the importation of communicable diseases, but they are unfavourable to the prevalence and multiplication of the infecting material when imported. Hence many diseases, as, for example, tubercle, cholera, malaria and dengue, are rarer at high than at low levels. But in a large number of instances a high altitude has proved no bar to the epidemic, or even the endemic, occurrence of disease; and not only those just named, but many other disorders that are most common at low levels have prevailed with intensity at great heights. One disease—verruga peruviana—is found only at altitudes exceeding 2500 feet above sea-level. The very obvious fact that a high altitude in the tropics may present conditions of climate and temperature and, consequently, of disease prevalence resembling those of the temperate or cooler zones is well illustrated by the behaviour of typhus fever. When that disease occurs in or near the tropics it is only or mainly at considerable heights, while in more temperate zones it prevails indifferently at high or low levels.

From the early days of medical observation the character of the soil has been thought to exert a powerful influence upon the prevalence and distribution of disease in man. The obviously unhealthy nature of many low-lying, marshy lands, the remarkable and apparently arbitrary distribution of some diseases, and the knowledge, which must have been acquired at an early period, that a dry porous soil is much healthier than a moist, impervious one, would all tend to encourage this view; and it was further strengthened by the belief, at one time universally held, that many diseases were due to "miasmas" or harmful emanations from the earth. The subject is one that has scarcely yet been finally worked out; but it would seem that what may be called the strictly geological characters of a soil have a less powerful influence upon disease distribution than its physical properties—its temperature; its dryness or moistness; its perviousness or imperviousness; its purity or impurity in regard to contained organic matter; and its relation to human dwellings. It has long been known, for example, that a damp and impervious soil is peculiarly favourable to the prevalence of consumption, lung diseases in general, rheumatism, and other affections; and there is good reason to believe that it is a leading factor in the distribution of diphtheria, and, possibly, of cancerous diseases. Cholera and typhoid fever have shown on many occasions that they are largely influenced by variations in the height of the water contained in the soil. Diarrhœal disease, particularly the diarrhœa of infants, has shown a marked parallelism with the rise and fall of the temperature of the soil. Yellow fever, typhus fever, and perhaps plague, Mediterranean fever, and other diseases, have seemed to require, or at least to be favoured by, the presence of organic matter, and particularly of animal pollution in the soil.

Other pathogenic micro-parasites would seem to spend one portion of their existence in the soil. The tetanus bacillus has a wide distribution in many varieties of soil; the bacillus of malignant œdema is found mostly in soil containing much putrefying matter, such as garden earth that has been recently manured; the anthrax bacillus, the actinomycosis fungus, the mycetoma fungus, would all seem to be capable of living in a saprophytic state for longer or shorter periods in or on the soil. The micro-organisms of typhoid fever, yellow fever, plague, and many other diseases are thought to possess the same power. The soil, further, plays an important part in the diffusion of some of the animal parasites to which man is liable. The chigo flea lives in or on warm sandy soils or on the earthen floors of human dwellings. Ankylostomum duodenale probably spends the larval stage of its life-history in moist earth, and is transferred to man either by means of food that has been contaminated by the earth, or by actual earth-eating (geophagy).

Finally, earth-disturbance on a large scale has been frequently followed by the appearance of certain diseases in an epidemic form. The diseases that have been observed to occur under such circumstances have been malaria, blackwater fever, ankylostomiasis, and that form of fever to which the name of typho-malaria or malarial typhoid has (probably erroneously) been given.

The exact influence of racial susceptibility or immunity upon disease-distribution is not always easy to determine. Differences of race are so commonly proved to be associated with differences in mode of life, in degree of "acclimatisation," in amount of exposure to infecting material, in readiness to take preventive measures, and the like, that it is often far from easy to say whether a difference in disease-incidence between any two given races should not be ascribed to one or other of these factors and not to a true difference in racial susceptibility. All that can be positively stated is that some races do seem to be more susceptible to certain diseases than others, while in a few instances a true racial immunity may with some reason be suspected. In the tropics, for example, the white races appear to be far more susceptible to malaria, liver abscess, scarlet fever and typhoid fever than the coloured races; while on the other hand they are much less likely to become subjects of elephantiasis, chyluria and other filarial diseases, the Guinea-worm, and many skin diseases, as, for example, craw-craw, tinea imbricata, and pinta.

The African negro is peculiarly liable to tubercle, small-pox, tetanus, and cerebro-spinal meningitis; and such diseases as ainhum, goundou, and sleeping-sickness, and such parasites as the filaria perstans, filaria loa, filaria diurna and some others are solely or almost solely confined to the negro race. On the other hand the negro enjoys a high degree of immunity from yellow fever and, to a less extent, from malaria. To the skin diseases named above as rare in the white races may be added yaws or framboesia, the frequency of which in any part of the world would seem to be determined largely by the depth of colour—or in other words by the amount of pigment in the skin—of the population. The Mongolian races have shown themselves remarkably susceptible to many of the infective diseases, both acute and chronic, and to ophthalmia and trachoma; and a few animal parasites, as for example distomum pulmonale and distomum sinense, seem to be almost peculiar to them.

That insanitary conditions on the one hand, and, on the other, man's conscious efforts to remove these and to prevent and control diseases of all kinds, have a very considerable share in determining the distribution of disease can scarcely be questioned. Sanitation and disease-prevention are both sciences of modern growth. They are practised to a very limited extent in the large majority of territorial areas of which the inhabited earth is composed, and are, for all practical purposes, entirely absent from not a few. In only a very small number of countries can they be said to have reached a high level throughout. In none have they attained the possible maximum. Their influence, therefore, upon disease-distribution is still far from being as powerful as it might be. They have not yet apparently succeeded in completely and finally removing from the face of the earth any of the known diseases to which man is liable; and it is easy to point to a considerable list of disorders—tubercle, syphilis, diphtheria, cancer and others—which, in spite of human effort, have, taking the world as a whole, become more instead of less rife in recent times. But on the other hand there is a vast and ever-increasing mass of evidence proving clearly that the control of many diseases is very largely in man's hands alone. A generally raised level of personal and communal hygiene and an ordered Public Health Administration in the more civilised countries have not only markedly lessened the general death-rate from all causes, but they have greatly reduced the power for evil of not a few special diseases, and all but extinguished some. Even in regard to the diseases named above as having become more frequent in recent times, it was necessary to add the proviso "taking the world as a whole," for three at least of the four named have become less prevalent or less fatal in those countries where serious efforts have been made to control them.

In England, where more persistent, more general, and more costly efforts to improve sanitation and control disease have been made than in any other country, it has been clearly shown that these measures as a whole have almost extinguished typhus fever, dysentery, and malaria; they have led to a marked reduction in typhoid fever and pulmonary consumption; and they have shown that such diseases as cholera and plague may for long periods together, and in spite of repeated importation of the infection, be kept at bay and at least prevented from gaining any serious hold upon the country.

Special health measures have succeeded in all countries in controlling special diseases. The draining of marshes and swamps has practically freed Holland and England from malaria; and similar works, combined with a war against the malaria-bearing mosquito, are leading to a striking diminution of the same disease in Italy, the West Coast of Africa, Cuba and elsewhere. Provision of a pure water-supply and care to prevent the consumption of fouled water have had an enormous share in controlling such diseases as cholera, dysentery, and typhoid fever. Careful inspection of meat has proved wholly successful in many countries in preventing trichiniasis. The muzzling order and other measures, particularly those concerned with the importation of animals from abroad, have almost extinguished rabies and hydrophobia in England. The introduction of vaccination by Jenner at the end of the eighteenth century led to an enormous diminution in small-pox in all those countries where it was practised; and the disease has almost ever since been less or more prevalent just in proportion as efficient vaccination and re-vaccination have or have not been carried out. Methods of protective inoculation against plague, typhoid fever, yellow fever and some other diseases have been recently introduced with promising results; and in regard to the first-named at least, the method has been so extensively and successfully practised in India and elsewhere as to rank already as an important factor in the local distribution of the disease. Syphilis and other venereal disorders have been controlled by judicious legislation and a healthier tone of public and private morals. Scurvy is no longer the scourge that it was at one time of gaols, barracks, and ships, and recent experience has shown that even an Arctic expedition may be successfully carried through without a single loss from this disease.

But when, on the other hand, no attempt is made to remove insanitary conditions, and disease is allowed to run its own course unfettered by any effort on the part of man to control it, the results are very different. Every one of the maladies named in the last two paragraphs, and a host of others—as ophthalmia, diarrhœal diseases of all kinds, erysipelas, puerperal fever, phtheiriasis and many other skin diseases, all the affections associated with animal parasites—are not only more common, but may literally run rife where man lives in the midst of filth and refuse, where drainage and a pure water-supply are unknown, and where organised public effort to control specific diseases is absent or impotent.

In this *résumé* of present-day knowledge as to the influence of different external factors upon the distribution of disease, the large class of "communicable" disorders has, of necessity, been most frequently mentioned. It is disorders of this class whose geographical distribution presents most points of interest and is of real service in explaining their nature and causation. In the diffusion of such diseases man's movements must always take a leading share. In the case of the chronic infective disorders, as syphilis, leprosy and (perhaps) tubercle, man's movements have apparently been alone responsible for their spread. In the case of the acute infective fevers this factor must also be an important one in their distribution; though in regard to many, as was pointed out above, it is probably not the only one.

Instances of the carriage of disease by man from one part of the earth's surface to another are innumerable—not only by the movements of individuals but by the movements of men in masses. Emigrants have carried leprosy and hydatids from Iceland to Canada and the United States. Traders and explorers have on countless occasions brought influenza, measles, syphilis, small-pox, and many another disease to communities where they were before unknown. Chinese labourers have been the principal agents in the distribution of leprosy in Australia, the Malay Peninsula, and the Far East generally. The great movements of peasants in Russia, who leave their homes in bodies and seek work in other parts of the country, have repeatedly been the means of spreading syphilis, ophthalmia, cholera, and other diseases. Beri-beri, plague, yellow fever, cerebro-spinal meningitis and many others have been transported by ships over thousands of miles, and often without losing any of their virulence by the way.

The movements of men associated with war have proved to be great factors in the distribution of disease. Here, no doubt, mere movement is not the only factor at work, but it has largely aided the other factors concerned. Typhoid fever, malaria, so-called typho-malaria, syphilis, measles, scurvy, small-pox, have all at some time or other been spread as the result of military movements. Typhus fever was widely diffused in Europe by the Napoleonic wars, and rapidly diminished in prevalence when those wars came to an end on the field of Waterloo. Leprosy was carried by refugees during the disturbances in Crete a few years ago to many parts of the Eastern Mediterranean. The Crusades were believed to have aided in the diffusion of more than one disease. It is, however, unnecessary to multiply instances further.

There is scarcely a war in ancient or modern times which does not furnish examples equally striking with those just quoted.

Pilgrimages to religious shrines have often aided in diffusing disease. The great annual flow of Moslem pilgrims to and from the western shores of Arabia has repeatedly been the means of bringing cholera to the Red Sea and of spreading it widely else-where; and there can be little question that this annual concourse of people also aids in distributing small-pox and other communicable diseases less striking and less dreaded than these. The same statement holds good for those gigantic gatherings of pilgrims at bathing and other religious festivals at Hardwar and elsewhere in India; the constant flow of Shiite pilgrims to the holy burying-grounds of Kerbela and Nedjef near the Turco-Persian frontier; and the annual gathering and dispersal of traders and peasants at the great fair of Nijni Novgorod on the Volga.

The geographical distribution of "communicable" diseases depends not a little upon the channel by which they are transmitted from man to man. This varies very widely. Some, as, for example, syphilis, gonorrhœa, ophthalmia, and possibly oriental sore, yaws, and other skin diseases require that the actual secretions of an infected person shall reach the whole or abraded skin or mucous membrane of a healthy person before the latter can contract the disease. The term contagious should be strictly limited to this class of disorder. In others the transmitted material is less gross, and, whether of the nature of a micro-parasite or not, is given off by the patient's body, exists in the air about him, and is carried to a greater or less distance by the air. In typhus fever this distance is perhaps only a few feet, in small-pox it may be measured by fractions of a mile. The part played by the air in the transmission of many other diseases has been much discussed and is still uncertain. Dust, which is merely a collection of particles, organic and inorganic, of "matter misplaced," may contain the germs of infectious disease or even fragments of the dried discharges of the contagious diseases, and the air carrying such dust may thus be the bearer of the disease. Ophthalmia, there can be little doubt, is often spread in this way; typhoid fever is thought by many to be a dust-borne disease; and in the aerial diffusion of small-pox it is reasonable to suppose that dust containing fragments of dried small-pox scabs is the main carrier of the infection. Malaria, at one time regarded as the type of an air-borne disease, is now known to be carried not by mere air-currents but by the agency of winged insects.

In the case of some diseases the infecting material is carried by water. Cholera, typhoid fever, and dysentery are all spread very largely, perhaps mainly, by this means. Goitre and urinary calculus have been thought to depend upon the consumption of particular kinds of water. The bilharzia disease, ankylostomiasis, Guinea-worm disease and other affections caused by animal parasites are certainly or probably spread by the agency of water. Several parasites of this nature pass the larval stage of their existence in the bodies of some minute fresh-water animalcule and gain access to man in one of two ways—either through his drinking water containing such infected animalcules or free larvæ, or from the larva-containing water coming into contact with the surface of his body.

Some water-borne diseases may be carried over considerable distances by this means. Cholera, the archetype of this class of disease, has on several occasions been spread over a great part of a large and populous city by a contaminated public water-supply ; and there is good reason to believe that such immense streams as the Elbe, the Volga, the Don, the Neva, and many other rivers have been the means of carrying the cholera virus over very long distances.

A large number of communicable diseases are transmissible by means of infected clothes, linen, merchandise and other fomites. The virus of small-pox, of scarlet fever, of measles, of plague, and of many other diseases may cling to the furniture, the bed-linen, the clothes, or any and every inanimate object that has been in the near neighbourhood of a patient suffering from the disease. In this position it may retain its virulence for long periods together, and if such infected articles are moved to other places they may carry the virus with them and reproduce the disease elsewhere. By this means infective diseases have been carried over much longer distances than they have ever been carried by the air or by water. It has probably been one of the most powerful means of distributing many communicable maladies; and there is reason to believe that in many of those instances of apparent carriage of infection by human beings from country to country or from hemisphere to hemisphere, the infecting material has really been carried in the clothes or belongings of the person,

and not in his tissues. This must always be the case when the
infection has been carried over very long distances—distances
which take longer to traverse than the incubation period of the
disease. It must equally certainly be the case in those well-known
instances where a person has come from an area where a particular
malady was epidemic to another area free from it, and where,
though not himself developing the disease, his arrival has been
followed by the appearance of the disease in other persons in
his immediate neighbourhood.

But man and his inanimate surroundings are not the only
means by which communicable disease is spread. The belief
that the lower animals are attacked at the same time as man by
epidemic diseases is almost as old as the history of disease itself.
In the early historical records of plague and pestilence nothing
is commoner than to find that a murrain among beasts and birds
or both prevailed at the same time as the human disease. Modern
observation has shown more precisely the true relation of the
lower animals to disease in man. It has shown that man
obtains the infection of some maladies, as, for example, anthrax,
glanders, farcy, foot-and-mouth disease, and hydrophobia, solely,
or practically solely, from animals already suffering from
them. Such diseases are, in fact, essentially animal affections,
and only attack man by accident. It has further shown that some
essentially human affections may accidentally attack the lower
animals, which thus become the carriers of infection. Human
influenza is believed to be transmissible to horses, cats, and other
domestic animals. Diphtheria, or a disease closely allied to it,
has been seen in birds. A form of infectious pneumonia has been
known to attack parrots and be transmitted from them to man.
Scarlet fever may perhaps affect cattle ; and it is still a debated
question whether the tubercular disease of cattle and other animals
is or is not transmissible to man. In at least one disease the
lower animals are not only capable of being attacked but appear
to be one of the principal means of diffusing the infection over
the earth's surface. The part played by the rat in the spread
of plague will be discussed in the chapter on that disease, where
it will also be shown that mice, monkeys, bandicoots, squirrels,
cats, and possibly marmots and other animals may all be carriers
of the infection.

Many of the animal parasites to which man is subject pass a

portion of their existence in one or other of the lower animals, and here the animal must be regarded as an essential, indispensable link in the transmission of the associated disease. Tænia solium and trichina spiralis complete their life-history in the pig; taenia mediocanellata in horned cattle; bothriocephalus latus in certain fish; tænia echinococcus—the parasite of hydatid disease—in the dog, sheep, or cattle. Such parasites can only pass to man where he is brought into relation with animals, living or dead, which are the subjects of them. Most of them gain access to his body through eating imperfectly cooked beef, pork, or fish containing them. Others, as for example taenia echinococcus and perhaps ascaris lumbricoides, probably escape from the animal in the excreta, and are conveyed to man in dirty drinking water or by contaminated hands or food.

Still lower in the scale of creation, many members of the insect family are found to act as the carriers of disease. Houseflies, mosquitoes, fleas, bugs, lice—any or every insect that may crawl on the body of a person suffering from an infectious disease or batten on his excreta, and then pass to the body of another person, may be the means of directly carrying the malady, through some portion of infecting material clinging to its legs or body. Or the insects may carry the infecting material to milk, or water, or solid food, which would then give the disease to any one swallowing it.

In other instances the infecting material, or in other words the parasite, actually enters the insect's body. This would appear to occur most usually, if not solely, in the case of diseases the parasite of which exists in the peripheral circulation, and the only insects capable of transmitting them are those which suck men's blood. The most dangerous insect in this respect would seem to be the mosquito. It has been shown to be the principal carrier of the malarial organism. A mosquito, though possibly of another species, is the extra-human host of the filaria sanguinis hominis (F. Bancrofti), and probably of many other varieties of filariæ; and strong evidence has recently been brought forward that certain mosquitoes may be active carriers of the yellow fever virus. Plague is also believed by some to be spread by the agency of insects; not, however, by the mosquito, but by fleas and other body-vermin found on man, on rats, or on other animals suffering from the disease. The cockroach has been suggested as the possible host of some of the tæniæ to which man is subject.

Finally, and as probably the lowest of all animals in the scale of creation that are capable of becoming the hosts of human parasites, must be named a fresh-water cyclops which has been shown to act in this way for the Guinea-worm; and it is far from improbable that later observation will show that some similar lowly organism is the extra-corporeal host of bilharzia hæmatobia and other human parasites.

In the case of those human diseases, the parasite of which passes half its life-history in some of the lower animals—whether a mammal, fish, insect, or crustacean—it is clear that the geography of the disease must be largely determined by that of the particular animal in question. There are in fact three essentials for the prevalence of such diseases—the parasite, the human subject, and the extra-human host. The geographical distribution of such diseases may differ from that of either of these essential elements taken separately; and in the absence of any one of them it must rapidly die out. The influence, further, of such external conditions as temperature, climate, elevation above sea-level, soil and season, upon the distribution of the disease must be somewhat more complex than in the case of other diseases; for such influence must act not only on the parasite itself, and the receptivity of the human subject, but also on the numbers and activity of the particular animal or insect which serves as the extra-human host.

This rapid survey of the forces at work in determining the geography of disease will, if it serve no other purpose, at least show that the distribution of diseases is a highly complex matter. In the case of each disorder to which man is liable, the areas of the earth's surface in which that disorder is met with, and its differing degrees of frequency in different parts of those areas must be regarded as the resultant of all, or of many of the factors, the most important of which have been discussed in this chapter.

AINHUM

Description. This disease consists essentially in the appearance of a constriction or furrow at the base of one or more toes. It is almost always the fifth or little toe; occasionally the fourth, and never any of the others that is affected. It is at first quite painless. The furrow becomes deeper and deeper, as if the toe had been tightly ligatured with wire or string; ulceration sets in; there is great pain and discomfort in walking, and the toe finally separates by gangrene or is removed by a surgeon or by the patient himself. The process may last several years; one case, in South Carolina, lasted for 50 years. Its etiology is quite unknown; of recent years some observers have adopted the view that it is only a form of leprosy, but this view is combated by others.

History. Attention was first called to the existence of this disease on the West Coast of Africa by Clarke in 1860, in a paper read before the Epidemiological Society of London. Later it was described by Corre, Crombie and other observers. But of the history of the disease itself almost nothing is known.

Geographical Distribution. *Europe.* The disease has never been met with in Europe. Ehlers has recorded an example of "ainhumoid" mutilation of the little toe by leprosy in a Greek, and shares the opinion of Zambaco Pasha and others that ainhum is only a form of leprosy. But this theory is still awaiting proof, and up to the present no certain case of ainhum has been seen in Europe.

Asia. In Asia the existence of the disorder has apparently only been demonstrated in India and China. In India ainhum is far from common, but is occasionally met with. Over thirty years ago a French naval surgeon recorded cases occurring in

Pondicherry among Indians of the Tamil race[1]. Elsewhere in India there are reports of cases from Calcutta, Bombay, Dacca, and Goa. Recently a case of ainhum, or of a condition closely resembling it, was observed in South Sylhet, Assam[2]. The patient in this instance was a Hindu washerman, a native of Sylhet; but the ainhumoid condition was complicated by the presence of marked keratosis of the palms of the hands and soles of the feet.

It is only within recent years that ainhum has been recognised in China. Hichins states that he has seen three or four cases of the disease at Ningpo, on the east coast of China, in the course of ten years[3]. Maxwell speaks of a case reported from Swatow, and also records another seen by himself in Shang-po in southern China early in 1899, in which the patient presented the characteristic affection of the little toe; but some 16 months later the man was found to present unmistakable symptoms of leprosy[4]. It is uncertain whether ainhum exists in the Pacific Islands.

Africa. Whether ainhum occurs to any extent in northern Africa is not known. In Egypt the only mention of its occurrence appears to be the record of certain cases at Cairo and Suez. At Massowa on the Red Sea coast it is said to be unknown[5]. In Algeria it has been seen (at Algiers).

The principal seat of ainhum is the West Coast of Africa. The larger number of records of the disease come from the Gold Coast, but it is found in many other parts of these shores, including Senegambia. The Yorubas call it "ayùn," the native word for a file; while the word "ainhum," commonly applied to the disease, is apparently derived from the language of the Nagôs tribe, and means "to saw." Other tribes call it "guduram." The exact distribution of the disease in this part of Africa is uncertain; but the Kroo negroes seem to be the most liable to it. One typical case, at least, has been seen in British Uganda, in eastern Africa.[6]

Ainhum is also said to have been seen among the Kafirs in South Africa, but fuller information on this point is wanting. On the island of Nossi-Bé, near Madagascar, and in Réunion

[1] Collas, *Archives de Méd. Navale*, 1867.
[2] Dalgetty, *Journal of Tropical Medicine*, March, 1900.
[3] *Medical Reports of the Chinese Customs.* No. 55 (1897–8).
[4] *Journ. Trop. Med.* Dec. 1900.   [5] Rho.
[6] Cook. *Journ. Trop. Med.* June 1, 1901.

cases of the disease have been met with. In the former island it is known by the native name of "faddiditi."

America. Ainhum has been observed in negroes in some parts of the United States, especially in the south. Thus there are records of occasional cases in North and South Carolina, in Louisiana, in western Virginia, and of one in Washington. It has also been observed in Philadelphia and even in Canada, in negroes coming from North Carolina.

It is present, but probably rare, in the West Indies; cases are known to have occurred in Trinidad, in St Thomas, and at Havana.

In South America Brazil would seem to be the only country where ainhum has been seen to any extent; and probably, next to West Africa, the disease is commoner in Brazil than anywhere else. Moreira[1] from personal observation says that nearly all the cases originate in the towns of Bahia, Rio de Janeiro, and Pernambuco, and in Buenos Aires in the Argentine Republic. In none of these places is the affection really a common one; and generally in Brazil it is thought that ainhum is now rarer than it used to be. This has been explained by the decrease in the number of African negroes among the inhabitants, and also by the supposition that the offspring of negroes, born in Brazil, may be less susceptible than their parents. It is perhaps commoner than it seems to be, for, as Da Silva Lima first pointed out, many of the affected negroes do not apply to a medical man for treatment, but either cut off the affected toe themselves or get a companion to do it for them.

In Georgetown, British Guiana, one observer is said to have seen as many as twenty cases[2].

Factors governing the Geographical Distribution. Ainhum is to some extent a race disease. The large majority of cases are found in the negro races. Hence it is mainly found where these races exist, as on the West Coast of Africa, in the coast towns of Brazil, and in the southern States of America. But while the negro race is most prone to it, it has also, exceptionally, been seen in Mongols in southern China, in Hindus

---

[1] *Monatshefte für prakt. Dermatologie.* Translation in *Journ. Trop. Med.* March, 1901.

[2] von Winckler, quoted by Moreira, *loc. cit.*

of pure descent in India, and in a Hindu-Malagasy half-breed in
Réunion.

The causation of ainhum is quite unknown. Its relation to
leprosy is uncertain, but the majority of observers regard it as a
distinct and separate affection. Traumatism may have something
to do with its immediate development, and in several cases it has
followed on a slight injury to the toe. The male sex is more
liable than the female. It has not been seen in children and
is rare in old men, the prime of life being the period at which
negroes are most likely to develope it. Heredity may have some
influence in the production of ainhum, and there are many instances
on record of several members of a family, of more than one
generation, suffering from it.

There is nothing to show that climate, soil, elevation, dirt, or
insanitation have any influence on the prevalence of ainhum. It
is found, it is true, most commonly in tropical and sub-tropical
countries, but this may be simply due to the fact that the races
liable to it are found in such countries.

BERI-BERI.

General Characters and Etiology. Beri-beri or Kak-ké is a rather widely distributed disease, of which two clinical types are usually described, the wet or dropsical and the dry or paralytic. Each form of the disease presents a rather complex but fairly definite group of clinical symptoms and signs, which the reader will find described in detail in any of the numerous recent treatises on tropical medicine. Although the clinical pictures of the two types of beri-beri are very different at first glance, they both represent a group of pathological conditions that are apparently the result of an inflammation of many peripheral nerves. What this multiple peripheral neuritis is caused by is at present unknown. It has been very generally held that it must be due either to some living organism, multiplying in the tissues of the patient, or to some toxine derived from micro-organisms growing in his surroundings. Pekelharing and Winkler have described certain bacilli and cocci as associated with the disease; Glögner, and later Fajardo of Brazil, have found a form of plasmodium in the red blood corpuscles of beri-beri patients analogous to the parasite of malaria. Innumerable other conditions—such as the consumption of decorticated rice, of damaged grain, of bad fish, and other imperfections in dietary, climatic conditions, malaria, ankylostomiasis, etc.—have been put forward as possible causes of beri-beri, but for the present the etiology of the disease remains uncertain. Perhaps the most widely held view as to its causation is that which connects the disease with the consumption of rice, but whether the poison is a chemical one or of microbial origin has not yet been determined[1].

---

[1] A spore-bearing diplobacillus has been found by Capt. Rost, I.M.S., in the rice sold in bazaars in an infected district, and also in the blood of 32 beri-beri patients. *Indian Med. Gazette*, Dec. 1900.

History. The oldest references to the existence of beri-beri are found in Chinese medical writings. The word *kak-kê* is said to occur in a work dating as far back as the year 200 B.C. The disease is also said to be mentioned in a Japanese medical treatise of the ninth century of our era. In the seventeenth century there is mention of it in the East Indies and in India. Its appearance in the Western Hemisphere is apparently an event of recent occurrence, as it was not until the year 1866 that it became epidemic in Brazil—a country which is now one of the principal endemic centres of the disease in the west.

Recent Geographical Distribution. *Principal Endemic Centres.* The geographical distribution of beri-beri is not a constant one. There are certain portions of the earth's surface, principally in or near the tropics, where the disease appears to be more or less permanently endemic. But, as the infection can be carried for long distances by the movements of human beings, it is frequently appearing in places hitherto free from it; and as, in some of these places at least, it finds conditions which favour its existence, the limits of its prevalence appear to be constantly widening.

Beri-beri was first recognised and described by Dutch physicians in the Dutch possessions in eastern Asia, and this part of the world is still one of the principal seats of the disease. A widespread epidemic in Brazil in 1866 drew increased attention to the malady, and since that time it has been closely studied in many countries where it has prevailed as an epidemic or become more permanently domiciled.

In the Malay Peninsula and the adjoining archipelago beri-beri is among the commonest of disorders. Its prevalence from year to year varies considerably, and in some years it becomes truly epidemic. It probably never disappears altogether. In the Straits Settlements it was unusually active in 1895, but has been somewhat less so since; in 1896 the number of cases admitted to hospital in the Settlements was 2057, in the next year exactly one more (2058), and in 1898 only 1329. A remarkable outbreak, however, was observed in the latter year in the Singapore Prison. Gaols are notoriously liable to be the scene of outbreaks of beri-beri, but this prison had been practically free from the disease from 1885 to 1897, only two cases being recorded in the whole period of thirteen years; at the end of 1897 one or two cases

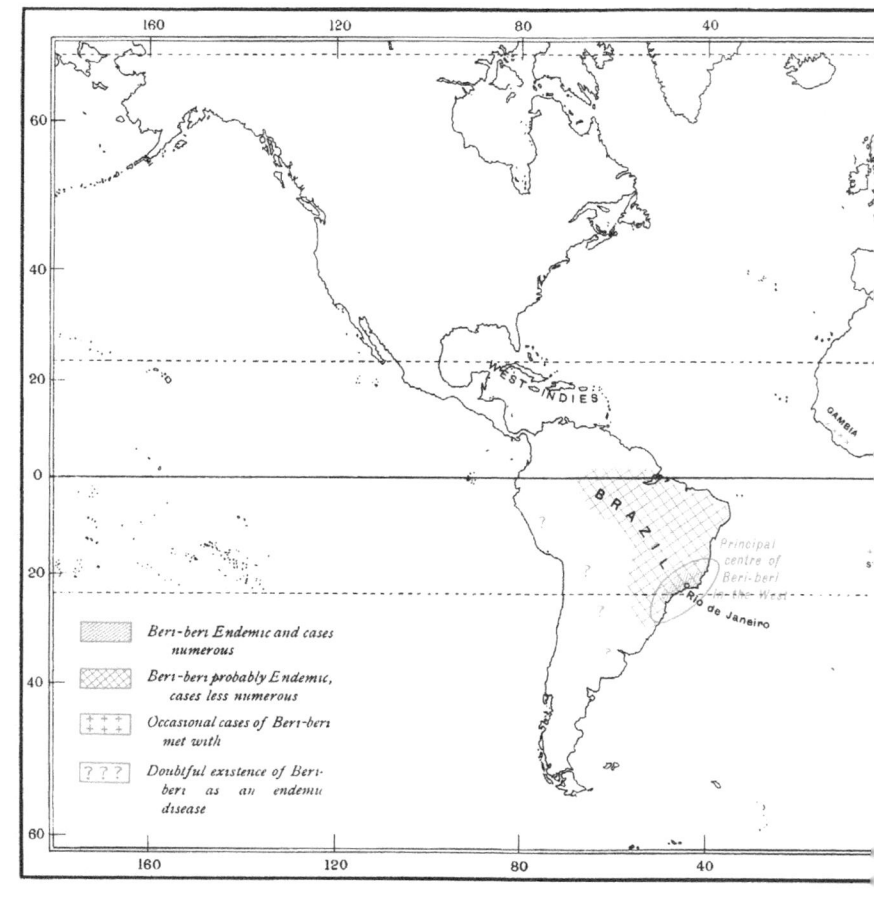

DISTRIBUTION OF BERI-BERI

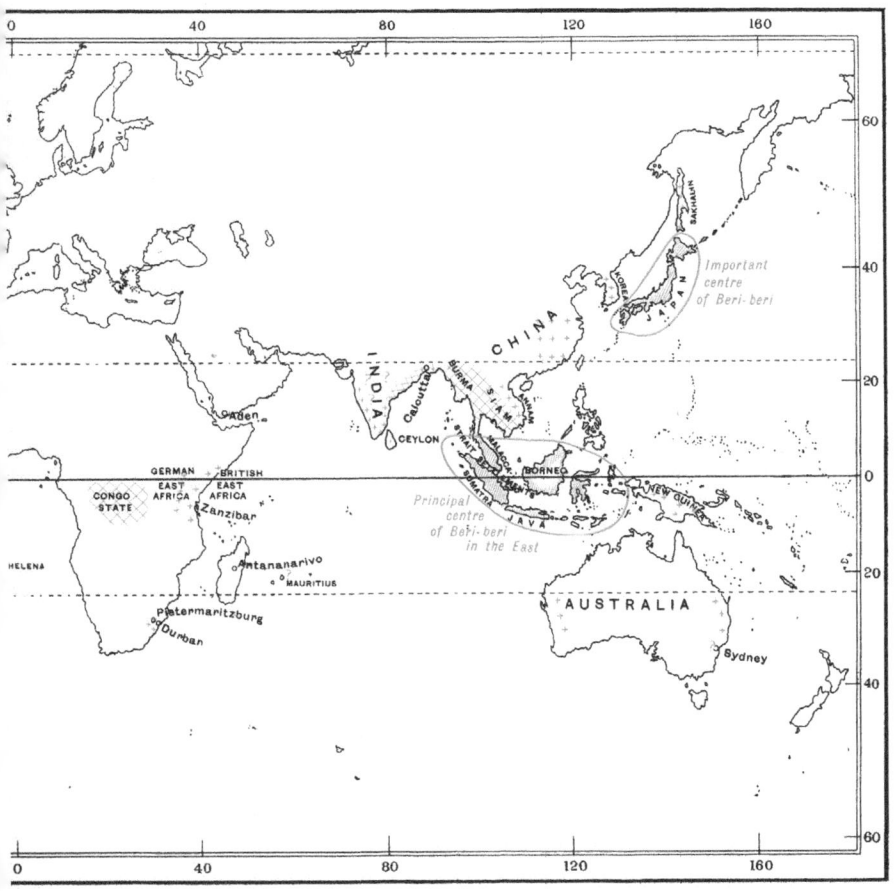

occurred, and in 1898 as many as 124 of the prisoners were attacked by the disease[1]. Similar instances have been recorded in other places.

Throughout the East Indian Islands beri-beri is endemic to a greater or less degree. Whether it is so in each of the numerous islands of the group is uncertain. It is undoubtedly common in Labuan, British North Borneo, Sumatra, Java, Celebes, the Moluccas, and in many other smaller islands; prevailing in some, as for example Borneo, specially on the coasts, in others becoming epidemic at times among labourers in the mines or on the plantations. Its degree of prevalence varies considerably. In the men of the Dutch army in the Netherlands-Indies (European and Asiatic) the recorded cases have diminished in recent years[2], but whether this indicates a lessening prevalence of the disease generally or has been due to increased precautions to prevent its development among the troops I am unable to state.

The evidence that beri-beri is a common disease in Siam, Annam, Tongking, and Cochin China is less conclusive. It is probably not a conspicuous disorder in Siam, as some recent writers on the prevailing maladies of that country make no mention of it. In 1892 or 1893, however, a severe epidemic occurred in a gaol at Bangkok and caused a high mortality. In Annam and Tongking it is said to be met with in the form of sporadic cases and also in localised outbreaks.

In China beri-beri is said to have been endemic for ages, but to be comparatively rare in the present day (Scheube). Manson states that it has been seen in Shanghai, Tsuchou, Wanchow, Foochow, Amoy, Swatow and Fatshan, and mentions several epidemics that have occurred in the past in the gaol and among the native population at Hongkong. It appears to be seen with great frequency in the last-named island; in 1899 alone as many as 197 deaths from this disease were recorded as occurring there among the Chinese population, the highest mortality being in the months of October, November, and December. In Swatow an

---

[1] In 1898, of 25,345 admissions to the hospitals of the Straits Settlements with 4139 deaths, 1329 admissions and 420 deaths were attributable to beri-beri (*Medical Report of the Straits Settlements*, for 1898).

[2] The number of cases of beri-beri in the Dutch army in the East in 1893 was 6174; there were 4994 in 1894; 5676 in 1895; 5780 in 1896, and only 2238 in 1897.

epidemic of beri-beri prevailed in 1895. It is certain therefore that it does exist in China, and the statement that it is rare there probably needs modification. The Chinese elsewhere seem to be specially liable to it; and in Calcutta, for example, the disease is practically confined to the Chinese inhabitants of that city.

Japan is almost as important an endemic centre of beri-beri as the Malay Islands or Brazil. The disease is said to be met with in all the islands of Japan, from north to south, and in the interior as well as along the coasts, but particularly in large, low-lying, damp, overcrowded cities. Further north the disease has been seen in the Russian island of Sakhalin, but it is doubtful whether it is endemic here[1]. It has also been met with in Formosa.

There is little mention of beri-beri in the Philippine Islands, but in New Guinea it is endemic in places. It is at times very prevalent in the crews of the pearling fleet in Torres Straits. In Australia beri-beri was until recently believed to be unknown, but lately a small epidemic occurred among a group of West Australian natives and also among Chinese on the eastern sea-board of Australia[2]. I find no mention of the disease occurring in New Zealand, but it is said to have been recently introduced into Fiji[3], New Caledonia, and the Society Islands, and perhaps some other islands in the Pacific.

Turning to the West it is found that in Burma beri-beri is believed to be endemic, and has in recent times been epidemic as far inland as Mandalay[4]. In India the Telugus from the East Coast and Northern Circars appear to be peculiarly liable to this disease; in 1897, out of a total of 137 cases of beri-beri recorded in the native army in India, as many as 135 occurred in Madras regiments, and nearly all the patients were of the race just mentioned. Less frequently the disease is observed on the Coromandel coast, in the Carnatic, and on the Malabar coast, and it probably exists in the interior, but to what extent is uncertain. It is constantly seen in Calcutta, where the Chinese

[1] In the Russian health statistics there is no mention of beri-beri.

[2] Molloy (*Trans. Intercolonial Med. Cong.*, Australia, 1892) suggests that beri-beri is endemic in Australia; and Graham, (Australian *Med. Gazette*, Nov. 1893) describes an allied disease as epidemic among the Chinese in Sydney. See also letter from Neil MacLeod, M.D., *Brit. Med. Journal*, Aug. 14, 1897.

[3] Probably by plantation labourers from Japan.

[4] M. J. Eyre, M.B., Lt.-Col. I.M.S. (*Indian Med. Gazette*, Jan. 1900), furnishes some details of recent prevalence of beri-beri in Burma.

community in particular suffer greatly from it, cases of the disease being daily treated at the Medical College and other hospitals. In Ceylon it is said to have been very common at one time, but to have now become very rare.

As south-eastern Asia is the principal home of beri-beri in the Old World, so is Brazil its most important seat in the New World. Since the great epidemic of the disease here in 1866 it appears to have remained more or less permanently established in the country, occurring in all parts of it, and at times becoming epidemic. In Rio de Janeiro beri-beri has recently been unusually active; in 1898 from 15 to 20 deaths from the disease were recorded in some weeks, and it was never absent from the mortality returns as a cause of death in a single week. It is said that "at first it was principally men in the navy who fell ill with beri-beri, next the soldiers on shore, chiefly those of the police, but now [May, 1898] we observe the sickness has in the same manner attacked private persons without exception of sex and social position[1]." So prevalent did the disease become in Rio de Janeiro on this occasion that it was mentioned, along with yellow fever, in the President's annual address to Parliament, as a matter requiring the attention of Government.

In Paraguay beri-beri is said to have been epidemic in the past; whether it exists there now, or in any other country of South America besides the Brazils, may be doubted. I find no mention of it as a cause of death in recent reports from Uruguay[2] and the Argentine Republic, but attention has been called to the occurrence in British Guiana of cases of multiple neuritis which present many points in common with beri-beri and would very probably be unhesitatingly classed as examples of that disease if met with in a country where it is known to be endemic[3].

Endemic Centres of lesser Intensity. There remain to be mentioned certain other portions of the earth's surface where beri-beri has been met with and is perhaps endemic, but where the disease

[1] *Public Health Reports* (Washington), 1899.

[2] T. W. N. Greene (*Brit. Med. Journal*, Jan. 9, 1898) states that cases are often seen in the British hospital at Monte Video, but they are mostly in men of the Brazilian navy. In 25 years he never saw a case outside hospital, either in the interior of Uruguay or in its capital, and therefore concludes that the disease is not endemic there.

[3] Fowler, *Journal of Trop. Med.*, May, 1900.

is apparently less widely prevalent than in either south-eastern Asia or the Brazils. It has been seen in many widely separate parts of the African continent. In German East Africa cases are occasionally recorded[1]; in British East Africa it apparently occurs only on the coast and as an importation from Asia. At Massowa on the shores of the Red Sea it is said to be unknown; but cases have been seen at Aden and also at Zanzibar. On the West Coast the disease is certainly observed in the Gambia, from one to four cases being returned in most recent annual reports of the Colonial hospital there.

In the Belgian Congo beri-beri is probably still more prevalent; it is said to be rare and benign among the white residents, but is more frequent among the natives. The labourers, both of black and yellow races, employed on the railway here suffered terribly from the disease. It raged severely among the native inhabitants in 1891-2, but has been much less prevalent since—a result which has been ascribed to an improved *régime*, and particularly to greater care as to diet[2]. In the Cameroons, on the other hand, beri-beri has not been seen[3].

To what extent beri-beri exists as an indigenous disease in South Africa it is not easy to say. A few cases are admitted in some years to the Durban Hospital, and in 1898 as many as ten came under treatment; two of which were in European residents of Durban who had never been outside the colony. In some other years the few recorded examples of the disease in Durban have been recognised as importations from elsewhere. At Pieter-maritzburg it is said to be far from infrequent, occurring not only among the coolie class but also among the European inhabitants. In Mashonaland beri-beri is sometimes seen in the natives but never in the white population.

Madagascar was the scene of an epidemic of beri-beri in 1866-7 which prevailed at Diego Suarez, in the extreme north of the island, and though the disease is said not to be endemic in the island generally, a few cases have been seen in the capital. A disease of a peculiar character, described as "acute anæmic dropsy," prevailed in Mauritius in 1878 and 1879; it presented

---

[1] Ref. in *Janus*, 1896–7, p. 389.
[2] *Congrès Nat. d'Hygiène et de Climatologie Méd. de la Belgique et du Congo*, 1897.
[3] Plehn, *Janus*, 1896–7.

some points in common with the moist variety of beri-beri, but differed from it in many important respects, "especially in the absence of paralysis, in its more evident transportability by human intercourse, by its attacking almost the whole population of certain districts, and by the much smaller mortality [in proportion] to the numbers affected." A similar disease is said to have prevailed in Assam and Lower Bengal in 1877–80, and the outbreak in Mauritius was ascribed to importation from India. Neither this disease nor true beri-beri is now endemic in Mauritius.

In St Helena, on the other hand, the disease has certainly been seen recently. In the (unprinted) hospital reports of the colony for 1897 and 1898, as many as 13 cases of beri-beri are included in the returns of the first year, and 8 cases with one death in those of the second. No details are furnished as to their source ; but, unless they were examples of imported infection and if the year in question was not an exceptional one, they would indicate a considerable degree of prevalence of beri-beri in the island, for in 1897 the number of cases of this disease exceeded those of any other, and in 1898 were only exceeded by cases of diseases of the skin, nervous, and respiratory systems. In the Island of Ascension also a serious outbreak, spoken of as a "recrudescence," of beri-beri occurred in 1898 among Kroomen stationed on the island [1].

It must be regarded as doubtful whether beri-beri exists as an endemic disease in any part of the North American continent or the adjoining islands, though it has frequently been imported to their shores. It is said to have been recognised (though whether as an indigenous or exotic disease is not stated) in Cuba [2], Guadeloupe, and other islands of the Antilles. A few cases are occasionally mentioned in the health returns of Jamaica and of Trinidad. Cases of a form of peripheral neuritis, supposed to be beri-beri, have been reported as occurring in fishermen on the North American coast.

Imported Cases of Beri-beri. There are many other countries and islands to which beri-beri has been occasionally imported, and without becoming even temporarily endemic has remained

[1] *Report on the Health of the Royal Navy* for 1898.

[2] Beri-beri is perhaps endemic in Cuba ; occasional cases of and deaths from the disease are observed at the present day at Havana ; and it was alarmingly prevalent at Matanzas in December, 1897.

entirely, or almost entirely, confined to the persons who have imported it. Cases of the disorder are frequently admitted to the Seamen's Hospital at Greenwich, the patients being usually from ships newly arrived in the Thames, but not infrequently the disease develops in the Asiatic crews of vessels which have been lying up in the river for several months. A very large number of instances of the carriage of infection of beri-beri by sea over long distances are on record, and in this way the disease has been imported to the coast of California, to New Jersey, to the Bermudas, to many English and probably many continental ports, to Cayenne in French Guiana, to South Africa, and to many other parts of the world. There would seem to be almost no limit to the distance over which beri-beri can be transported by ships, and sea-carriage is apparently the principal means by which the disease has been and is being diffused over the earth's surface.

Ship Epidemics.- Ships, indeed, under certain conditions, seem to be peculiarly favourable, not merely to outbreaks of beri-beri among the crews, but to its persistence for very considerable periods, not only in the men originally attacked but even after a complete change of crew. Such outbreaks, which are apparently often associated with, if not due to, conditions of dirt, over-crowding, bad ventilation, and an imperfect dietary, were perhaps more common in the past than they are at the present day. In former years quite one-fourth or more of the *personnel* of the Japanese navy were, it is said, annually attacked by the disease. Examples of quite recent occurrence are also not wanting. Thus in December, 1896, the barque "Lodestar" arrived at Falmouth from Rangoon with a cargo of rice. During the voyage, when near St Helena and already 92 days at sea, two men developed symptoms of beri-beri, and later the whole crew were attacked by the disease and three died. Subsequently this barque was sold to German owners, was re-christened the "Steinbok" and entirely remanned, mostly by Norwegians and Swedes, shipped in Amsterdam. On January 27th, 1898, she sailed from Java for New York with a cargo of sugar; she touched at Bermuda in June, and reported that on the voyage the whole of her crew, with one exception, had developed symptoms of beri-beri. It is important to add that as this boat had visited Java, where this disease is notoriously prevalent, the second outbreak may have been due to a fresh infection, and not to persistence of infection from the first outbreak. Another notable instance was that of the s.s. "Nour-el-Bahr" in the coastguard service on the Arabian coast of the Red Sea. She arrived in the Camaran roads on September 21st, 1898, and in the middle of November twelve of the crew, all Indians, fell ill with beri-beri. This was a new boat, freshly arrived from Genoa where she was built, and the crew of 33 (28 of whom were Indians) had left Bombay seven months before the appearance of the disease. In this instance, unless it was of local origin, the infection must have remained latent for seven months after its importation from Bombay. In the same year a remarkable outbreak of beri-beri occurred on a Brazilian man-of-war, the "Benjamin Constant." She was sent with a crew of 250 officers and men to the north of Brazil on a voyage of instruction, and in the harbour of Para beri-beri broke out. She went on to Pernambuco, and during the voyage ten men died of the disease. In the harbour of Pernambuco eight more men died, and 10 officers and 65 men of the crew were affected by the malady.

In some of these and similar instances of outbreaks of beri-beri on board ship the insanitary conditions already briefly mentioned have seemed to contribute powerfully to the development of the disease. Manson has described in lurid detail the hot, steamy atmosphere in which the crews of some ships live, and the "sodden state" of the dark, damp, overcrowded, unventilated places they sleep in. Coming from a warm climate into cooler British waters they light fires, block up every means of ventilation, and bring about "a very good imitation of the tropical conditions the germ of beri-beri requires for its development." "In other words," it is added, "these lascar sailors create an incubator on a large scale, which, should it chance to contain a beri-beri germ, quickly becomes extensively infected and lethal."

Epidemics in Institutions. It is a well-recognised fact that the inmates of schools and gaols in the countries where beri-beri is endemic are particularly liable to suffer from the disease. The example of the Singapore prison has already been quoted, and a no less striking instance is found in a severe and fatal outbreak which occurred in the Rajahmandry gaol in the Madras Presidency in the year 1898[1]. Labourers in mines and on plantations, or

[1] Report on Sanitary Measures in India in 1898–99, p. 21. *Parliamentary Paper*, Cd. 397.

employed in large engineering operations, also appear to be peculiarly prone to its attacks. At Banka and Billiton, in the Eastern Archipelago, the tin-mining population has frequently suffered from beri-beri. In Borneo, where beri-beri is regarded as essentially a jungle-disease, coolies labouring on the estates are frequently attacked by it. In the Belgian Congo the labourers employed in constructing the railway suffered terribly from the disease. In Japan it is said to be particularly prevalent in the mining districts of Kiushiu.

In like manner bodies of troops have from time to time been attacked by beri-beri. At Atjeh in Sumatra the troops have suffered severely from it. In the Japanese army it was very prevalent in former years.

It is probable that in all these instances overcrowding and bad hygiene play an important part in fostering the disease.

Finally the occurrence of remarkable outbreaks of beri-beri in lunatic asylums has to be mentioned, as opening up some difficult and hitherto unsolved problems in the etiology and epidemiology of the disease. A well-known example is that of the Tuscaloosa Asylum, in Alabama, U. S. A. In this institution, which usually contains some 1200 inmates, a few cases of beri-beri developed at the end of 1895; they were not severe or apparently of an infectious character. But in the very hot and dry summer of 1896 the disease became epidemic in the asylum, 71 cases in all, of which 21 ended in death, being recorded. The larger number of the patients were white persons, but negroes were also attacked, and in them the disease was more severe and fatal.

An analogous outbreak occurred in the Richmond Lunatic Asylum at Dublin in the years 1894–97. The disease first appeared here in the autumn of 1894, and by the end of the year it had attacked no less than 174 persons, and caused 25 deaths. The building was much overcrowded at the time, and there seemed to be little doubt that this had in some way contributed to the outbreak. In 1896 there was a fresh epidemic of the disease in the asylum ; it was somewhat milder than the first, and was the cause of only 115 attacks and 8 deaths. In the following year a third outbreak occurred, and in this no less than 246 persons suffered from the disease and 11 died. The cause of this, as of other asylum epidemics of beri-beri, has hitherto remained obscure. About the same time smaller outbreaks of the disease occurred in the Suffolk County Asylum at Melton in the winters of 1894–5 and 1896–7 ; and in 1895 a similar epidemic was reported from the Arkansas State Asylum at Little Rock, Arkansas State. A few cases of an analogous disease have also been observed in at least two German lunatic asylums[1], and in a French asylum at Saint Gemmes-sur-Loire (Le Dantec).

[1] Conolly Norman, F.R.C.P.E, *Brit. Med. Journal*, Sept. 24, 1898.

BLACKWATER or HÆMOGLOBINURIC FEVER.

General Characters and Etiology. Hæmoglobinuric Fever or Blackwater Fever is by some regarded as a severe form of malarial infection, by others as a distinct and separate disease. It is characterised by fever; the passage of dark, almost black, urine; bilious vomiting; severe lumbar pain, and other symptoms. It usually attacks European residents in those countries where it is endemic in the second or third year after their arrival. It is one of the most deadly diseases that have to be feared by the European compelled to reside in tropical Africa, or in the other countries where it prevails. Its relation to malaria and to the tick fever of cattle will be briefly discussed later. Its exact etiology is unknown. Blackwater fever is essentially an endemic disease, but it can take on epidemic characters both in countries where it is permanently endemic, and occasionally in other countries where it is not usually met with.

History. The disease was first described by French naval surgeons, who observed it in the island of Nossi-Bé, off Madagascar. Comparatively little is known of the history of the disease itself. It is possibly only of recent introduction into India, as the earlier writers on Indian diseases make no mention of it. In many other parts of the world where it is now known to exist attention has only quite recently been called to it. But there is nothing to show that the fever had not existed in some of these countries before, remaining unnoticed, or at least unrecorded. On the other hand, the disease certainly seems capable of appearing, under special conditions, in countries believed to be hitherto free from it. As an example may be mentioned the outbreak among labourers employed in cutting the canal through the Isthmus of Corinth.

Recent Geographical Distribution. *Europe.* In Europe hæmoglobinuric fever has occasionally appeared in the form of localised prevalences, usually for short periods and under special circumstances. It has been recognised in Sicily and also in the Roman Campagna. In Sardinia it has been seen, and in 1885 twenty cases of the disease developed in prisoners in a gaol at Castiadas. During the construction of the canal through the Isthmus of Corinth it prevailed as an epidemic among the labourers; and elsewhere it has been associated with soil-disturbance on a large scale. It has also been seen in Rhodes and other islands in the Greek Archipelago. A few cases have been observed in England and France in persons returning from Africa or other places where the disease is endemic.

Asia. In Asia the disease is not common. It is, however, occasionally met with in India, China, and the eastern Peninsula, although some observers have denied its occurrence here. Thus it is said to be far from uncommon in Assam and the Duars. It is not unknown in other parts of India, but the evidence for its existence is scanty, and it would appear to be a rare disease here. Many of the leading writers on Indian diseases (as, for example, Chevers, Maclean, Fayrer, and others) make no mention of this fever. But Firth and Notter have described some doubtful cases as occurring at Mian Mir, Amritsar, and Meerut[1]. Recently some cases, apparently free from doubt as to their nature, have been seen in the Darjiling Tarai, or swampy forest-belt at the foot of the eastern Himalayas, and one case was seen at Sylhet[2]. In the Dutch possessions in Malaysia the disease first came under observation after the Atjeh war of 1874–78. Cases occurred at various places in Java, Celebes, and Borneo. Its occurrence at Singapore is doubtful.

In French Cochin China it is said to have been occasionally seen. In China it must be very rare. Manson never saw a certain case in Amoy, Hongkong, or Formosa, and similar negative evidence is forthcoming from others from Shanghai, Hongkong, and north and south Formosa. No mention of the

---

[1] *Army Medical Department, Appendix to Report for* 1885.

[2] Baldwin Seal, *Journ. Trop. Med.*, Feb. 1899, p. 179.

disease has apparently been made by any of the writers in the Medical Reports of the Chinese Imperial Maritime Customs during the ten years ending in 1899. In the Chinese army, on the other hand, hæmoglobinuric symptoms were said to have been frequent in 1885 on the borders of Tongking and the Chinese province of Kwang-si, and a few cases have been seen at Fatshan.

Australasia. In Australasia and the Pacific Islands blackwater fever is also either unknown or extremely rare, and generally throughout the Far East the disease is either absent or very exceptionally met with. In New Guinea, however, cases of the disease have been seen.

Africa. In Africa, on the other hand, the reverse is the case. The tropical and subtropical portions of this continent are the principal homes of hæmoglobinuric fever, which in many places is the cause of a very serious degree of mortality among European residents. While in Algeria it is said to be quite unknown[1], it is particularly prevalent along the West Coast from the Senegal to the Kuanza. It is endemic in the valleys of the Senegal, the Gambia, the Niger, and the Congo Rivers. At Bathurst (Gambia) in 1898, of four deaths among the Europeans, two were due to hæmoglobinuric fever. In the French Sudan it is particularly severe. At certain French settlements on the Gaboon and on the Gold Coast, 38 per cent. of the European residents were, a few years ago, annually attacked by the disease; and on the Upper Senegal 28 per cent., at Cazamance and on the Rio Nuñez 15 per cent., at Cayor 8 per cent., and at St Louis and Goree from 1 to 3 per cent. of the European inhabitants, according to the same estimate, annually suffered from an attack of this fever[2]. At Sierra Leone and at Cape Coast Castle Europeans are no less liable to it; and at the last-named place it is the principal cause of mortality. In the Congo Free State it is a prolific source of invaliding and death. One estimate has placed the mortality from this cause in a group of Congo Free State employees between the years 1878 and 1892 at from 11 to 12 per cent. of the total mortality from all causes. Of 23 deaths among Baptist Missionary Society missionaries on the Congo six were due to this malady. In 1896 it became epidemic in Boma. At Chisamka, Angola, in Portuguese West Central

[1] Brault.
[2] Ref. in *Trans. Epidem. Soc.*, N. S., vol. XII., p. 123.

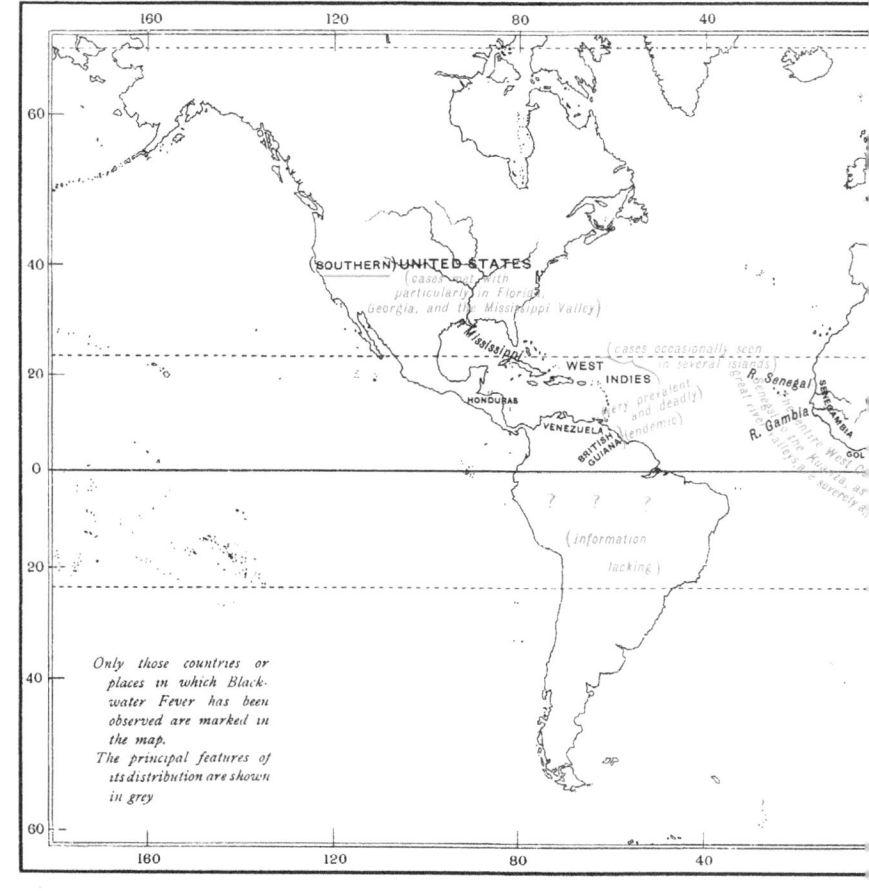

DISTRIBUTION OF BLACKWATER FEVER

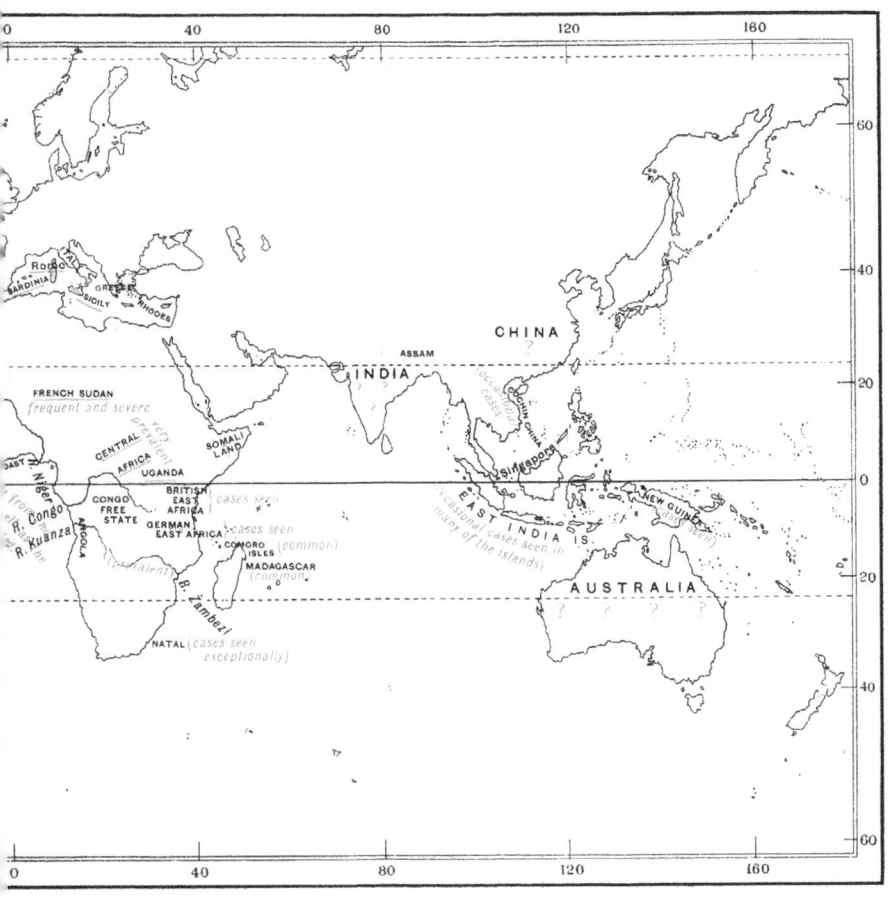

Africa, cases of the fever have been observed at a spot 300 miles from the west coast and 400 miles south of the Congo[1].

In British Central Africa blackwater fever is a frequent disease, and about 10 per cent. of the European residents, it is estimated, are attacked by it. In this part of Africa it is said to be more common on elevated than on low-lying ground, but the reverse appears to be the case elsewhere. It is known in British East Africa, in Uganda, in German East Africa, and at least one case has been seen in Somaliland[2]. In Uganda it is thought to be increasing in frequency.

It is frequently seen on the banks of the Zambesi and the shores of Lake Nyassa[3], and unmistakable cases of the disease have been met with in the Shire highlands. It has been observed in Mashonaland, though it is not common there, and even as far south as the colony of Natal it has, though exceptionally, been seen.

It is common in Madagascar, not only in the lower-lying regions, but also in the more elevated district of Antsianaka, where it prevails especially in the cool season. It is also common at Nossi-Bé (where the disease was first described as a separate entity); and at Mayotta in the Comoro Islands it is very prevalent.

North and Central America. In North America hæmoglobinuric fever has been met with as an indigenous disease in the southern States, especially in Florida, Georgia, and the States in the basin of the Mississippi river. In the West Indies it is sometimes seen, and in Cuba it is not infrequent; in Jamaica it is less so, and one observer there saw three cases only in the course of twenty years; occasionally a single case of the disease is recorded in the annual reports of the Island Medical Department. In Martinique and other French West Indian possessions it is said to prevail; but in St Vincent, St Lucia, and other British possessions in the West Indies it is probably not common, as no mention of it is apparently found in any of the recent annual health reports of these islands. In regard to Trinidad, however, Dr Bennett, the Government Medical Officer, has himself seen one case of blackwater fever and has heard of two others in the island; while in 1901 five fresh

---

[1] Massey, *Journ. Trop. Med.*, Feb. 15, 1901.

[2] Crosse, *Epid. Soc. Trans.*, N. S., vol. XVIII., p. 114.

[3] W. Poole, *Journ. Trop. Med.*, Jan. 1899.

cases of the disease were recorded there.   In British Honduras also cases have been met with.

South America.   In Venezuela this form of fever is said to be exceedingly prevalent and particularly deadly in the plains, but less common in the highlands[1].

The disease is apparently endemic in British Guiana ; several sporadic cases are mentioned in the Surgeon-General's report for the year 1898-99 as occurring in Cotton Tree district and in other parts of the colony.   In Brazil it has not been met with, and with regard to the other portions of South America there is an absence of information which may be held to indicate that the disease is rare if not unknown.

Factors Governing the Geographical Distribution. The difficulties surrounding the diagnosis of hæmoglobinuric or blackwater fever, the uncertainty as to its specific nature and causation, and the confusion which of necessity has arisen between this disease and other analogous diseases, such as paroxysmal hæmoglobinuria, renders any exact account of its geographical distribution almost impossible.   Our knowledge of its true position in the category of disease is at present imperfect. It is only in comparatively few countries that its existence has been determined with accuracy.   These have been named above, and their position is shown in the accompanying map.   But it is far from certain whether the disease does not exist, actually or potentially, in many other countries besides those named.   If there is any truth in the theory (which will be referred to later) that hæmoglobinuric or blackwater fever in man is the same disease, and due to the same cause, as "redwater fever" or "tick fever" in cattle, then it is clear that the geographical distribution of this disease, or at least of the micro-parasite which produces it, is wider than is at present believed.   Australia and South Africa, for example, would have to be mentioned with other countries as important homes of the disorder.   But until further light is thrown upon this subject it would serve but little purpose to treat the two diseases as one and the same.   It is not known that they are due to the same parasitic cause, and even if that were proved, there is still an absence of proof that in those countries such as Australia, South Africa, and Texas, where the

---

[1] Ackers (of Caracas), quoted by Manson, *Epid. Soc. Trans.*, N. S., vol. XII., p. 126.

parasite is widely spread in animals, it affects man to any extent. It is better, therefore, to confine the present consideration of the disease to those countries where it is known to exist in man.

The limits of prevalence of blackwater fever, though they cannot be stated with absolute accuracy, appear to be approximately as follows. In Europe it has not been seen as an indigenous disease further north than Italy. In Asia, Assam and (perhaps) the southern portion of China appear to represent the most northerly limits of the affection. In the Western hemisphere it has not been recognised north of the "Southern United States." In the southern hemisphere hæmoglobinuric fever has not been observed further south than the latitude of Natal, and there only as an exceptional, and perhaps imported disorder. In general terms the utmost limits of latitude in which this disease has been proved to exist may be said to be the 45th parallel to the north and the 35th parallel to the south of the equator.

Briefly, then, hæmoglobinuric fever prevails principally in the tropical and subtropical portions of Africa and in Madagascar, and to a less extent in the corresponding area of North, Central, and South America, and the West Indies. It is rare but not unknown in India and the Far East, and has been seen in certain parts of Europe, principally on the shores of the Mediterranean. It is mainly therefore a disease of warm or hot climates.

It prevails principally, but far from solely, in low-lying regions. In Africa it is very frequently seen along the valleys of the great rivers, such as the Senegal, the Zambesi, the Congo, and the Niger; and in America the Mississippi valley is mentioned as a region where the disease is common. But it is also met with at considerable heights, as for example in the Shire highlands and on the high plateaux of Madagascar, while, as already pointed out, in Venezuela it occurs in the high lands, though it is more common and assumes a severer form in the plains. In its endemic areas it is variable in its prevalence. "It may not be seen for years in a district and then numbers of cases may occur within a short time."

Its relation to race is uncertain. Most authorities are agreed that it attacks Europeans far more than the natives; but the late Dr G. F. Reynolds stated that in Western Africa it attacked natives more frequently than Europeans[1]. In the Belgian Congo old

---

[1] *Journ. Trop. Med.*, 1899, p. 145.

Africans frequently suffer from it. In the Antsianaka district of Madagascar, on the other hand, the Hovas, who are as prone as the white residents to all forms of malaria, are said to be immune to hæmoglobinuric fever[1]. It is to be noted also that some Europeans appear to be immune to the disease, and may live for long periods in places where it is endemic without contracting it. Estimates vary widely as to the proportion of European residents in such places who may be expected sooner or later to suffer from it, but one good observer places it as low as only 10 per cent.[2] When Europeans are attacked it is generally during the second or third year of their residence.

Recently the analogy between blackwater fever and paroxysmal hæmoglobinuria in man and "Texas fever," "redwater fever" or "tick fever" in cattle has attracted much attention. The relationship of blackwater fever to paroxysmal hæmoglobinuria is uncertain, but Sambon has shown that the symptoms and morbid anatomy of the two diseases are practically identical. Without going so far as to regard them as one and the same disease, he points out that the high mortality of blackwater fever (in which alone it seems to differ from the other) may be due to the fact that it usually attacks "the wrecks of severe tropical malaria." The same author is of opinion that blackwater fever is probably due to a protozoal parasite which, if not identical with, is very similar to, that which produces Texas fever in cattle. This disease, which has been shown to be associated with, if not caused by, a blood parasite to which the name *Pyrosoma bigeminum* has been given, has been recognised not only in Texas, but in South Africa, Australia, Roumania, on the shores of the Danube, and in the Campagna near Rome. The fever described by Koch as prevalent in German East Africa is probably of the same nature. It is believed that the parasite gains access to the blood of the cattle through the bites of the cattle-tick, *Boöphilus bovis*; it has, at any rate, been shown that animals never become infected unless they have fed in a field previously infected with this tick. In South Africa, Edington has success-fully practised a method of prophylactic inoculation against the disease in cattle, and a similar method has been recently tried,

---

[1] Laffay, *Archives de Méd. Navale*, Oct., 1899.

[2] Crosse, circular on *Blackwater Fever* for the Medical Department, Brit. Central Africa.

with satisfactory results, in Queensland.   In this colony the disease was first imported in 1894; it has spread, in consequence of the absence of restrictive measures, far to the south of the Tropic of Capricorn, but has not yet obtained a footing on the dry western plains of Australia, and is (or was in 1898) confined to the more humid eastern seaboard[1].   The results of inoculation were most successful in this colony.   Dr Tidswell reported in 1899 that 17,960 head of cattle had then been inoculated with the blood of animals that had already passed through an attack of the fever, and it had been found that, whereas before the introduction of this practice the mortality had been between 60 and 70 per cent., it fell afterwards to between 3 and 5 per cent.[2]

Whether hæmoglobinuric fever in man is due to the same organism as the redwater fever of cattle is uncertain.   Proof of the transmission of the organism from the cattle-tick to the human host is at present wanting.

The exact relationship of hæmoglobinuric fever or blackwater fever to malaria has not, as already pointed out, been finally determined.   It appears, however, to be no longer open to doubt that the condition of hæmoglobinuria with fever frequently occurs in malarial subjects in whom the malaria parasite is present; that it may also occur in persons who have never suffered from malaria and in whom the malarial parasite cannot be found; and that its geographical distribution corresponds to that of malaria only to the extent that it is found in many notoriously malarial regions on the earth's surface, but in far from all, and that there are wide tracts or whole countries in which malaria even in severe forms is endemic where hæmoglobinuric fever is unknown or extremely rare.   It seems probable that persons who have suffered or are suffering from malaria are more liable than others to attacks of the disorder.   On the West Coast of Africa, where blackwater fever is the most fatal type of fever met with, it has been observed that "the more malarious a district is, the more malignant and fatal is the type of blackwater fever met with there[3]."

[1] See a paper on "Queensland's Progress" by Sir Horace Tozer, Agent Genl. for Queensland, read at the 3rd Ordinary Genl. Meeting of the Royal Colonial Institute.

[2] Report by Dr. Frank Tidswell to the New South Wales Government, *Journ. Trop. Med.*, May, 1899.

[3] G. F. Reynolds, *Journ. Trop. Med.*, Jan., 1899.   Dr Reynolds himself died from blackwater fever the month after his paper appeared.

In the Belgian Congo old Africans the subjects of chronic paludism are said to be specially liable to attacks of blackwater fever[1].

In their seasonal relations the two diseases have also much in common. In some parts of the West African coast both malaria and blackwater fever are least prevalent during the rainy season and in the dry season, and most so during the periods of transition from one of these seasons to the other. But this parallelism between the two diseases is far from being universal, and at Lagos cases of blackwater fever are said to occur all through the year and quite independently of the presence or absence of malaria[2].

If hæmoglobinuric fever is in truth only an intensified form of malaria it is at least remarkable that, as just pointed out above, its geographical distribution should be so much more circumscribed than that of the various other forms of fever included under the general term "malaria." Ordinary "malaria" is widely prevalent in India, China, and the Far East generally, and yet, as already shown, blackwater fever though met with in these regions is a rare disease there. From both English and French possessions in Eastern Asia multitudes of invalids return yearly to Europe, many suffering from "malaria" in one of its numerous forms, yet cases of hæmoglobinuric fever are practically unknown among them. Among invalids returning from Africa, on the other hand, the disease is far from uncommon.

Like malaria, blackwater fever is said to have appeared concurrently with earth disturbance. Laffay has recorded this fact in connection with the prevalence of the disease in Madagascar. Crosse also records that the fever has become much more prevalent in recent years in the Niger Territories and this he attributes to the turning up of virgin soil for coffee and other plantations. Before 1886 the disease was unknown in those regions, and it was only after the introduction of extensive coffee plantations that it began to prevail. " It is significant," this author adds, "that our first three gardeners died from blackwater fever, and that for some considerable time cases only occurred near the plantations; and as plantations became more common, so the disease spread to the other stations

---

[1] *Congrès National d'Hygiène et de Climatologie Médicale de la Belgique et du Congo*, 1897.

[2] Strachan, *Journ. Trop. Med.*, Feb., 1900.

in the territories[1]." The occurrence of blackwater fever among the labourers employed in constructing the canal through the Isthmus of Corinth has been already mentioned as an example of the association of this disease with soil disturbance on a large scale.

A few years ago it was suggested that blackwater fever is nothing more than the result of quinine poisoning. This view was first propounded by the Italian observer, Tomaselli, and has been supported by no less an authority than Koch. But the arguments on which the theory is based are scarcely convincing, and an abundance of observations have been published tending to disprove it.

[1] *Guy's Hospital Gazette*, Oct., 1898.

CALCULUS (STONE IN THE URINARY PASSAGES).

General Characters and Etiology. Stone may form in any part of the urinary tract. Hence renal, ureteral, vesical and urethral stones are all met with. The chemical constituents vary widely. Uric acid, urates, oxalates, various phosphates, calcium carbonate, cystin, xanthin, and concretions of blood and fatty matters are the principal varieties of substances found in urinary calculi. Two or more of these are often seen together in the same stone.

The mode of formation of a stone is uncertain. In some a foreign body forms the nucleus; in others the ovum of a bilharzia worm; in the majority no definite nucleus is discoverable, and it is surmised that excess of earthy constituents in the renal secretion, combined with an habitually concentrated condition of the urine, leads to deposit of one of the salts named, and that this forms a nucleus around which fresh deposits are successively formed, leading to the production of a stone. The matter is, however, still a subject of controversy, and it is probable that the true pathology of calculus is less simple than it appears. Gout is probably a strong predisposing factor in the production of some kinds of stone.

Geographical Distribution. *Europe.* Urinary calculus is met with to some extent in all European countries. In England it is not a particularly common disorder; in Scotland it is rather more frequent, but in Ireland it is exceedingly rare. In England the Eastern and Southern counties and the West Riding of Yorkshire formerly furnished a larger number of cases of the disease than other counties. I have no recent information in regard to France, but the tendency of the disease has been to decrease there during the last century. In Holland it was at one time very frequent, but is now less so.   In Germany stone is, or was several decades ago, rare, except in a small portion of the Duchy of Altenburg, and in parts of Old Bavaria and Upper Swabia.

In Denmark, Norway, Sweden, Iceland, and Finland stone is very rare.   In Russia, on the other hand, while infrequent in the extreme north, the disease is very common in many of the central governments.   The recently published returns do not give the figures for each government separately, and the territorial groups for which statistics are available are unfortunately, so far as concerns European Russia, more of administrative than of geographical interest.   It will, however, be useful to give the figures for the principal divisions of both European and Asiatic Russia. I have worked out the proportions from the reports of cases treated both in and out of hospitals in each of the years named :—

Cases of Urinary Calculus per million inhabitants.

| Year | European Russia | | | | | Asiatic Russia | | |
	Governments with Zemstvos	Governments without Zemstvos	Polish Provinces	Baltic Provinces	The Don Cossacks	The Caucasus	Siberia	Central Asiatic Russia
1893	113	39	64	194	87	39	20	10
1894	113	40	62	224	91	49	16	12
1895	123	43	79	220	98	44	21	14

The first two columns of this table comprise the greater part of European Russia.   I cannot account for the great difference in the frequency of stone between those governments with and those without the system of local self-government known as the *Zemstvo.* It is true that the second column includes a greater proportion of governments in the north of the country than the first ; but the Baltic provinces are also in the north, and they return more cases of stone than any other part of Russia.   The only statement, indeed, in regard to European Russia which these figures justify, is that stone is very common on the shores of the Baltic, less so in most other parts of the country, still less so among the Don Cossacks, and least so in Russian Poland.

Concerning the rest of Europe it seems that calculus is very rare in Switzerland, but not infrequent in Hungary, Italy, and in parts of Turkey and Greece. Recent information in regard to Spain seems to be lacking.

Asia. The figures given in the table on page 55 seem to indicate that calculus is very much rarer in all parts of Asiatic Russia than in European Russia. The returns from the former, however, in regard to this as to all other diseases, are much less complete and reliable than those from the latter.

In Asia Minor, Arabia, Syria, Mesopotamia, Persia, and Afghanistan, urinary calculus is remarkably common. But of all Asiatic countries India is the one in which stone is most frequently seen. It is, indeed, probably more prevalent here than in any other part of the world. The urate and oxalate of lime stones are the varieties most often seen. Rare in Madras, they are extraordinarily common in the Punjab, and even more so in Hyderabad in Sindh. Freyer states that "broadly speaking, stone is mainly confined to the Punjab, North-West Provinces and Oudh, the upper part of Bombay (Sindh and Gujerat) and, to a lesser extent, Central India—that is, chiefly to the great alluvial plains watered by the Indus and upper half of the Ganges[1]." In Farther India stone is found to be common in Bangkok, and still more so in some other parts of Siam (Gowan). It is rare, on the other hand, in the Malay Peninsula.

In China stone is seen principally in and near Canton; it is extremely frequent here, but in other parts of the empire it is only occasionally seen.

In Australia, New Zealand, and the Pacific Isles, calculus is said to be a very rare disorder.

Africa. The natives of Upper and Lower Egypt are very subject to urinary calculus. In Abyssinia, in Morocco, and in Algiers the disease is also somewhat common; it is less so in Tunis; and in Nubia, Uganda, and the great zone of tropical Central and Western Africa it is remarkably rare. Information as to its prevalence in South Africa seems to be incomplete.

---

[1] Brit. Med. Association, 1901.

America. Rare in Alaska, stone is comparatively common in many parts of Canada. In the United States it is, or was, most frequently seen in Kentucky, Tennessee, Virginia, Ohio, the North of Alabama, and perhaps Missouri[1]; while in the rest of the States it appears to be comparatively infrequent.

The data for Central and South America are extremely scanty, but they seem to indicate that calculus is rare throughout, with the possible exception of Entre Rios and perhaps certain parts of Brazil.

Factors concerned in the Distribution. Climate can scarcely be regarded as having any very direct influence upon the frequency of stone. The disease is very common in tropical India, in the temperate zones of Europe, and in the decidedly cold Baltic provinces of Russia. It is, however, rare in the far north of both hemispheres. If climate has any effect on the production of stone, it is probably an indirect one. In India, for example, the area where stone is common is, according to Freyer, "characterised by the scantiness of its rainfall as compared with the rest of India." "It is further characterised," he adds, "by great alterations in temperature, being intensely hot in summer and cold in winter. These two conditions—the scanty rainfall and intense heat—promote excessive perspiration, most of the fluids of the body thus passing off by the skin and leading to concentration of the urine. It is reasonable to suppose that this condition would be more likely to facilitate the deposition of crystals in the urinary passages than that prevailing over the rest of India where the rainfall is heavy and the temperature equable, thus checking the perspiration and leading to dilution of the urine. It is very suggestive that even in the stone-producing region itself the incidence of stone is, as a rule, in direct proportion to the scantiness of the rainfall and the intensity of the heat." Similar conditions, however, prevail in the Caucasus, Central Asiatic Russia, and elsewhere, without being associated with great frequency of stone.

Calculus occurs on every variety of soil, and the belief formerly held that it was commoner on chalk formations than on others does not seem to be any longer tenable.

The nature of the food usually eaten does not seem to have

---

[1] Gross, quoted by Hirsch.

any influence on stone formation. Animal and vegetable eaters are found to be equally liable. The water drunk is, however, more liable to suspicion. The presence of lime salts in drinking water has long been regarded as predisposing those who drink it to stone. The evidence on this point is extraordinarily contradictory and inconclusive, and there is not room here to discuss it. Briefly it would seem that, while excess of this constituent in water has often been found associated with prevalence of stone, it is by no means necessary to such prevalence, for stone may occur with frequency in regions where the water is free from lime salts.

That stone and "endemic hæmaturia" undoubtedly occur with great frequency in the same parts of Africa has long been known; but the explanation has only been forthcoming since the discovery of the bilharzia hæmatobia and its relation to the hæmaturic disease. In the endemic areas of this disease stone is found to be frequently produced by deposits forming round a bilharzia ovum in the urinary passages.

The female sex is much less liable to urinary calculus than the male. The effect of race is uncertain; but the pure-blooded negro enjoys a remarkable, if not absolutely complete immunity from the disease.

CANCER or MALIGNANT DISEASE.

Under the term "malignant growths" or "cancer" is included, it need scarcely be said, a considerable variety of different pathological conditions. It is very possible that some of these differ from the others entirely in their mode of production, and it may well be that a fuller knowledge of their nature will demonstrate the impropriety, from a scientific point of view, of grouping together phenomena so distinct as, for example, epithelioma and sarcoma. It has however become the custom among those who prepare mortality or disease statistics to form a special group under some such heading as that given to this chapter, and in many instances no attempt is made to differentiate the statistics of the various kinds of growth included under this heading. The principal forms of growth thus grouped together are the varieties of carcinoma, sarcoma, and epithelioma, which, while probably of "developmental" origin, all present that character of "malignancy" which it is unnecessary here to define more fully. This group of growths has a very wide geographical distribution, but like most other forms of disease their distribution is far from even over the earth's surface, and in some happy countries, such as the Faroë Islands and Arabia, they are said to be almost, if not quite unknown.

Recent Geographical Distribution. *Europe.* In England cancer or malignant disease annually causes between twenty-six and twenty-seven thousand deaths, and there has for many years past been a steady and almost unbroken rise not only in the total number of deaths from this cause, but also in the proportion which it bears to the population. The Registrar-General, in his report upon the mortality in England and Wales for the decennium 1881–1890, pointed out that the death-rate from cancer or malignant disease had risen from 384 per million living of all ages and both sexes in 1861–1870, to 468 in the succeeding decennium, and to 589 in the decennium 1881–1890. This rise was partly accounted for by the transference, during the years 1885 to 1890, of a considerable number of deaths from other groups to this group in consequence of further inquiries made into the cause of doubtful deaths, in accordance with a system which was not in practice in previous years. It was, further, partly accounted for by improved methods of diagnosis, a certain proportion of deaths formerly ascribed to other causes being now more correctly attributed to malignant disease. But after these sources of fallacy have been eliminated, a large difference between the figures for the earlier and the later periods remains and can only be explained as due to a really increased prevalence of this form of disease. The annual figures, published since the year 1890, have shown that this increase is maintained. In the years 1891 to 1900 the annual mortality from this cause per million living has steadily risen from 692 in the former year to 828 in the latter. During the 28½ years ending in 1897 the average increase in mortality from malignant disease had been as high as 3·1 per cent. per annum in males, and 1·9 per cent. per annum in females, but the proportional mortality still remains considerably higher in females than in males. The area of greatest prevalence of malignant disease before 1890 was situated in and around the counties of Huntingdon and Cambridgeshire, and consisted of the districts of Stamford, Bourn, Spalding, and Holbeach in Lincolnshire, Oundle and Peterborough in Northamptonshire, and the entire counties of Huntingdonshire and Cambridgeshire, with the exception of the registration districts of Caxton, Linton, and Newmarket. In 1881–1890 this area had a mean population of over 300,000, and its crude cancer-mortality rate was 859 per million, or 46 per cent. higher than the general English rate. In the preceding decennium the excess over the general English rate had been 44 per cent. After còrrecting for age and sex distribution, and taking the death-rate for age-groups above 35 only, it was found that the highest cancer rates were in Huntingdonshire, Cambridgeshire, Sussex, and North Wales; and the lowest in Monmouthshire and Derbyshire.

In Scotland the mortality from cancer or malignant growths increased from an average of 580 per million per annum in the decade 1881–90 to 730 in each of the years 1895 and 1896, and to 770 in 1897. In Ireland a very marked rise has also been observed, the cancer mortality in the years 1881–90 being equivalent to a ratio of 410 per million, while in the preceding decade it had been only 340 per million inhabitants. In 1900 it has risen to 600.

In France there is also reason to believe that cancer is increasing. The official statistics, however, include "cancer *and other tumours*" under a single heading and it is possible that the term "other tumours" may include some new growths of a benignant character. The deaths from these causes in French towns of over 10,000 inhabitants steadily rose from an average of 840 per million in 1887–1890 to 1050 per million in 1898. These figures referring to urban populations only cannot, of course, be compared with those for the British Isles, which apply to urban and rural populations together.

Other European States in which cancerous diseases prevail to a considerable extent appear to be the Low Countries, Germany, Norway, Sweden, Switzerland, Greece, and Austria. In all of these countries there is reason to believe that they are increasing in frequency.

In Holland this group of diseases has become considerably more prevalent in the past decade; in the year 1889 the total deaths registered from this cause were 3411, and the ratio per million of the population was 750; but in each succeeding year (with one slight exception) the mortality has steadily increased, until in the year 1898 the number of deaths from these causes was 4685 and the ratio per million of the population was 930, or considerably higher than that observed in the English "cancer-area" already described.

In Germany cancer is said to be somewhat less common than in England or in Sweden, but is yearly increasing in frequency.

In Denmark, at least in the urban population, deaths ascribed to malignant disease have also increased in number in recent years. There have been considerable fluctuations in the annual returns under this heading, but on the whole the tendency is towards a regular increase. The deaths from cancer of the stomach rose from 299 in 1892 to 385 in 1896 and 377 in 1897; those from cancer of the uterus from 106 in 1892 to 118 in 1897; those from cancer of the female breast, in the same period, from

32 to 66, and those from cancer in other parts from 400 to 493. Deaths from cancer in all these forms, taken together, rose in the six-year period from a total of 837 to 1054. This indicates a very high degree of mortality, and in this portion of the Danish population deaths from cancer exceed those from all the principal acute febrile diseases taken together. The last-named figure represented a mortality in the urban population of over 1300 per million.

A similar increase has been observed in Norway. In the quinquennium 1881–1885 cancer yearly caused an average of 945 deaths, or 5·6 per cent. of the mortality from all causes; in 1886–1890 these figures rose to 1083 for the average annual number of deaths from the disease, and 5·84 for the percentage ratio. In the four following years the total deaths from cancer were respectively 1224, 1278, 1405, and 1449, and the percentage ratios to total deaths from all causes were 6·44, 5·9, 6·39 and 6·2 respectively; the last-named figure being equivalent to a mortality-rate of about 700 per million living.

In Sweden the death-rate from cancer in 1886–7 was no less than 950 per million.

The returns of cancerous diseases in the Russian Empire are very incomplete, and no figures are in fact available to enable an opinion to be formed as to the relative frequency of these maladies. Each successive year, however, certain groups of "cases" and of "deaths" from these causes are recorded in two tables dealing with diseases treated "inside" and "outside" the hospitals respectively. The figures in these tables do not represent more than a small proportion of the total cases of and deaths from the various diseases included in them, but it is probable that they represent more or less nearly the same proportion each year, and it is therefore interesting to note that so far as they go they seem to represent a marked increase in the mortality from cancer, and a great increase, followed by a slight fall, in its prevalence in recent years. Thus this special group of cases of cancer rose from an average of 48,506 in the years 1887–92, to 70,005 in 1893, then fell to 65,025 in 1894, rising again to 69,329 in 1895. The group of deaths, on the other hand, rose from an average of 3,408 in the years 1887–92, to 3,942 in 1893, 4,092 in 1894, and 4,529 in 1895. These figures indicate a very marked rise in the period in question; and if the total numbers of cases and deaths have followed the same course as the numbers in these special groups, it would appear that cancerous diseases have recently become much more frequent in Russia as they have elsewhere.

In Austria cancerous diseases are said to be far from rare, but I have not succeeded in obtaining recent statistics for their degree of prevalence. Switzerland is also spoken of as a country in which this class of disease is particularly fatal. In Bohemia cancer is believed to be on the increase; in Prague, at least, cases of the disease are said to be more commonly seen than formerly[1].

In Servia cancer would appear to be comparatively rare. In the year 1894–5 only 156 deaths were ascribed to this cause in a population of 2,312,484. The mortality rate per million living was therefore 67; and the deaths from "cancer" were only 1·92 per cent. of the total deaths from all causes. In European Turkey all forms of malignant growths are seen, but no statistics as to their frequency are available for comparison with other countries, and I cannot learn that they are either exceptionally common or the reverse. In Greece cancer is frequent, and is said to have become more so in recent years.

In Italy cancer gives rise to a considerable annual mortality. In the decennium ending in 1896 there was an increase in the number of deaths returned under this heading, the mortality figure having risen from 428 per million living in 1887 to 496 in 1896. The rise was not quite an unbroken one, and it will be seen that the mortality from this cause in Italy was in 1896 about the same as it was in England four decades ago. Cancer is distinctly more prevalent in England and *à fortiori* in Sweden than it is in Italy.

In Iceland malignant growths are very rare, and in the Faroës they are said to be quite unknown.

Asia. For the greater part of the Asiatic continent no accurate information as to the prevalence of malignant disease is available, and with one or two exceptions it is only possible to assert in regard to Asiatic countries that cancer is believed, on more or less good authority, to prevail in a greater or less degree. For many parts of the continent no information of any kind is to be obtained. The following are the principal features of the distribution of this class of disease in Asia so far as is known

[1] Dr Skaliçka at Conference of Czech Naturalists and Physicians, 1901.

at present, but later and fuller knowledge may very probably throw a different light upon the subject.

In Persia cancer is said to be either very rare or almost unknown. In the Caucasus, on the other hand, it is perhaps far from uncommon; in the returns of the Tiflis municipal hospital patients of all nationalities are mentioned as suffering from this form of disease, Russians and Armenians being apparently more liable than persons of other races[1]. Throughout the Arabian plateaux and in Syria cancer is said to be extremely rare, if not unknown. In Asia Minor it occurs, but as to its relative frequency information is lacking.

Hitherto it has been generally thought that the natives of India were comparatively free from malignant growths, but this opinion was probably based on insufficient knowledge of the facts. Dr Leonard Rogers, Professor of Pathology in the Medical College, Calcutta, states that the records of autopsies performed in the College Hospital furnish figures which "show that malignant diseases not only occur in natives of India, but may be said to be common among them." Buchanan, on the other hand, does not think that cancer or other malignant disease is by any means so common among natives of India as in Europe, and he adds that among British and native troops, and among prisoners (for all of whom accurate statistics exist) there are very few admissions to hospital recorded for these diseases. It is to be noted, however, that these statistics only refer to a limited class of the population in regard to sex, age, and occupation, and are no true indication of the frequency of these affections. From a recent correspondence in the columns of the *Lancet* it may be gathered, from numerous rather conflicting statements, that while cancerous growths do occur in India, both among Hindus and Mohammedans, they are probably considerably less frequent than in many European countries.

In Kashmir malignant disease is met with, but with no great frequency, in spite of the presence of many rivers which annually overflow their low-lying banks—conditions which have elsewhere been believed to favour the prevalence of cancer. Epithelioma of the thigh is seen here in men, produced by local irritation, and cancer of the female breast is also met with.

[1] *Statistics of Caucasian Pathology.* By Dr I. I. Pantiukhof. Tiflis, 1898 (in Russian).

In Ceylon cancer showed a marked increase in the early years of the last decade, the registered deaths from this cause rising from 129 in 1891 to 235 in 1893; but there has since been a diminution, and in 1897 the deaths from cancer were only 177, indicating a mortality ratio of about 54 per million.

In Farther India and the Malay Peninsula malignant disease is not common. Gowan, during four years' residence in Siam, states that he occasionally saw cases of osteo-sarcoma, and also of epithelioma, but had never come across a case of cancer of the female breast. Another observer, Rasch, states that he never met with sarcomatous disease in Siam[1], and in the Laos country cancerous disease generally would seem to be equally rare[2].

In the East Indian Islands this form of disease must also be rare. Nieuwenhuis never saw a case of either carcinoma or sarcoma in the natives of Borneo, and he adds that in consequence of the scanty clothing worn by the people it would be difficult to over-look cases were the disease at all common[3]. It is, however, known there, and in Java cases of carcinoma have been oc-casionally seen by at least one observer, among the hill tribes of that island[4].

The Chinese appear to be more liable to malignant growths than other eastern races, and immense sarcomatous tumours have been described by writers in the Chinese Customs Medical Reports, as observed by them in natives of the country. Of the relative frequency of this form of disease in the greater part of China, however, little appears to be known with accuracy. Maxwell has recorded, in the southern province of Fokien, the occurrence of most of the forms of malignant disease, but is of opinion that there are "many fewer cases than at home."

In Corea cancer is said to be far from common; the forms most often met with are myeloid sarcoma of the jaw and epithelioma of the tongue and mouth[5].

Australasia. In New Guinea cancer must be very rare, at any rate among the natives. Sir William MacGregor states that only one case of the disease has been seen in a native of British

---

[1] *Janus*, 1896–7, p. 445.
[2] Hanson, *Pacific Medical Journal*, Jan. 1902.
[3] Ref. in *Janus*, 1899, p. 422.
[4] Kohlbrügge, *Janus*, 1897–8, p. 221.
[5] *Chinese Customs Med. Reports*, No. XLVIII. 1894.

New Guinea, and that it was in a person who had lived for many years more like a European than an islander[1].

In Australia, on the other hand, cancer is prevalent to a considerable degree, and in some parts of the continent it is increasing in frequency. In Queensland the mortality from this cause was 284 per million inhabitants in 1893; in the following year it rose to 358; in 1895 it was 417; in 1896, 392; and in 1897, about the same (391). In Western Australia, however, though the total number of deaths from cancer rose on the whole (not without marked annual fluctuations) from 20 in 1889 to 55 in 1898, the estimated population in the same period increased in a much more rapid proportion (from 43,698 to 168,129); consequently the ratio of cancer deaths per million inhabitants fell from 458 per million in the first year of the period to 327 in the last. It is obvious that to a great extent the increase of population in this, and also in other divisions of Australia, has been due to immigration, and the figures indicating the mortality from cancer or any other disease have proportionally less value. A large number of the deaths occurring in these ten years may have been in persons who developed the malignant growth elsewhere, and it is impossible to judge, from these figures alone, whether the conditions obtaining in Western Australia are favourable or otherwise to the prevalence of malignant disease.

The same limitations must be borne in mind in regard to the presence and recent history of cancer in South Australia. Here, with a larger, but less rapidly growing population than in Western Australia, the crude death-rate from this group of diseases increased considerably in the same decade. Thus the total annual deaths returned under this heading rose from 133 in 1889 to 184 in 1898; but as the estimated population during the same interval only rose from 311,112 to 362,897, the cancer mortality had increased from 427 per million in the first year of the period to 507 per million in the last. The returns recently published for 1900 indicate a still further rise, to 590 per million.

In Tasmania the deaths from cancer in the successive years 1894 to 1897 were equivalent to 500 per million in 1894, 470 per million in 1895, 580 in 1896 and 480 in 1897. It is stated in the Report of the Central Board of Health for 1897 that "during the

---

[1] Annual Report on Brit. New Guinea for 1897-8.

last ten years there has certainly been no marked tendency towards an increase in the prevalence of this disease in Tasmania," and these figures bear out the statement. It will be observed that the cancer mortality in Tasmania is about equal to that in Southern Australia, but above that in Western Australia or Queensland.

In New Zealand, on the other hand, the mortality from this group of diseases has increased considerably in the same period. In 1889 the deaths under this heading were 260, and there has been an almost unbroken increase in this figure up to the year 1898, when it was 471. The population in the same period rose from 616,057 to 743,463, and the mortality rate from cancer therefore rose from 422 to 633 per million. In New Zealand cancer would seem to be decidedly prevalent—more so than in Western Australia, Queensland, Southern Australia, or Tasmania, but still a considerably less fertile cause of death than in England, Sweden, or Holland. In New Caledonia the disease is said to be very uncommon.

Africa. Throughout the African continent cancer is very much less frequent than in Australia or in many parts of Europe. This class of diseases is according to some authorities very rare in British East Africa, but it has recently been stated that in Uganda cases of malignant disease (especially of sarcoma of the jaws) are not rarely seen, but that on the whole these diseases are not nearly so frequent as at home, and that they run a slower course[1]. In general the black races of Africa appear to suffer remarkably little from, though they are not entirely immune to, malignant growths. Livingstone never met with a case among the Makalolo or the Barotse on the Upper Zambesi. In Abyssinia, one observer has, however, described cancer as rather frequent, and in Algeria cancerous growths of all kinds are common, both in natives and Europeans[2]. In Western Africa cases of cancer have been exceptionally seen; Plehn mentions two cases of sarcoma and two of osteo-sarcoma as occurring in natives in the Cameroon district. On the Gold Coast, and adjoining coasts, cancerous diseases are said to be rare. At Lagos, during a practice of fourteen years' duration Dr Johnson saw only five cases of cancer

[1] A. R. Cook, *Journ. Trop. Med.*, June 1st, 1901.

[2] Brault, who states that epithelioma "*pullule* en Algérie sur les Européens," while sarcoma is often seen among the natives.

in native patients, and in each case the sufferer had lived as Europeans live. Similarly in Bechuanaland and in Mashonaland malignant disease seems to be decidedly uncommon.

In Mauritius cancer is also far from common, and it is said to cause only one-eighth of the mortality which arises from the same cause in England. Both carcinoma and epithelioma are met with on the island.

North and Central America. Malignant disease is prevalent to a considerable extent in both Canada and the United States. In British Columbia it was the cause of 22 deaths in Victoria in 1896, out of a population of 20,000. This was equivalent to a mortality of 1100 per million living; but being the mortality of a single city only, where possibly many deaths from cancer occur in persons drawn from elsewhere, this figure is not to be compared with those already given for whole countries. In New Brunswick malignant disease is also the cause of a considerable mortality; and in Nova Scotia it is not only on the increase, but is grouped with tuberculosis as causing together with that class of diseases no less than 40 per cent. of the total number of deaths at the principal hospital in Halifax.

In the United States generally malignant disease is prevalent, and is, in many places, distinctly increasing as a cause of death. In Massachusetts this increase has been extraordinarily great and comparatively steady. In the year 1856 the cancer mortality in that State was only 190 per million, but in 1895 it had risen to a maximum of 700. In the first twenty years of that period the mortality figure was 300, in the second 560. Or taken in five-year periods the death-rate from cancer per million inhabitants in Massachusetts, during the past forty years, has been as follows: 230, 270, 330, 370, 450, 540, 590, and 640. Part of this increase is recognised as due to improved knowledge and more accurate methods of diagnosis, but this cannot possibly account for a rise of between 200 and 300 per cent. in the cancer mortality in forty years. In seven of the principal cities of the United States for which statistics are available an equally striking increase has been observed in recent times. The greatest rise has been in San Francisco, where the deaths from this cause, which in 1866 had been only 165 per million inhabitants, were no less than 1036 per million in 1898. In Boston the cancer death-rate nearly trebled itself between 1863 and 1887; it then fell slightly, but has again increased in the last few years. In the seven cities of New Orleans, Philadelphia, Boston, New York, San Francisco, St Louis, and Baltimore the mortality from cancer rose from 354 per million living in 1870 to 664 per million living in 1898[1]. On the whole the Southern States are said to be less affected than the Northern.

Of the prevalence of cancer in the West Indies little is known. It is probably not common.

In Central America cancer is, at least in some States, comparatively rare. This appears to be the case in Guatemala, where in 1894 the deaths from this group of diseases in a population of 1,431,506 were only 60. The cancer mortality for that year was, therefore, only 42 per million, a rate which, in comparison with that of many European countries, is a very low one.

South America. In some South American countries malignant disease gives rise to a considerable annual mortality. Trustworthy figures as to its frequency here are, however, for the most part lacking. In Brazil it is common in some provinces; in that of Buenos Aires the annual deaths returned under this heading are particularly high. In Guiana cancer is not very prevalent. In Uruguay it is a frequent cause of death; in 1897 401 deaths from malignant disease were recorded in this Republic, which is equivalent to a mortality of 477 per million of the population. Cancer accounted for 3·28 per cent. of the mortality from all causes.

Factors governing the Geographical Distribution of Malignant Growths. From the above rapid summary it will be observed that cancerous diseases are comparatively common in the following countries:—In Great Britain, France, Holland, Sweden, Norway, Austria, Switzerland, Greece, and to some extent Germany; in Australia; in New Zealand; in China; in some parts of Canada; generally throughout the United States; and probably also in Brazil, Uruguay, and some other parts of South America. They are rare or unknown, on the other hand, in Iceland and the Faröe Isles, in Servia, in Persia, in Arabia, in Corea, in New Guinea, generally throughout the entire continent of Africa, and in some Central American States, as, for example, Guatemala.

---

[1] *American Journal of the Medical Sciences*, Feb. 1900.

The conditions which determine the geographical distribution of malignant disease are not fully known. From what is known as to the etiology of the various forms of growths grouped under this heading it is probable that racial proclivities have a considerable share in determining their prevalence. The black races appear to enjoy a remarkable immunity from cancerous tumours. The yellow races are more prone to suffer; and in the Chinese malignant growths of great size are sometimes met with. But it is the white races that are most liable to their development. The Jews have been thought by some to be rather less liable to cancer than other races, but the evidence on the point is conflicting.

The precise influence which climate, soil, elevation, humidity, and other local conditions have upon the prevalence of cancer is also unknown. Certain countries and certain special areas in those countries have been shown to be associated for long periods together with a high mortality rate from cancer. Generally speaking, but not without exceptions of considerable importance, these cancer areas are low-lying, damp, often on clay or other retentive soils, and watered by streams or rivers. High-lying, dry districts on the limestone, chalk, or other porous soil, enjoy a relative, but far from complete immunity. On the other hand, cancer has been found prevalent at considerable elevations, and in districts where the soil was entirely chalk. It is fully recognised that areas where cancer is common and areas where it is rare may adjoin; and it has been pointed out that even the opposite banks of the same river may show a marked difference in the rate of mortality from malignant disease. In a certain number of instances a succession of cases of cancer has been reported from one and the same house; and it has been thought that in such houses there must be some local condition or set of local conditions peculiarly favourable to the development of this class of growths. But whether these are not instances of mere coincidence, or whether in truth such "infected" houses exist is still open to doubt. The term "cancer-house" has been sometimes employed to designate a house of this nature.

Perhaps the most striking feature in the recent history of cancerous diseases has been the tendency they have shown to increase in many parts of the world. Although the increase may in some instances be apparent rather than real, and due partly at any rate to greater accuracy of diagnosis and registration, it has in many instances been far too marked to be accounted for entirely in this way; and it is now very generally admitted that diseases of this group have become really more frequent in recent years than they were formerly. The European countries in which an increase in cancer prevalence has been most clearly shown appear to be the following:—England, Scotland, Ireland, France, Holland, Denmark, Norway, and (perhaps) Russia. No less striking—perhaps even more striking—than the increased frequency of cancer in Europe, has been the simultaneous increase in this class of diseases in Australia, in New Zealand, in some parts of Canada, and in the United States. In the States the recent rise in cancer mortality has, indeed, been more marked, perhaps, than in any other country. A large number of theories have been advanced to account for this strong tendency on the part of cancerous diseases to become yearly more and more widely prevalent and fatal; but no quite satisfactory explanation of the fact has as yet been found. Some have even questioned whether it is indeed a fact, but the evidence of the increase in cancer prevalence is unfortunately too strong to admit of serious question.

It has been frequently suggested that cancer and malaria are antagonistic diseases, and that where the one is common the other is absent or rare. The figures given in this chapter, if compared with those quoted in the chapter on malaria, will be found to lend considerable support to this view. Cancerous diseases are certainly very common in the northern European countries—Norway, Sweden, and Denmark, where malaria, on the other hand, is very rare or unknown. They are frequent and increasing in Holland and England, both at the present day almost free from malaria. In Australia, New Zealand, Tasmania, Canada, and the United States cancer is remarkably prevalent, while malaria is, in at least many of these countries, very rarely seen. The reverse is also found to hold good in several instances, and in many highly malarious countries, such as Persia, Siam, the Laos country, many of the West Indian Islands, Corea, New Guinea, Mauritius, and many parts of Africa, malignant growths are notably infrequent. The most marked exceptions to the apparent rule have been the rarity of both diseases in Iceland, the Faröe Islands, and the central plateau of Arabia; and the frequency of both in Italy, and perhaps in China and Brazil.

When the statistics of smaller areas are compared the inverse ratio of prevalence between the two diseases is also to some extent observed. Kruse has published (in the *Münch. Med. Woch.*[1] 1901, No. 48) tables showing the mortality from each in the various provinces of Italy. Though quoted by him as leading to conclusions adverse to the theory in question, these tables seem to an unbiassed reader to show in a rough but decided manner that, whether large or small areas be taken, cancer is found to be more or less prevalent as malaria is more or less rare.

The diminution of malaria in many European countries in recent years and the increase of cancer in the same countries are facts which may also indicate this relation of the two diseases. Some have gone so far as to suggest that the antagonism is a really fundamental pathological antagonism, and that the inoculation of a cancerous subject with the protozoon of malaria might be tried as a cure for intractable malignant growths. Further evidence is needed before this relationship of the two diseases can be accepted. It may show that no such antagonism between the two exists in individual cases. It will more probably show that the two are antagonistic only in the sense that there is some general condition, or set of conditions, the presence of which is favourable to the prevalence of one disease and adverse to that of the other. What these conditions are is at present quite uncertain.

[1] Ref. in *British Medical Journal*, March 1, 1902, where the tables are quoted in full.

CEREBRO-SPINAL FEVER

(EPIDEMIC CEREBRO-SPINAL MENINGITIS).

General Characters and Etiology. Epidemic cerebro-spinal meningitis, or cerebro-spinal fever, has been recognised as a specific fever since its appearance in southern Europe some seventy years ago; the Germans have given it the name of 'cerebral typhus'; and these various titles indicate that the most prominent feature of the disease is a severe affection of the nervous system. The disorder is, in brief, a specific fever in which the principal lesion is an inflammatory condition of the membranes of the brain and spinal cord—just as in enteric fever the principal lesion is an affection of the intestinal glands, in plague, of the lymphatic glands or lungs, and in cholera, of the mucous lining of the bowels. The pathological changes are associated with the presence in the tissues of a micro-organism, first demonstrated by Weichselbaum, who gave it the name of *diplococcus intracellularis meningitidis*.

History. This disease has always shown a distinct tendency to recur in epidemics of longer or shorter duration; and according to Hirsch four principal epidemic periods may be differentiated. In the first period, 1805—1830, the disorder occurred in the form of isolated outbreaks in different parts of Europe and, to a greater extent, in the United States. In the second period, 1837—1850, there were widespread epidemics in France, Italy, Algiers, the United States, and Denmark. In the third period, 1854—1875, the disease was diffused through a great part of Europe, the adjoining portions of Asia, the United States, and some parts of Africa and South America. The fourth period

was described as "a return to merely casual epidemic outbreaks,
or to more or less considerable groups of cases here and there
within its former distribution area." This latter description
applies to some extent to the behaviour of the fever during
the past two decades; but in quite recent years there has been
again a tendency to renewed activity in the disease; and to
the four periods described a fifth might perhaps be justly added.
Jaeger[1], indeed, is of opinion that there is now (he wrote in 1899)
"an epidemic period of this disease," and he points to its
wide prevalence all over the United States, in Germany, in
France, and in Greece.

Recent Geographical Distribution. *Europe.* In
England cerebro-spinal fever is a very rare disorder, and in
the past only occasional and slight epidemics have been recorded.
In Scotland it appears to be almost, if not quite, unknown, and
in Ireland, with the exception of some well-known epidemics
among the recruits of the Royal Irish Constabulary in certain
Dublin barracks, the disease has also been rare. Generally
the British Isles have been far less affected by this malady
than most of the other European countries. France, on the other
hand, has been the scene of frequent epidemics, occurring
especially among troops in barracks; and the same is true of
Germany, where a very large number of outbreaks have been
reported within the last four years. In Holland and Belgium
the fever has been as rare as in England. In Denmark it is
probably commoner; from 5 to 34 deaths from this cause have,
at least, been annually reported between the years 1892 and
1897 in the Danish urban population alone. Sweden has been
rather frequently visited by it; the northern limit of the disease
in this country is stated to have been the 63rd parallel of
latitude (Hirsch). It is met with in European Russia, where it is
no very great rarity; in the years 1893—1895 the number of
registered cases of the disease in different parts of that country
varied between 7 and 50 per million inhabitants—the lowest
figures occurring in the territory of the Don Cossacks and the
highest in the Baltic provinces and Poland. In Austria-Hungary
cerebro-spinal fever is said to have been comparatively rare in
the past, and in Spain and Portugal it is perhaps still rarer.

---

[1] *Deutsche Med. Woch.*, July, 1899.

In Italy, on the other hand, especially in the southern provinces and in Sicily, the disease is common; it was epidemic there in the years 1893 and 1894, causing a total mortality throughout the peninsula equal to 66 and 41 per million respectively in the two years. In the Balkan peninsula generally this fever has not been frequently seen, but neither in Turkey nor Roumania is it altogether unknown, and in Greece a certain number of cases occur every winter and on rare occasions epidemics.

Asia. Epidemic cerebro-spinal meningitis would seem to be a very much less common disease in Asiatic than in European Russia. The number of cases registered in the three years 1893 to 1895 in the Caucasus varied between 4 and 6 per million inhabitants per annum, the average for the six preceding years having been 6 per million. In Russian Central Asia the corresponding figures had been from 0·5 to 1 per million in the three years named, and an average of 1 per million for the six preceding years; while in Siberia the returns showed from 2 to 10 cases per million in the triennial period, and an average of 6 per million in the six preceding years. These figures are only a remote approximation to the truth, but they show that the disease is not altogether unknown in any of the great geographical divisions of the Russian Empire.

At least two epidemics have been recorded in Asia Minor; the first at Magnesia, near Smyrna, in the year 1869, and the second in Smyrna itself in the following year. Two years later a few cases of the disease occurred in Jerusalem, and in 1874–5 a slight epidemic was reported from Persia. Whether more recent epidemics have occurred in these countries I am not able to say.

In India this malady is by no means of rare occurrence. It especially attacks the inmates of Indian gaols, and certain gaols have been the scene of repeated small outbreaks of the fever[1]. The history of the disease in this country is incomplete, but apparently it has only been known there within the last score of years. The first mention of it in the health reports of the Indian Government is found in the Sanitary Commissioners' Report for 1881. Between that year and the year 1892 the

---

[1] As for example the Alipur Central Gaol, Calcutta, in which no less than eleven outbreaks of the disease occurred between 1885 and 1898.

statistics of the Indian army and of Indian gaols furnished a total of 333 cases and 259 deaths[1]. None of the three Presidencies escaped, and epidemics were also reported from Burma, the Andaman, and the Nicobar Islands. The principal outbreaks, however, were in Bengal and Bombay, and the Bengal gaols suffered especially. Cases also occurred in the Bombay and Lahore gaols; and in part of the Kangra district of the Punjab the malady is said to have been "very prevalent every year," and to have caused an epidemic of some severity in 1894. How far the general population in India is affected by this disease it is impossible to say. It is apparently rarely seen in Ceylon: only one death from it was recorded in the island in the year 1895, one in 1896, and none in any other year between 1890 and 1897.

The evidence for the existence of cerebro-spinal fever in China is excessively scanty. Coltman[2], who spent many years in the country, expresses his belief that the disease is not unknown there, his own child having suffered from an attack; but he was unable to find any reports of it from other sources. I have come across no mention of it in the reports of the medical officers in the Chinese Customs Service; and if this fever exists in China it is probably very far from common.

Australasia. The disorder is probably exceedingly rare in Australia. Until quite lately no report of its occurrence there was known, and even in recent years the only references to it I have found have been the records of one death from this cause in Queensland in 1897, of one death in South Australia in 1892 (the only death of the kind in this colony in the ten years 1889 to 1898), and of the entire absence of this disease as a cause of death in Western 'Australia in the same decade. Nor were any deaths from the fever recorded in New Zealand during the same ten years,—a fact which may be taken to show that, if not unknown, it must be excessively rare and mild in that colony.

Of the existence or frequency of this disease in the Pacific Islands little appears to be accurately known. One case was recorded as occurring in a Polynesian inhabitant of Fiji in the

---

[1] Moorhead, *Transactions of 1st Indian Medical Congress*, 1894.

[2] *The Chinese: their Present and Future ; Medical, Political, and Social.* By Robert Coltman, Jnr., M.D., 1891.

year 1897, but beyond this I have met with no recent record of the disorder. A small epidemic, however, occurred there in the year 1885[1].

Africa. Whether this disease exists or occasionally causes epidemics in Egypt and the Sudan is uncertain, but it seems very probable that some outbreaks of a malady hitherto called by another name are really of the nature of cerebro-spinal fever. Slatin Pasha in his interesting record of captivity in and escape from the Sudan has referred to certain outbreaks of fever which attacked the native population of the Ghezireh and the Sudanese troops stationed near Omdurman. These outbreaks were spoken of as "typhus," but from the recent report of Capt. H. E. H. Smith, R.A.M.C., it seems that the epidemics in question, which prevail mostly in the autumn in the neighbourhood of the Sudanese capital, are probably of the nature of epidemic cerebro-spinal meningitis[2].

In Abyssinia, Tripoli, and Tunis there are, so far as I am aware, no records of the existence of this disease. In Algeria many epidemics of it have occurred in the past; it was apparently introduced to the colony by French troops in the year 1840, and prevailed rather severely there in 1841–2, 1844, 1845–6, and with special severity in 1846–7. A later outbreak occurred here in 1868.

On the West Coast of Africa there appear to be almost no records of the occurrence of cerebro-spinal fever, with the single exception of a remarkable outbreak which recently occurred at Cape Coast among a number of coolies brought from Mombasa in East Africa for duty with the Ashanti Field Force[3]. This was regarded as the first known outbreak of the disease on this coast, but the evidence was uncertain. The carriers, it seems, came in three transports; the first batch of 1500 men were apparently free from the complaint, but among the second batch of 1500 some were probably already affected by it, and after their landing the disease spread to a great many other persons. From these facts it must be regarded as an example of imported infection, but whether from East Africa—whence

[1] *Trans. Epidem. Soc. London*, N.S. Vol. VII. Sir W. MacGregor has recently spoken of the appearance of the disease in Indian coolies in Fiji.

[2] Ref. in *Indian Medical Gazette*, Sept. 1899.

[3] *Journ. Trop. Med.*, Nov. 1900.

these transports originally sailed—or from elsewhere there is nothing to show. With the exception of this epidemic, the only other record of cerebro-spinal fever on the Atlantic shores of Africa is the bare mention of the disease among the maladies known in the Cameroons[1].

In South Africa, on the other hand, there is good reason to believe that cerebro-spinal meningitis is a much less rare complaint. How long the disease has existed in this part of the continent is uncertain, but it was apparently in 1883 that it began to prevail widely in Cape Colony, and in the following year a certain belt of the colony was visited by an epidemic of some severity. It subsided in 1890, but in that year a small outbreak was observed in Natal, mostly among natives; of the fourteen cases known to have occurred on this occasion nearly all ended in death. More recently there is mention of a severe epidemic of the fever at Tijgerberg in the northern part of Cape Colony in 1897; and in the same year an outbreak of some severity occurred at Malmesbury in the same colony, followed in 1898 by a less extensive recurrence.

North and Central America. The North American continent has in the past been more repeatedly and extensively visited by cerebro-spinal fever than any other. Twenty years ago Hirsch wrote that "hitherto the headquarters of the disease, both as regards wideness of diffusion and severity of the outbreak, have been the United States, where its area extends from Canada to the Gulf of Mexico, and from the Atlantic to the Prairie States of Minnesota and Iowa." In a footnote the same author adds that he has seen no accounts of the disease in Georgia, Florida, or Arkansas, but that he is not prepared to conclude from this absence of evidence that these States have escaped it.

This frequency and wide diffusion of the fever in the western continent are still among the most striking features in its epidemiology. For many years past repeated outbreaks of the disease have occurred in widely separated regions, but usually in villages and in country districts. Recently some of the larger towns have also been affected. In 1893 an epidemic appeared in New York. In 1896, 1897, and 1898 epidemics occurred in Boston, and in 1898 cases of the disorder were recognised in

---

[1] Albert Plehn, *Janus*, 1896–7.

Baltimore and other towns. In the latter year cerebro-spinal fever prevailed—usually, it is true, in a mild form—in twenty-seven of the States and in the District of Columbia. Prof. Osler, in his address to the West London Medico-Chirurgical Society in 1899, drew attention to the renewed diffusion of the disorder in the United States, and described it as the fourth successive "wave" or epidemic period of the disease in the States in the nineteenth century (the fourth epidemic period referred to by Hirsch and spoken of above did not apparently affect America).

Canada is perhaps less widely troubled by this disorder than the States; but it is said that in Alaska cerebro-spinal fever is very commonly seen, and it is even spoken of as one of the many dangers inseparable from gold-mining in that remote and inhospitable country.

In Mexico and the Central American States nothing seems to be known in regard to the presence or absence of cerebro-spinal fever. I find no mention of it in recent reports from British Honduras. But in some of the West Indian Islands it is certainly known. In Jamaica, for example, it must be very far from uncommon. At Half-way Tree in that island as many as 12 cases of the disease occurred in the year 1897–8, all of which ended in death. A few cases were also recorded in the same year in the eastern district, and in the health reports for the year 1898–9 the absence of any cases of cerebro-spinal fever in the returns from all parts of the island is noted as very exceptional, and as being the first occasion for many years in which such a thing had happened [1].

South America. Whether epidemic cerebro-spinal meningitis occurs in any part of South America I am not in a position to say. The absence of positive records can scarcely be taken to prove that the disease does not exist there, as it is among the lesser known disorders, and may be easily confused with other cerebro-spinal affections of an inflammatory character. In the past it was rumoured that cerebro-spinal fever had been prevalent in Brazil and in Monte Video in the year 1840, but the rumour was unconfirmed.

Outbreaks at sea. There remain to be mentioned certain

[1] It is noteworthy that among the recent epidemics of this disease in Calcutta was one which occurred in the depôt of the Jamaica Government Emigration Agency in that city, in the latter part of 1900.

outbreaks of this disease which have been reported as occurring on ships at sea. One occurrence of this nature has already been mentioned above, in the instance of the three transports which arrived at Cape Coast from Mombasa in East Africa, with some of the coolie passengers already showing symptoms of cerebro-spinal fever. Another example recently recorded is that of the s.s. " Clyde," an emigrant ship carrying a large number of persons from Calcutta to Georgetown in British Guiana. In the former instance, that of the three transports at Cape Coast, the published facts throw no light on the original source of the infection. The disease does not appear to have shown itself until towards the end of a long voyage, and the majority of cases occurred after the ships' arrival. But on board the " Clyde " the first case developed three days after the ship left Calcutta, and as it is well known that cerebro-spinal fever is no rarity there it may be presumed that the emigrants in question brought the infection on board with them when they embarked in that city. The " Clyde " appears to have left Calcutta on January 24th, 1900, and arrived in Georgetown on May 9th. The outbreak was confined to the occurrence of four cases, between January 27th and March 19th. An interval of no less than forty-one days occurred between the third and last cases.

An outbreak presenting remarkable analogies with the above occurred on another emigrant ship, also carrying coolies from Calcutta to Georgetown in British Guiana. This boat, the " Elbe," left Calcutta on October 18th, 1900, and reached Georgetown on January 5th, 1901, and during the voyage 17 cases of cerebro-spinal fever with 14 deaths occurred. The first patient developed symptoms the very day of embarkation, only 9 or 10 hours after leaving the coolie depôt in Calcutta, and as other cases of the disease had already occurred there, there can be no doubt that the outbreak on board was due to an infection imported from the depôt.

Factors concerned in the Geographical Distribution of Cerebro-spinal Fever. It cannot escape notice that the facts as to the geographical distribution of cerebro-spinal fever over the earth's surface are very imperfectly known. This is no doubt true of many other of the diseases dealt with in these pages, but it is more especially true of the one now under consideration. Some of the reasons for this have just been pointed out in referring to the absence of information concerning this disease in South America. Cerebro-spinal fever is a malady that is not always easy to differentiate from ordinary idiopathic, traumatic, or tubercular meningitis, or from several other pathological conditions of the brain and spinal cord and their coverings. The aid of bacteriology in the diagnosis of a given case has only recently become possible, and the bacteriologist and the means for making a bacteriological inquiry are not always at hand. It is therefore practically certain that the area of prevalence of this, as of many other diseases, is much wider than the known records of its occurrence would imply.

Its known area of prevalence is, however, a wide one at the present day, and is probably always becoming wider. With the (uncertain) exception of South America it is found in all the great continents. Its northerly limit varies in the two hemispheres; in Sweden it is said to cease at the 63rd degree of latitude, and it is absent from Iceland and the Faroë Islands; but in Alaska it exists in the region of the gold-fields, which are on the very borders of the Arctic circle. In the south it is common at the southern extremity of Africa, and an occasional case is seen in Australia.

The title, so frequently given to this disease, of epidemic cerebro-spinal meningitis expresses one of its most interesting features—its tendency to occur in the form of epidemics. The epidemics vary very widely in their extent and intensity, and in not a few instances outbreaks have occurred of so limited a character—the disease attacking a mere handful of people—that the word "epidemic" is quite out of place in connection with them. In some instances observers have even been hardy enough to diagnose single isolated cases of disease as of this nature, but in general terms epidemicity is one of the main characters of the fever. The epidemics are usually curiously irregular in their relations both to space and time. They do not as a rule pass steadily through a country from one end of it to another, like a cholera or an influenza epidemic. They rather tend to occur as a number of isolated outbreaks, here and there, in different parts of a country, the separate outbreaks showing little or no relation with each other. It is, indeed, usually impossible to trace any connection between them. So also in regard to time, an individual outbreak often shows no definite course, with a distinct beginning, a rise to a maximum degree of prevalence, and a distinct decline to extinction. An epidemic may linger on for months or even a year in a given place, and it may recur at irregular intervals in the same village or district, showing that the infection is really—at least for the time being—endemic in that particular spot.

The exact manner in which the infection is spread is not known, but this tendency to recur in certain places seems to show that the virus of the disease can exist for longer or shorter periods outside the human body. This "saprophytic" existence of the microparasite is rendered still more probable from a study of those repeated outbreaks of the disease which have been observed in such institutions as barracks and gaols, where it would seem that the infection clings to a building for long periods together. Some of these instances have been already referred to. The most striking have been met with in Indian gaols. Thus for example, in the Alipur central gaol, in Calcutta, the first known epidemic of this disease occurred in 1885, and in the next decade nine more outbreaks were recorded in the same gaol; then after an interval of three years an eleventh outbreak—confined to three cases—occurred (in 1898)[1]. Of similar import were the facts connected with the outbreak of cerebro-spinal fever at sea, in the s. s. "Clyde," already alluded to. That forty-one days should have elapsed between the 3rd and 4th cases in this outbreak can only be explained by supposing that the infection clung in some way to the vessel itself or its contents during that period. In this respect the behaviour of the disease presents not a little analogy with that of beri-beri.

If cerebro-spinal fever is directly infectious from person to person—in the sense that scarlet-fever and measles are in ordinary parlance "contagious"—it is so probably in a low degree. In most epidemics no connection can be traced between individual cases. Sometimes several persons are attacked more or less simultaneously; and, in like manner, it has occasionally been observed that several isolated outbreaks in districts at a considerable distance from each other have arisen at or about the same time. Both these observations seem to show that the

[1] *Indian Medical Gazette*, June, 1899.

disease is spread less by direct transmission of a virus from one person to another than by the exposure of several persons or communities to a common source of infection.

There seems nevertheless to be good reason for believing that the virus of cerebro-spinal fever can be and is carried from place to place by means of human beings. Hirsch has quoted a number of instances in which the movements of troops have seemed to contribute to the spread of the disease. Its appearance in Algeria many years ago was thought to be due to an importation of infection by the French, and three examples have already been quoted above (the transports at Cape Coast, the "Clyde" and the "Elbe") in which the disease was apparently carried long distances by sea.

The relations of this disease to *temperature* are briefly as follows. The actual height of the thermometer is without influence in its production. It can exist equally well in tropical, temperate, and sub-arctic climates. The view formerly held that one of the most important determining causes of an outbreak was a low temperature was based on imperfect knowledge of the area of distribution of the disease. A disorder which can cause severe epidemics in Indian gaols, among the poor of Jamaica, and on the tropical coasts of Western Africa cannot in any way be dependent for its causation upon absolute cold. The recent history of the malady has in fact thrown much new light upon its relations to temperature and to many other external conditions. How far this apparent change in the behaviour of the disease is due to the mere absence or imperfection of the earlier records for many countries, or how far it is a real change it is not easy to say. But it seems certain that in the last two decades cerebro-spinal fever has spread to, and become widely diffused in, countries where formerly it was unknown. Sixteen years ago Hirsch wrote that "hitherto it has been confined really to temperate and subtropical latitudes; it has not penetrated into the cold zone except to a very slight extent; while the tropics and the whole of the southern hemisphere have escaped it altogether." Some time, however, before those words were published the disease had already appeared in many parts of India, and the recent outbreak in connection with coolies for the Ashanti Expedition, referred to above, seems to show that it exists or can exist in tropical Africa.

In Southern Africa also the prevalence of the disease seems to date only from tolerably recent times. It may be accepted, therefore, as beyond question that the area of prevalence of this fever is now a much wider one than formerly, and that conclusions based on our present knowledge of its distribution may be upset by later behaviour of the disease.

For the present it may be stated, in regard to its relation to temperature, that while absolute cold is certainly not necessary for its production, a relative degree of cold seems to be of some importance. In temperate climates epidemics have most often occurred in the cold season of the year, and changes of temperature seem to favour their development—possibly by their effect on the individual, whose resisting powers may be lowered by a fall in the thermometer, or possibly by some more indirect action.

The distribution of cerebro-spinal fever shows little definite relation to elevation above sea-level or to the nature of the soil. Wetness of the soil appears to be without influence, unless perhaps in an indirect manner by rendering the individual more susceptible to the virus. Racial differences have perhaps some importance in determining the geographical distribution of the disease, but to what extent is uncertain. All races have suffered from it, but it has been thought that negroes are particularly susceptible. How far this increased susceptibility is really due to racial difference and how far to difference in habits or in exposure to infection is unknown. It would appear that among the determining factors of an epidemic of this disease the character of the surroundings of the individual plays a very important part. The frequency with which children or persons of early adult life alone suffer from it, and the fact that attacks are exceedingly liable to occur in individuals who are in a state of fatigue or exhaustion from some unusual or excessive degree of labour or physical exercise, or after exposure to cold and wet, all tend to confirm this statement. So also does the tendency shown by the disease to become epidemic in towns, and especially in the poorest quarters, where dirt, overcrowding, and general insanitary conditions are found at their worst. In this respect cerebro-spinal fever shows a certain analogy with yellow fever and with dengue.

CHICKEN-POX (VARICELLA).

Chicken-pox being a mild disorder, rarely resulting in death, mortality returns are of little value as indications of its prevalence. In the case of many countries it is difficult to find any mention at all of the disease, as owing to its comparatively trifling nature it is often not referred to at all by writers on the diseases prevailing in those countries. Most writers on the geographical distribution of diseases ignore chicken-pox completely.

History. Chicken-pox was for long regarded as a mild form of small-pox. The name "varicella" is as old as the Arabian writer Rhazes. Fuller, in 1730, and later Heberden, in 1767, established the difference between the two diseases.

Recent Geographical Distribution. *Europe.* The disease is probably more or less common in all European countries. In England and Wales it causes one or two hundred deaths in most years, but some of these are believed by the Registrar-General to be due to small-pox. This confusion between small-pox and chicken-pox is doubtless common to other countries. The official returns for most other European States that I have consulted make no mention of this disease. It is undoubtedly known in Russia, where it is called "Windy-pox" (*vietriannia ospa*); it is not infrequently seen at the hospitals in St Petersburg and other large towns, and is probably common in the country. At Constantinople it is more or less constantly present and often of a severe character. In Malta many cases of chicken-pox are registered in some years.

Asia. In many parts of Asia chicken-pox is far from rare. It is said to be very common in India[1]. In Rajputana it is

---

[1] Buchanan, *Journ. Trop. Med.*, Sept. 1899.

occasionally epidemic. In Ceylon it appears to be frequently seen; in 1895, 350 cases of the disease were recorded in the island; in 1896 the recorded cases rose to 579; and in 1897 they were 508 in number. It is probably rarer in Malaysia; Kohlbrügge never saw it in Java[1]. A few cases, however, of a mild type occurred among the Dutch troops in the Indies in 1897.

In China chicken-pox is probably very prevalent. It is frequently seen in Shanghai[2], and is one of the commonest of contagious disorders in the Chung-king district[3]. It was epidemic at Kiukiang in the spring of 1898[4]. It is also known in the Fo-kien province.

Australasia. Chicken-pox is seen sometimes in Fiji. In Australia it is far from uncommon, occasional cases being recorded in recent reports from Queensland, New South Wales, Victoria, Western and South Australia.

Africa. In Africa chicken-pox is certainly known on the West Coast, and is very common in the south. Of its prevalence elsewhere little seems to be known. I find mention of the disease in recent reports from Sierra Leone, Lagos, and the Gold Coast. In Uganda it is very common. In South Africa also it is very prevalent; in the health-reports of the Cape and other colonies for recent years cases or epidemics of chicken-pox are frequently mentioned as met with not only in many parts of Cape Colony, Namaqualand, Griqualand West, the native territories of Tembuland, the Transkei, and Pondoland, but also in Johannesburg in the Transvaal, where it appears to have been frequently seen before the war. The indigenous races of South Africa seem to be very susceptible to the disease, and it sometimes prevails very severely amongst them.

It is certainly met with in Mauritius.

America. In North America chicken-pox is an endemic disease, at times becoming epidemic. Frequent mention of it is found in reports from Ontario, Manitoba, and other Canadian provinces. It was epidemic in Montreal in 1897. It is probably also fairly common in the United States, where occasional outbreaks of the disease are observed. It is not unknown in the West Indies, cases being sometimes recorded in Jamaica.

[1] *Janus*, 1897–8, p. 221.   [2] *Chinese Cust. Med. Reports*, No. 41.
[3] *Ibid.* No. 45.   [4] *Ibid.* No. 56.

Factors governing the Geographical Distribution of Chicken-pox. Chicken-pox is undoubtedly a widely-spread disorder. Its exact distribution cannot, it is true, be determined for the present, owing not only to the absence of information in regard to its existence in many countries, but also to the uncertainty of diagnosis which often surrounds this disease, and the ease with which it may be confused with mild or modified small-pox. Still, in spite of these difficulties, there is a sufficiency of evidence to show that chicken-pox prevails under a great variety of conditions and in countries widely apart. It can certainly exist just as well in the tropics as in temperate climates, though it is perhaps more common in the latter.

Chicken-pox is essentially a disease of child-life, especially attacking children under five years of age. It sometimes takes on epidemic characters, but such outbreaks are usually more or less localised in character, and show no regularity in time of recurrence or in the number of persons attacked.

CHOLERA.

Cholera is a disease which for long periods, and in some instances permanently (if our limited historical knowledge will permit the use of such a word in regard to any disease), remains endemic in certain countries, at varying intervals spreads more or less widely elsewhere, and from time to time becomes pandemic, extending over a great portion of the inhabited world.

The Endemic Centres of Cholera. It has long been customary to regard Lower Bengal—the flat, marshy, malarial country of the Sundarbans and the delta of the Ganges—as the endemic home of cholera, whence extensions of the disease to other parts of India and thence to other countries were to be always traced. It is certain, however, that at the present day cholera is endemic, in the fullest sense of the term, in many other parts of India, in Farther India, in China, and perhaps in Japan.

In India itself the region mentioned above is, doubtless, a most important area of endemic cholera ; it is perhaps the oldest, it is certainly among the worst. But it is not the only one. From the discussion on cholera at the first Indian Medical Congress in 1894, and from the annual returns of the Sanitary Commissioners with the various Governments of and in India, it is clear that cholera is endemic in many other parts of the country. It is certainly so in the Bombay Presidency. Capt. Herbert, I.M.S., stated at the Congress that the difference between Bombay and Bengal in regard to the endemicity of cholera is one of degree only, and that the same might probably be said of Madras and Berar. If British India, this observer added, could be divided into endemic and non-endemic areas of cholera, the latter would only include Sindh and the Punjab, and even in the easterly districts of the Punjab a mild type of the disease is endemic. It is certain that in most years—in *all* recent years—cholera prevails more or less in most portions of British India, including Assam. There are years of great prevalence, such as 1877, 1887, 1891–2, and 1897, and there are periods of slight prevalence, but it is rarely entirely absent from any province, excepting Sindh. How far in these "cholera years" the excessive prevalence in any given province is due to an importation from Lower Bengal or elsewhere, or to a revival of local infection is uncertain. The factors governing its spread are to some extent known. In all the Indian provinces, and even in the great endemic area of Lower Bengal, where the virus may be supposed to be permanently and widely present, carriage of infection by human beings and its diffusion by contamination of water-supplies are constantly observed. It has been pointed out also that an appearance of western extension from Bengal is sometimes brought about when no such extension has really taken place, because the seasonal commencement of cholera activity is early in Lower Bengal, occurring before the hot weather has far advanced, and it is delayed more and more as one travels westward and the annual rainfall diminishes, until in the Punjab the ordinary epidemics come during and towards the end of the rains. In the areas of endemic prevalence, and of epidemic extension in India, localised outbreaks of great violence are not infrequent.

In Farther India cholera is as truly endemic as in India proper. Lombard speaks of Bangkok as "un véritable foyer de choléra." In the plains of Cambodia cholera appears invariably to become epidemic in the hot season. In Annam it is said to begin to prevail each year with the rains. It is constantly epidemic, if it be not truly endemic, in Cochin China. In Lower Tongking it is admittedly endemic (Rey) and frequently severely epidemic. In the Straits Settlements cholera seems to have been prevalent every season in recent years, and it may be doubted whether the infection is on each occasion newly imported—though on many occasions it has been thus imported by coolies from China[1].

That cholera is endemic in some parts of China no longer admits of doubt. At Shanghai it is spoken of as "more or less

---

[1] I am informed, however, by Prof. W. J. Simpson that the infection is "almost always" so imported.

endemic, making an annual appearance generally toward the end of summer[1]." At Chinkiang an epidemic of cholera usually decimates the native population in the hot season[2]. At Hankow and at Ichang, both on the Yangtze-kiang, the disease appears each year in the hot months[3]. At Pakhoi it prevails in most years between July and September[4]. At Lung-chou (near Pakhoi) it has been epidemic in most recent years[5].

From the evidence of the medical officers in the Chinese Customs service it appears then that cholera is truly endemic in the places just named, if in no other parts of China. Manson also states that the disease is endemic in Canton. It is scarcely possible to suppose that the annually recurring epidemics in all these· places are due to successive introductions of infection from India or elsewhere, and the conclusion is unavoidable that they are merely a seasonal revival of an endemic disease, just as happens in Lower Bengal and other parts of India. As cholera in its endemic homes in India occasionally causes severe and widespread epidemics, while in other years it is quiescent, so also in China there are "cholera years" and years of little cholera. At Ichang the memorable epidemic years were 1850, 1864, 1883, and 1892. In 1892 cholera was extremely prevalent in many parts of China, particularly in the Yangtze-kiang valley. In 1895 epidemics, in some places of "appalling" violence, prevailed throughout a great portion of the country; in Peking alone cholera caused over 50,000 deaths. It is noteworthy that while 1892 was a great "cholera year" in India, 1895 was not, though in the previous year the disease had been more than usually active there.

Epidemic Extensions of Cholera to Countries near the Endemic Centres. While the endemicity of cholera in the countries already named may be regarded as fully established, there are some other countries or islands in the Far East where the disease is at times endemic, at times absent altogether. In these countries cholera has been imported from India or China, has given rise to severe epidemics, and has remained for the time being truly endemic, for a longer or shorter period : it has then disappeared and apparently remained absent altogether until a

---

[1] Cox, *Chinese Cust. Med. Reports*, No. 55.
[2] Lynch, *Ibid.* No. 50 and No. 54.
[3] *Ibid.* No. 41 and No. 44.   [4] *Ibid.* No. 42.   [5] *Ibid.* No. 50.

fresh importation has started the cycle again. This has happened repeatedly in Ceylon and Japan. In Ceylon large numbers of cases of cholera occur every year, but the disease is apparently not permanently endemic in this island. The infection is repeatedly imported from India. A violent epidemic occurred in Ceylon in 1892, and each year since that date a very large though diminishing mortality from cholera has been annually reported from the island.

In Japan among recent epidemic years may be mentioned the years 1877, 1882, 1886, and 1890. These outbreaks were of very great violence, and were followed by recrudescences of an infection for the time endemic in the island. It may indeed be questioned whether cholera is not now as truly endemic in Japan as in China. A recent writer in the Sei-i-Kwai Medical Journal has called special attention to the fact that since 1877 "*cholera has appeared every year in Japan*" (the italics are his). In Corea cholera has been frequently imported.

In some of the East Indian islands local extensions of cholera have also been repeatedly observed, though some of these islands, as for example Amboina, are said to have enjoyed a remarkable immunity from the disease. Java experienced a severe epidemic in 1901, and the disease is possibly truly endemic here. In German New Guinea cholera is said to have been introduced by a steamer in 1896, and to have become for the time being endemic there, as also in New Britain and in the Bismarck Archipelago, though it had not (in 1897) spread to the British Solomon Islands[1].

To the west of India similar occasional extensions of cholera are not infrequently observed. Afghanistan, Persia, and Arabia are all instances of countries which, while taking a prominent part in the great pandemics of cholera, are also visited by the disease at other times. The infection on these occasions appears to be always imported from India, and either a serious epidemic may follow, or the disease may become truly endemic for a longer or shorter period. Cholera was in this manner practically endemic in Persia from 1852 to 1861 and from 1865 to 1872.

Cholera has often spread very widely from its eastern habitats without becoming truly pandemic. It has several times broken out among the pilgrims at, or *en route* to, the Moslem Holy Places in

---

[1] Colonial Report for the British Solomon Islands, 1897.

Arabia. In 1823 it spread as far as Astrakhan at the mouths of the Volga, but did not produce a European epidemic, possibly because of the stringent measures taken by the Russian authorities. There is in fact no true dividing line between a widespread epidemic and a pandemic, the terms being merely conveniently used to indicate a wide or a very wide (but never truly *universal*) extension of the disease outside those countries where it is more or less permanently present.

It would appear that the virus of cholera finds in certain countries conditions suited to its existence for very long periods, and perhaps, as in Lower Bengal, permanently. In others the conditions favour the existence of an imported virus for a considerable number of years, after which the disease disappears altogether and remains absent until it is again imported. It is possible that the countries named above—such as Farther India and China,—which are now endemic homes of cholera, may not remain so permanently. How long the disease has already been endemic in these is uncertain, but in some—*e.g.* China—it has almost certainly been endemic for the last twenty or thirty years. When the infection is carried still farther from India, as for example to Europe, it usually dies out after from one to five years of irregular and varying prevalence. It is probable, however, that India is the true original nursery ground of cholera, whence the disease has been carried to the other countries where it is now endemic.

The Great Pandemics of Cholera. To Europeans the greatest interest attaches to those occasional extensions of cholera which from time to time visit not only Europe, but a very considerable portion of the inhabited earth's surface. These so-called pandemics have occurred at irregular intervals of ten, twenty, or more years, and have not always followed the same course or covered the same ground. What causes this disease, or influenza, or plague, thus occasionally to spread widely throughout the world after long periods of inactivity is not known, and any explanation of the phenomenon is largely based on speculation. It is not due to any meteorological conditions alone; and the views formerly current as to an "epidemic constitution" of the atmosphere, or as to the influence of great earth changes as manifested by earthquakes, inundations and the like, are no longer entertained as explanations of the pandemic extensions of disease. The true relation of the micro-organisms associated with these diseases to their causation is also as yet not sufficiently established to justify the assertion that some acquired virulence and power of diffusion and resistance to outside influences on their part will fully explain these occasional pandemics, though it is very probably an important part of the explanation.

In regard to cholera it is established, practically beyond a doubt, that carriage of the infection by human beings and its diffusion by water are important factors, but probably not the sole factors, in the extension of the disease. The carriage of infection by human beings is probably the principal explanation of the fact that cholera, in its pandemic extensions, follows the great routes of trade and travel. Its diffusion by water explains the fact that cholera follows the course of streams and rivers, and appears to spread more rapidly along the great lines of human communication when these are also water-ways (streams or rivers) than when they are not.

The pandemics of cholera have always started from India. They have usually been preceded by a year or years of unusual activity of the disease in India, and sometimes as for example in 1891-2 by its great prevalence throughout the Far East. The diffusion of cholera westward into Europe has been frequently accompanied by a diffusion eastward into China, Farther India, and elsewhere. The disease, as just stated, has almost invariably followed the great routes of trade and travel. The principal routes traversed by cholera in the past have been the following :—(1) From India to Afghanistan, Persia, Russian Central Asia, the Caspian, the Caucasus, Astrakhan, the Volga or shores of the Black Sea or both, and so to Europe generally (as in 1830 and 1892). (2) A more northerly route, through Cabul, Bokhara, and Khiva, to Orenburg on the Ural river, and so to European Russia (as in 1829). This and the first route are sometimes followed in one and the same epidemic. (3) From India to China, and thence probably through Tibet, to Kashgar, Khokand, Yarkand, and Bokhara, and so to the west (as in 1841–44). (4) From India to the Red Sea, Arabia, Egypt, and thence to Syria, Palestine, Asia Minor, and through these or directly from Egypt to Turkey, and up the Danube to Europe generally (as in 1864–5). (5) From India to the shores of the Persian Gulf, Mesopotamia, Asia Minor, and thence to Europe.

It is important to note that on several occasions these routes have been combined; the disease following, at or about the same time, two or more of the great lines of communication in the East.

Without entering into the controversial question of the early history of cholera epidemics, it will be useful, and indeed essential to an understanding of the geographical distribution of the disease, to give a rapid summary of the course followed by the known epidemics of cholera in the past century.

Historical Summary of the great Cholera Epidemics or Pandemics. The first wide extension of cholera outside India appears to have taken place in 1819 and following years. In 1819 the disease spread along the coasts of the Malay Peninsula; in 1820 it was in Siam, and was terribly rife in China: in 1820–22 it spread to Oman and Muskat in Arabia, to several places on the Persian Gulf, and through Mesopotamia and Persia to the Caspian Sea. In 1822 it was again rife in Mesopotamia, and was largely spread as a result of the war between Persia and Turkey. In that and the next year it passed up through the Caucasus and reached Astrakhan, but—thanks to the energetic measures put in force there, or to some other cause—failed to spread into Europe[1]. It reached Syria and Asia Minor, but did not invade Egypt. In 1823 this epidemic apparently came to an end; but in the meantime (in 1819) cholera had been imported by a ship from India into Mauritius and Bourbon (now Réunion) in the south, causing a serious epidemic in the former, and in 1820 it appeared in Zanzibar and Somaliland.

The next extension of cholera began in 1828. In that year cholera was in Arabia. It probably spread in 1829 through Cabul to Khiva, Herat, and Bokhara, and thence to Orenburg in Russia. It was probably carried to Orenburg either by caravans from Bokhara or by nomad Kirghiz from there or from Khiva. In 1829–30 it was epidemic in Persia. In 1830 it spread up the Volga, just as it did more recently in 1892, and soon invaded the greater part of European Russia. In that and the following year it was very rife in many parts of Russia, and

---

[1] In Astrakhan 371 cases of cholera, with 192 deaths, occurred in 1823; and at Krasnoe-Yar, 20 miles away, 21 cases, with 13 deaths (Arkhangelski).

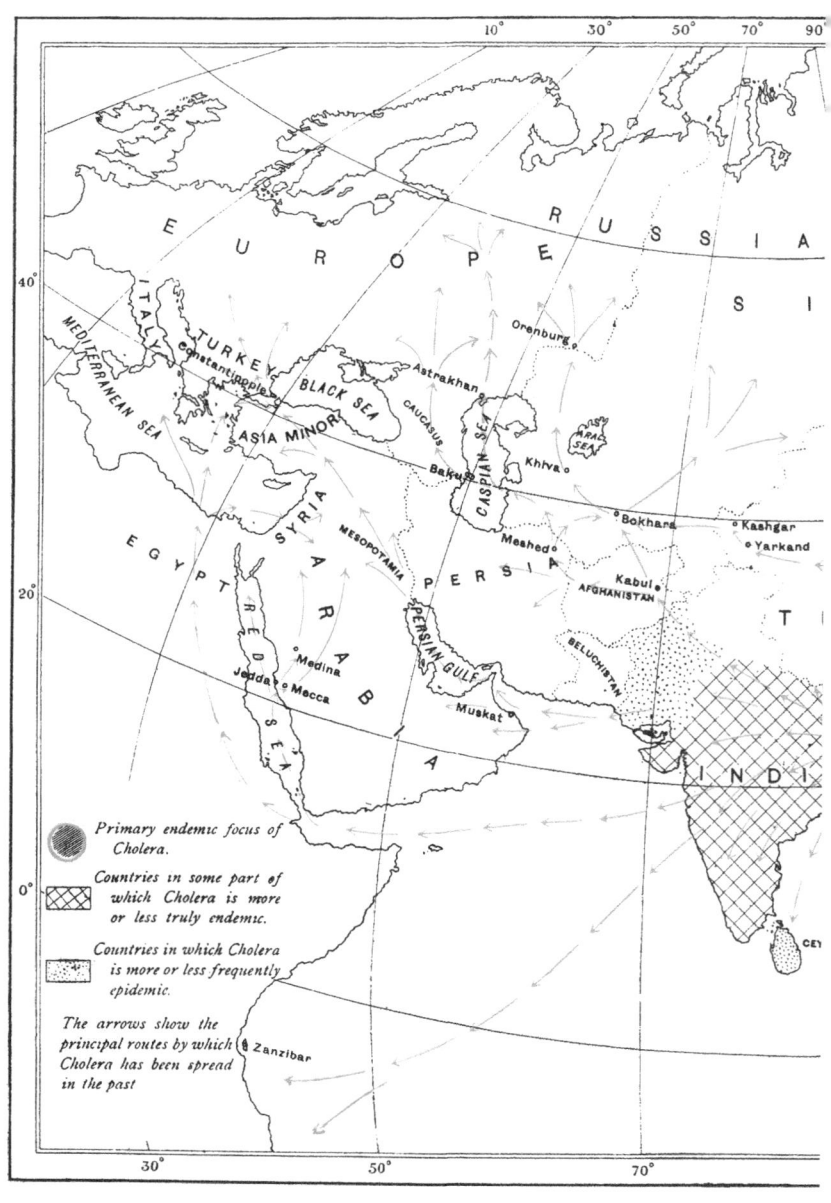

CHART SHOWING THE PRINCIPAL ROUTES BY

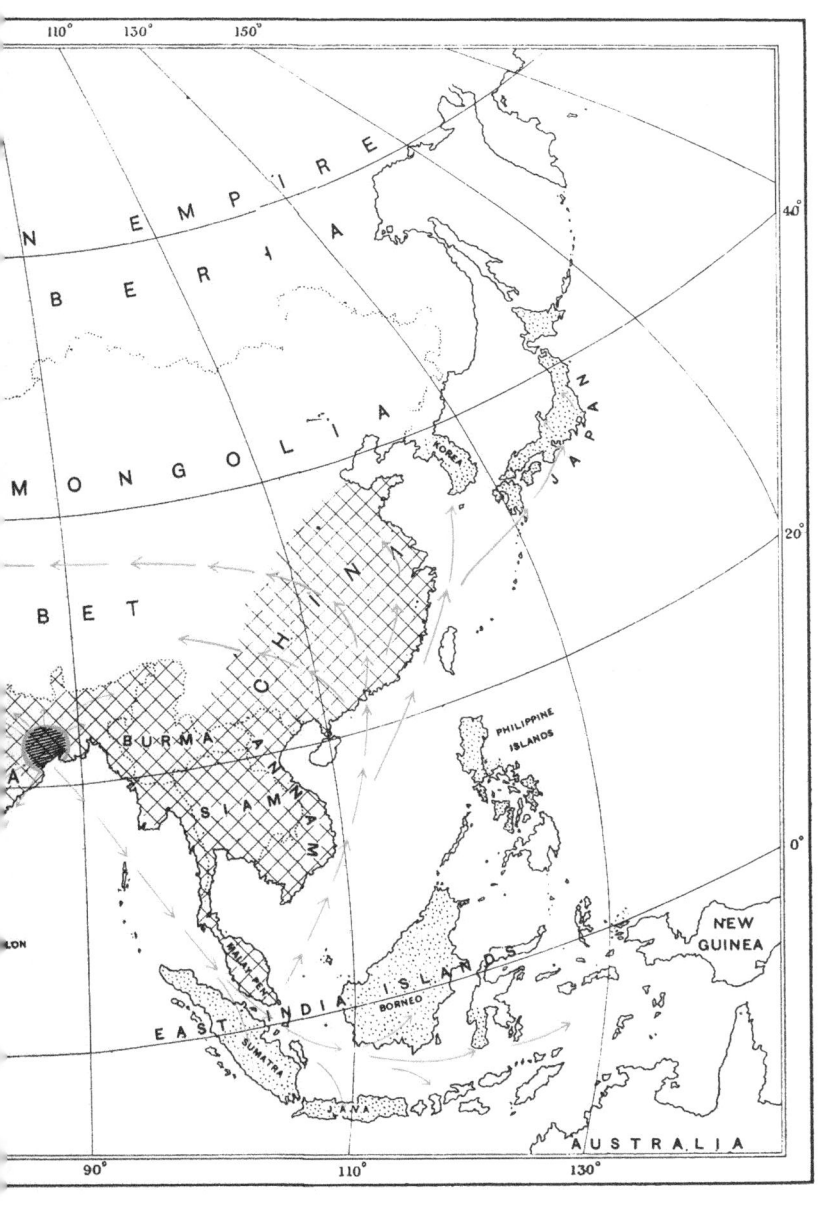

WHICH CHOLERA HAS SPREAD IN THE PAST

particularly among the troops engaged in the Russo-Polish war. In 1831 cholera also spread widely in Arabia, and caused the death of nearly one-half of the pilgrims to the Holy Places of El Islam. Asia Minor and Egypt also suffered. In Europe few countries escaped. Turkey, Bulgaria, Austria, Germany, France, Italy, Spain, Portugal, Norway, Sweden, and the British Isles were all invaded in 1830–32, and from the latter it spread to America. In America cholera broke out in 1832 at Quebec, Montreal, New York, and Philadelphia, spread along the St Lawrence and the shores of Lakes Ontario and Erie, and finally appeared in almost all parts of the United States, in Mexico and in Cuba. In the Old World cholera prevailed in 1836–7 in Palestine, in Somaliland, in Zanzibar, and in Algeria, as well as in many parts of Europe, where it had remained more or less endemic since its first introduction in 1830.

The year 1840 marked the commencement of a third extension of cholera. In that year the disease spread along the coasts of the Malay Peninsula, and was introduced by British vessels into China, where it raged severely in 1841–3. In 1842 cholera appeared in the north of Burma, apparently by extension from China. In 1844 it was rife in Kashgar, Yarkand, Khokand and Bokhara, and in all these places it appears that the infection was originally derived from China; possibly it was carried along the great trade-route from China through Tibet to these Central Asiatic provinces. In 1844–45 cholera was epidemic in Afghanistan and in Central Asia, and it would seem that after the westerly extension from China just described the infection took a south-easterly turn from Bokhara back through Afghanistan to the Punjab. The original westward course was, however, also continued, and before the end of 1845 cholera appeared in Persia, and in 1846 in Mesopotamia, and to a disastrous extent among the Haj pilgrims in the Hedjaz; it also appeared at Aden and in Oman. In 1847 it prevailed over most of the districts already named, and from Persia spread to Asia Minor, the Caucasus, Astrakhan, and the Volga basin, to many places in central and southern Russia, and to Constantinople.

The year 1848 was one of the worst " cholera years " on record. The disease burst out at Mecca and Medina, ravaged Wallachia and Moldavia, and spread all over Russia, Finland, and Sweden. In 1848 and particularly in 1849, few countries in Europe escaped its ravages. It is noteworthy, however, that in Austria and Spain cholera did not spread to any great extent, and that Portugal and Greece remained quite free. Almost the whole of England suffered, but the behaviour of the disease was very capricious in this country. In 1849 Tunis, Oran, and Algiers were invaded, and in 1848 and 1849 epidemic cholera was more or less rife in Canada and the United States. In 1850 Egypt and the whole African sea-board of the Mediterranean and some islands in that sea were visited, and the disease spread over Mexico and California, Cuba and Jamaica. In 1851 isolated outbreaks occurred in Poland, Silesia, and Pomerania, and again in Cuba and Jamaica, and for the first time cholera appeared in Grand Canary Island. The disease appears to have then almost, if not quite, died out in the West for a time. But the respite was very brief and cholera was in the meantime very active in the East.

In Persia and Mesopotamia it was widely epidemic in 1851, and in 1852 spread thence to Russia through the Caucasus and Astrakhan. This new invasion of the south-east of Russia began before the remains of the former epidemic had quite ceased in Polish Russia on the west. At this time and until 1861 cholera was truly endemic in Persia and Mesopotamia. In 1853 it was more or less active in Russia, Denmark, Norway, England, Germany, Holland, Bessarabia, Moldavia, Wallachia, Piedmont, Barbary, and Portugal, and for the third time America was invaded. In 1854–5 cholera was pandemic, raging in Europe and in North America as well as in many parts of Asia. A few cases occurred in Dutch Guiana in 1854; it was epidemic in British Guiana in 1857, and in 1855 and 1858 it was widely epidemic in Brazil. In 1857–8 the disease quite died out in the greater part of Europe and America, only to reappear in many places in 1859. In 1858 cholera was raging on the shores of the Red Sea. Mauritius had been invaded a second time in 1854, and was the scene of three subsequent epidemics between that year and 1862.

In 1860 a great revival in cholera activity was recorded in India; in 1862 the disease was severely epidemic in Peking, Taku, Tientsin and other places in northern China, and in 1864 the first signs of a renewed westward extension were observed. From Sindh cholera spread to Persia, Hadramut, Yemen, and Somaliland. In 1865 it raged at Makalla and Mocha, and spread into the Hedjaz

Abyssinia, and Egypt, where it caused a terrible mortality. Thence it passed to some places on the Dardanelles, to Cyprus, to Syria, Palestine, and Asia Minor. From Constantinople cholera passed up the Danube into the heart of Europe, and also into the Caucasus, which on all former occasions had been invaded from the East and not from the West. The Russian shores of the Black Sea became infected, and finally Italy, France, Spain, Portugal, Malta, Gozo, and England, all suffered from more or less extensive outbreaks. Imported cases occurred also at New York, and in Guadeloupe the disease became epidemic.

In 1866 cholera prevailed widely in most of the places affected in the preceding year, and extended in addition to St Petersburg, Bavaria, Saxony, Prussia, Belgium, and Holland, and in fact over the greater part of Europe. It was largely diffused by the operations connected with the campaign in Prussia, Austria, and Italy. In this year cholera again attacked the Mecca pilgrims, was active in Mesopotamia, Persia, and the Caucasus; and in England reappeared in Southampton and caused an epidemic in London and over the greater part of the country. From England the disease again spread to America, was widely carried through the United States by the movements of recruits, and also appeared in Cuba. In 1867 it was still present in Galicia, Switzerland, Dalmatia, Montenegro, Albania, Sardinia, and in a few places in Russia; it had died out in Arabia, Egypt, Mesopotamia, and the greater part of Turkey in Asia. In 1868 cholera completely disappeared from Europe; but in Africa it spread in 1867 through Abyssinia to Central and Southern Africa, appearing on the East Coast in 1868 and reaching Zanzibar in 1869. In 1868 it broke out for the first time in history on the West Coast of Africa, in the French settlement of St Louis, in Senegal, whither it was believed to have been imported by Moorish trading caravans from north-eastern Africa. It thence spread to Bathurst, and southward along the West Coast and into the interior. In 1867–8 cholera was also epidemic in Nicaragua, Honduras, Brazil, the Argentine Republic, Paraguay, Uruguay, Bolivia and Peru, and the Pacific Coast.

Meanwhile, in the East cholera had again spread from India into Afghanistan, into Kashmir, and into Persia in 1867. In the two following years it was widely epidemic in Persia, and in 1869 it broke out at Kief in western Russia, and in Nijni-Novgorod after the great Fair.

The year 1870 marked the commencement of a most intense cholera prevalence in Europe, and during the next three years it was estimated that not less than· one million lives were sacrificed to the disease. It was extremely rife in Russia and in Prussian Poland in 1870. In 1871 it broke out in Asia Minor and in Constantinople, and in 1872 in many parts of Austria-Hungary, the Roumanian Principalities, Prussia, and Belgium. It was several times imported into England, but no epidemic resulted. In 1873 cholera began to subside in Europe, but remained in Russia, in the Netherlands, and in Munich until 1874.

In the meantime cholera had spread in Central Asia, invading Balkh in 1871, and Khokand, Tashkent, Bokhara, and other places in 1872. In 1870–1 it also spread to Arabia and Nubia, again attacking the Mecca pilgrims. In 1873 the disease appeared in New Orleans and in many places in the Mississippi valley.

There now followed a period of eight years, during which cholera was apparently confined to its endemic homes in the East[1]. But in 1881 and 1882 the disease was again terribly rife among the Mecca pilgrims and widely spread through the Hedjaz. In 1882 it was epidemic in Bokhara and Afghanistan, and from 1882 to 1884 was widely spread throughout Egypt; in 1884 it appeared at Toulon, Marseilles, Genoa, Naples, and other places in Italy, where it remained until 1886–7. Spain was lightly visited by cholera in 1884 and severely in 1885.

Another great western extension of cholera began in the year 1892. In 1891 the disease had been unusually active in India, and in 1892 it caused one of the highest mortality-rates on record in that country. From India it spread through Afghanistan to Persia, crossed the Russo-Persian frontier, and was then carried by the Transcaspian railway both eastward and westward. Eastward it passed to Samarkand, Tashkent, and thence to the greater part of Siberia; westward it reached the Caspian. It soon crossed that sea, and invaded the Caucasus on the one hand and Astrakhan on the other. It rapidly spread up the Volga and more slowly along the shores of the Black Sea, until the greater part of European Russia was more or less infected by the disease.

---

[1] Hirsch, however, appears to be in error in stating that "from 1875 to the end of 1880 it [cholera] had shown itself at no point of the globe out of India." It was severely epidemic in Japan in 1877–79; and was present in several parts of China during the five-year period in question.

In the same year it also became severely epidemic at Hamburg, in Germany, and in some parts of Austria, Hungary, Turkey, and Arabia; and isolated cases were recorded in England, France, Belgium, Holland, Denmark, Norway, Italy, Servia, Morocco, and the United States.

In 1893 the disease was again rife in Russia, especially in the Ukraine in the south-west, while the Volga provinces and the Caucasus in the south-east, which had been most severely affected in 1892, suffered but little. Cholera also prevailed to some extent in many parts of France, Holland, Belgium, and Germany (where Hamburg was again the scene of an epidemic). Hungary, Austria, Bosnia, and Roumania were all invaded. It was more or less prevalent in many parts of Italy, and in some places in Spain, Tunis, Algeria, Senegambia, and the Canary Islands. The Haj pilgrims suffered from a terrible outbreak of the disease, which, according to an estimate of the Grand Sherif of Mecca caused 50,000 deaths. Imported cases again occurred in the United States and also in Brazil. A considerable number of strictly limited outbreaks of cholera occurred this year in England.

In 1894 the disease was still active in many parts of Russia, especially in the Polish and other western governments. In Germany cholera was widely prevalent; in Galicia there was a very fatal epidemic; in Turkey, Belgium, Holland, and Hungary the disease was present, and it had not quite disappeared from France. In 1895 it was again epidemic in Volhynia, Podolia, Kief, and other parts of western Russia, in Galicia, in European and Asiatic Turkey, in Morocco, in Egypt, and in Japan. Finally in 1896 cholera disappeared from Russia, and apparently from Europe; but it was widely epidemic in Egypt. In all these years the infection was frequently imported to England, but without giving rise to any epidemic. In 1899 the malady appeared at Bushire, and at Basra and some other places on the Persian Gulf. As usual, the outbreak here occurred at the time of the date-harvest, when thousands of people are congregated together under the worst possible conditions of sanitation. Finally, while these pages are passing through the press, cholera has caused a considerable mortality among the pilgrims to the Hedjaz, and in many parts of Egypt; while it is reported to be raging in Manchuria and at numerous places on the Russo-Chinese frontier.

Map of European Russia and the Caucasus,
showing the distribution of Cholera in 1892 (from the beginning of the epidemic
to December 1st).

The shading in the accompanying map represents a varying ratio of cases of cholera per 100,000 inhabitants. The numbers inserted in the different governments and provinces correspond to the numbers in the following table:

No.	Name of Government or Province	Cases of cholera per 100,000 inhabitants	No.	Name of Government or Province	Cases of cholera per 100,000 inhabitants
I.	Daghestan	3912	23.	Podolia	148
II.	Terek	3384	24.	Perm	150
III.	Stavropol	2115	25.	Yaroslavl	104
IV.	Kuban	1944	26.	Riazan	90
V.	Baku	1607	27.	Tula	84
VI.	Kars	1468	28.	Orel	71
VII.	Zakataly	1309	29.	Poltava	70
VIII.	Erivan	1308	30.	Tchernigof	70
IX.	Elizavetpol	1079	31.	Radom	99
X.	Tiflis	749	32.	Moscow	68
XI.	Kutais	6	33.	Kielce	52
			34.	Volhynia	49
1.	Astrakhan	2494	35.	Kostroma	31
2.	The Don Cossacks	1812	36.	Olonetz	27
3.	Saratof	1721	37.	Vladimir	19
4.	Samara	1556	38.	Warsaw	17
5.	Simbirsk	1028	39.	Mogilef	16
6.	Voronezh	894	40.	Płock	12
7.	Orenburg	746	41.	Livonia	8
8.	Tambof	733	42.	Minsk	8
9.	Lublin	536	43.	Pskof	6
10.	Penza	483	44.	Lomza	8
11.	Kursk	397	45.	Grodno	5
12.	Kazan	380	46.	Tver	3
13.	Ufa	353	47.	Novgorod	2
14.	Viatka	343	48.	Piotrkow	1
15.	Kief	336	49.	Vologda	1
16.	St Petersburg	290	50.	Kaluga	0·6
17.	Nijni Novgorod	280	51.	Courland	1
18.	Ekaterinoslav	258	52.	Smolensk	0·3
19.	Kharkof	255	53.	Vilna	0·5
20.	Taurida	215	54.	Vitebsk	0·1
21.	Kherson	219	55.	Siedlec	211
22.	Bessarabia	194			

The Caucasus & Transcaucasia [bracket covering I.–XI.]

The only parts of Russia which escaped the epidemic altogether in 1892 were the governments of Archangel and Esthonia; the Polish governments of Suvalki and Kalisz, and the Principality of Finland. These are consequently omitted both from the table and the map.

Factors governing the Geographical Distribution of
Cholera. It will be seen that in its epidemic or pandemic spread
westwards cholera has been mainly confined to the northern
hemisphere. The occasions on which it has crossed the equator
have been the following:—in 1820, 1837, and 1869, it appeared
in Zanzibar[1]; in 1819 and four times between 1854 and 1862
it was epidemic in Mauritius; it has recently appeared in New
Guinea, a few degrees south of the equator; and it was epidemic,
in one or more years between 1855 and 1868, in Brazil, Paraguay,
Uruguay, the Argentine Republic, Bolivia, and Peru.

In considering the geographical distribution of cholera it is
more easy to name those countries where the disease has never
appeared than those where it has at some time prevailed. Some
doubtful sporadic cases only have occurred in Iceland (in 1860).
The Faröe Islands have never been visited. In Norway Bergen
has been the most northerly limit of cholera. Finland has almost
entirely escaped its ravages. No other European country has
quite escaped the disease. But in saying this it is not intended
to imply that all parts of those countries have been the scenes of
cholera epidemics; on the contrary, there are many towns and
many districts of considerable size which have always been
fortunate enough almost or quite to escape the disease; as, for
example, Rouen, Versailles, Lyons, Sedan and the Vosges
district in France; Baden and Würtemburg in Germany; several
towns in Russia, and so forth. The disease is notoriously
capricious in its behaviour, though possibly a fuller knowledge
will in time enable us to explain phenomena that now appear to
be capricious. In Siberia the northerly province of Yakutsk
seems to have escaped cholera in all the past epidemics of the
disease; Kamtchatka has also never been visited, and generally
in Siberia the northerly limit of cholera appears to have been
about the 60th parallel of latitude. In Europe, on the other hand,
the disease has on many occasions been mildly epidemic considerably to the north of this, *e.g.* in the Archangel government.

---

[1] Isolated cases of what is described as "apparently a not very rapid form
of cholera" were seen by Pruen in Equatorial Africa; epidemics of cholera he
never saw or heard of there. There is no other evidence to show that true
cholera exists in this part of the world, and the cases described were perhaps
cases of an acute diarrhœal disorder other than cholera. *The Arab and the
African*, S. T. Pruen, M.D., London, 1891.

In the western hemisphere, Greenland, Newfoundland, and the northern portions of Canada have never been visited by cholera. Some parts of Central America, the Bermudas, and several West Indian Islands have been equally fortunate. In South America, Venezuela and Chile have also escaped hitherto, and, so far as the absence of positive records justifies the assertion, the same is true of Patagonia and the Falkland Islands. The infection has been several times imported to the shores of South Africa (one of the most recent instances occurred in Durban in 1890), but has never gained a footing there, and Australia and New Zealand seem to be equally unsuited for the spread of cholera, or have been equally successful in dealing with imported infection. The complete immunity of the Andaman Islands, so close to the principal home of the disease, is even more remarkable.

That cholera is largely spread by the movements of human beings is shown firstly by the part that pilgrimages have taken in its development, secondly by the fact that the course and rapidity of its diffusion have often varied with the course and rapidity of the usual means of human communication, and thirdly by the disease always following the great routes of trade and travel. The commencement of the last pandemic of cholera in 1892 was associated with the movements of a vast body of pilgrims dispersed from the great bathing festival at Hardwar in northern India. This and other fairs have on many previous occasions played a similar part, though it is important to add that they have in many years been attended by equally large numbers of pilgrims amongst whom cholera has appeared without being followed by any wide extension of the disease. This happened in 1891, when three deaths from cholera occurred in pilgrims to Hardwar, yet no epidemic resulted, although in the following year the same measures of precaution seemed powerless to prevent the development there of one of the greatest and widest epidemics of cholera ever known. The dispersing pilgrims in 1892 carried the infection with them and became foci of the disease in many parts of India. That they were the sole factor in starting the epidemic must however be doubted. Cholera had been unusually prevalent in India in the previous year without their taking any part in its diffusion, and it had already been unusually active in the North-West Provinces in 1892 before the Hardwar Festival (an exceptionally numerously attended one, though less so than that of 1891) of that year was held. It is probable therefore that the movements of the Hardwar pilgrims only contributed to an extension of cholera which would have occurred in any case.

The great annual pilgrimage to Mecca and Medina has on many occasions been not only the scene of terrible outbreaks of cholera, but the means of dispersing the disease far and wide; and smaller pilgrimages, such as those to the shrines of Meshed in Persia, and the many local fairs and festivities in India, have frequently aided in spreading the disorder. So also the movements of large bodies of peasants in Russia, the dispersion of traders and others after great fairs such as that of Nijni-Novgorod, and the movements of troops in war-time (in this case aided by other conditions than mere movement) have all powerfully contributed to the diffusion of cholera. But a full consideration of all the facts will usually lead to the conclusion that the disease is not solely spread in this manner.

The same remark applies to the diffusion of cholera by water. That the disease can be, and is spread in this way to an immense extent, and perhaps more than by any other means, can scarcely be denied. The last epidemic furnished innumerable examples of this. In Russia such great rivers as the Volga, the Neva, the Vistula, the Dnieper, and the Don all became seriously contaminated by the cholera virus in 1892[1], and at Hamburg the Elbe was the principal distributor of the infection. In London in 1854 the disease was diffused by a contaminated public water-supply, and the same thing has been observed over and over again, not only in countries where cholera has become epidemic, but in India and even in Lower Bengal itself, where, if anywhere, cholera is permanently endemic. But this spread of the infection by water—no more than its diffusion by human beings—is not the whole explanation of the behaviour of the disease. There are phenomena—particularly the quiescence of cholera for several years in its endemic homes, and its wide extension outside them in other years under apparently identical conditions—which cannot be explained by these considerations alone, and which involve problems too wide and too controversial to be discussed here.

[1] "The Spread of Cholera by Water." By Frank G. Clemow, M.D., D.P.H. *Transactions of the 8th International Congress of Hygiene and Demography.* Buda-Pest, 1894.

Cholera, it will be seen, remains endemic for very long periods only in tropical countries. In some subtropical countries, such as Persia and Mesopotamia, it has remained endemic for a longer series of years than in the temperate countries of Europe or America. That local conditions, and not climate alone, have some share in determining this behaviour of the disease, is, however, clear from the fact that cholera has remained endemic in Russia for a longer series of years than in other European countries with warmer climates. Warmth is, in fact, not the only condition needed for cholera prevalence, though it is an important one. This is shown by the rarity and mildness of the disease in northern Europe and its absence from northern Siberia. The influence of warmth was remarkably demonstrated by the distribution of cholera in Russia in the year 1892. The intensity of the epidemic decreased with something like mathematical regularity as the disease travelled from south to north, and though other factors,—such as a loss of virulence in the cholera virus as it travelled further, and an increasing preparedness on the part of the authorities to meet and deal with the disease—may have been at work, it is scarcely possible not to believe that a diminishing temperature was the most important factor. The map on p. 100 shows, by means of shading, this remarkable distribution of the disease in Russia in 1892. Many other considerations—especially the subsidence of cholera in Europe in the winter and its recrudescences in summer—prove further the great influence of warmth upon its prevalence. The disease has, however, exceptionally prevailed in quite cold weather, and, as stated above, it has been seen as far north as Archangel and, doubtfully, in Iceland.

An immense number of other influences are at work in determining the geographical distribution of cholera. Race is probably of no great consequence, but the black races seem to be more susceptible than the white. Elevation is of importance, low-lying, alluvial countries seeming to suffer much more than those more elevated; but this rule is not without marked exceptions. A country or district which has suffered severely in one year is often visited mildly or escapes altogether the next year; but exceptions to this rule are also observed, not only in Europe, but also in India. Soil, meteorological conditions, the degree of intelligence and civilisation of the people, and many other local conditions all play a considerable part in the distribution of cholera. To what extent the disease can always be kept at bay by well-directed measures is perhaps uncertain. Certainly a high level of Public Health administration, combined with great care in preventing the spread of imported infection, has in a vast number of instances proved successful in saving a country from a cholera epidemic. These instances have, however, all occurred in countries far removed from the original source of the disease in the East, and it has yet to be proved that they would be equally successful in countries nearer to India.

Lastly, perhaps the most important factor of all in the geographical distribution of cholera is distance from the endemic homes of the disease. Those countries and islands which are farthest removed from India, and have consequently less constant and direct communication with her, escape altogether or are visited rarely. Those countries, on the other hand, such as Farther India, China (where for the present cholera is as truly endemic as in India), Persia, Mesopotamia and Central Asia, between which and India there is a constant stream of intercommunication, are the most liable to epidemic recurrences. It is more than probable that this relative immunity to cholera of countries distant from the source of the disease is due, not only to their less constant and direct communication with India, but also to the greater distance over which the virus has to be carried, with consequently increased chances of its dying on the way or greatly losing its virulence.

DENGUE.

General Characters and Etiology. Dengue, Dandy, or Breakbone Fever is a highly infectious febrile disease, characterised by severe rheumatoid pains in the joints and limbs, and in some cases by a cutaneous eruption of varying character and duration. The cause of this disease is quite unknown, but it presents all the characters of a specific infective fever, and is very possibly of microbial origin. The microbe, if it exists, has not, however, been discovered. The malady can apparently remain for long periods as an endemic, but it is seen most characteristically as an epidemic, or, at irregular intervals, as an almost pandemic disorder.

Geographical Distribution. Dengue appears to be mainly a disease of tropical and subtropical zones, but it can occur and give rise to epidemics of some severity in comparatively temperate countries. Whether dengue is to be regarded as permanently endemic in any part of the world appears uncertain, the records of the disease dealing almost entirely with recurring epidemic prevalences. In this respect it presents many analogies with influenza, both diseases occurring from time to time in the form of localised outbreaks, and at intervals spreading with extraordinary rapidity over very large areas of the earth's surface. Dengue has, however, at no time become truly pandemic in the sense that influenza has on many occasions ; though the area which it has covered in some epidemics has been of very considerable size, as for example in 1870—73, when it was seen as far west as the southern United States, and as far east as China and Formosa. Even in its widest epidemic prevalences, however, it does not spread very far outside tropical and subtropical latitudes.

Moreover it retains a much more sporadic and capricious character than influenza. In the years just named, for example, dengue was present, between the easterly limit of Formosa and the westerly limit of the United States, solely in Burma, India, China, the East Indies, Arabia, Aden, the East African coast, Zanzibar, Mauritius, and Réunion, whereas influenza, in its epidemic prevalences, usually attacks every country, and sometimes every inhabited place in every country (with the exception of isolated communities) between the extreme limits of its extension.

Dengue resembles influenza further, not only in its tendency to attack a very large proportion of the inhabitants of a place in which it breaks out, but in the extraordinary speed with which these large numbers of persons develop the disease; so that—just as in the case of influenza—it is common to read of "nearly the whole" or of "three-fourths" of a community being "simultaneously" attacked. It is probable that these expressions are not to be taken quite literally, but it is certainly true that the disease does attack very large numbers of persons, and that it may do this in a very short period of time. Trustworthy estimates have placed the proportion of a community attacked at as high a figure as two-thirds of the population (at Mauritius in 1873), at three-fourths of the population (at Smyrna and at Amoy), and even higher (at Jaffa and St Thomas). In some of the annual outbreaks of this disease at Beirût in Syria I am informed on excellent authority that almost the entire population are attacked.

Recent Endemic Prevalence in certain Countries. If dengue is truly endemic at the present time in any part of the earth's surface it is probably so only in certain tropical and subtropical countries. In India it is occasionally seen among the natives, but I am aware of no trustworthy statistics proving its degree of prevalence in the Indian peninsula. Many epidemics of the disease, on the other hand, have been recorded here in the past. In Egypt dengue is said, on good authority, to be met with in a sporadic form every year during the damp months of autumn, dying out in December. In Beirût, on the Syrian coast, dengue would appear to be truly endemic, and every year, usually in the autumn, an epidemic of greater or less severity occurs. From time to time epidemic outbreaks of the disease are also reported from Tripoli, where an outbreak of this kind occurred as recently as November, 1901.

It is possibly also endemic in East Central Africa; Pruen observes that it occurs there sporadically (he ascribed it to the use of impure drinking-water); and another traveller in East Africa, Scott-Elliot, states that "in the colony zone, at a height of 5000 to 7000 feet, there is a curious kind of rheumatism or 'dengue fever,' which affected many of my men[1]." It is also said to be seen sporadically on the West Coast, at Senegambia[2].

In the West Indies there is some reason to believe that dengue is more or less permanently endemic. It is sometimes seen in Cuba; it was epidemic there in 1897, at the same time as yellow fever, but as Cuba was at the time under exceptional conditions owing to the war, and as dengue, also accompanied by yellow fever, was epidemic in Texas at the same time, this outbreak was quite possibly due to an imported infection and not to excessive activity of an indigenous disease. In the Bermudas dengue is said to occur sporadically from time to time. It is sometimes seen also in British Guiana.

Epidemics or Pandemics of Dengue. Historical Summary. The epidemic extensions of dengue have been marked by great irregularity, both in regard to time and to the countries attacked. The history of the disease is far from completely known, as it has only been distinctly differentiated in the last century, and is even now not always easy of diagnosis from other disorders. But so far as recorded facts enable us to follow its history the principal epidemics in the past have been the following :—

In 1779 dengue was prevalent in Batavia and Egypt; in 1780 in Spain, Zanzibar, India (the Coromandel Coast), and possibly in Philadelphia. In 1818 it broke out at Lima. In the years 1824–28 there was a wide prevalence of the disease; it spread over a great part of the Ganges delta, extending as far inland as Berhampur; it appeared on the Coromandel Coast, at Gujerat and Rangoon; it (probably) raged at Zanzibar; it prevailed from 1826–28 in the south of the United States, in the West Indies, and at Carthagena and Bogota in South America. Whether any of the countries intervening between the wide limits here named were affected by the disease at this time there is apparently no evidence to show.

[1] *A Naturalist in Mid-Africa.* G. F. Scott-Elliot, M.A., F.L.S., etc. London, 1896, p. 189. Cook, on the other hand, never saw a case during four years' residence in Uganda.

[2] Two cases of dengue were admitted to the Colonial Hospital, Gambia, in 1898; none in 1897; one in 1890. See the Colonial Hospital Reports.

In 1835 a limited outbreak of dengue was seen in Arabia; in 1836 in India; in 1837 in Bermuda; in 1844 in India; in 1845 in Egypt; in 1847 in India; in 1845–48 in Senegambia; in 1845–49 in Rio de Janeiro; in 1847–56 repeatedly in Tahiti; in 1850 in the southern United States; in 1851 in Mauritius and Réunion, and at Callao and Lima; in 1853–54 in India; in 1854 in the southern United States and in Havana; in 1856 in Senegambia and at Benghazi in Tripoli; in 1860 in Bermuda and in Martinique; in 1863 in Bermuda; in 1864 in Cayenne; in 1865–67 in Senegambia, in Spain, and in Teneriffe; in 1868 in Port Said.

What is spoken of as a third pandemic of dengue began in Zanzibar in 1870; the scattered portions of the earth's surface affected by the disease in that and the following three years have been already enumerated. In India, during this epidemic, dengue spread from Bombay, Calcutta, and Madras along the lines of communication to most parts of the east and west coasts of the peninsula and inland to the foot of the Himalayas.

More recent limited outbreaks of dengue have occurred at Benghazi in Tripoli in 1878; on the coast of the Gulf of Mexico in 1880; in New Caledonia in 1884–85; in Fiji, for the first time in 1885; at Port Said in the same year, and widely throughout Egypt in 1887. In 1889 dengue spread to both shores of the Ægean Sea and to most of the islands of the Greek Archipelago excepting Crete. Beirût, Jaffa, Smyrna, Constantinople, Salonika, and Athens were all visited, and the epidemic is said to have spread all over Asia Minor and the Lebanon. Still more recently there is mention of an epidemic at Amoy, on the coast of China. Canton had also been the scene of an outbreak at some earlier date. In 1895–96 dengue was twice epidemic in Cochin China[1]. In 1897 an epidemic of the disease broke out in Queensland and caused as many as 97 deaths. Four cases of this disease had been recorded in the preceding year, but before 1896 dengue had apparently been absent from the colony for some time[2]. In that year the malady, as already stated, was epidemic, together with yellow fever, in Cuba and in Texas. In 1898 seventeen cases of dengue were admitted to the Durban Hospital, Natal, so that probably there was an epidemic there in that year. In November, 1901,

---

[1] Nogué, *Arch. de Med. Nav. et Colon.* Dec. 1897.
[2] Reports of the Registrar-General. 1896 and 1897.

the disease became epidemic in Beirût and Tripoli, in Syria, and a few isolated cases occurred a little later in Constantinople. Finally, in British New Guinea dengue has recently been imported and has attacked the natives, but without doing much harm[1].

Factors governing the Geographical Distribution of Dengue. The known geographical range of dengue is a restricted one in regard to latitude. The disease appears to be confined to a belt running round the world, the northern limit of which may be placed in Europe at about the 41st degree (Constantinople); in Asia not higher than the 33rd degree (Northern India), and in America at about the 40th degree (Philadelphia). The southern limit of the belt would seem to be about the 30th degree of south latitude in Africa (Durban); the 25th or 26th degree in Australia (Queensland); and the Tropic of Capricorn in South America (Rio de Janeiro). Within this belt, which is practically the tropics plus a strip of the earth's surface on either side of the tropics, dengue has been able at times to prevail widely in both hemispheres.

That temperature is one of the leading factors in defining the range of dengue is therefore exceedingly probable. A warm or hot atmosphere seems to be essential for its prevalence. This is further shown by its behaviour in individual outbreaks. Almost without exception the onset of cold weather has always put an end to an epidemic of dengue. The only apparent exceptions have been certain epidemics in the West Indies, which occurred in cool weather; but the term "cool" must be taken as having but a relative significance here, as the minimum temperature at the time was as high as 64° (Hirsch). Temperatures near the freezing-point are said to arrest an epidemic completely.

Certain other conditions seem to be also necessary for the prevalence of dengue. Dirt, insanitation, and overcrowding have appeared to favour its development. Towns have usually suffered more severely than rural districts, and particularly those portions of the towns affected where the general conditions of hygiene have been most defective. In this respect, as Hirsch and others have pointed out, dengue presents some analogy with yellow fever. A further point of analogy is found in the tendency dengue has shown to occur along coast-lines and to avoid penetrating far into

[1] Sir W. MacGregor.

the interior of countries. A low elevation above sea-level is apparently more favourable than a high one.

This disease appears to be almost entirely independent of such conditions as the race, age, or sex of the community attacked. All ages and both sexes are liable to it, and an apparent immunity observed in certain races in some outbreaks has not been maintained when the same races have been exposed to its attacks elsewhere. A similar indifference is shown by dengue to the nature of the soil on which it prevails, whether as regards its physical or its geological characteristics.

The clinical and other characters of this disease point to its being one of the acute infective fevers, but the microorganism associated with it, if there be one, has not yet been discovered. That the disease is infectious from person to person is practically certain. Corré and others have quoted a large number of instances in which dengue has been spread by the arrival of a ship with cases on board. Instances are also known in which a person has visited an infected locality and on his return thence introduced the disease to a place hitherto free from it. The fact that medical men and nurses have also rather frequently suffered from its attacks may be held to be further proof of the considerable degree of infectivity possessed by the disease. It is still a moot point, just as in the case of influenza, whether this is the only or the principal manner in which dengue is spread when it becomes epidemic, or whether the great rapidity of its diffusion and the occasional occurrence of what seems to be a simultaneous outbreak in a large proportion of a community do not require some other explanation. It is probable, however, that the rapidity and the "simultaneity" of these epidemic outbreaks have been somewhat exaggerated, and that dengue is, at any rate to a very great extent, spread from person to person.

In some epidemics animals have been believed to suffer from an analogous disorder at the same time as human beings, but no proof has ever been adduced that this disease can in fact attack the lower animals.

DIABETES MELLITUS.

General Characters and Etiology. The general characters of diabetes mellitus are too well known to require detailed description here. The causation of diabetes is uncertain. Hereditary influences, temperament, racial predisposition, all perhaps have some share in it, and there can be no question that disturbances of the nervous system are largely concerned in its production. Some have hazarded the suggestion that this disease, like so many others, will prove to be of micro-parasitic origin. But no facts in support of this suggestion have as yet been brought forward.

History. The oldest known mention of this disease seems to occur in one of the early Hindu writings—the Ayur Veda of Susruta. References to diabetes, or at least to glycosuria, are frequent in the writings of the Græco-Roman and Arabian physicians, and throughout medieval times. Towards the end of the eighteenth century the study of the disease was placed on a scientific basis by the work of two English physicians, Dobson and (later) Rollo. In recent times much light has been thrown on the nature and cause of diabetes by the investigations of Pavy, Naunyn and others.

Geographical Distribution. *Europe.* Diabetes mellitus is increasing in frequency in the British Isles. In England and Wales the mortality from this cause has steadily risen from an average of 57 per million in the years 1881–90 to 86 per million in 1900. Among the specially liable group of males above 45 years of age the mortality-rate is very much higher. Thus in the years 1881–90 the diabetes deaths among persons between 45 and 55 was 134 per million; between 55 and 65, they were 282; between 65 and 75 they were 397; and over 75 they were 314

C.   8

per million. In Scotland diabetes is somewhat rarer than in England, and in Ireland it is still less prevalent.

In France diabetes also, appears to have become more frequent in recent years. In Paris, at least, there has been a steady rise in the mortality from this cause in the past four decades; in 1891 the death-rate rose to 140 per million of the population.

Although statistics relating to the prevalence of this disease in other European countries are largely wanting, there is evidence from many of them to show that diabetes is not a rare disease. When statistics are available they tend to show a rise in the mortality from this disease similar to that observed in the British Isles and in France. Thus in Denmark the deaths from diabetes in the urban population rose from 48 (about 60 per million) in the year 1892 to 87 (over 100 per million) in 1897. In Norway the diabetes death-rate is about equal to that of England and Wales. In Russia, in the three years 1893, 1894, 1895 there was a steady rise in the number of cases of diabetes treated both inside and outside the hospitals. Cases of, and deaths from the disease were reported in those years from all parts of European Russia, including the Baltic and Polish provinces, and the Don Cossack territory.

In Italy diabetes seems to be rarer than in England or in Denmark, but here also its frequency is increasing. Between the years 1887 and 1896 the deaths ascribed to it in the whole of Italy continuously rose from 17 per million in the first year to 27 in the last.

Diabetes undoubtedly exists in Turkey in Europe. In Constantinople it is seen with some frequency; but I am unable to satisfy myself that it is of more common occurrence here than elsewhere in Europe, in spite of the unmeasured consumption of sweetmeats of all kinds by the inhabitants and the sedentary, inactive life led by the majority.

Asia. Diabetes is certainly met with in the Caucasus and Transcaucasia, though it is probably not a common disease here. The returns do not give the separate figures for the flat Caucasus proper and the hilly regions of Transcaucasia. From Russian Central Asia a few cases of diabetes are yearly reported, but the returns from these provinces are very imperfect, and it is impossible to state whether the disease is common or rare here. The same remark equally applies to Siberia. The disease statistics for the Siberian provinces are for the country as a whole, and not for each province separately.

With regard to Arabia, Syria, Asia Minor, Mesopotamia, and Persia, medical literature affords no information as to the presence, absence, or frequency of this disease.

In India, on the other hand, diabetes is known to be a disorder of very considerable frequency. The Hindus here suffer much more than the Mohammedans. It is said to be particularly common among the educated Bengalis. The Jains, whose only animal food is milk and butter, escape it, and so also do the Sadhus, Yogees, and Chorbays of Muttra, who live almost entirely on sweets[1]. In Ceylon, also, diabetes would seem to be as prevalent as in India. Whether the disease exists in any of the East Indian islands is uncertain. It is said to be unknown among the hill tribes of Java[2]. It is occasionally seen in China[3].

Australasia. In Australia diabetes is probably a rare disorder. Occasional cases are seen in the hospitals of New South Wales; and in Western Australia it is the cause of a few deaths in most years. In New Zealand on the other hand this disease must be very much more common than in Australia, and here, as in Europe, it is becoming more frequent. In 1889 the deaths from diabetes in New Zealand were only 14 (about 22 per million); in 1898 they had risen to 48 (about 64 per million). It is therefore nearly as common in this colony as in the mother country. Nothing seems to be known concerning the existence of diabetes in the Pacific Islands.

Africa. Diabetes certainly occurs in Egypt, in Algeria, and in Morocco. It is relatively common in Tunisia, particularly among the natives; the Mussulman and Jewish population suffer from it here much more than the Christian inhabitants. On the other hand diabetes appears to be absent from the pathology of Western and Central Africa, and is seen but rarely in Madagascar.

America. Osler states that in America diabetes is a rarer disease than in Europe. The last census gave a death-rate of only 38 per million for this disease. It is however increasing in frequency in the United States, as in Europe. In Mexico it has been seen and is perhaps common. In some of the West Indian

---

[1] *Transactions 1st Ind. Medical Congress,* 1894.
[2] Kohlbrügge, *Janus,* 1897–8.   [3] Coltman, *op. cit.*

islands, as Jamaica and Grenada, an occasional death from it is recorded.

For South America the evidence in regard to this disease is conflicting, and scarcely justifies any more definite assertion than that it occurs, but with great rarity, in Brazil and Peru, and that it is perhaps unknown in British Guiana.

Factors governing the Distribution of Diabetes. This disease occurs indifferently in tropical, sub-tropical, and temperate climates. It is apparently commonest in the highly civilised countries of Europe and in India. It is, perhaps, a disease of civilisation. Luxury, a sedentary life, a neurotic temperament, and town life all seem to favour the production of diabetes in the individual, and these are found at their highest development in the most civilised countries. Mental strain, worry, anxiety, "nerve-wear" also serve to predispose to diabetes. All these factors increase as civilisation advances, as the city population increases at the expense of the rural population and as the "struggle for life" becomes keener, and to this may perhaps be attributed in part, if not wholly, the recent increase in frequency of this disease in so many countries.

Race has probably not a little influence upon the prevalence of diabetes. The Jewish race has shown a remarkable tendency to suffer from it, not only in Europe, but in the United States and in Algeria. The negro race is not exempt, and in a series of 77 cases recorded by Futcher in America as many as one-tenth of the patients were negroes. In India, as stated above, the Hindus suffer more than the Moslems, while the Jains and some other races escape.

The belief has often been expressed that an exclusively or mainly farinaceous diet is largely responsible for the production of diabetes. The prevalence of the disease in India and Ceylon to some extent supports this view. But there are some considerations of weight against it. Thus in many other countries outside India the natives live practically on nothing but farinaceous food and yet do not develope diabetes; and in India itself the Jains under similar conditions escape it, and so also do Hindu widows, who never touch animal food[1].

The male sex is very much more liable to be the subject of

---

[1] Koïlash Ch. Bose, I. M. S. *Transactions 1st Ind. Medical Congress,* 1894.

diabetes than the female, and the disease is mainly—though not exclusively—one of advanced years. Children have occasionally suffered from it, and even infants under one year of age have been known to develope it.. There is nothing to prove that insanitary conditions favour its prevalence ; and in London and elsewhere it has been shown to occur more often in the well-to-do than in the poorer classes of the community.

DIARRHŒAL DISORDERS.

1. THE DIARRHŒA OF INFANTS, CHOLERA NOSTRAS, SUMMER DIARRHŒA, ETC.
2. SPRUE (PSILOSIS).
3. HILL DIARRHŒA.
4. "CARIBI," "EL BICHO," OR "INDIAN SICKNESS."

The Principal Varieties of Diarrhœal Disorders.
Diarrhœa is, it need scarcely be said, a symptom of many morbid conditions, and not as a rule a disease in itself. It is unnecessary to enumerate the very large number of causes which may give rise to loose evacuations. Many of these causes are local, accidental, or temporary in their character, and, as such, it would be purposeless to attempt to discuss the geographical distribution of the symptoms to which they give rise. In most countries diarrhœal diseases contribute largely to the sum of annual mortality, but it is usually impossible to distinguish in the returns what proportion of the deaths is due to diarrhœa of local, accidental, or temporary causation, and what to the more specific forms of diarrhœa which will alone be discussed here.

Some forms of diarrhœa may be conveniently regarded as almost constituting specific diseases in themselves, inasmuch as they are probably the result of certain constant causes or groups of causes, they show distinct seasonal or climatic variations, or they bear definite relations to elevation above sea-level, or to the age of the persons attacked. Among these is the diarrhœa of infants, which is one of the most prolific causes of infantile mortality. Another is the so-called cholera nostras, or summer diarrhœa, from which many people suffer in the summer months, and which is probably in part due to the consumption of unripe or over-ripe fruit, or of fruit in excess. A third form is the so-called hill diarrhœa, met with at high elevations in India and elsewhere. Sprue or psilosis is a fourth form of affection of the alimentary tract in which diarrhœa is a more or less prominent symptom. "Caribi" or "Indian sickness," "El Bicho," and "epidemic gangrenous rectitis," all probably the same disease, form yet another distinct variety of diarrhœal disorder.

1. THE DIARRHŒA OF INFANTS, CHOLERA NOSTRAS, SUMMER DIARRHŒA, ETC.

In most countries a very large number of children die in the first years of life from diarrhœal troubles. To what extent these troubles are due to one specific cause and may therefore be regarded as a specific disease cannot be determined. Diarrhœa is such a constant symptom in infantile diseases, it is so easily brought about by teething, by improper feeding, neglect, and other temporary and accidental causes, that unquestionably much of the so-called *cholera infantum* or infantile diarrhœa is due to no one specific disease. But the remarkably definite relation which the mortality of infants from diarrhœa shows to variations of temperature seems to indicate that a considerable proportion of it is due to a definite cause or group of causes which varies in activity with the variations in temperature, and that in this sense a certain proportion of the deaths from so-called infantile diarrhœa may be justifiably regarded as due to a more or less definite disease or group of diseases.

It is probable that the deaths from infantile diarrhœa form a very notable proportion of the deaths from diarrhœa generally recorded in the vital statistics of many countries. It is, however, rarely possible to determine accurately what this proportion is, and the returns for "diarrhœa" as a whole, or for "dysentery and diarrhœa" are alone available. Although infantile diarrhœa contributes largely to the sum of these returns, the diarrhœa of elderly persons, and *cholera nostras*, together with the diarrhœa from temporary or accidental causes already alluded to, form the majority of the recorded cases. Dysentery will be dealt with in a separate chapter.

Recent Geographical Distribution. *Europe.* In England and Wales the deaths from "dysentery and diarrhœa" fell from an annual mean of 1076 per million in the decade 1861–70 to one of 935 in the next decade, and to 674 in the years 1881–90. The fall has not been maintained, for though the mortality from these causes was as low as 358 per million in 1894, it rose to 902 in the following year, and has since been respectively 564, 871, and 727 in the years 1896, 1897, and 1900. Dysentery being all but extinct in this country, almost the whole of the deaths in this group may be regarded as due to diarrhœa. These deaths are vastly more numerous in densely-crowded urban populations than elsewhere, and hence the highest mortality ratios are constantly returned from such districts as those of London, the East and West Ridings of Yorkshire, Lancashire, Leicestershire, Nottinghamshire, Staffordshire, and Warwickshire. In the East Riding the mortality from this cause has in some years been as high as from 1500 to 2000 per million.

In Scotland the deaths from the same causes fell from an annual mean of 453 per million in 1888–90 to 360 per million in 1896; in 1897 the ratio rose again to 510 per million.

In the larger French towns (those with over 10,000 inhabitants) deaths from "diarrhœa and gastro-enteritis" have varied in the last decade between 1550 and 2020 per million inhabitants; while "*maladies cholériformes*" caused in the same period a mortality varying between 20 and 270 per million. The higher figures under the latter heading fell in the cholera years of 1892 and 1893.

In Belgium the mortality ratio from "cholera" (? nostras), "enteritis and diarrhœa" together rose considerably between 1870 and 1897; in the year 1897 it was nearly 1250 per million inhabitants. In Holland, on the other hand, the mortality ratio from "diarrhœa" steadily fell from about 450 per million in 1889 to about 110 per million in 1898. In Germany infantile diarrhœa causes an excessively high mortality, particularly in large and densely populated towns, and in some, as for example Berlin, the death-rate from this cause has increased with the growth of the population. In Bavaria and Swabia very high death-rates from "cholera nostras" (which here appears to include all forms of diarrhœa except dysentery) are annually recorded.

In Denmark the deaths from "cholerine and acute intestinal catarrh" vary considerably from year to year; in each of the years

1896 and 1897 the mortality from this cause in the principal Danish towns was rather less than 1300 per million living. In Norway the deaths from "diarrhœa acuta, cholera nostras, and enteritis" showed some tendency to increase between 1881 and 1894; the mortality from these causes in the latter year was about 540 per million. The corresponding mortality ratio in Sweden in the years 1891–96 was between 300 and 400 per million per annum.

Diarrhœal diseases are excessively rife in European Russia. The official statistics of that country now return deaths from "simple cholera" only; and this term appears to include all forms of diarrhœal disease excepting Asiatic cholera and dysentery. Among the statistics of a large number of registered cases of various diseases, the term "gastro-intestinal catarrh" also appears, and exceedingly high figures are always inserted under this heading. There can be no question that diarrhœal disorders are among the commonest of all maladies in Russia. They are closely associated with the frequent use of polluted water-supplies, with the nature of the usual diet of the peasantry (which consists largely of sour black rye-bread, fermented cabbage-soup, salted herring and raw cucumbers, melons and other fruit), with excessive alcoholism, and generally with the low level of knowledge among the bulk of the population, leading to neglect of sanitation, improper feeding of infants, and other harmful conditions. A form of hæmorrhagic diarrhœa is said to be very common in children in southern Russia, owing to the large amount of raw vegetables and fruit eaten by them. Among Bashkir children, who consume this dangerous diet to a less extent, this form of diarrhœa is rarer.

In Austria-Hungary, and particularly in Hungary, "cholera infantum" and "cholera nostras" give rise to high annual death-rates, and in Switzerland the enteritis of infants also causes a large sacrifice of infantile life. Throughout Italy diarrhœal diseases (*enteritis, diarrhœa, colitis*, etc.) are prolific causes of death; the mortality from them rose, with some fluctuations, from 3159 per million in 1887 to 3683 per million in 1895. In Spain and Portugal diarrhœal disorders are widely prevalent, and no doubt the causes of their prevalence are largely the same as those already mentioned as acting in Russia. In most, if not all of the countries of the Balkan Peninsula diarrhœa, and particularly infantile diarrhœa, is extremely common, especially in the summer months. In Constantinople these disorders are the cause of a considerable annual mortality.

The distribution of diarrhœal disorders in Europe is to a very great extent determined by temperature. The warmer the temperature of the summer months, the more prevalent and fatal do they become. Hence they are comparatively uncommon in the north and excessively common in the south. In Iceland diarrhœa is but little seen, and in the Faröe group it is only moderately frequent. In Malta and other Mediterranean islands enteritis is often very prevalent in the summer.

Asia. The same law holds good to a considerable extent in Asia. Diarrhœa is, however, met with as a common disease throughout the greater part of Siberia; though probably more rife in the south than in the north. It is certainly extremely prevalent in the Caucasian provinces, where gastro-intestinal catarrh is the cause of more deaths than perhaps any other single disease. In the Central Asiatic provinces of Russia, with their high summer temperature, and where sanitation is as yet of necessity of a primitive and oriental character, diarrhœal disorders prevail to an enormous extent. The same statement appears to be equally applicable to Syria, Mesopotamia, and Persia. In Asia Minor these ailments are exceedingly common. At Trebizond severe dysenteric forms of gastro-intestinal catarrh are often observed. In Arabia diarrhœal diseases are among the most frequent of maladies. The annual mortality from these affections among the pilgrims to the Hedjaz is enormous; it appears to be largely attributable to the very polluted water of the wells which form the sole water-supply on the long caravan routes between the ports of Jedda and Yanbo and the holy places of Mecca and Medina. This water is said to be frequently extremely foul, and to have in many instances a strong smell of sulphuretted hydrogen.

The prevalence of diarrhœal diseases in India will be again briefly discussed in the chapter upon dysentery, as in Indian statistics dysentery and diarrhœa are grouped together. In India generally the deaths from diarrhœa are very numerous, and are highest in the months of August, September, and October, that is to say in the latter half (or more) of the rainy season. They have risen enormously during the recent years of drought and famine. It is of interest to note that in 1891 and 1892 when cholera was extraordinarily rife in India and became pandemic elsewhere, other diarrhœal disorders were showing no unusual activity.

In Ceylon diarrhœa at times causes acute epidemics, and on some occasions at least (*e.g.* in 1892 and 1897) these have been traced to the consumption of rice made from paddy which had been kept for use as seed, and had therefore become stale.

Throughout Farther India and China, in the East Indian Islands, and in Japan, diarrhœal affections are among the commonest of diseases, attacking both the indigenous and foreign inhabitants of these countries and islands. In China diarrhœa causes a vast amount of mortality in children and adults.

Australasia. In Australia diarrhœa would seem to be quite as frequent a cause of death as in many European countries. In Queensland "diarrhœa" and "enteritis" taken together caused a mortality of between 1000 and 1200 per million per annum between 1893 and 1897. In New South Wales, Victoria, and Western Australia, these diseases appear to be somewhat less common. In New Zealand, in the years 1889–98, this group of disorders gave rise to an annual mortality of from 300 to 500 per million per annum. In most of the islands of the Pacific diarrhœa is a frequent disorder, especially among children, and in the warm season.

Africa. In Northern and Central Africa diseases of this class are commonly met with. In the equatorial portion of the continent, both in the centre and on the east and west coasts both natives and Europeans suffer largely from diarrhœa. It is mentioned as causing considerable trouble to the members of almost every exploring expedition that has ventured into the interior of the Dark Continent. It is prevalent in Morocco, Algeria, and Tunis; it is extremely common in Egypt. In the colonies along the western shores of Africa it is rife, and here, as well as in Central Africa generally, it is said to be particularly prevalent after the first rains have washed surface and other pollutions into the streams from which drinking water is taken. It is frequently associated with carelessness in the disposal of human excreta, which are often allowed to pollute watercourses. It is largely avoided in many European colonies where precautions are taken to ensure a pure water-supply. On the West Coast of

Africa diarrhœa appears to be as common on the sea-board as inland, but on the east it is said to be both more frequent and more severe on the interior plateau than on the coast.

In many parts of South Africa diarrhœa is very prevalent. At the Grey Hospital, for example, in King William's Town this is in some years, next to typhoid fever, the commonest disease that comes under treatment, and at many other of the colonial hospitals cases of diarrhœal disorders are exceedingly numerous. Diarrhœa was very prevalent in many places in 1897, owing to the drought of that year; at Aberdeen, at Aliwal North, and in the Jamestown sub-district it was epidemic. At Worcester a severe outbreak of diarrhœa caused an enormous increase in mortality among the natives in 1898. During the recent military operations in South Africa many of our men suffered severely from this disorder; and it is noteworthy that among the forms of diarrhœa observed was one associated with acid, frothy, liquid, green stools, and much scalding about the anus, which was believed to be brought about by the large amount of sand swallowed in the dust-storms which frequently visited the camps.

In Madagascar, in the Seychelles, and in the Mascarene Islands (Mauritius etc.) diarrhœa is very common.

North and Central America. The tendency for diarrhœa to become more prevalent and severe as warmer latitudes are approached is well seen in North America; but there are not lacking important exceptions to the rule. Thus in Greenland diarrhœa is said to be the chief sickness in the winter months, and a form of cholera has even been epidemic here and in Labrador in the past. In other parts of Canada diarrhœal disorders always help to swell the annual mortality, and the diarrhœa of infants appears to be prevalent to a considerable degree. Davidson states that the provinces along the Atlantic coast suffer less than others, and that Quebec shows an extremely high death-rate from diarrhœal diseases. They appear to have been less conspicuous causes of death here in more recent years, thanks probably to the same care as to water supplies that has so reduced the death-roll from enteric in the Quebec province. These diseases also seem to be less prevalent or less fatal now in Manitoba and British Columbia than they were some years ago.

In most parts of the United States diarrhœal disorders are

very common, and cause a high annual mortality. In Massachusetts this group of diseases, which here includes "dysentery, cholera infantum, diarrhœa, cholera nostras, and enteritis," caused an annual mortality of 2111 per million inhabitants in 1856–75, and of 1834 per million in 1876–95. Of the diseases named cholera infantum is the most important, and this alone has caused an annual mortality varying between 640 and 1580 per million living (of all ages) in the years 1856–95. Throughout the States generally diarrhœal disorders are a cause of a greater mortality than in England, cholera infantum being especially frequent and fatal. These diseases are certainly more prevalent in the States than in the higher latitudes of Canada.

In Mexico and throughout Central America diarrhœal disorders are extremely prevalent. Frequent mention of them is found in the annual reports on the health of British Honduras, where the general infantile mortality, doubtless partly through their agency, is very high. In Guatemala "diarrhœa" alone accounted for 1292 deaths (about 900 per million inhabitants) in 1894.

Reports from most of the West Indian Islands seem to indicate a marked frequency of diarrhœa. In Jamaica a form of septic diarrhœa is one of the most prevalent diseases at Port Royal. In many of the Leeward Islands, and in Trinidad, Grenada, Barbados, and elsewhere, diarrhœal disorders are very frequent. To what extent the deaths from this cause in Mexico, Central America, and the West Indies are deaths of infants is not clear.

South America. In the tropical portions of South America diarrhœal diseases are no less prevalent than in most other tropical countries. In the Guianas they are the cause of a considerable annual mortality. Throughout Brazil diarrhœa is a common disease, but appears to be less prevalent in the southern provinces of São Paulo, Paraná, Sta Catharina and Rio Grande do Sul than in the warmer provinces further north. In Uruguay the deaths from "enteritis," "infantile gastro-enteritis" and "entero-colitis" together caused, in 1897, 63·15 per 1000 of the deaths from all causes, and a mortality of little short of 1000 per million inhabitants. In the Argentine Republic diarrhœa is certainly a common disorder, and it is probable that it is more or less prevalent in the western countries of the South American continent.

Factors governing the Geographical Distribution of Diarrhœal Disorders. The causes of diarrhœa are practically ubiquitous, and consequently, as the above brief summary of facts shows, the condition itself is ubiquitous. On the whole, diarrhœal disorders of all kinds are more common, and to some extent more severe, in hot countries than in warm, and in warm countries than in cold. The mean annual temperature of a given portion of the earth's surface is, however, of little importance ; and in temperate climates the most important factor in the prevalence of diarrhœal diseases seems to be the height of the temperature in the warm season of the year alone ; hot summers in any country being almost invariably associated with diarrhœa prevalence, although cold winters may reduce the annual mean temperature in those countries to a comparatively low level.

In England it was long ago shown by Ballard that it is the temperature of the soil and not that of the air which seems to determine the spread of diarrhœa ; he demonstrated that " the summer rise of diarrhœal mortality does not commence until the mean temperature recorded by the 4-foot earth-thermometer has attained somewhere about 56° F., no matter what may have been the temperature previously attained by the atmosphere or recorded by the 1-foot earth-thermometer." It is probable that a like though not quite identical statement might be made concerning most countries. It appears, however, that the temperature at which diarrhœa begins to be prevalent is not everywhere the same. It is less a question of absolute than of relative temperature. The records of earth-thermometers are not available for most countries, but there is probably a rough annual parallelism between their records and those of the air-thermometer ; and it is certain that in many temperate countries diarrhœal diseases begin to be more than usually prevalent not when the air-thermometer records a particular temperature common to all, but when the summer heats, whatever they may be in the given place, begin to make themselves felt. In New York, for example, the temperature in May is about 54°, and in Berlin about 55·4° F. ; diarrhœa is then but slightly active ; in June the average temperature rises in both cities to 61° and the diarrhœa mortality increases seven-fold. In Melbourne similar facts are observed. But in Queensland, where the average winter temperature is about 60°, diarrhœa shows no special activity at that temperature, and it is only when the mean monthly temperature rises to nearly 70° that the diarrhœa mortality markedly rises[1].

The other causes which determine the geographical distribution of diarrhœal diseases are too numerous to discuss at length. Many forms of diarrhœa are the result of neglect of sanitary measures, ignorance, improper food, excessive indulgence in certain foods and alcohol, and other quite preventable causes. Race appears to have no very striking influence, all races showing a liability to diarrhœa if exposed to the causes of it. Acclimatisation is undoubtedly possible, and the waters of some rivers, as for example the Neva in St Petersburg, always cause diarrhœa in new-comers, though taken with impunity by persons accustomed all their lives to drink of them.

2. SPRUE (PSILOSIS).

General Characters. Sprue, or psilosis, is now recognised as a distinct morbid condition, in which an inflamed, bare, aphthous and eroded condition of the mouth and tongue, much wasting and anæmia, and certain other features which will be found described in most text-books, are, in addition to a definite form of diarrhœa, the most prominent symptoms. Its relation to hill-diarrhœa is uncertain, but as the latter is much more common in men than in women, while sprue shows no such relation to sex, it is possible that they are separate diseases. The cause of sprue is as yet unknown.

Recent Geographical Distribution. This disease is probably essentially tropical or subtropical in origin, but as in some instances the symptoms do not develope until the subject has left the tropics for a considerable time it is actually met with in temperate countries. In England and on the continent of Europe cases of sprue have been seen, but the patients have always been persons who have lived in the tropics and almost certainly contracted the disease during their previous residence in some country where it is endemic. Southern China and the southeast of Asia generally seem to be the principal homes of sprue. In China, at Wenchow, it is spoken of as "one of the most formidable tropical diseases" which are met with, and in Hong-

---

[1] Stawell, *Intercolonial Medical Journal*, March 20, 1899.

kong it is of somewhat frequent occurrence. In Cochin-China a form of chronic diarrhœa is spoken of, a residuum of dysentery, which may perhaps be sprue. The disease is also said to be known in North China and Japan. Some doubt has been entertained as to whether the disease is met with in India. Some have asserted that "both primary sprue and secondary sprue, following on dysentery, acute enteritis," and similar bowel troubles are common among the natives of India. Giles, on the other hand, doubts if true psilosis is ever seen in that country ; and Moorhead states that it is rare in India[1]. In the Malay Peninsula it is certainly met with. It is also mentioned as occurring in the Straits Settlements, but is probably not very common there, only two cases of sprue being recorded in the hospitals, prison hospitals, and lunatic asylum at Singapore in the year 1898. Manson speaks of it as "especially common" in the countries already mentioned, and also in Java, Manila, Ceylon, tropical Africa, and the West Indies. Probably the term "especially common" is to be understood in a relative sense, as the disease is scarcely mentioned by some writers on the most prominent maladies of these countries.

Sprue is probably fairly common in some of the West Indian Islands ; it is said to have caused "many" deaths at St Michael in Barbados in 1898. It would appear also to be not unknown as far south as Natal, in South Africa, as two cases of psilosis were among the admissions to the Durban Hospital in 1898, but whether these patients had contracted the disease elsewhere is not stated.

Further knowledge as to the etiology, and greater certainty in the diagnosis of sprue, are needed before its geographical distribution can be adequately determined. Some (e.g. Buchanan) have regarded famine diarrhœa as essentially the same disease as sprue, and if this be so, its distribution must be much wider than it is at present generally believed to be. A tropical or subtropical climate and a high degree of atmospheric moisture seem to be conditions favourable to the occurrence of the disorder.

[1] *Trans. 1st Indian Med. Congress*, 1894.

3. HILL DIARRHŒA.

A form of diarrhœa is not uncommonly seen in Europeans who have resided for some time in the plains of India and thence proceeded direct to a hill-station. It is probable that changes of temperature, of barometrical pressure, and of other conditions, acting on the alimentary canal of subjects already rendered susceptible to such changes by a prolonged residence in the plains, are the principal causes of this malady. How far it is to be regarded as a specific disease is uncertain. Sir J. Fayrer and other writers have treated it as identical with sprue, but it has already been mentioned that there are certain considerations which point to their being different diseases. Hill-diarrhœa usually takes the form of the passage of loose colourless stools, as a rule in the morning, between the hours of 3 and 11 a.m., together with flatulence and other dyspeptic symptoms. It generally yields in a few days or weeks to treatment, but in some cases it may necessitate a return to the plains, when it usually ceases at once.

Further information is needed before the geographical distribution of hill-diarrhœa can be accurately stated. Up to the present most of the published descriptions of it have come from India. The conditions which seem to be essential to its development are certainly found more commonly in India than elsewhere. Simla, Kasauli, Chakrata, Sabathu and Dugshai, all of which are in the western Himalayas, are said to be particularly affected by hill-diarrhœa (Davidson), but it is also occasionally seen at Mount Abu, Darjiling, and other hill-stations.

Outside India an identical or analogous form of "hill-diarrhœa" has been seen in South Africa, and even in Europe, and it is probable that this affection has a wider distribution than is generally believed.

4. "CARIBI," "EL BICHO," OR EPIDEMIC GANGRENOUS RECTITIS.

Under these names a disease characterised by a rapidly spreading phagedæna of the rectum, spreading upwards from the anus, or starting in the colon, is met with in Guiana, Venezuela, Brazil, and probably also in Fiji. In Guiana it is called "Caribi" or the "Indian sickness." In 1856 it caused a serious epidemic

C. 9

among the Indian tribes of that country. Whether it still exists or recurs there is not clear; I have come across no mention of it in recent reports on British Guiana. Manson quotes a full description of the disease as seen in Venezuela, where the natives ascribe its occurrence in children to the habit of chewing the green, tender stalks of unripe maize, of which they are very fond on account of their sweetness. It appears to attack the lower animals—principally fowls, but also dogs and calves—in Venezuela.

In some of the marshy districts and low humid localities in the tropical parts of Brazil an exactly similar disease is described as occurring, and it is here called " El Bicho." Dwellers in towns where putrid emanations abound also suffer from it; and the Indian population when they leave their mountains to reside in the hot and humid plains are subject to its attacks.

Whether the analogous disease seen in Fiji is the same as this is uncertain. Corney describes it as an epidemic gangrenous stomatitis with an analogous condition of the rectum, and states that it is the most fatal of all the diseases he has met with in the South Seas[1]. No mention of gangrenous stomatitis is made in the brief references to "Caribi" and "El Bicho," and it is therefore impossible to assert definitely that they are the same as the Fiji malady. It would be a remarkable fact if one and the same disease should be met with in, and apparently confined to, certain tropical portions of South America and an island in the Pacific five or six thousand miles away.

---

[1] *Trans. Epidemiol. Soc.* 1887–8.

DIPHTHERIA.

General Characters and Etiology. Diphtheria is essentially a specific fever, probably caused by the entrance of a microbe or microbes into the tissues, with local production of a peculiar membranous inflammation and ulceration, usually, but not invariably, in the throat and upper respiratory passages, and the absorption of powerful toxins which may leave behind the acute attack severe or slight paralytic symptoms. The disease thus briefly defined has a very wide distribution, and as a cause of general mortality it ranks prominently among the acute infective fevers. The Klebs-Löffler bacillus is generally regarded as the specific bacillus associated with diphtheria, but some believe that the disease is the result of a combined infection with this bacillus and other micro-organisms, of which certain of the pyogenic streptococci and staphylococci appear to be the most important.

The true relation of "croup" to diphtheria is uncertain. In most countries the two are dealt with together in the public health returns.

History. Diphtheria is probably a very ancient disorder. Hirsch mentions (doubtful) references to a disease that might have been diphtheria in the Talmud, and in the Hippocratic collection. Later Greek medical writers make mention of a throat affection which may with greater certainty be regarded as diphtheria. In medieval and recent times there are numerous accounts from many parts of the world of epidemics of malignant sore throat, and there seems no good reason to doubt that many of these were epidemics of diphtheria.

It was not, however, until the early part of the century just closed that the true specific character of diphtheria was recognised and established, mainly by the labours of the French physician

Bretonneau, who published his treatise in 1821. The bacillus associated with the disease was first described by Klebs in 1883, and was first cultivated by Löffler in 1884. Ten years later the "serum" treatment of diphtheria was introduced. The effect of this treatment has been to diminish to a very marked extent the fatality of the disease, and this result has been reflected in many countries by a diminished general death-rate from diphtheria. When therefore, as in most instances, death-statistics are alone available, caution is necessary before accepting a lower diphtheria death-rate occurring in any country within the last 7 or 8 years as proof of lessened prevalence of the disease.

Recent Geographical Distribution. *Europe.* In England and Wales diphtheria is the cause of a considerable, but for the present diminishing, annual mortality. The recent history of the disease in this country has been somewhat remarkable. In the decennium 1861—1870 diphtheria caused an annual mortality of 185 per million living; in the next decennium this figure fell to 121; but in the ten years 1881–90 it rose again to 163. In 1891 there was a further increase to 173, in 1892 to 222, and in 1893 the diphtheria deaths were no less than 318 per million of the population. This was the highest rate since registration was introduced with the exception of the years 1858 and 1859, when the diphtheria death-rate was 339 and 517 respectively. Since 1893 there has been a decline, and the deaths from diphtheria in the years 1894—1900 have varied between 245 and 291 per million. It is probable that this diminution is associated with the introduction of the serum treatment. Diphtheria has long shown a marked tendency to prevail in certain districts. As long ago as 1884 the areas of highest diphtheria mortality in England and Wales were defined as follows :—the largest has its base in the south-eastern counties of Sussex, Hampshire, Surrey, and Kent, and extends upwards through Middlesex, Hertfordshire, Essex, Cambridgeshire, and Bedfordshire, and it occasionally includes Lincolnshire, Norfolk, and Nottinghamshire; the second and smaller area has as its nucleus the whole of North Wales and the adjoining county of Shropshire, and tends to spread through Herefordshire and other counties bordering on Wales, in some years extending into South Wales, Monmouthshire, and even across the Bristol Channel into Somersetshire. Since the year 1884 these areas of special diphtheria prevalence have each year

maintained their reputation. The first or south-eastern diphtheritic region is more constant than the other, which is sometimes
called the western or Welsh diphtheritic region.

The tendency for diphtheria occasionally to become epidemic
is well seen in England and Wales, and each year a certain number
of localised epidemics are investigated by the medical officers of
the Local Government Board.

The figures quoted above for England and Wales refer to
diphtheria alone, but it is probable that many of the deaths
returned as due to croup are in reality due to diphtheria. Taken
together, the mortality from the two diseases has fallen from
a maximum of 389 per million in 1893 to 281 in 1897, which was
the lowest mortality from these diseases since the year 1891.

In Scotland the mean annual mortality from diphtheria was
214 per million in the years 1881–90; this figure has steadily
fallen in succeeding years to 180 in 1895, 160 in 1896, and 140 in
1897. In Ireland the mean mortality figure from this disease in
the years 1881–90 was only 70 per million inhabitants. In the
next decade it rose to 75, and in 1900 to 81.

In France the diphtheria mortality has markedly diminished
in the last 15 years. In the towns with over 10,000 inhabitants,
the deaths from this disease averaged 610 per million population
in 1886–90; the ratio fell to 540 in 1891, and 510 in each of the
two following years. In 1894 there was a more marked fall, and
the decrease has been maintained until in 1897 and 1898 the ratio
was only 120 per million inhabitants. It can scarcely be doubted
that this recent fall in diphtheria mortality in France has been
due in part, though probably not wholly, to the introduction of
the serum treatment.

In Holland the diphtheria mortality has markedly decreased
of recent years; between 1889 and 1892 it had been about
stationary at 140 or 150 per million living, but in 1893 it rose to
200. Since that year it has almost uniformly declined, until in
1898 it was only 90 per million. The mortality from croup,
which is returned separately in Holland, has also steadily fallen
from a maximum of 210 in 1889 to a minimum of 50 per million
in 1898.

In Belgium there has been an equally marked decline in the
mortality from diphtheria and croup. Since 1881 the two diseases
have been grouped together, and the average yearly number of

deaths caused by them in the decennium 1881—1890 was 4216 (in the two previous decennia it had been 5843 and 4761 respectively) but in 1895 the deaths from these causes, in spite of an increasing population, had fallen to 2383, and in 1897 to 1734, or 266 per million inhabitants.

In Germany the mortality from diphtheria and croup is high and exceeds that from any other of the principal zymotic diseases. In 1894 the deaths from these causes amounted to 1308 per million inhabitants, a rather lower figure than the mean of the preceding seven years (1887—1893) which had been 1383 per million, but much higher than the corresponding figures for the British Isles, Holland, or Belgium.

In the separate portions of the German empire the mortality from diphtheria has shown very great fluctuations, with a tendency to decrease in some States, but in others to increase. Thus in Prussia in the nine years 1886 to 1894 the total deaths have varied between a minimum of 36,160 (in 1891) and a maximum of 55,401 (in 1893), and there has been no uniform increase or decrease in the mortality for any consecutive period. In Saxony the first year of the period was the highest (with 6483 deaths), and on the whole the tendency has been towards a diminution in the number of deaths from this cause. In Würtemburg, on the other hand, the later years of the period showed a marked increase; the deaths in the years 1893 and 1894 being considerably more than four times as numerous as in the years 1887 and 1888.

In Bavaria and in Swabia in the ten years 1889—1898 diphtheria was a prolific cause of sickness. In each of these countries it occupied the third place in the list of fatal diseases.

In Denmark, at any rate in the urban population, the deaths from diphtheria and croup have very markedly diminished in recent years. From diphtheria alone the number of registered deaths in the urban population uniformly fell from 676 in 1892 to 198 in 1897; those from croup similarly diminished from 346 to 84 in the same period; and those from both diseases from 1022 to 282. The diminution, which is somewhat less striking when the figures for the whole Danish kingdom are taken, cannot be wholly ascribed to the introduction of the serum treatment, as it commenced before this form of treatment began to be practised.

Diphtheria has been seen in the past in Iceland.

In the Faröe Islands it is now a rare disease. At the end of the decade 1871—1880 it was prevalent there to some extent, but throughout the next decade it diminished greatly, until in 1889 and 1890 not a single case of the disease was reported. In each of the four succeeding years, however, a small number of cases of, and deaths from diphtheria and croup occurred in the islands.

In Norway the mortality from diphtheria has remained comparatively stationary in recent years. In the quinquennium 1886-90 there had been a considerable rise in the number of deaths from this cause over the preceding five-year period, the number of deaths per annum in the two periods being respectively 1025 and 1648; but in 1891 the number fell to 1592, in 1892 to 1230, and in 1893 and 1894 it was respectively 1469 and 1438. The deaths from croup have diminished very greatly, from an average of 369 in 1881—1885 to only 140 in 1893 and 182 in 1894. The ratio of deaths from both together was in 1891 approximately 880 per million, and in 1894 approximately 800. These diseases have therefore been much more prevalent recently in Norway than in England and Wales.

In Sweden diphtheria and croup appear to be slightly less prevalent than in Norway. The mortality from these causes together was 815 per million per annum in the seven years 1880-86; it fell to 566 in the next period of seven years, but again rose in 1894 to 735 per million.

In Russia in Europe diphtheria and croup prevail to a remarkable extent, though less so than in Hungary, Croatia and Slavonia, or Servia. In the years 1893 to 1895 the annual mortality from these causes in European Russia was 1470 per million living, which was higher than the mortality in any other European country with the exception of those just named. Both diphtheria and croup are very widely spread throughout the country. In the extreme north, however, they are rare, and in the government of Archangel with a population of over 350,000 inhabitants, only 4 cases of diphtheria and no deaths were registered in 1893, 3 cases with 2 deaths in 1894, and 8 cases with no deaths in 1895. No part of European Russia is free from the disease, and since the year 1894 it has become very widely epidemic. In 1881—1884 there was a similar epidemic prevalence of diphtheria in Russia, but the case-mortality then was nearly twice as high as it is now since the introduction of the serum treatment.

In Switzerland diphtheria and croup are much less prevalent than in some other countries. They showed great fluctuations in the ten years ending in 1897. In a group of 15 towns with a population exceeding 10,000 the highest annual number of deaths in that period was 365 in 1890, the lowest 152 in 1897. The marked and steady diminution since the year 1894 may be in part due to the introduction of the serum treatment, but it is noteworthy that almost equally striking diminutions in the mortality from these causes had occurred in the preceding years before this form of treatment was practised. The rates which the deaths from diphtheria and croup in these 15 Swiss towns bore to the population were approximately 730 per million in 1890 and 250 per million in 1896.

In Austria the mortality from the two maladies under consideration remained remarkably steady in the years 1889 to 1894 inclusive, each year something over 28,000 deaths being ascribed to them. These diseases are a particularly high cause of mortality in that country; in the seven years 1880 to 1886 they caused an annual number of deaths equivalent to 1484 per million inhabitants; in 1887—1893 the mortality fell to 1271 per million, and in 1894 it was slightly lower. In Hungary and Fiume, on the other hand, the mortality from diphtheria and croup rose from 1129 per million per annum in 1880—1886 to no less than 2135 in 1887—1893. This latter figure has in Europe been exceeded only in Croatia and Slavonia—where in the same two successive seven-year periods the mortality from the two diseases rose from 1596 to 2510 per million living per annum—and in Servia, the figures for which country are given below. Here diphtheria is the cause of an even higher mortality than in Croatia and Slavonia.

For Spain trustworthy figures appear to be wanting, but there is reason to believe that diphtheria and croup are prevalent there to a considerable extent. In Gibraltar they are annually the cause of a large number of deaths, and of recent years have been particularly active.

A very marked decline in the mortality from diphtheria alone has been observed in Italy of recent years. It fell from 835 per million in 1887 to 409 in 1890, and after remaining, with slight fluctuations, about this latter figure, it further fell to 239 per million in 1895 and 205 in 1896. The mortality-rate from diphtheria and croup together in Italy fell from 956 per million in 1887 to 501 in 1890, and to 298 in 1896.

Of the prevalence of diphtheria or croup in the countries of the Balkan peninsula it is impossible to speak with statistical accuracy, with the single exception of Servia. In that kingdom diphtheria is the cause of a higher mortality than in any other European country for which definite figures are available. In the years 1887 to 1893 the annual mortality from this cause (deaths from croup being probably included) was equal to 2378 per million of the population; but in 1894–5 the diphtheria deaths rose to no less than 4295 per million, the actual figures being 9934 deaths in a population estimated at 2,312,484. The deaths from this cause in Servia in that year far exceeded those from any other disease or group of diseases and constituted as many as 15·58 per cent. of the deaths from all causes.

Diphtheria is by no means rare in European Turkey. In Constantinople the disease is known, and is at times epidemic; it was particularly prevalent there in the early part of 1898. In many villages of the Salonika province and in Salonika itself diphtheria has become endemic in recent years. In Kavala it recently (1901) caused a rather severe epidemic.

In Malta diphtheria is an annual factor in the mortality of the island. It has become less common in recent years than it was formerly. In 1885 as many as 536 cases of the disease were recorded in the island; in the following year only 192 cases occurred, and then a steady and almost uniform decline in the number of cases followed, to a minimum of 6 cases only in 1896. In 1897 8 cases, and in 1898 14 cases of the disease were recorded. But an epidemic occurred in 1901, in which out of 172 cases 52 proved fatal.

In Cyprus the disease occasionally becomes epidemic, as for example in 1889.

Asia. Throughout Asia generally diphtheria appears to be less prevalent than in Europe, the only marked exception to this rule occurring in the northern plains of the Caucasus.

In the Caucasus and Transcaucasia diphtheria has in recent years prevailed with very varying degrees of severity in different parts of the country. In the years 1887 to 1892 the extremely mountainous province of Daghestan was practically free from the disease, and in the more or less mountainous governments of

Baku, Elizavetpol, Tiflis, and Erivan low mortality rates from diphtheria were recorded. On the other hand in the flat steppes of the Terek, the Kuban, and Stavropol, which together constitute the Caucasus proper, an excessively high degree of diphtheria prevalence was observed in those years. In the years 1893—1895 a somewhat similar distribution of the disease was maintained. In the year 1895 the mortality from diphtheria alone in the government of Stavropol was no less than 4300 per million (approximately), and similarly high figures were observed here in the preceding years.

In Central Asiatic Russia diphtheria is met with throughout but is not excessively prevalent. In the provinces of Askhabad, Samarkand, Viérnoé and Tashkent, the mortality from this cause is particularly low; it is higher in Turgai, and in Uralsk diphtheria was the cause of a high and increasing mortality in the years 1893 to 1895. Among the Bashkirs, on the borders of European and Asiatic Russia, diphtheria is said to be practically unknown.

Throughout Siberia the malady is commonly met with except in the extreme north, and is at times particularly fatal. In the northerly government of Yakutsk it is, however, almost unknown; in 1887—1894 not a single case was registered, and in 1895 only 5 cases with 1 death. Elsewhere in Siberia the returns of diphtheria were, for the most part, not high in each of the years 1887 to 1895. In the convict island of Sakhalin 17 deaths from this cause occurred in 1895, a mortality of 654 per million. In Siberia as a whole the recorded annual mortality from diphtheria in the period 1887—1895 has oscillated about 100 per million inhabitants; but here, as also in Central Asia, the returns are far from accurate, and it is probable that a large number of cases yearly escape record.

In Asia Minor diphtheria is far from rare. In 1898 an epidemic of the disease broke out at Malali, a village 8 hours from Trebizond, and severe epidemics have been recorded elsewhere in the peninsula. At Erzerum it is occasionally seen, and in Mitylene, Samos and other islands off the coast. The disease has also been met with in Baghdad (*e.g.* in 1885). In Persia it certainly exists and is seen from time to time at Teheran. In Syria it is far from rare. It was severely epidemic at Aleppo in the summer of 1900, while in Tripoli and Jaffa it is not un-common.

In India, and generally throughout the tropical and subtropical portions of Asia, diphtheria is apparently exceedingly rare. It is not mentioned in the reports on the health of the troops in India and is probably not endemic in the peninsula. Cases have however, been reported from time to time from Calcutta, Darjiling and elsewhere.

In Ceylon 7 deaths from this cause were registered in 1896, and 2 in each of the years 1897 and 1898.

The disease is almost unknown in Farther India. It is not mentioned by some recent writers on the principal diseases of Siam, while one states that it is unknown there. In the Malay Peninsula it is very rare; in the Straits Settlements only 5 cases of the disease were recorded in 1897, and 3 in 1898.

It is probably equally infrequent in the Malay islands; in Java it is said to be quite unknown.

In China, on the other hand, diphtheria, though a rare disease, is undoubtedly met with. Until comparatively recently it was believed that diphtheria did not occur in that country, but in the last decade evidence of its existence there has accumulated. Isolated cases of the disease have been mentioned by many writers in the Chinese Customs Medical Reports,—at Newchwang, at Pakhoi, at Chung-king and elsewhere. Cases have also been reported from Tientsin, Peking, Chinan-fu, P'ang Chuang and Ching-chou. It is said to be much more fatal in the native Chinese than in European residents. At Shanghai it is "no longer an interesting rarity" and appears to be increasing in frequency. In Formosa it is rare, but in 1891 a small epidemic of diphtheria prevailed in the natives of Tamsui in the north of the island, and caused the deaths of 20 children. In Hongkong, and generally throughout the southern part of China, diphtheria appears to be still rare,— more so than in the north of the country.

In Japan diphtheria is a far more prevalent disease than in China. In 1893 and 1894 cases occurred in all parts of the country, excepting in Okinawa Ken, and "its fierceness" was said to be "increasing year by year[1]." By this was probably meant that the severity of the disease in individuals was increasing; the actual numbers of cases and deaths being less in 1894 than in 1893. The figures in the two years were 5726 cases

[1] *Annual Report of the Central Sanitary Bureau attached to the Home Department of the Imperial Japanese Government.* Tokyo, 1897.

with 3205 deaths in 1893, and 5308 cases with 2903 deaths in 1894 in a population of approximately forty million. The mortality in each year was consequently between 70 and 80 per million inhabitants. This would indicate a degree of prevalence of diphtheria in Japan about the same as that of the disease in Holland or Ireland. It is much less than the corresponding figure for the adjoining island of Sakhalin, which has already been given.

Australasia. Diphtheria has not of recent years been a very conspicuous disease in Australia. It is, however, found more or less all over the continent and in some parts appears to be increasing. In the following figures diphtheria and croup are, apparently, dealt with together.

In Queensland only 42 deaths from diphtheria were recorded in 1897, or less than 90 per million of the population. This figure must be received with caution, as it indicates deaths only, and relates to a period subsequent to the introduction of the serum treatment. In New South Wales diphtheria has occupied a far from conspicuous place in the recent death-returns. In Victoria also the disease is apparently not very prevalent, though in recent years it has shown signs of increasing. In Melbourne and other towns it has at times become epidemic within the last decade. In 1895 the recorded cases in that city (which had a population of over 450,000) were 69, and the deaths 8; in 1896 the cases were 98 and the deaths 10; and in 1897 the cases were 139 and the deaths 7. These figures indicate an increasing prevalence with a diminishing case-mortality, which may have been due to the disease becoming milder, or to a wider use of the serum treatment. At Horsham, Victoria, a very severe epidemic of diphtheria prevailed between July 1898 and June 1899. In South Australia the recorded deaths from diphtheria rapidly diminished from 174 in 1890 to only 21 and 22 in 1896 and 1897; the figure for 1898 was 38. As the population in that period considerably increased, the fall in diphtheria mortality was very striking, and the more so as it is only since 1894 that the diminution could even in part be ascribed to the introduction of the serum treatment. In Western Australia, on the other hand, the reverse has been the case, and here a slight increase in the mortality from diphtheria was observed in the same period; in 1890 and 1891 only 3 deaths from this cause were recorded in each year, but in 1897 and 1898 the deaths rose to 26 and 34 respectively. As the population more than trebled itself in that period the rise is less in reality than it seems. The mortality here is still low in comparison with that in England and Wales.

In Tasmania 114 cases of diphtheria with 13 deaths were recorded in 1896 and 150 cases with 12 deaths in 1897, in an estimated population (on December 31st, 1896) of 166,113 persons. In recent years the case-mortality has fallen very considerably owing to the wide use of the serum treatment. In 1896 the disease was recorded as being somewhat widely epidemic in Hobart.

In New Zealand a marked fall in the recorded deaths from diphtheria was observed between the years 1892 and 1898. The 195 deaths from this disease recorded in 1892 exceeded considerably the deaths in the preceding three years, but then a steady fall set in, until only 45 deaths were recorded in 1898. In the later years the fall must in part be ascribed to the introduction of the serum treatment, but as a decrease had set in before that treatment began to be used it is probable that there has really been a diminished prevalence of diphtheria in New Zealand in the years in question.

Africa. In some countries along the northern shores of Africa diphtheria is not only a comparatively rare disease but when it occurs it usually assumes a benign form.

In Tunis arrangements for the bacteriological diagnosis of diphtheria have existed since 1894, and it has been clearly demonstrated that the disease is a great rarity there. It was most prevalent in 1894, when 61 deaths from this cause were recorded. The cases met with are usually remarkably mild and are often only recognised by the subsequent occurrence of paralytic sequelæ. In Algeria, on the other hand, diphtheria and croup together were, at least prior to 1890, a cause of very high mortality.

Throughout Central Africa generally diphtheria is, so far as is known, a rare disease. Fuller and more careful observation, particularly with the aids to diagnosis which bacteriology alone affords, may show that it is more prevalent than is at present believed. One observer (F. Plehn), by the discovery of the bacillus in 14 negroes in the Cameroon district, has established the existence of the disease in Western Africa although it had been stated before on good authority that diphtheria did not occur there.

In British East Africa it has also been stated that diphtheria is unknown, but fuller knowledge may possibly show that this statement is equally incorrect. One observer, however, during four years' residence in Uganda never saw a true case of the disease[1].

As the cooler portions of the African continent are approached diphtheria gradually becomes a more conspicuous disease, though in no part of South Africa is it so prevalent as in many European countries. In British Bechuanaland it was epidemic in 1891–92; and about the same time it prevailed in Basutoland. In the Annual Reports on the Public Health of the Cape of Good Hope for the years 1897 and 1898 mention is made of small numbers of cases of diphtheria in many places throughout Cape Colony. Thus in the Aliwal North district several cases occurred about the same time in 1897, in places widely scattered; in the Jamestown district and at Colesberg one or two cases were recorded; and at Barkly East an outbreak of the disease prevailed at the end of 1896 and beginning of 1897. The coloured population is not entirely free from it, as shown by the occurrence of one or two cases among them at Murraysburg in 1898. In the native territories of Tembuland, the Transkei, and Pondoland diphtheria is, according to the evidence of natives, frequently seen, but the statement requires confirmation. It was prevalent in and around Wynberg in 1898. It is probably rare, if existent at all, in Mashonaland.

In Mauritius diphtheria is endemic, but the mortality it causes is very low, being only about 25 per million inhabitants. It occurs only in the form of isolated cases and is said never to become epidemic. It is here more a disease of the country than of towns, and of the cooler inland districts than of the coasts. The whites suffer most and the Indian residents the least.

North and Central America. Diphtheria is met with throughout the greater part of Canada with varying degrees of frequency. In many of the provinces it is exceedingly prevalent and one of the most prolific causes of death of all diseases. In Canada the returns for diphtheria and croup are dealt with together. In Quebec it appears, from the Report of the Board of Health of

---

[1] Cook, *Journ. Trop. Med.* June 1, 1901.

that province for the year 1896, that diphtheria is the most frequent of all contagious diseases met with in the province. It is constantly present in one place or another, sometimes as isolated cases, sometimes as an epidemic, sometimes mild, and sometimes very severe. It is one of the highest factors in the mortality of the province. In the seven years ending in 1896 there had been no general epidemic of the disease, but many partial epidemics, and on the whole it had shown a tendency to diminish. In Ontario diphtheria is also widely prevalent; the mortality caused by it has been practically stationary since 1887, and is far higher than that from scarlet or enteric fever. In New Brunswick the disease would appear to be scarcely less prevalent, and occupies a leading position among the causes of mortality throughout the province. In Nova Scotia diphtheria is said to be "the most prevalent and fatal of all germ diseases"; it usually occurs in the form of scattered epidemics. In Manitoba the disease has become more frequent in recent years; in the four years 1894 to 1897 the recorded cases in the province were 83, 112, 223, and 235 respectively.

Diphtheria is in most years a factor in the mortality throughout the United States. In 1897 the only States in which no deaths from this cause (croup being here included with diphtheria) were recorded, at least in the urban population, were those of Arkansas and New Mexico, both in the south. In the towns of Arizona, which adjoins New Mexico, only 6 deaths, in those of Mississippi 6, in those of North Carolina 3, in those of South Dakota 2, and in those of Virginia 9, were registered in the same year. States, on the other hand, in which the diphtheria mortality in the urban population was high in that year, were those of New York, Pennsylvania, Massachusetts, and Maryland. In Massachusetts diphtheria has been more prevalent of recent years than formerly; in the 20-year period 1856—1875 the deaths from this cause (diphtheria and croup together) in the urban population were equivalent to 720 per million living; but in the period 1876—1895 they had increased to an equivalent of 960 per million. In the State of Michigan, in the neighbourhood of the great lakes, there have been marked fluctuations in the prevalence of diphtheria; in 1881 it caused a mortality in the cities and towns of that State of 1226 per million, but this figure has since more or less steadily diminished, and in 1895 the deaths from this cause

were only 260 per million of the population; in 1897 the mortality
had risen again to 492 per million. In Ohio about the years
1883 and 1884 diphtheria was so prevalent as to be spoken of
as a "terrible scourge."

In general the prevalence of diphtheria in the United States
appears to be less in the south than it is in the north. In Central
America the disease is of incomparably less frequency, and, it
would appear also, of less severity than further north. In Mexico
it is both rarely met with and is benign in character.

In the West India Islands diphtheria is a rare disease. It is
scarcely mentioned in the annual health reports of the various
British possessions in this group. A few mild cases occurred
in Trinidad in 1898; 3 cases in the southern Port of Spain
district, 5 in that of Arima, and 2 in that of Gran Couva. One or
two cases were observed in Jamaica in 1897. Sporadic examples
of "diphtheritic sore-throat" occurred in St Vincent in 1896.
Elsewhere in the West Indies the disease has appeared in recent
years only as single cases or has been entirely absent. In the
Bermudas one case was observed in 1898.

South America. In British Guiana diphtheria is exceptionally
met with. Three cases of the disease with one death were
registered in the colony in 1898. In Brazil diphtheria is said to
have been endemic since 1860, but not to rank as one of the
fatal diseases of the country, while croup is the cause of only a
small mortality. Diphtheria is also met with in Peru.

In the Uruguay Republic diphtheria was the cause of 144
deaths in 1897, out of a population of some 800,000. In the
Argentine Republic the disease is also frequently met with; it was
particularly prevalent here in the years 1889 and 1890; in 1897
diphtheria and croup together caused 353 deaths in Buenos Aires
province, of which number 105 were returned from the northern,
77 from the central, 170 from the southern, and 1 from the
Patagonian region of the province.

**Factors governing the Geographical Distribution
of Diphtheria.** It will be seen that diphtheria is almost a
world-wide disorder. It is however mainly a disease of the
temperate zones, and becomes comparatively rare as the tropics
or the poles are approached. In Europe it has been most
prevalent recently in Germany, Norway, Sweden, Central and
Southern Russia, Austria-Hungary, and Servia, while it has been

very much less so, or at any rate less fatal, in the British Isles, Holland, Belgium, Denmark, Switzerland, and Italy. In Asia it has been extraordinarily active in the plains of the northern Caucasus, and has caused severe epidemics in some parts of Asia Minor; it has also become prevalent to some extent in Japan. For the rest of the continent it may be said that diphtheria is a comparatively rare disorder, particularly in the extreme north (Northern Siberia) and in India, Farther India, and the East Indian islands. In Africa, it is generally a rare disease, and is only found with any marked degree of frequency in the southern, or temperate portion of the continent. Similarly in the New World, diphtheria is but little seen in the tropical and sub-tropical belts—including Mexico, Central America, the West Indies, and Guiana, while in the northern United States, in Canada, and in Uruguay and Argentina it is of frequent occurrence.

Diphtheria bears a close resemblance to scarlet fever in its distribution. Both are mainly diseases of the temperate zones, but both can occur either sporadically, or even epidemically, in tropical and even in sub-arctic regions.

It may justly be concluded then that temperature is a powerful factor in determining the distribution of this disease, and that extremes of heat or of cold are unfavourable to its prevalence. Heat is probably more unfavourable than cold, for while diphtheria is found as a severe endemic in no tropical country, it is endemic and a cause of high mortality in such cold countries as Norway and Sweden. This conclusion is further justified by the seasonal relations of the disease. Diphtheria in many temperate countries is found to be most active in the autumn and winter, and least so in the warmer summer months.

Altitude and the nature of the soil are probably factors of considerable importance. Low-lying areas are generally believed to be more favourable to diphtheria prevalence than high altitudes. This has been doubted by Hirsch and others, but a general survey of the recent history of the disease certainly tends to support the statement. A striking illustration has been furnished by the behaviour of the disease in the Caucasus in the past few years. The highest mortality of any country for which definite statistics are published has occurred in the flat, low-lying plains of the Caucasus proper, while in the high mountainous provinces of Transcaucasia, on the other hand, there has been a strikingly low diphtheria mortality. The comparative infrequency of the disease in Switzerland also lends support to the view that diphtheria prefers low altitudes. The rule, however, is not without important exceptions, such as the occurrence of epidemics on the Himalayan slopes, and the recent high degree of diphtheria mortality in Norway.

A damp soil is also very commonly regarded as favouring the endemic prevalence of the disease, and probably a continued residence in a damp dwelling is of even greater importance. The influence of race has never been fully determined. All the great divisions of the human family, including pure Mongols and full-blooded negroes, seem to be capable of becoming subjects of the disease, though probably both their susceptibility to attack and their power of recovery when attacked vary greatly. In China, for example, the disease is said to be much more intense and fatal in natives than in European residents, while in the United States the white races suffer much more than the black.

The disease shows definite age relations. It is mainly a disorder of childhood, and is commonest between the ages of three and twelve. Diphtheria is acutely infectious, and it is now generally admitted that it is largely spread among children between these ages by means of their attendance at schools. The late Sir Richard Thorne showed very clearly how powerful "school influence" might be in the propagation of the disease. As he pointed out it brings together in the closest relations the members of the community at the most susceptible age, and not always under the best hygienic conditions, while the habits of kissing, of transferring sweets from mouth to mouth, the joint use of drinking-cups and the like are obviously ready means of spreading the infection[1]. Sir R. Thorne further believed that the special conditions of aggregation of children in schools might act in a still more serious way, by "manufacturing" as it were "a form of disease of particular potency for spread and for death." Elsewhere he expressed the same idea in other terms, and spoke of "the possible occurrence of what may perhaps be looked upon as *the progressive development of the property of infectiveness*." The relation of the ordinary sore throats which sometimes precede a diphtheria epidemic to the true diphtheritic affection is uncer-

---

[1] Milroy Lectures, 1891.

tain, but it seems possible that, given certain conditions, the intensity and infectiveness of the virus of the former may be gradually increased until it is capable of producing, in place of a simple, almost non-infective inflammation, the highly infective condition known as diphtheria. Space will not permit of a fuller discussion of this suggestion here. The problem presents close analogies with that of the true relation between cholera and the simple diarrhœas which are so often observed about the commencement of a cholera epidemic ; or that of the exact relation between pestis minor and true plague.

Whatever the explanation may be, diphtheria does from time to time cause epidemics not only in schools but in larger communities, such as whole villages and towns ; and there is reason to believe that it also occurs epidemically over still wider areas, affecting whole countries and even whole continents. Hirsch has shown that these cycles of epidemic or pandemic prevalence are extremely irregular both in time and in the areas covered. The recent history of the disease seems to show that in most temperate countries it is more or less widely endemic, and that its fluctuations from year to year or from decade to decade, while often very wide and striking in any one country, are by no means the same for all countries. In many European countries, for example, there has in recent decades been a marked decrease in diphtheria prevalence, while in others there has been a no less marked increase, or the disease has been stationary. The decrease has been most noticeable in France, Belgium, Denmark, Switzerland and (especially) Italy. In Hungary on the other hand there has been a marked increase, while in Germany, Norway, Sweden, Russia, and Austria the disease has been either stationary, or has fluctuated irregularly.

DYSENTERY.

General Characters and Etiology. It is customary, in the case of many countries for which statistical data of disease prevalence exist, to group together dysentery and diarrhœa under a single heading. Inasmuch, however, as the two conditions are probably totally distinct, it would serve little purpose to discuss their geographical distribution together. Dysentery is now generally regarded as etiologically a specific disease, or, perhaps, a group of allied diseases, of which one form is associated with, and possibly caused by, the *amœba coli.* Diarrhœa, on the other hand, is a symptom of a large number of pathological conditions, and is due to a vast variety of different causes. In the present chapter dysentery will, so far as is possible, be dealt with alone.

While, as just stated, the cause of one form of dysentery is very possibly the *amœba coli*, it is not yet certain that all forms are due to this or to any one invariable cause. Some have ascribed it to the action of the *bacterium coli commune* (Celli and Fiocca) ; others to the action of streptococci (Jancarol) ; while Flexner of Philadelphia claims to have isolated a group of bacilli, allied to the colon bacillus, which he believes to be the specific cause of the disease, and to be probably identical with those isolated previously by observers in Japan. Some have believed that the pathological condition to which the name of dysentery is applied is only a manifestation of malaria, and they have pointed in support of this view to the frequent occurrence of dysentery in malarial subjects, and to the undoubted fact that in many malarious countries dysentery is particularly common, and *vice versâ.* But a study of the geographical distribution of the two diseases shows clearly that they are far from constantly found together, and that there are many parts of the earth's surface where the one is common and the other unknown. The same statement may also be made in regard to the distinction between dysentery and diarrhœa, already alluded to. Dysentery, however, has this in common with many forms of diarrhœa, that it appears to be largely associated with impurities of water-supplies, and that when pure water is supplied to a given community in which either dysentery or these forms of diarrhœa have been prevalent, they have to a great extent, if not entirely disappeared.

There are many forms of dysentery, and it is probable that the term is often used in a very loose manner. Too little is yet known of the intestinal flora and fauna and their relation to the disease to make the presence or absence of any of them a test of the specific nature of any dysenteriform disorder which may be endemic or epidemic in a given place or country. Pathologically, the disease is distinct and characteristic enough. But it is open to question whether, in many of the countries where dysentery is spoken of as prevalent, the presence either of true dysenteric ulceration of the intestine or of the *amœba coli* or the other organisms named above has been demonstrated ; and it is, therefore, possible that in some of these countries other diarrhœal diseases have been regarded as dysentery.

The term tropical dysentery is very frequently used, but true dysenteric ulceration of the bowels, together with the discovery in the bowel contents of one of the organisms mentioned, has been observed in many extra-tropical parts of the world ; and it would seem desirable to avoid the use of this term, if it is intended to be synonymous with acute dysentery. Chronic forms of dysentery are met with in countries where it is not endemic, in persons who have contracted the disease in countries where it is endemic.

Recent Geographical Distribution. *Europe.* The diminution of dysentery in Europe in recent times has apparently gone hand in hand with improvements in sanitation and hygiene, and it is more than probable that the principal factors in this reduction of the disease have been the improvements in water-supply and drainage.

In the British Isles dysentery is now almost entirely a thing of the past. Cases of the disease are occasionally seen, but usually in persons who have contracted it abroad, or in inmates of asylums. It appears to be an undoubted fact that true dysentery is endemic in many of our larger institutions for the insane, and it is said to attack especially the feeble and debilitated and to show no predilection for any particular form of insanity, or for any age or either sex. The heading " diarrhœa and dysentery " still remains in the Registrar-General's Reports, but it is probable that almost the whole of the figures grouped under it are accounted for by various diarrhœal diseases other than dysentery.

In Europe generally dysentery is far more prevalent in the south than in the north. In many northerly countries it is almost, if not quite unknown. In Norway and Sweden, for example, no mention of it is found in the statistical returns of disease. In Denmark, also, only from o to 6 deaths from it were annually recorded in the urban population in the years 1892–97 inclusive. In the Low Countries it is somewhat more common, but is decreasing as a cause of mortality. Holland has returned from 10 (the lowest) to 41 (the highest) deaths from dysentery in the years 1889–98. In Belgium the deaths from the disease steadily fell from a mean of 881 per annum in the years 1871–80 to only 264 in 1897 ; the last figure was equivalent to a ratio of about 40 per million.

That the disease cannot exist in northern latitudes, however, it is impossible to assert. On the contrary, there is proof that it has in times past been epidemic—even malignantly so, as for example in 1860,—in Iceland, and is now occasionally seen there in a mild form. It is also met with in the far north of Russia, and some deaths from the disease are yearly reported from the town and government of Archangel. Throughout European Russia generally dysentery is more or less endemic.

In the countries of Central Europe dysentery is more common than in the north. In Germany it is said to be most prevalent in the Prussian districts of Bromberg, Oppeln, Gumbinnen, Dantzig, and Lüneberg. In Dantzig it is endemic, and at times takes on an epidemic character ; it was very prevalent in 1895, when over one thousand cases of the disease were recorded. It is said to be more common on the plains than in the higher lands.

In Austria-Hungary the malady seems to be still more common ; it is said to be most fatal in Galicia, Bukowina, and Dalmatia, and least so in Upper and Lower Austria and Salzburg, where, on the other hand, diarrhœa is said to be most prevalent. The deaths from dysentery (*Rühr*) in Austria in 1896 were at the rate of about 90 per million inhabitants[1], and in Hungary, where the disease appears to be still more prevalent, they were at the rate of about 230 per million in 1897.

In Switzerland, on the other hand, the disease is said to be unknown. In Italy and the rest of southern Europe it is rather prevalent. It seems to be increasing in Italy, where it is said to occur most often in the south of the peninsula. In Spain, also, where dysentery is frequent, it is particularly so in the southern provinces. It is perhaps in the countries of the Balkan peninsula that dysentery is most prevalent in Europe. It is spoken of as one of the most fatal diseases in the low lands of Roumania and Bulgaria, and as generally diffused throughout Eastern Roumelia and Macedonia. In Servia it caused a mortality of about 680 per million in 1894; it prevails here especially in low-lying regions, and is said to be comparatively rare in the hilly districts of this and of the other Balkan countries. It is probably common in Turkey; it is certainly frequently seen in Constantinople and in some parts of the Salonika province, and it is endemic but mild in character in many parts of Greece.

It is met with also in many of the islands of the Mediterranean, such as the Ionian Islands and Crete. The French Marine Infantry suffered considerably from the disease in the latter island in 1896–7.

Asia. In the Caucasus dysentery is frequently seen; in Tiflis the Tatars and Russians are said to suffer from it more than other races.

The disease is not unknown in Siberia, and in Vladivostok and along the Russo-Chinese frontier it seems to be more or less endemic. Deaths from this cause are yearly reported from all the Siberian provinces without exception. They are comparatively numerous in the south, but are reckoned by units only in the far northerly province of Yakutsk.

In Persia and the adjoining parts of Turkey in Asia dysentery certainly exists. At some places, such as Suleimaniyé and Khanikin on the Turco-Persian frontier, true acute dysentery seems to be endemic. In Teheran the European inhabitants sometimes suffer considerably from the disease. At Basra, at

[1] In 1885–87 the mean mortality from dysentery in Austria was as high as 440 per million (Davidson).

the head of the Persian Gulf, grave forms of the disorder are not uncommon.

It is known in Arabia, and one of the commonest causes of death among the Moslem pilgrims at Mecca and Medina is a chronic form of dysentery which, no doubt, is not wholly of local origin. In Syria the disease is met with at Jaffa, and at Aleppo it is the most prevalent of all maladies.

In India the returns of dysentery and diarrhœal diseases are grouped together. Both groups of disorders are very prevalent, nor is this surprising in view of the polluted character of the water frequently drunk by the natives, and their habits of constantly bathing in, and drinking, the very foul water of tanks, rivers, and streams.

From the returns of the European and native armies in India some idea may be gathered of the frequency and distribution of these diseases in different parts of the country. In the decennium 1886–95 the annual admission-rate for "dysentery and diarrhœa" in both armies was 30 per 1000 strength and the death-rate 0·73, or 730 per million. In most of those years Bengal and Orissa gave the highest ratios for dysentery. In the years 1891–95, on the other hand, the highest ratios were returned from the Burma coast district, Bengal-Orissa, Burma (inland), and South India following in the order named. Low dysentery-ratios were recorded on the West Coast, in the Indus valley, in Central India, and in the Hills. In the Rajputana States the disease is neither common nor severe. The native troops suffer considerably more than the European troops in India generally, but they are much better able to withstand an attack of the disease, the case-fatality rate in natives being less than one-fifth of what it is in Europeans. Dysentery is a common disease of gaols in India, and prevails especially when these institutions are at all overcrowded. The famine of the past few years, with its consequent ills—particularly starvation and the ingestion of improper articles of food—has enormously increased the death-rate in the general population. In each of the years 1895 and 1896 the recorded deaths in India from dysentery and diarrhœa were at the rate of about 1100 per million, but in 1897 the ratio rose, in consequence of the famine, to 1860 per million, and has probably been still higher since. The recorded deaths from these causes in 1897 were 403,833 in number. What proportion of these were due to dysentery alone it is impossible to say.

In Ceylon dysentery is exceedingly common, and from time to time becomes epidemic in certain parts of the island.

In Farther India both dysentery and diarrhœal diseases are also exceedingly prevalent. In Siam dysentery is the most fatal of all diseases, in some years causing more deaths than cholera, malaria, and enteric fever together; it is very frequent in the Laos country, and breaks out every year in the plains of Cambodia. In French Cochin-China dysentery accounts for no less than 28 per cent. of the mortality from all causes, paludism coming second and causing 24 per cent. At Saigon the disease is said to have diminished greatly since the water-supply has been improved. In Annam and Tongking it is very widely prevalent, attacking both natives and European residents. In the Malay Peninsula and Straits Settlements dysentery of a mild type is frequently seen.

In many parts of China dysentery is rife. At Tientsin it prevails in most years. In Ningpo a mild type of this disorder is, next to malaria, the most common disease in the native population. At Hankow it becomes prevalent every year, and at Shanghai cases are frequently recorded. In the Government Civil Hospital at Hongkong the disease is one of the commonest that comes under treatment. At Lung-chou it is not infrequently seen; and in Canton it is a common disorder. All round the neigh-bourhood of Wuhu, on the Yangtze-kiang, it is said to be very rife. In a word dysentery is found, more or less, throughout China; it appears to be met with in the north (at Peking for example) as well as in the south, but perhaps to a less extent, and in Corea it becomes widely prevalent in the summer and autumn of most years.

In Japan dysentery is endemic and "gradually extends its dominions year by year." In 1893 there was a very severe epidemic of this disease, and as it was spoken of as the "severest epidemic ever experienced" it is clear that others must have preceded it. The recorded cases were no less than 167,305 and the deaths 41,282, figures which were more than double the corresponding returns of the preceding year. The disease pre-vailed over the greater part of the archipelago. In the following year it was very nearly as rife, and the numbers of cases and deaths were but little less than in 1893, places which had escaped in the first year being attacked in the second. The recorded deaths in 1893 were equal to about 1000 per million inhabitants, but it is probable that they were below the truth. Since that year there has been an annually recurring epidemic of greater or less severity in Japan, at least until the year 1898.

In many of the islands of the Malay Archipelago dysentery is known, but its degree of prevalence varies very widely in the different islands. In Java it is endemic and was at one time a most serious cause of mortality in the European troops of the Dutch army in the Indies, but improved sanitation and particularly the digging of wells with good water have greatly diminished its activity. Stokvis states that between 1869 and 1878 the deaths from this cause in the Dutch troops were no less than 13 per 1000 per annum: the first well was sunk in 1875, and already in 1879–83 the mortality from this cause had fallen to 4·2; still later in 1884–88 it fell to the low figure of 0·7 per 1000. It is interesting to note that in the report on the Dutch troops in the Indies in 1897 tropical dysentery is spoken of as very rare, and only 3 cases in Europeans and 6 in natives were recorded in the year, out of a strength of over 17,000 of the former and over 24,000 of the latter. In Borneo dysentery is neither common nor severe; at Labuan 2 or 3 cases are admitted to the colonial hospital each year. In Celebes the disease is somewhat prevalent on the coasts. Ternate, Amboina, and Timor suffer only to a small extent. In the Philippine Islands dysentery is met with, but of a mild type.

Australasia. In British New Guinea dysentery is the most serious disease among the inhabitants. Sir William MacGregor, in his annual report for 1897–98, stated that a few years before dysentery was practically unknown here, but it had been imported in an epidemic and contagious form within the preceding three or four years, and, like all new diseases when introduced among an aboriginal people, it had raged with very great severity and caused many deaths. It had been carried all along the coast, but did not seem to have extended inland.

In many parts of Australia dysentery is endemic, and at times epidemic. It appears to be decreasing in Queensland, where the mortality from this cause steadily fell from 484 per million in 1893 to 134 per million in 1897. It was formerly very much more prevalent[1], but now it is spoken of as "not a frequent or fatal disease in adults, except when it from time to time takes an epidemic form from some cause more or less ascertainable, when it is generally confined to a special locality in the vicinity of the generating cause." Polluted water appears to be the most frequent source of origin. In Victoria the disease is not very common, but in 1896 it became widely epidemic in a severe form in the neighbourhood of Leigh. In New South Wales dysentery is seen, but not very frequently.

In South Australia the returns for dysentery and diarrhœa are grouped together, and the degree of prevalence of the former cannot be ascertained from them alone. In Tasmania dysentery is said to be less common than in Australia. In New Zealand it is far from frequent; only 15 deaths from this cause were registered in the colony in 1898. In Australia and New Zealand the degree of prevalence of dysentery appears to decrease more or less regularly as the cooler regions of the south are approached. In the former it is said to be by no means rare in the aborigines.

In many of the Polynesian Islands dysentery is a conspicuous disease. In Fiji it accounts for more admissions to the colonial hospital at Suva than most other diseases. The malady seems to have been brought to these islands with the advent of a European population, and on some plantations it has been known to become epidemic in a most malignant form, carrying off almost every labourer on the estate. Sir W. MacGregor has recorded the instance of a certain plantation on which 115 labourers were employed, every one of whom, with the single exception of a boy, was carried off by dysentery. In New Caledonia it is met with, but is not of a severe type. In Samoa and the Sandwich Islands dysentery is among the commonest of all diseases.

Africa. Along the northern shores of Africa dysentery is more or less prevalent. In Morocco it is seen principally along the coasts. In Algeria it is said to be one of the most fatal diseases of the country, especially in the province of Oran[2]. In Tunisia it is endemic, but of a benign character.

[1] The mean mortality from dysentery in Queensland in 1883—1888 was no less than 1158 per million (Davidson); in 1897 it was about one-ninth of this.

[2] The excess of dysentery in the Oran province is ascribed to the large amount of sulphates of soda and magnesia in the drinking water—an explanation which would seem rather to apply to an excess of ordinary diarrhœa than of true dysentery.

Few parts of the western shores of Africa seem to be free from dysentery, and in many places it is one of the most serious causes of illness and death. In Senegal it is the most fatal disease in Europeans; it appears to be rife throughout the province, and has been, at least in part, ascribed to the use for drinking purposes of the waters of the Senegal river, which are extensively polluted by human dejecta[1]. In Lagos dysentery is one of the most frequent of diseases. It is very prevalent in the Cameroon district. On the Gold Coast dysentery accounts for a large number of the annual admissions to the colonial hospital. It is said, however, that the disease is uncommon among Europeans in this colony, owing to improved water supply, and that, though chronic dysentery is fairly common among the natives, true acute dysentery is probably extremely rare. It is very common on the Upper Congo in European residents, but in the neighbourhood of the Cataracts it is said to be less so than elsewhere, as a result of the greater purity of the water here[2]. Among the natives dysentery and diarrhœa are rife throughout the Belgian Congo, prevailing especially after the first rains have fallen and washed the surface and soil impurities into the streams. This characteristic of the disease is observed throughout the greater part of Central Africa. Pruen, whose experience lay chiefly on the east side of the continent, writes of dysentery, typhoid fever, and diarrhœa that they all prevail most "when the showers first moisten the refuse lying about and thus cause its decomposition, and then wash the decomposing materials into the nearest stream."

On the eastern side of the African continent dysentery is widely prevalent. In Egypt it appears to be one of the most prolific causes of death. The British troops in the Sudan in 1898 suffered considerably from it after the Omdurman campaign; and in the Dongola campaign of 1896 it attacked large numbers of our men. In Abyssinia it is one of the most prevalent of all diseases. It is noteworthy on the other hand that at Massowa the Italian army of occupation suffered but little from it.

In British East Africa it is common, and its prevalence is said to be largely associated with impure water-supplies and with

[1] *Le Progrès Médical*, June, 1898.
[2] Congrès National d'Hygiène et de Climatologie Médicale de la Belgique et du Congo, *Compte Rendu*, 1897.

excessive indulgence in alcohol. It prevails in a severe form in the country between the Lualaba and Lomami rivers in Central Africa, and in a milder form on the Upper Zambesi. Along the shores of the Nyassa and Tanganyika lakes and on the highlands to the east it is endemic. It is noteworthy that while dysentery is more common but less fatal on the interior plateau than on the coast, diarrhœa is both more common and more severe in the interior than on the coast.

Throughout the greater part of South Africa dysentery is a very prevalent disorder. In Cape Colony and in Bechuanaland it is among the most frequent of diseases ; large numbers of cases are annually treated at the provincial hospital at Port Elizabeth, and at the Frere hospital, East London, dysentery and diarrhœa on some days account for 50 or 60 per cent. of all the patients seen. At Murraysburg acute dysentery prevailed at the end of the year 1898, and many cases of the disease were treated at Vryburg in the same year. In the native territories of Tembu-land, the Transkei, and Pondoland it is the most common disease among the natives. In Natal dysentery is said to be even more prevalent than in Cape Colony, and to be more frequent on the coasts than in the interior. Throughout South Africa generally dysentery is believed to occur to a great extent independently of diet, and to be rather dependent upon climate. True dysentery seems to have been one of the many forms of diarrhœal disorders from which our troops suffered in the recent South African campaign. In Mashonaland a mild type of the disease is said to be common at certain seasons.

In Madagascar dysentery is endemic on the coast line, and is met with commonly, but in a benign form, even on the central table land, as at Antsianaka. The Hova troops suffered considerably from it during the French invasion a few years ago, and the Hovas are at all times very liable to its attacks. In the Seychelles it is the most frequent and fatal of all diseases. In Mauritius it is only surpassed by malarial fevers as a cause of death ; in the first half of 1898 it accounted for 10 per cent. of all deaths in the island ; it is also prevalent in Rodriguez and Réunion.

In the islands off the west coast of Africa the disease is met with. In Madeira it was at one time the cause of destructive epidemics, but is now usually mild and tractable. In the Cape

Verde Islands it is severely prevalent along the coasts, and it is also frequent in Fernando Po and in St Helena.

North and Central America. The limit of prevalence of dysentery extends far to the north in the western hemisphere, and the disorder is met with near to, if not actually within, the Arctic circle. Dysentery is said to be seldom seen in Greenland, but it is by no means rare in Labrador, among both Eskimos and Europeans. Of the prevalence of the disease in other parts of Canada little can be said with accuracy. I have found no mention of it in recent health reports from the various Canadian provinces. It appears, however, that the disease is known there, and is moderately prevalent among the Indian tribes.

Of the distribution of dysentery in the United States complete information is also lacking. In the past it is said to have "frequently assumed an epidemic form over larger or smaller areas of the States, and certain cycles of years in a given locality are marked by a high dysentery mortality." Judging from the returns of the disease in the United States army in the years 1839–59, dysentery would seem to have been more prevalent then in the southern than in the northern part of the Republic, and at the present day Osler states that the disease is still commoner in the southern cities than in those of the north. In Baltimore it prevails every summer and has on several occasions been epidemic. The returns for Massachusetts have recently shown a very marked and continuous reduction in the mortality from dysentery. Thus in 1856–75 the annual mean mortality from the disease in this State was 510 per million; but it fell in the next period 1876–95 to only 160 per million. In the early years of the first period dysentery was an epidemic disease of the summer months in Massachusetts, prevailing more or less in almost every city and town, but recently it has lost its epidemic character, and the hope is expressed that at no distant date it may become extinct. Some cases of true amœbic dysentery have been recorded in Kansas City, Missouri, and an "epidemic bloody flux" has been observed as far north as the State of Michigan.

In Mexico dysentery is widely endemic on the west coast, and is present, but less frequent, on the east coast; in the past, it has been on at least one occasion (1848) the cause of a violent epidemic here.

In British Honduras dysentery is frequently seen, especially in the rainy season; it is noteworthy however that in the flat, alluvial swamps of Stann Creek district it is rare. In Guatemala it is common, and in 1894 it was the cause of 809 deaths, or nearly 570 per million inhabitants. It is endemic in San Salvador, not only in the low lands, but also at considerable elevations. In Nicaragua it is severely endemic, and in Costa Rica it is next to malaria the most common disease along the coasts, though less frequent in the interior.

Dysentery is met with in many of the West Indian Islands. In Cuba it is endemic, and during the recent troubles in that island has been epidemic to a truly terrible extent. It was estimated that in the Spanish army alone "enteritis and dysentery" were the cause of as many as 12,000 deaths, but what proportion of these were due to dysentery alone is not stated. In Jamaica dysentery is also endemic; at Port Royal it is one of the most prevalent of all diseases. In Guadeloupe and Martinique it is common, and generally throughout the Leeward and Windward Islands. In Grenada, St Lucia, and Barbados it is a prominent cause of death; in Trinidad it is prevalent in some districts in a peculiarly aggravated form, but it is noteworthy that in the town of Arima the disease practically disappeared after the introduction of a good water-supply. In Puerto Rico a mild epidemic of dysentery occurred after a great hurricane in August, 1899.

South America. Scarcely any portion of the great South American continent seems to be free from dysentery. In British Guiana it is one of the principal diseases seen, and large numbers of persons suffering from it are annually treated in the public hospitals; the Indian tribes in the interior are also said to be attacked by it. In Dutch Guiana and in French Guiana the malady is also frequently met with. In Brazil it appears to be common throughout the country. Davidson quotes a number of authorities to the effect that it is one of the severe endemic diseases of the Pará province in the north, that it is from time to time epidemic in the Amazon valley, and is of rather frequent occurrence in the provinces of Maranhaõ, Piauhi and Parahyba, in the south. Epidemics have also occurred at Rio de Janeiro, and in Pernambuco the disease is seen, but not commonly.

Dysentery can scarcely be very common or fatal now in Uruguay; in 1897 it caused only 9 deaths in the whole republic, or less than 12 per million inhabitants. It is said however that it is not rare along the Uruguayan coasts, and in the past has given rise to very severe epidemics. It appears to be more prevalent and fatal in the Argentine Republic, and particularly in the northern provinces. In the province of Buenos Aires the deaths ascribed to dysentery in 1897 were 95 in a total population of about one million. The disease is also endemic in Paraguay.

On the western side of South America dysentery is widely prevalent. In Ecuador it is common in the rainy season. In Peru it is endemic along the coasts, and is said by Hirsch to occur in the interior, at such elevations as 8000 feet and even 13,000 feet (at Cerro Pasco) above sea-level. Dysentery is met with in all parts of Chile.

Factors concerned in the Geographical Distribution of Dysentery. The causes and conditions which determine the present geographical distribution of dysentery are no doubt numerous and complex. Climate has a considerable influence, and the disease is on the whole more prevalent in warmer than in cooler countries. It is now, for example, far commoner in Southern Europe than in Central, and in Central than in Northern Europe; it is more frequent in the south of Italy than in the north, and the same is true of the Spanish peninsula and of France. In the southern hemisphere it is found to diminish in like manner, both in Australia and New Zealand, as latitudes more distant from the equator are approached; and in the southern parts of both Africa and South America it appears to be less rife than in the tropical portions of those continents. Dysentery would also appear, to some extent, to increase in severity as well as in its degree of prevalence as the equator is approached. But both generalisations are not without important exceptions. The disease was for example in times past exceedingly prevalent in many parts of Northern Europe; it has been malignantly epidemic in Iceland —an all but arctic island, and it is not unknown in Greenland.

Elevation seems to have some influence on its prevalence; low-lying, marshy districts appearing to be the most favourable to its activity. But there are numerous exceptions to this rule; in the elevated table-land of Madagascar, for example, it is quite common, and, as just stated, it has been seen at a height of 13,000 feet in Peru. Race has perhaps some influence on its distribution, but no race seems to be immune to it.

Local conditions, of which polluted water-supply is probably the most important, play a leading *rôle* in determining the distribution of dysentery; and many other examples are known, beside those already quoted, of the diminution or disappearance of the malady in a town or district after a pure water-supply has taken the place of an impure one. Thus at Saigon, in French Cochin China, and at St Louis de Senegal, on the west coast of Africa, the establishment of a good water-supply led to a great decrease in the prevalence of dysentery. In many isolated places in the French colonies the use of Chamberland filters has been followed by equally happy results, and similarly in the Fiji Islands as soon as the houses of Europeans were provided with rain-water stored in metal tanks dysentery became very much rarer than it had formerly been. Measures of sanitation generally, particularly such as tend to drain and purify the soil, have aided in diminishing or extinguishing the disease in many parts of the world where it had previously prevailed.

The distribution of dysentery has varied much from time to time. Its disappearance from certain parts of Europe in comparatively recent years, and its continued diminution in many countries and increase in others, have already been alluded to. In some parts of Europe in the past, as for example Sweden, Iceland, and Germany, and in many countries in Asia, Africa, and America in the present, dysentery has taken on epidemic characters, but as a rule it remains more or less constantly prevalent in those districts in which it is endemic. Of the cause of these epidemics, or of the considerable fluctuations in endemic prevalence, little is definitely known. The disease, however, often shows marked seasonal relations in places where it is endemic.

Dysentery does not appear to be transmitted like smallpox from person to person by direct infection or contagion, and consequently its distribution is not influenced by the movements of emigrants or traders. The cause of it can, apparently, be carried in drinking water, hence outbreaks have occurred on board ships supplied with polluted water; in no other sense can it be called a sea-borne disease. It does, however, sometimes appear in countries previously free from it, as for example in New Guinea in the case mentioned; and unless its appearance is due in such instances to an acquired virulence of an indigenous and hitherto harmless amœba or bacterium or whatever else is the cause of the disease, it can only be explained by an importation from elsewhere of a something essential to its production. In what manner dysentery can be thus carried from one country to another, however, is not clear.

Military operations, particularly in countries where dysentery is endemic, have frequently favoured the activity of the disease. In the French invasion of Madagascar; in the war in Uruguay in 1867; in the recent troubles in Crete; in the Spanish-American war in Cuba; in the Peninsular War; in the Crimean War; in the French operations in Tongking; in the recent war in South Africa, and in many others, dysentery has caused considerable, and in some instances, terrible havoc among the combatants of one or both sides.

It has already been stated that dysentery has many points in common with malaria, but that the geographical distribution of the one does not correspond with that of the other. As examples of places where malaria is absent but dysentery prevalent may be mentioned the Seychelles Islands, Peyta in Peru, New Caledonia, Barbados, and perhaps Hawaii. To these may be added the island of Mauritius before the appearance of malaria there in 1866.

ERGOTISM.

Characters and Etiology. Ergotism is the name given to a group of symptoms which follow on the consumption of ergot. Ergot is a substance produced in the grains of rye by a fungus, the *Claviceps purpurea*. In some years a very large proportion of the rye crop may be affected with this parasite, and in peasant communities where rye is largely eaten ergotism may, under these circumstances, become seriously epidemic. There are two principal clinical varieties of the disease, the gangrenous and the spasmodic. In both the illness begins with vomiting, diarrhœa, abdominal pain and cramps—in brief, with all the symptoms of irritant poisoning. To these are added other symptoms—tingling and itching of the skin, giddiness, headache, muscular pains and twitchings, and general depression. In the gangrenous form gangrene occurs in some peripheral parts of the body, as the toes, fingers, or, less commonly, the ears or nose. In the spasmodic form, severe and painful spasms of the muscles occur. Mental symptoms usually supervene, such as delirium, melancholia, or dementia, and the disease frequently ends in death.

History. It is thought that many of the medieval epidemics described under the name of "Saint's Fire" (*ignis sacer, ignis S. Antonii*) may have been of the nature of gangrenous ergotism. Hirsch has tabulated a long list of such epidemics from the year 591 A.D. to the year 1879. The larger number of outbreaks of gangrenous ergotism have occurred in France and Spain; and of convulsive ergotism in Germany and Russia. These epidemics appear to have been much more frequent in former centuries than in recent times.

Recent Geographical Distribution. With the exception of a small outbreak in a New York prison early in the last century and the occurrence of occasional cases in the Caucasus and Siberia there seems to be no mention of ergotism outside Europe.

In Europe the disease has been most frequently seen in France, Germany, Russia, and Sweden. In the British Isles there is no record of ergotism during the past three centuries. In France, on the other hand, there were nine such epidemics in the 17th century, seven in the 18th, and three in the 19th. It is noteworthy that in France the disease has usually taken the gangrenous form. It has been most often seen in the upper and lower basins of the Loire and in the Rhone valley. In Belgium there was a small epidemic in 1845–6. In Holland, on the other hand, as in England, the disease has not been seen for the past three centuries. In Germany there were eleven epidemics in the 16th century, ten in the 17th, twenty-one in the 18th, and fifteen in the 19th. The majority have occurred in the north-west and north-east portions of the country, only a few in the south-west and Bohemia, and none in the central regions. The overwhelming majority of outbreaks of ergotism in Germany have been of the spasmodic form.

In Sweden ergotism was not infrequently seen, at least up to the year 1867, particularly in the southern part of the country. In Norway on the other hand the only mention of the disease seems to relate to a small epidemic there in 1851.

In Russia, where rye-bread is the principal food of the peasants, ergotism has on several occasions become epidemic. It has been seen in many central and southern governments, but apparently not in the extreme north. The most recent epidemic of the kind seems to have been one that occurred in the north-easterly government of Perm in the autumn and early winter of 1895. In Russia, and no doubt elsewhere, there is great difficulty in persuading the peasantry that the diseased grain and the symptoms of ergotism stand in the relation of cause and effect, and in inducing them to take precautions to prevent the continuance of an epidemic. The Russian name for the disease, "*zlaia kortcha*" or "malignant spasms," may be taken to indicate that the nervous, rather than the gangrenous form is the commoner in that country. Among the places mentioned as scenes of ergotism epidemics between 1832 and 1864 is the government of Tomsk, in Siberia. In Finland there have been at least two epidemics (in 1840–4 and in 1862–3) and, as just stated, cases have in recent years occurred in the Caucasus and Siberia.

Austria-Hungary seems to have escaped this disease, with the exception of an outbreak in Transylvania in 1857. Switzerland and Italy have been the scenes of a few epidemics in the past; while Spain seems to have been quite free from ergotism during the last three centuries. The disease is occasionally seen in some parts of Turkey but is not frequent.

Factors concerned in the Distribution. The relation of the ergot-parasite of rye to the disease in man has been proved beyond possibility of doubt. The parasite attacks other grasses, but it is said to be poisonous only when growing in rye or possibly in bromus-grass. Under certain conditions of season and weather the parasite rapidly multiplies, and on some occasions a very high percentage of the rye grains may be affected by it. In the outbreak in Perm, just alluded to, as many as 17 per cent. of the grains were found to be diseased in some districts.

The disease is most common in countries where rye bread, or other food-stuffs made from rye, are commonly eaten. The peasantry naturally suffer most. Unhygienic conditions have seemed to favour the development of the malady, and several epidemics have been recorded in insanitary prisons, foundling hospitals, and similar institutions. Children have frequently suffered from it, but infants at the breast are said to escape its attacks.

The diminution of ergotism in recent times has been attributed to an improved knowledge of agriculture, to more careful cultivation of the crops, to greater facilities for providing the poorer classes with good and wholesome food-stuffs, and finally to increased cultivation of the potato.

The distribution of the two types of ergotism is remarkable. As already stated the gangrenous type has almost invariably been seen in France and Spain, the convulsive in Germany and Russia. The explanation of this is uncertain. It seems probable that each set of symptoms is due to a particular poison; thus the gangrenous form is thought to be due to the action of sphacelinic acid, the convulsive to that of cornutin. It may be assumed, therefore, that the conditions of soil and latitude in the more southern countries tend to the production of the acid in the growth of the ergot-parasite, while in more northerly latitudes the cornutin is in excess. Possibly racial differences in the peasantry of these countries have also something to do with the development of the two types of the disease.

ERYSIPELAS.

General Characters and Etiology. The symptoms of erysipelas scarcely require description here. It is an acute, and highly infective inflammation of the skin, believed to be caused by the *Streptococcus erysipelatis* (or *S. pathogenes longus*).

History. Erysipelas is mentioned by the oldest writers on medicine. It was formerly more common than it is now, and there are records in the past of many epidemics of the disease in hospitals, lying-in homes, and other institutions in Europe and America. Hirsch has tabulated a long list of epidemics of a peculiarly malignant type of erysipelas which have occurred frequently in the United States since the year 1822.

Recent Geographical Distribution. *Europe.* Erysipelas occurs in all European countries. It is said to be especially frequent in Iceland and the Faröes. It must be a common disorder also in Norway, where in the years 1881 to 1894 it is stated to have occupied a higher place in the mortality returns than either rheumatic fever or syphilis. It appears to be prevalent throughout European Russia, but is somewhat less common in the Don Cossack territory, and remarkably less so in the Polish provinces. In Germany erysipelas is said to be much less frequent in the northern plains than in the south. It appears to be very common in Bavaria and Swabia. In England and Wales erysipelas causes a yearly mortality of between 30 and 40 per million, but twenty years ago the annual mortality ranged from 80 to 90 per million. In Switzerland, at least in the urban population, the corresponding figure has varied, in the ten years 1888—1897, between 21 and 49 per million. In Italy the average mortality has been placed at 150 per million. Erysipelas occurs in Turkey in Europe, but I cannot discover that it is more frequent there than elsewhere.

Asia. Erysipelas is certainly not unknown in Asia Minor, Syria, or Persia. In India I have seen very typical cases of the disease in Bombay and Calcutta, but in some parts of India, as, for example, Rajputana[1], it is said to be extremely rare. There is mention of its occurrence in Ceylon, and cases are sometimes treated in the Singapore prison. It is said to be unknown among the hill tribes of Java; in Siam it is rare (Rasch); in China very rare (Coltman); and in Japan perhaps equally so.

Australasia. In Australia erysipelas would seem to be rarer than in Europe; but of its occurrence or frequency elsewhere in this part of the world little seems to be known.

Africa. Erysipelas occurs in Egypt, Tunis, Algiers, and other countries, but whether frequently or not I am unable to say. It is rare in British East Africa, and also in Madagascar. In Réunion, on the other hand, Pélissier (in 1881) had seen many cases of the disease. Plehn has seen one typical case of erysipelas in the Cameroons.

America. Epidemics of erysipelas have occurred repeatedly in Greenland, and also in Alaska. In many parts of Canada the disease is seen with some frequency. There is recent mention of it in Health Reports from Manitoba, Nova Scotia, and elsewhere. In the United States erysipelas seems to be a commoner disorder than in England. It is said to be most prevalent on the prairies and south-west central plains, and to be of frequent occurrence in the Cordillera region, but is less common on the Atlantic and Pacific coasts, and generally in the country east of the Mississippi. The historical epidemics of a malignant type of the disease in the States have been already mentioned.

In Mexico erysipelas is infrequent. There is recent mention of its occurrence in British Honduras. In Guatemala it is perhaps not a rare disease; it was the cause of 215 deaths there in 1894 (about 150 per million inhabitants). Occasional cases are seen in the West Indies. In British Guiana 19 cases of erysipelas were reported in 1898; I do not know whether this was an exceptional or average number. For the rest of South America recent information upon this disease seems to be wanting.

Factors concerned in the Geographical Distribution. The information hitherto published as to the frequency, or even

---

[1] Adams.

as to the existence, of erysipelas in many parts of the world is so imperfect, that much caution is needed in discussing the causes which have led to its distribution over the earth's surface. It is not absent from any of the great geographical zones—tropical, sub-tropical, temperate, and sub-arctic. It is perhaps more common in cool or temperate climates than in hot ones, but even this rule is not without exceptions. It is possibly more fatal in the cooler regions of the earth than in the warmer.

In its seasonal relations erysipelas has shown a tendency to prevail most in the cooler half of the year. The mortality from the disease shows a rough tendency to vary inversely with the amount of rain-fall. Cold east winds have been thought to favour its occurrence. It is found at all altitudes and on all varieties of soil and geological structure. All races suffer from it, including the negro, who was at one time thought to be immune to it.

The disease is highly infectious, but it seems to require the pre-existence of some open wound, some small abrasion or discontinuity of skin, or some absorbing surface (such as that of the uterus after child-birth or a patch of eczema) before it can be set up. Hence it is a less "infectious" disorder than, for example typhus fever or influenza. In hospital wards, and particularly in surgical wards, it was at one time a frequent source of serious epidemics, the infection spreading rapidly from case to case and from ward to ward, often with disastrous results. Such "hospital epidemics" of erysipelas have been known to occur in hospitals which were above reproach in regard to cleanliness, structure, ventilation, absence of overcrowding, and general hygiene. But the larger number of them, as also the continued recurrence of sporadic cases of the disease, have been associated with some hygienic defect, of which overcrowding, imperfect ventilation, imperfect cleanliness, and (in some instances) the escape of foul air from sewers or refuse heaps into the ward, have been the most important. It can scarcely be questioned that the introduction of the antiseptic and aseptic methods of treating surgical cases has had an enormous share in lessening the frequency of hospital erysipelas.

The mode of origin and even of transmission of erysipelas is still a matter of some doubt. Cases of so-called idiopathic erysipelas outside hospitals are by no means rare, and it is the exception, rather than the rule, in such cases to learn of a definite exposure to infection from a person already ill with the disease. The bacteriology of the disease has not yet been finally worked out, and the relation between *Streptococcus erysipelatis* and the pyogenic streptococci is still uncertain. The morphological and cultural differences between them are so slight that it has seemed possible that the former might be no other than one of the latter rendered unusually virulent by certain external conditions. What those conditions are, or whether under any conditions a comparatively harmless pyogenic coccus can acquire the power of producing erysipelas (not as a rule a suppurative affection) is quite unknown.

The relationship between erysipelas and puerperal fever is also as yet a matter of uncertainty ; on the whole, however, the evidence points strongly to their affinity at least, if not to their identity. Certainly women in child-birth run enormously increased risks of puerperal fever, if attended by persons the subject of erysipelas. Moreover, doctors and midwives and nurses attending on puerperal fever patients have been known to develope erysipelas, while the proportion of deaths from the latter disease in the newly-born children of women with puerperal fever is unusually high. All these facts point to the conclusion that there must be some very close relationship between the two maladies.

GOITRE AND CRETINISM.

Goitre, or enlargement of the thyroid gland, and cretinism, or a peculiar form of congenital idiocy, are two pathological conditions that are most conveniently dealt with together. Although final proof of their exact relation to each other is perhaps wanting, it is the general opinion of most observers that they are very closely allied the one to the other, and that they are probably both due to the same cause or set of causes; while strong evidence of their kinship is found in the fact that large numbers of cretins are themselves goitrous, and that equally large numbers are the offspring of goitrous parents. The geographical distribution of the two conditions may therefore well be discussed together.

History. Goitre has been known to exist as an endemic disease in the Alps at least since the time of Pliny. Marco Polo makes mention of its occurrence in Yarkand and other parts of Central Asia in the 13th century; and throughout later medieval and more recent times there are frequent references to goitre as occurring in many European countries and in some other parts of the world. It is probable that most of the great centres of goitre mentioned in the following paragraphs are of very considerable antiquity. Cretinism is apparently not referred to by any writer before the 16th century.

Recent Geographical Distribution. *Europe.* In the British Isles neither goitre nor cretinism can be regarded as common forms of disease. There are, however, certain parts of the country in which goitre at least is endemic, and the most important centre of the disease appears to be in the north and centre of England, "in the valleys which drain the Pennine range of hills, east and west, from the Border as far south as the Peak of Derbyshire[1]." The magnesian-limestone districts of the last-named county have long been regarded as peculiarly favourable to the development of goitre, and " Derbyshire neck " is a name not infrequently applied to the disease. Cretinism, on the other hand, is much more rarely seen than goitre in England ; but it is said to be occasionally met with in Derbyshire, and in the dales between Lancashire and Yorkshire. In Scotland and Ireland both forms of disease seem nowadays to be exceedingly rare. In some parts of Wales goitre appears to be somewhat frequently met with.

In France goitre is said to be far from infrequent in some parts, notably in Savoy and the Hautes-Alpes. Epidemics of the disease have been recorded as occurring among soldiers in France, but these have been regarded as "probably the result of forced marches through goitrous districts, combined with scarcity of food," and "in such districts the practice of carrying loads on the head seems largely to increase the ratio of the disease[2]." In the valleys of the Hautes-Pyrénées, in the mountainous parts of the department of the Rhône, and in the southern valleys of the Upper Auvergne goitre and cretinism are, or were some years ago, most commonly seen (Hirsch). In Germany the Black Forest is said to be an important seat of these diseases, which occur particularly in the lower valleys and are said to be unknown at the higher points of the forest. In Bavaria, in Würtemberg, in Baden, and in Hesse scattered centres of goitre and cretinism are met with. The plain of North Germany and of the Netherlands is, on the other hand, said to be quite free from both forms of disease, and in Belgium they seem to be rare. Sweden and Norway and Denmark also appear to suffer but little from either.

In European Russia goitre is perhaps rather commoner. Some years ago the shores of Lake Ladoga were regarded as peculiarly goitrous districts, and possibly they are so still. But the most reliable recent evidence points to the east and north-east of European Russia as specially liable to this form of disease. From

---

[1] *The Goulstonian Lectures*, 1899. "On the Pathology of the Thyroid Gland," by Prof. G. R. Murray, M.A., M.D. Hirsch, on the other hand, on the authority of numerous writers in the eighteenth and first half of the nineteenth century, believed goitre was commoner " in the southern and midland counties than in the northern and mountainous districts."

[2] Dr Pugin Thornton, in Quain's *Dictionary of Medicine*.

the conscription returns of the Russian army for the years 1893, 1894, and 1895, it seems that the governments between the Volga and the Urals are the most goitrous. Thus in the government of Ufa the number of conscripts rejected on account of goitre was in 1893 two, in 1894 five, and in 1895 nine (out of 17,000 to 22,000 conscripts). In that of Samara the corresponding figures were 6, 2, and 3 (out of 20,000 to 27,000 conscripts). In that of Perm they were 10, 10, and 17 (out of 20,000 to 27,000 conscripts); in that of Viatka they were 3, 4, and 2 (out of 24,000 to 32,000 conscripts); and in that of Kostroma the numbers rejected were as high as 6, 12, and 23 in the three years in question (out of only 10,000 to 14,000 conscripts). Elsewhere, with the exception of some of the Polish governments, the numbers of goitrous conscripts were very much lower. It would seem from these figures that in the governments of Kostroma and Perm goitre finds more favourable conditions for its endemic existence than in any other part of European Russia, though in comparison with such countries as Italy, Austria, and Spain the figures cannot be held to indicate a very high frequency of the disease in any of the districts mentioned.

In Austria both goitre and cretinism are very common; occurring especially along the banks of the Danube, and particularly in the provinces of Styria and Carinthia, on the slopes and valleys of the Carpathians, and in Transylvania, Galicia, and Hungary. The valley of the Drave on the other hand is said to be quite free from both diseases. Cretinism is particularly frequent in these countries, and in 1896 the official statistics contained returns of 10,086 cretins; they are most frequent in the provinces of Styria, Carinthia, and Salzburg, where their numbers per 100,000 of the population were 215, 254, and 265 respectively. In Dalmatia, on the other hand, the ratio fell to 30.

Switzerland has long been regarded as peculiarly favourable to the existence both of goitre and cretinism, and both forms of disease are found with frequency in many parts of the country, but especially in the Vallais. In Spain and in Italy both goitre and cretinism are said to be of very frequent occurrence. In the former country the southern slopes of the Pyrenees are, or were some years ago, more especially affected by them, while in Italy the district of Aosta and the valleys of the Alpine chain which traverses Piedmont and Lombardy are mentioned by Hirsch as being particularly goitrous. It appears certain that in Italy, Spain, Switzerland, and Austria, both goitre and cretinism are more widely prevalent than in the more northern European countries.

Of the degree of frequency of goitre in the countries of the Balkan peninsula little is accurately known. In the valleys of Montenegro and perhaps in other valleys of the Balkan mountains cretinism is met with to some extent. In some of the mountain districts of Roumania (Moldavia and Wallachia) goitre is endemic, while cretinism is met with sporadically. Goitrous persons are occasionally seen in Constantinople, but the disease is not a common one here. It is remarkable that the Constantinople *hamals* or street porters, who carry astoundingly heavy weights on their backs, and not infrequently pass a band round the forehead, so that part of the weight is borne by the head and neck muscles, are rarely, if ever, the subjects of goitre.

Goitre exists in Greece, but only in two villages in the mountains of Phthiotis, and it is rarely complicated with cretinism.

Asia. The existence of goitre on the European side of the Urals has been pointed out above. The valley of the Ural river, which for some distance divides Europe from Asia, is also described as very favourable to the development of goitre. Kandaratski[1] states that the disease is widely prevalent all along the river valley, and that the Russian inhabitants suffer from it no less than the Cheremiss, Permiaks, Voguls, and Tatars who dwell in the same region. In the Caucasus goitre is said to be frequently seen in some valleys, but so far as may be gathered from the numbers of conscripts rejected from the Russian army for this cause it would seem to be mainly in the governments of Kutais and Tiflis in Transcaucasia that the disease is at all common, while in the plains of the Caucasus proper (Stavropol, Kuban, and Terek), north of the mountain range, it appears to be unknown. In many parts of Siberia, notably in the valley of the Lena and in the Altai mountains, goitre is endemic.

In Asia Minor and in Mesopotamia centres both of goitre and cretinism are met with. Syria, Arabia, and the table-land of Persia have all been said to be quite free from both diseases. Hirsch,

---

[1] At the 7th Conference of Russian Physicians, held in Kazan, 1899.

who quotes authorities for those countries, also cites a paper by Burnes communicated to the Calcutta Medical Society in 1835, in which Bokhara is added to the list of countries free from goitre and cretinism. It is, however, certainly not so at the present day. On the contrary, goitre is frequently met with in all parts of Bokhara, but especially in Kalai-Khumba, the capital of Darwaz. Here, it is said, on the authority of the Bek of the province, that there is scarcely a family in which some cases of the disease are not met with, and the same statement is made in regard to Karatag, where goitres of enormous size are common (Grekof). Cretinism is also not rare in Bokhara.

The Turkomans of both Russian and Chinese Turkestan appear to suffer to a remarkable extent from goitre. In Yarkand, Khokand, Kashgar, Kargalik, Utch-Turgan, and other places goitres of great size are very frequently met with[1]. Contiguous with this vast Central Asiatic area of goitre-prevalence are a number of districts of northern India where the disease appears to be no less common. Hutcheson[2] describes it as widely prevalent throughout the northern part of the peninsula, but as specially localised in certain well-defined areas, extending from the Brahmaputra to the Indus. It is found in some Himalayan valleys from end to end of the great range. In the Tarai district it is very common, and in Nepal, particularly in the neighbourhood of Khatmandu, the capital, as many as forty per cent. of the population are said to be affected with the disease. Eastern Bengal, Assam, and some parts of the North-West Provinces and the Punjab are also not wholly free from it. In Rajputana and generally throughout the plains of India goitre appears, on the other hand, to be rarely seen.

In Farther India goitre seems to be much less commonly prevalent than in India proper. In Burma the affection is said to be unknown in Arakan and practically so in Pegu, though a considerable number of cases are met with in the Irawadi and Tenasserim divisions[3]. In the Kwala Kangsa district of Perak in the Malay States, a mountainous limestone area, goitre is endemic. In Siam the disease is, according to Rasch, exceedingly

[1] Raemdonck, *Report of Leprosy Conference in Berlin*, 1897.
[2] Brig. Surgeon Lt.-Col. G. Hutcheson, M.D., *Transactions of the First Indian Medical Congress*. Calcutta, 1894.
[3] Davidson.

rare. In some of the East Indian islands, on the other hand, both cretinism and goitre seem to be quite common affections. Thus Kohlbrügge observed them frequently among the hill tribes of Java, and in many parts of Sumatra their comparative frequency is well recognised.

In Northern China goitre appears to be widely endemic. In some villages it has been estimated that one-sixth of the inhabitants are affected by the disease. Western Sze-chuan and Yünnan in the south-west are also highly goitrous provinces, and in Tibet there appears to be equally good reason for believing that this form of disease is widely endemic.

Australasia. Whether goitre or cretinism exist in the Pacific Islands and in Australasia generally is doubtful; and in the absence of positive statements to the contrary it may be accepted that Australia, New Zealand, and the host of islands in their vicinity are practically free from both forms of disorder.

Africa. Comparatively little is known of the existence or frequency of either of the diseases now under discussion in the greater part of the African continent, at least at the present day. Hirsch quotes a number of authorities (mostly of not very recent date) for the statement that Lower Egypt, the Abyssinian basin, the East and West Coasts, and the littoral of Algiers are free of these diseases, while goitre is endemic on the Abyssinian plateau, in a few places in Sennaar, on the slopes and in the valleys of the Atlas (*e.g.* in Kabylia), in the mountainous parts of Morocco, and to a very considerable extent in the basin of the Niger and in the valleys of the Greater Sudan. In the centre of Africa goitre is probably of rare occurrence. Pruen did not recollect to have seen a single case during his sojourn in the interior, though on the coast mild cases were not infrequent—the waters here being impregnated with salts of lime. Cameron, on hearsay authority, believed that the inhabitants of the country round Lake Kassali, in the south of the Congo State, suffered greatly from goitre. Cook says it is common in Uganda.

In the Azores and in Madagascar goitre, and perhaps cretinism, are known to exist; but in Madeira the former is very rare and in Mauritius it seems to be unknown.

North, Central, and South America. While goitre is not unknown in some parts of the continent of North America, and is even very prevalent here and there, it appears to be less common than in either Europe or Asia, and it would also seem to have diminished very considerably within recent years. Hirsch, writing some years ago, gave a long list of districts in Canada and the United States where the disease was endemic—beginning as far north as the banks of the Saskatchewan river in Hudson Bay Territory. He pointed out, however, that already from the beginning of the nineteenth century a diminution in the extent and frequency of goitre in this continent had been observed, and it appears that this process has been continued up to the present day. In 1893 Prof. Osler[1] wrote to a large number of medical practitioners in the localities said by Hirsch to be goitrous, and ascertained that in several of these districts goitre had become very rare if not quite extinct. At present the malady appears to be by no means uncommon in the province of Quebec, and in Montreal cases are not infrequently seen in hospital practice. In Ontario it is very prevalent in the limestone regions at the end of Lake Ontario[2] and is common in parts of Michigan. In the States of Alabama, Virginia, and Vermont, where goitre was formerly common, it is now very rare. Cretinism is thought by Osler to exist nowhere as an endemic disease in any parts either of the States or Canada, though sporadic cases of it are at times met with.

In Mexico and New Mexico, on the other hand, cretinism is said to be still endemic, while goitre was within the last half century excessively prevalent throughout a vast area extending through Mexico, Guatemala, San Salvador, Nicaragua, and Costa Rica, and down the Cordilleras as far to the south as Chile. This zone of goitre-prevalence has been compared with those of the Alps in Europe and the Himalayas in Asia[3]. The frequency of the disease increases in this Central and South American zone from north to south, and the widest diffusion of goitre is found on the eastern slopes of the southern Cordilleras, in the states of the Argentine Republic. The river basins of the mountainous parts of Paraguay are also goitrous, and throughout a great portion of Brazil the affection is commonly met with, though cretinism

[1] *American Journal of the Medical Sciences*, Nov. 1893.
[2] In the asylum of Kingston, Ontario, of 600 inmates no less than 288 were affected by goitre. No cases of cretinism existed in the asylum.
[3] Hirsch.

on the other hand is unknown as an endemic disease in Brazilian territory.

Factors determining the Geographical Distribution of Goitre and Cretinism. It will be observed that the distribution of goitre over the earth's surface is an exceedingly wide one. The disease has been observed in the far north and in the tropics, and appears to occur quite independently of all conditions of climate, season, or weather. It is apparently uninfluenced by race—the North-American Indian, the European, the Chinese, the Himalayan Indian, the Turkoman, and the white and negro inhabitants of Brazil, Chile, and Central America all suffering from it under certain conditions. It is, indeed, rather in local circumstances and conditions that the causes of goitre and the main factors in its geographical distribution are to be sought. What these circumstances and conditions are is not fully known. Altitude above sea-level appears to have but slight influence. Goitre is found, it is true, more often among highlanders than dwellers in the plains, and its chief endemic areas are in great hill ranges such as the Alps, the Himalayas, and the Cordilleras. It is also rare on table-lands, very rare on low lands, and seldom found on sea-coasts[1]. But to all these general statements there are very marked exceptions. Thus while goitre has been found as high as 10,000 to 13,000 feet above the sea in the Cordilleras it is also met with as truly endemic in the low-lying Tarai at the foot of the Himalayas, in the northern plains of France, and in the flat valley of the Danube.

The most important local conditions which are believed to influence the prevalence of goitre in a district are connected with the geological character of the soil and the nature of the water-supply. There is not space here to discuss this question in full, but briefly it may be said that from a very large variety of sources evidence is forthcoming that goitre is mainly found where people live on a magnesian-limestone soil, and where the water drunk is impregnated with the various salts of lime and magnesia. A further development of this view is found in the suggestion that it is not by this "dolomitic" character of the water drunk that the existence of goitre (and, it should be added, cretinism, as both are believed to be due to the same cause) is to be explained,

[1] Hirsch says it has never been found hitherto on coast margins, but cases of goitre have been seen on the east coast of Africa, *vide supra*.

but by the presence of sulphide of iron or of copper, both of which are more likely to be found dissolved in dolomitic waters than in others. There are, however, many difficulties in the unqualified acceptance of these views. For while a dolomitic soil is often found associated with these diseases it does not seem to be essential for their development. Thus goitre is found on the sand and mud-banks of Eastern Bengal and Assam, on the laterite of the Dacca and Maimensingh districts, on the calcareous soil of the Trans-Gogra districts of the North-West Provinces, on the alluvium and detritus of the Tarai, and on different kinds of soil in the Punjab[1]. Then again in New Zealand, where large masses of magnesian limestone lie exposed in the northern island, goitre is, or rather was some decades ago, wholly unknown[2]. Further, while goitre is found among people in the Ballia districts of the North-West Provinces of India whose water-supply is not of dolomitic origin, it is unknown at Mussuri and Naini-tal, where the water-supply is of this nature. For the present then the question of the part played by the chemical constituents of the soil and of drinking-water in the production of goitre and cretinism must remain an open one. Recently the suggestion has been made that these diseases are not due to the action of any special chemical substance contained in soil or water, but are, like so many other diseases, of bacillary origin. Lustig and Carle found a bacillus which they named the " bacillus liquefaciens " in waters taken from districts in which goitre prevailed, but they were unable to prove definitely that any relation existed between this bacillus and the disease.

While both goitre and cretinism are essentially maladies of an endemic character, the former has on rare occasions become distinctly and undeniably epidemic. A large number of instances of epidemics of goitre have been recorded in the past in France, and not a few in Germany and elsewhere. A remarkable outbreak of the disease occurred among Russian troops in the Turkestan campaign in 1877, and a similar occurrence among recruits in Brazil is also recorded. In India in 1886 and in 1893 goitre was most unusually prevalent in the Himalayan and other districts where it is always more or less present, and in 1894 a

[1] Hutcheson, *loc. cit.*

[2] Thomson, quoted by Hirsch. The reference, however, is to the year 1840.

sudden outbreak of the disease occurred in the native contingent of the Duffla Field Force. This outbreak affected the troops in a clearance in the virgin forest in a district notorious for malaria, and among a people where goitre was common. The soil was described as damp and swampy, and a dense fog prevailed at the time, allowing the sun to be seen for only four or five hours daily. The troops were only there for ten days, yet nearly every man developed the disease with the most alarming rapidity[1].

Finally, while cretinism is only found in districts where goitre prevails, there are, on the other hand, vast endemic areas of goitre where cretinism is very rare or almost wholly unknown. In many goitrous districts the lower animals, such as dogs, cats, sheep, pigs, horses, and mules, have been observed to suffer from the disease.

[1] Hutcheson, *loc. cit.*

GOUNDOU, ANAKHRE, OR GROS-NEZ.

General Characters and Etiology. Goundou, Anakhre, and Gros-nez are names given to a curious and interesting disease which is met with on the West Coast of Africa and in some other parts of the world. Its principal characteristic is the appearance of symmetrical swellings on either side of the nose. It usually begins soon after childhood, but may attack adults. The appearance of the swellings is in most cases preceded by headache and a discharge of blood and pus from the nose. These symptoms last some months, and then disappear, but the swellings continue to increase, until they ultimately interfere with vision and may destroy the eyes. The tumours appear to consist merely of a thin hollow shell, formed of the nasal process of the superior maxilla and adjacent bones. The swellings are oval, with their long axes directed downwards and outwards.

The cause of the disease is unknown. Maclaud[1] has suggested that it is due to the larvae of some insect finding their way into the nostrils; but, as Manson has pointed out, the symmetry of the growths is difficult to account for on this hypothesis. Braddon regards it as apparently due to a chronic sub-periosteal inflammation involving the nasal process and perhaps the orbital plate of the superior maxilla and the nasal bone[2].

History. Goundou seems to have been first described in a paper read by Prof. A. MacAlister before the Royal Irish Academy on December 10, 1882, upon what were termed the horned men of Africa. Later descriptions of the disease were published by Lamprey in 1887[3], and by Maclaud in 1895[4]. Within the past

---

[1] *Archives de Méd. Navale*, Jan. 95.   [2] *Journ. Trop. Med.* May 15, 1901.
[3] *Brit. Med. Journ.* Dec. 10, 1887.   [4] *Loc. cit.*

two years cases of apparently the same disease have been reported from China, Sumatra, the Malay States, and Uganda, but it is not clear that the disease is a novelty there. Indeed of the true history of goundou practically nothing is known.

Recent Geographical Distribution. *Europe.* In Europe goundou is wholly unknown.

Asia. In Asia it has been seen in Sumatra, in the Malay States, and in China. The number of cases recorded up to the present is, however, exceedingly small. In Sumatra itself one case has been seen in a Malay woman[1]. In the Malay States three cases have been observed, two of which were in Boyanese (natives of Sumatra), and the third in a Malay[2]. In China one quite typical case has been seen in Shangpoo, in southern China, the patient being a Chinese woman[3].

Australasia. The disease appears to be quite unknown in Australasia.

Africa. In certain parts of the African continent, on the other hand, goundou is found with considerable frequency, and particularly on the West Coast. In some villages of the Ivory Coast as many as one or two per cent. of the population are affected. Maclaud believed that the disease was confined to the riverine districts of the Lower Komoë, and that, if found elsewhere, it was probably only in persons who had at some time or other lived in those districts. Brault also states that the disease is particularly common in villages in the districts of Beltié and Krinjabo ; in Indenié, Attié, Morénou, Baoulé, and Esikasso, and in short the whole of the Lower Komoë. But though this region is the principal centre of goundou it is certain that the disease occurs in other parts of West Africa. Lamprey has described three cases in Fantis, of whom one came from the Wassan territory, one from the Gamin territory, and one was a visitor to Cape Coast Castle. Renner has recorded a case in the village of Kissy, on the right bank of the Rokelle or Sierra Leone river[4]. This patient was born in the village and had never left it, and as no other cases had been observed in the village the disease in this instance must almost certainly have been of local origin. Quite recently, too, evidence has been adduced that goundou is not wholly confined even to the western half of the African

[1] *Journ. Trop. Med.* Aug. 1900, p. 11.   [2] *Ibid.* May 15, 1901.
[3] *Ibid.* Dec. 1900.   [4] *Ibid.* Jan. 15, 1900.

continent, for two cases of it have been seen in Uganda, though whether of local origin or imported is not stated[1].

America. Finally the only mention of goundou on American soil seems to be the record of a single case observed in a West Indian negro child in 1894. Dr Strachan, who described the case, stated that he had seen two similar examples, and had often observed a "ridge" in this part of the face in West Indian negroes[2]. He suggested that this feature in West Indian negroes might be an example of atavism referable to some tribal peculiarity of the original West African stock.

Factors which govern the Geographical Distribution of Goundou. This disease has up to the present been observed solely in tropical countries, and it may therefore be presumed that warmth is either directly or indirectly essential to its development. Beyond this very general statement little can be said as to the influence of other physical conditions, and no observations have yet been published as to the relation of the disease to soil, atmospheric moisture, elevation, or the general level of sanitation and personal hygiene. Race is probably a factor of importance, as the overwhelming majority of cases have been seen in African negroes and the cases in the West Indies were in negroes of African stock. But the recent evidence, summarised above, of the occurrence of the disease in Malays, East Indians, and Chinese, shows that the negro family is not the only branch of the human race liable to it. A very similar affection having been observed in a chimpanzee[3], it is just possible that the lower animals may share with man the susceptibility to goundou.

---

[1] Cook, *Journ. Trop. Med.* June 1, 1901.
[2] *Brit. Med. Journ.* Jan. 27, 1894.   [3] Maclaud, *loc. cit.*

GOUT.

General Characters and Etiology. Gout has been
defined as "a nutritional disorder, one factor of which is an
excessive formation of uric acid, characterised clinically by attacks
of acute arthritis, by the gradual deposition of urate of soda in
and about the joints, and by the occurrence of irregular constitutional symptoms[1]." The characters of an acute attack of "the
gout" are too well known to need description, as are also the
general features of chronic gout and of the so-called "gouty
diathesis." An immense variety of affections have at one time
or another been regarded as of gouty origin, but, while the term
is a convenient one, it would be difficult to define its exact
scientific meaning. The cause of gout is unknown; the disease
is essentially a disturbance of tissue metabolism, to which many
factors predispose, and in which the production of uric acid in
excess is one of the most striking features. But the excess of
this substance is in no sense the cause of the disease, for it is
found in many other conditions in no way resembling gout.

History. Gout or podagra is one of the oldest of known
diseases. From the time of Hippocrates to the present day it
has been mentioned by almost all the medical historians, including
the Greeks, the Romans, the Arabians, and the later medieval
writers. No doubt the term podagra in early times included
many other affections than would now be strictly classed as
gouty, and this may perhaps account for the apparently greater
frequency of the disease in many countries in former centuries
than at the present day. But there is good reason to believe
that true gout has become very much more rare in recent years
than it used to be. To what extent this diminution is to be
ascribed to changes in the habits of the civilised nations will be
referred to again later.

[1] Osler.

Recent Geographical Distribution. *Europe.* Gout is met with in almost all European countries. Apparently, however, it is particularly common in the British Isles and in Germany, and this has been attributed to the large amount of alcoholic drink consumed in those countries. The disease, at least in its acute form, is incomparably less frequent at the present day than it was a hundred years ago. It is still far more commonly seen in England than in either Scotland or Ireland. A very notable diminution in the prevalence of gout in recent times has also been observed in the Low Countries and in Switzerland. In Russia gout is seen with some frequency in the cities, but among the vast mass of the population the disease is certainly not a common one, in spite of the large quantities of fermented liquor drunk by them. In Lapland, Iceland, and the Faröes, gout is said to be unknown. The disease occurs in European Turkey, but comparatively rarely. In Servia it is perhaps more common; in 1894–5 gout was the cause of 0·9 per cent. of all the deaths in that kingdom. In Greece gout was formerly rare, but I am informed by Prof. Hadji Michali of Athens that it is daily becoming more frequent.

Asia. The distribution of gout in Asia is very imperfectly known. There is no mention of it in recent health returns from the Caucasus, Siberia, or Russian Central Asia. It is very doubtful if it exists in Asia Minor, Syria, Mesopotamia, Arabia, or Persia. In India it has been seen, both in natives and Europeans, but as a comparatively rare disease. In Ceylon and Assam it is said to be quite unknown. It is probably very rare, if it occurs at all, in the East Indian Islands. No case of gout has been observed among the hill-tribes of Java[1]. The disease is also very infrequent in the Malay Peninsula. Gowan saw one case in a native during four years' residence in Siam. An occasional case is seen in the hospital at Penang[2]. There seems to be no positive evidence of the existence of the disease in China or Japan.

Australasia. In Australia on the other hand gout certainly occurs and apparently with some frequency. In Queensland occasional cases are recorded in the Annual Health Reports. In New South Wales a certain number of cases come under hospital treatment in some, if not in all years, and in both South Australia

---

[1] Kohlbrügge.
[2] *Straits Settlements Medical Reports.*

and West Australia a few deaths from gout usually appear in the annual mortality returns. There seems to be no mention of gout in any of the Pacific Isles, but this can scarcely be taken as proof that it does not exist there.

Africa. Gout is apparently a rare disease in the African continent, but information upon the subject is at present very imperfect. It is said to be unknown in Egypt, and in all the countries along the northern shores of Africa. The West Coast was also at one time believed to be exempt from gout; but this is apparently not the case at the present day. In the Colonial Hospital, Gold Coast Colony, five cases of gout were treated in 1897, and unless this was an exceptional occurrence it would seem to show that gout is not particularly rare here. Some decades ago it was said to be very common among the Hovas on the central plateau of Madagascar.

America. The references to the occurrence of gout in Canada and the United States are extremely scanty. It is seen in many of the cities of the United States, but is probably less common here than in England. Gout is also exceedingly rare in Mexico, Central America, and the West Indies. In South America it is said to be infrequent in British Guiana and almost unknown in Brazil and Peru, but of somewhat frequent occurrence in Chile.

Factors concerned in the Geographical Distribution of Gout. It will be seen from the above summary that the facts concerning the distribution of gout over the earth's surface are very imperfectly known. So far as they go they warrant the general assertion that this disease can occur in almost all latitudes, but that it is principally found in the temperate zones and is rare or unknown in most tropical countries. Exposure to cold and damp seems to favour the production of gout in the individual, and this may be part, though it can scarcely be the whole, of the explanation of the comparative rarity of the disease in warm latitudes.

While the actual nature and cause of gout are unknown, it is nevertheless certain that heredity has a large share in its production. Excessive indulgence in luxurious food, especially if combined with the consumption of much alcohol and a sedentary life, seems undoubtedly to predispose to gout. It is believed that the egregious excesses of this nature which were indulged in in the early days of the Roman Empire were the main reasons of the extreme frequency of the disease at that time. It is also believed that the greater moderation in eating and drinking which obtains now as compared with the very different habits of our forefathers explains the diminished frequency of gout in recent times. The part that over-consumption of strong drink plays in the production of the disease is uncertain. It is an undoubted fact that heavy drinking in better-class society is not now the rule, as it formerly was, and that the average man is not in the habit, as his great-grandfather possibly was, of consuming from one to three bottles of port after dinner each night, and being carried to bed by his servants, unconscious from alcoholic poisoning; and it is very probable that this change in social habits has had a large share in reducing the frequency of gout. But it is equally undeniable that alcohol is still consumed to an enormous extent, and that though individuals in the richer strata of society may drink less than formerly, the community as a whole, at least in some countries, as for example, England and Germany, drink more, and the yearly bill for alcoholic drinks in most European countries is an increasing one.

Probably gout is influenced as much, or more, by the kind of fermented liquor drunk as by the quantity. The heavy beers consumed in England and in Germany are said to be much more conducive to gout than the lighter beers drunk in America. Strong, sweet, red wines have always borne an evil reputation for gout. The spirits distilled from some grains, as for example whisky and vodka, are probably less nocuous in this respect; they are at least drunk to an enormous extent in Scotland, Ireland, and Russia, and gout is comparatively rare in these countries.

The relation of gout to race is not wholly certain. It is far commoner in the Aryan races of the European continent than in any others; but it has been seen in African negroes, in the natives of India, Siam, and the Straits Settlements, in the Hovas of Madagascar, and in the Creoles of South America. In India the Mohammedans suffer from it occasionally, while the Hindus are said to escape it entirely. Probably social habits, mode of life and other factors are of much greater importance in predisposing to gout than any variations in racial susceptibility.

INFLUENZA.

History. The earliest reference to the occurrence of influenza epidemics is probably to be found in Hippocrates, and relates to the year 412 B.C. (*circa*). In 591 A.D. Bishop Gregory of Tours mentions an epidemic which was probably one of influenza. Other epidemics are spoken of by the chroniclers as occurring in the years 877, 889, and 927. Then after a long pause came the great outbreak of 1173. The records of the thirteenth century are silent as to influenza, but there were many outbreaks in the fourteenth—notably in 1323, 1327, and 1386[1]. From the fifteenth century onwards the references to influenza epidemics are not only much more numerous but are also much more definite than the earlier ones. Hirsch, Theophilus Thompson and others have given complete lists of the principal epidemics and pandemics of this disease from medieval to recent times, and the story of the latest pandemic is told in detail in a subsequent portion of this chapter. These successive outbreaks of influenza have shown no definite regularity either in the intervals between them, their duration, or the area they have covered.

Innumerable names have at different times been suggested for this disorder. The term "influenza" as applied to the disease was first used by an Italian writer, Piero Buoninsegni, whose Florentine history was published in 1580. It is the one now most generally employed. The etiology of influenza is uncertain. A minute bacillus has been observed in the secretions of influenza patients and is thought by some to be characteristic of the disease. It was first described simultaneously by Pfeiffer, Kitasato, and Canon in 1892.

---

[1] The early history of influenza is told with greater detail in a paper on "Epidemic Influenza" read by the author before the Society of Med. Officers of Health in 1890 and published in *Public Health* (April, 1890).

General Epidemiological Characters. In discussing the geographical distribution of influenza it will be impossible to ignore the striking epidemiological characters of the disease, which seem to distinguish it from most other epidemic disorders. No other disease appears to spread so rapidly, to attack almost simultaneously so large a proportion of the people exposed to it, or to extend in epidemic form over so wide an area of the globe. In these respects dengue is the only other malady which at all resembles influenza, and it will be recalled that when in 1889–90 the last epidemic of influenza swept over a great part of the globe many people regarded it as a form of dengue. But dengue has never prevailed as an epidemic over so wide an extent of the earth's surface as even approximately to justify the use of the word "universal" in regard to its prevalence, and this has certainly on many occasions been the case in regard to influenza.

But while influenza differs from most other diseases in this important respect, there are other characters which it shows in common with such typical epidemic diseases, for example, as cholera or plague. Like cholera and like plague, influenza appears to remain for long periods together endemic in certain countries or regions, where from time to time it becomes epidemic; and, like them, at irregular intervals it spreads from these centres, or from one of these centres, and becomes more or less widely pandemic. On some of these occasions, it would seem, the disease extends, often in a remarkably short space of time, over very nearly the whole inhabited globe. Countries which have been quite free from influenza for many years and which are visited by it on these occasions may then become for the time endemic homes of the disease; and for a considerable period may be ravaged by recurring epidemics of this very troublesome, and (in its total results on the community) very fatal malady. The epidemics of influenza during the past three winters in London, Paris, Madrid, Constantinople and many other places were sufficient to demonstrate that after twelve years of irregular pandemic prevalence the disease had lost none of its vigour, or of its power to increase the general mortality to an extent which, had the cause of the increase been plague or cholera, would have caused very great alarm.

Whether influenza is permanently endemic in any part of the world must for the present remain uncertain. Before 1889, when the latest pandemic extension of the disease began its course, it was apparently endemic, and had perhaps been so for some time, in several parts of the earth's surface. Thus influenza, or a disease allied to it, is said to have been common in China before the year named, and to account for its existence there a theory was advanced that it was the result of some colossal earth-commotion or gigantic inundation in that little-known country. It was even suggested that some catastrophe of this sort might have raised a cloud of dust particles which, like that following the great Krakatau eruption of 1883, might have swept round the world and so produced the pandemic of influenza of 1889 and following years. The more sober observations of the medical officers in the Chinese Customs service, however, lend no sort of support to this view, and there is, indeed, no reason to believe that the pandemic in question started in that country. At the same time it appears probable that influenza was locally endemic in some parts of China before 1889; and in 1888 a mild outbreak of the disease occurred in Hongkong[1].

In like manner, as will be shown later, influenza was endemic in many parts of Russia before the pandemic "wave" began its course. In India a few cases of "influenza" were annually recorded for many years before 1889. So also in the Solomon Islands, in Fiji, and in Trinidad, a disease apparently identical with influenza had for long been endemic. Elsewhere influenza or a disease closely allied to it had from time to time been observed. The well-known "strangers' influenza" occurring in St Kilda, in Iceland, in other isolated islands, and among isolated communities such as certain Indian tribes in North America and in Brazil, which is said invariably to follow the arrival of a stranger from the outside world—even though the new-comer be himself in health and come from a country free from influenza—is a phenomenon which is established beyond doubt, but which has not yet been explained. An outbreak of this kind was recorded in the month of May 1889 as occurring among the Indians, half-breeds, and whites in Athabasca and other districts of British North America. It is noteworthy that at the same time influenza was epidemic in Greenland and in Bokhara, but none of these outbreaks seem to have had any

---

[1] Cantlie. *Medical Reports of the Chinese Imperial Maritime Customs.* No. 42, p. 37.

connection with the great epidemic extension of the disease which began in the following autumn.

The relation which these temporary—or possibly in some instances permanent—endemic centres of influenza bear to the pandemic extensions of the disease is uncertain. It appears, however, to be exceedingly probable that the latter do have their beginning in one of these centres. That it is always the same one must be doubted. The history of the disease, indeed, shows almost conclusively that this is not the case; though probably several of the past pandemics have begun in or near the same region of the world as that which started in 1889. It has, however, proved impossible to assert, even approximately, where most of the past pandemics began. Their early history has been usually wrapped in obscurity, and though their course has in some instances been traced back over considerable distances, it has never been possible to assert definitely that they began in one particular district or even in one particular country. The reason of this is obvious; the disease has almost invariably come from the East, and the further back it is traced towards the rising sun, the greater becomes the difficulty of obtaining accurate information.

To some extent these considerations are true in regard to the great pandemic which began in 1889. But some few years ago, a careful study of a large number of Russian papers and reports[1] led me to the conclusion that the area of origin of that pandemic lay somewhere on the borders of European and Asiatic Russia, and that the disease spread thence eastwards, westwards, and southwards. I propose to dwell in some detail upon the ascertained facts which led me to that conclusion, as they were derived from comparatively little-known and not easily accessible sources, and appear to justify a more definite statement as to the place of origin of this pandemic than has ever been possible in the case of any of its predecessors. The following are the principal facts

---

[1] These consisted principally of numerous articles in Russian medical papers, various brochures and pamphlets, the annual reports of the Medical Department of the Russian Ministry of the Interior, and a valuable report on the influenza epidemic in the Russian army, issued at the end of 1891 by the Army Medical Department in St Petersburg. The latter report was based upon no less than 11,000 documents, obtained in reply to a circular sent to every army medical inspector in the empire in Nov. 1889.

in connection with the early period of the outbreak of influenza in Russia in the year 1889[1].

Course of the Last Pandemic. Before the outbreak influenza had unquestionably been endemic in many parts of Russia. In the official returns for 1887 and 1888 there was scarcely a government from which some cases of, or deaths from the disease were not reported. It appears to have been particularly frequent in the governments lying on or near the Gulf of Finland, and not least so in the city of St Petersburg itself. In 1889 localised epidemics of *grippe* occurred in the government of Moscow in January, February, April, and June. In the Tchernigof government, influenza broke out in the town of Borzna, and caused a sharp epidemic, in the month of February. In Nijni Novgorod there were three periods of prevalence of the disease in the year 1889—the first in March and April, the second in August, and the third (*the* epidemic) in October and November. In Kazan cases of the disease occurred in the early months of the year. It is unnecessary to multiply instances further, and it may be asserted with confidence that influenza was endemic in many parts of Russia prior to the autumn of 1889.

But at that period the disease suddenly took on a new character. Throughout the length and breadth of the country it spread with great rapidity and with extraordinary intensity, and crossing the frontiers of Russia gradually extended its ravages over almost the entire inhabited globe. The relation of this rapidly spreading form of the disease, which attacked in many places an overwhelming proportion of the population, to the localised, endemic form just described as existing previously in Russia is uncertain. But it appears probable that in these phenomena we are dealing with two different forms of one and the same disease—that in the steppes of western Siberia in the autumn of 1889 influenza acquired extraordinary powers of diffusion and of virulence, that it spread thence in all directions, and that its early course was to some extent obscured and rendered difficult to follow by the existence here and there of the disease in its mild, endemic, non-diffusive form. That

[1] For detailed references to the very large number of Russian authorities consulted for the facts contained in the text the reader is referred to two articles by the author, on "The Recent Pandemic of Influenza, its Place of Origin and Mode of Spread," published in *The Lancet*, Jan. 20, and Feb. 10, 1894.

influenza did begin earlier in one part of Russia than another, and that it did definitely spread as from a centre is demonstrated by a study of the dates at which the earliest cases of the great epidemic occurred in different places. I am able to reproduce here a map which I published some years ago in *The Lancet*, showing the results of a study of this nature. The dates of the earliest cases of the disease are inserted against the names of all the principal towns of Asiatic, and many of those of European Russia, and in following these dates a very accurate idea is gained not only of the course taken by the epidemic, but also of its place of origin.

In retracing the lines of the epidemic it is found that they converge in a region lying between 60° and 70° east longitude, and about the 55th or 56th degree of north latitude. Within this region the earliest cases of influenza appear to have occurred in Petropavlovsk (Akmolinsk province) and in Tcheliabinsk (Orenburg government). In each of these towns the epidemic was said to have begun in the end of September. The next earliest cases occurred in Tiumen (Tobolsk government) on October 2nd, and in Tomsk on October 7th[1]. Nowhere else in European or Asiatic Russia were cases registered at so early a period (with the exceptions already mentioned, where cases of the endemic disease were set down as early cases of the epidemic). The authenticity of these dates appears to be unquestionable; and they are further supported by the fact that in each of the towns named the epidemic attained its highest point in October, and was declining, if not over, before the month was at an end—that is to say before the epidemic had begun in St Petersburg and the west of Europe.

Tcheliabinsk was at that time the terminus of a line of railway which ran from Samara through Ufa, and which has since been continued as the Trans-Siberian Railway; the town lies to the east of the Ural Mountains, though just within the limits of European Russia. Petropavlovsk lies entirely in the steppes, and at the time of the epidemic was connected by roads with the great postal road, which was then the main route of communication across Siberia. This road, known as the great Moscow Tract, is marked by the thick line across the accompanying map. There

---

[1] All the dates given in this chapter in connection with the outbreak in Russia are according to the Old Style.

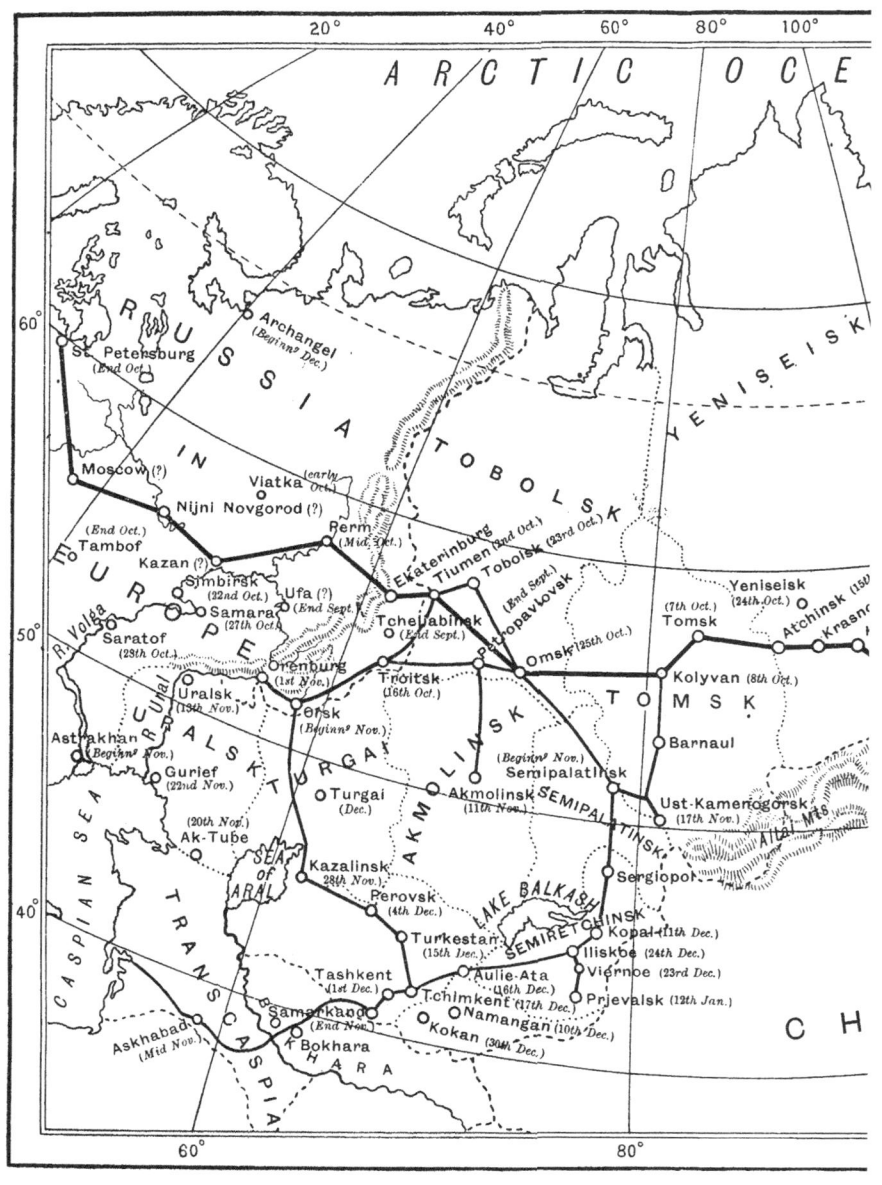

CHART SHOWING THE EARLY COURSE O

(The dates are those of the earliest kno

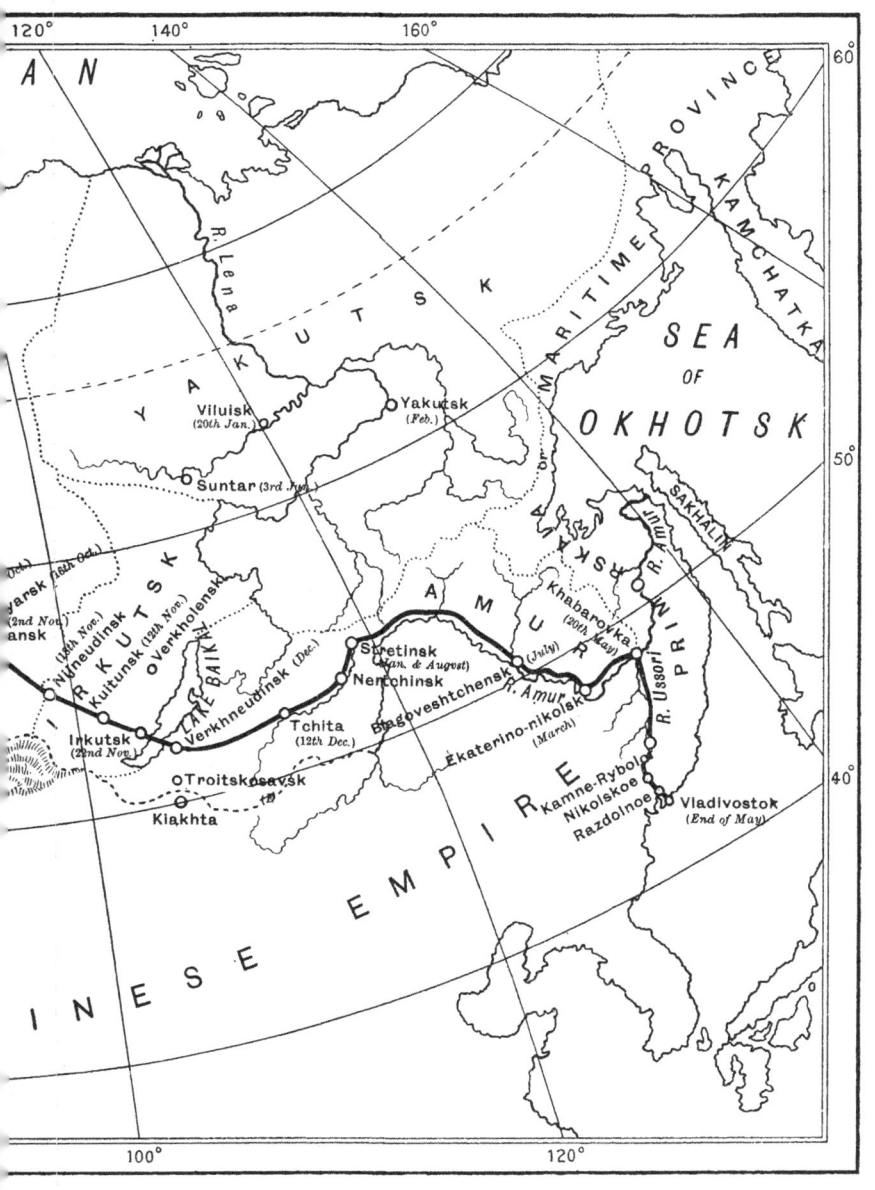

F THE INFLUENZA PANDEMIC OF 1889-90

wn cases in each of the towns named)

was then railway communication as far as Tiumen, an important junction where several lines of communication by rail, road, and river diverged. This was the first town on the Moscow Tract in which influenza appeared, the earliest cases occurring here on October 2nd.

The region defined above as the birthplace of influenza in 1889 lies in the northern part of the Kirghiz steppes. The large district covered by these steppes is divided into five provinces—namely, Uralsk, Turgai, Akmolinsk, Semipalatinsk, and Semirét-chinsk. The inhabitants are mostly Kirghiz, belonging to three great divisions—the Great, Middle, and Lesser Hordes. Their mode of life is nomadic in the summer, but as winter draws near they select a suitable place for their *zimovka*, or winter quarters, where they either pitch their felt tents (*kibitki*), or construct *zemlianki* or earth huts for themselves and their cattle. Their principal occupation is the rearing of cattle, and an enormous trade in live-stock is carried on. The chief market is in the town of Orenburg, but a considerable amount of business in cattle, hides and leather is done at Petropavlovsk.

The suggestion that the influenza epidemic arose among these nomad tribes was first made by the compilers of a valuable report upon the epidemic issued by the Russian Army Medical Depart-ment, and it scarcely admits of question that a large number of considerations go to bear out the suggestion. Unfortunately there is no positive evidence of the existence of influenza among the Kirghiz before it appeared in Petropavlovsk and on the Moscow Tract. Beyond the facts stated all is conjecture. It is certain, however, that the places where the earliest cases of influenza occurred were either in the steppe country or closely bordering upon it, and there was a clear progression of the epidemic in different directions in lines which radiated from the steppes. Whether the cattle, possessed in large numbers by the Kirghiz, suffered in any way from the disease, or from a disorder at all like it, is unknown. The Kirghiz themselves call influenza "*gangei*" and also "Chinese fever." The latter name might be held to point to China as the place of origin of the disease, but, apart from the absence of direct evidence, there are other considera-tions which negative the idea. The towns earliest affected by influenza are more than a thousand miles from the Chinese frontier. The province of Tomsk has a short strip of frontier occupied entirely by the Altai mountains, a formidable barrier to any epidemic disease. The provinces of Semipalatinsk and Semirét-chinsk, which also border upon the Chinese empire, were not invaded by influenza until November and December respectively, or much later than the appearance of the disease in Petropavlovsk. Further, the epidemic spread from north to south in these provinces, and not from south to north as it must have done had it been imported from China.

The radiating lines of the epidemic may be briefly traced as follows. The disease broke out, as stated above, at Petropavlovsk in the end of September, 1889. Thence to Tomsk is a distance of about 1000 miles, which could be travelled by post in about six days. Influenza broke out in Tomsk on October 7th. Nine days later it appeared in Krasnoyarsk, some 370 miles further east along the great Moscow Tract. It did not reach Irkutsk until November 22nd, or 46 days later than its appearance in Tomsk. Beyond the great Baikal Lake its progress was even slower. Tchita was invaded on December 12th, Strétinsk not until January. The road at that time ended at Strétinsk, and passengers for Vladivostok, Russia's naval station on the Pacific, had to descend the Amur by steamer to Khabarovka, and thence to ascend the river Ussuri as far as Kamne-Rybolof, whence there was a post-road to Vladi-vostok. Influenza broke out in the last-named town at the end of May, 1890, and there is some interesting evidence of the importation of infection here, not by road from the west, but by steamer. The disease broke out immediately after the arrival of the first steamers of the Russian volunteer fleet from Odessa, on the opening of the navigation season of 1890. On the convict island of Sakhalin the like fact was observed, the earliest cases being among the keepers of a certain lighthouse immediately after the arrival off the lighthouse of the steamer *Baikal*.

From Vladivostok influenza passed along the rivers Ussuri and Amur, reaching Blagovéshtchensk in July and re-infecting Strétinsk in August. There was thus in Eastern Siberia a double current in the course of the epidemic, firstly from the west, and secondly from the east.

The epidemic followed a more rapid course to the west of its place of origin, but apparently its progress was not more rapid than that of an ordinary traveller. The journey from Petropavlovsk to St Petersburg, *viâ* Tiumen, by rail to Perm, thence by the rivers Kama and Volga to Nijni Novgorod, and thence by rail through Moscow, could be accomplished in eleven or twelve days. From Tcheliabinsk, through Ufa and Samara the journey could be completed in some six days. Between the date of the appearance of influenza in either of these towns and the date of the first cases in St Petersburg at least three weeks elapsed. In the towns along the first route, influenza manifested itself in Perm in the "middle of October," in Kazan in the "middle" or "end of October," in Nijni Novgorod in the "beginning" or "middle of October," and in Moscow about the 18th of the month. Of towns along the second route, Ufa and Samara are the most important; in Ufa the epidemic passed unnoticed as such, but there was a great increase in cases of illness in the town in the early days of October and even at the end of September; in Samara influenza appeared on October 27th.

Influenza seems also to have followed the course of certain rivers. Along the Ural river the disease appeared in Orenburg on November 1st, in Uralsk on the 13th, and in Gurief at the mouth of the river on the 22nd. Along the Volga it appeared in Simbirsk on October 22nd, in Samara on the 27th, in Saratof on the 28th, in Tsaritsyn at the end of the month, and in Astrakhan at the beginning of November. The influenza infection therefore travelled down these rivers in the direction of their current, and the same was observed in the river Lena in the Siberian government of Yakutsk; while the cholera infection, it may be recalled, was in 1892 carried up the two first-named rivers against their current. This difference was obviously due to accidents of geographical position of the towns on the rivers, and not to anything inherent in the nature of the two diseases. In the one case towns on the upper reaches of these rivers, in the other towns at their mouths, were nearer to the place of origin of the epidemic, and were consequently attacked earlier in the course of the disease.

The Caucasus was not invaded until November, and some parts of it not until towards the end of that month or the early days of December. The Transcaucasian Railway is said to have assisted in the spread of the disease from Tiflis. Further west, Rostof-on-Don was invaded on November 16th; Tambof at the end of October; Kief on November 1st; Kharkof on the 5th; Poltava on the 23rd; Vilna on the 10th; Warsaw on October 29th.

It would serve no useful purpose to quote the dates of invasion of other Russian towns. There is complete and more or less rapid railway communication over all this part of the country.

More interesting evidence may be derived from the provinces of Central Asia where there are no railways, except in Trans-caspia, where the roads are few and communication slow, and where, therefore, the epidemic could be more easily followed. In Central Asia influenza appears to have passed from north to south and from west to east, and to have followed the post-roads. It seems to have been carried either from Petropavlovsk or Orenburg to Orsk, where the epidemic began in the early days of November. In Kazalinsk (457 miles to the south in the province of Turgai) the epidemic began on November 28th; in Perovsk (233 miles further along the same post-road) on December 4th; in Turkestan (208 miles) on the 15th; and in Tchimkent (103 miles) two days later. At Tchimkent the road joins another great line of communication running east and west. This is a post-road, beginning at Samarkand on the west (then the terminus of the Transcaspian Railway) and running through Tashkent, Tchimkent, Iliskoé, and other towns to Kopal, thence turning northwards to Sergiopol and Semipalatinsk, where it is joined by other roads from Omsk and Kolyvan on the Moscow Tract. Of the actual course of influenza in Transcaspia and along the Transcaspian Railway little is known with certainty. In Askhabad the disease appeared about the middle of November. Among the troops stationed there the first case was on the 15th. How and whence the disease was introduced into Askhabad there was no evidence to show; but from this important town the infec-tion spread eastwards; it appeared in Kaachka on December 1st, in Samarkand at the end of November, in Tashkent on December 1st, in Tchimkent (as already stated) on the 17th, and in Iliskoé on the 24th. In Viernoé, on a branch road from Iliskoé, the first cases were observed on December 18th, and in Prjevalsk, further along the same road, on January 12th. Another route followed by the infection appears to have been southwards from either Omsk or Kolyvan on the Moscow Tract to Semipalatinsk, where it appeared in the beginning of November; thence on the one hand to Ust-Kamenogorsk, where the first cases occurred on November 17th, and, on the other, through Sergiopol (where there is said to have been no epidemic at all, either among the troops or the civil population) to Kopal, where the malady appeared on

December 11th, and where the road meets that already traced from Samarkand through Iliskoé.

It has appeared to me to be useful, at the risk of seeming tedious, thus to describe in detail the early days of the last great pandemic of this disease, as it establishes in a manner more exact than was possible in the case of any previous pandemic both the approximate place of origin, and the rapidity of spread from the place of origin. Outside Russia it will be impossible here, for sheer lack of space, even if it were necessary, to trace the course of the disease in the same detail.

Westward and southward of Russia influenza spread in the winter of 1889–90 to almost, if not quite, all the countries of Europe. In England, it will be recalled, the epidemic appeared shortly before Christmas, 1889. The Faröe Islands were visited by the disease at a later period; Iceland was attacked in July and August, 1890; and whalers reported that influenza appeared on the shores of Greenland at some date not definitely stated[1].

In Asia, the spread of the disease has been already described in detail for the whole of Siberia and of Russian Central Asia. In Arabia, influenza broke out at Hodeida on the Red Sea coast early in April. In Persia, on the other hand, the disease was prevailing early in January. In India, it appeared as an epidemic early in 1890. In Bombay and Calcutta cases were first noticed in the month of February, the Health Officer of the latter city reporting that the disease was imported there from Bombay. In the Jalaun district of the North-West Provinces the epidemic began even earlier—in the middle of January. The course of the disease in India seems to have been very irregular, but in general terms the epidemic among European troops stationed in India began in March, reached its height in April and then began to decline, very few cases being reported in the latter half of the year. In many places the general population suffered earlier than the troops, while in others the reverse was the case. Burma was affected rather later than India proper. In Ceylon the outbreak began as early as the first week in February,

---

[1] *The Faröe Islands.* By J. Russell-Jeaffreson, F.R.G.S. London, 1898. Whether Greenland shared in the great epidemic extension of influenza subsequent to the autumn of 1889, or whether the statement of the whalers referred to the outbreak of the disease there in May 1889 (already alluded to in the text) is uncertain.

and was apparently caused by the arrival of H.M.S. *Himalaya* with many cases of the disease on board. The earliest local cases occurred among the pilot's boatmen in Colombo harbour.

Farther India seems to have been attacked not much later than India. In the Straits Settlements influenza appeared towards the end of February; it was most active in Singapore, but Penang and Malacca also suffered considerably. The East Indian Islands seem to have escaped with a less severe epidemic than the places just named; in Java a mild outbreak only of the disease occurred, while in Labuan, off Borneo, there was no epidemic of any kind in the year 1890.

In China influenza appeared in the neighbourhood of Swatow in the latter end of February, 1890, attacking first the native population. Hongkong is said to have been affected still earlier. At Chinkiang the epidemic began towards the end of March, and raged violently among the Chinese population. Ichang was attacked some time in April; Hoihow was affected "in the winter and early spring"; Hankow in April (it was believed that the infection spread to this city from Shanghai). In Wuhu influenza appeared in March, and Foochow was visited about the same time. Ningpo was not affected until the autumn, and in some other places the disease did not break out until even later. Korea seems to have escaped altogether. Formosa was less fortunate, the disease appearing in Tamsui and Kelung in April and May, 1890. Japan, on the other hand, is said to have been invaded by the epidemic as early as February of that year.

The whole of Australia was visited by the epidemic. In Victoria cases of influenza were first reported towards the end of March, and the entire colony suffered considerably. In New South Wales and South Australia the beginning or middle of March was the date to which the commencement of the epidemic was ascribed. In Queensland and Western Australia it appears to have begun about a month later. New Zealand and Tasmania were attacked by the disease some time during the month of March.

Returning to the west, the course of the epidemic on the African continent may be very briefly traced as follows. In Egypt, influenza was observed in Cairo and other parts of Lower Egypt as early as November, 1889; in December many places elsewhere were already attacked by the disease, as, for example, Alexandria, Port Said, Damietta, and Ismailia. In January and February outlying places removed from the great centres were attacked, and the disease invaded Upper Egypt, reaching Assouan and Wady-Halfa on the Nile only in March[1]. Along the northern shores of Africa Algiers is said to have first felt the effects of the epidemic in the first or second week of January, 1890, and in Tunis the earliest cases were reported about the same time. On the West Coast influenza seems to have been introduced to Sierra Leone early in March, where it broke out after the arrival of two steamers with cases of the disease on board. In the Gambia there was a very severe epidemic, which caused a high mortality among the natives in the early months of 1890. Lagos and the Gold Coast were apparently affected somewhat later, and in Senegal the epidemic was at its height in the month of July. In the Cameroon district, on the other hand, influenza is said not to have become epidemic at any time.

South Africa was visited by the disease even earlier than many places on the West Coast. Thus in Cape Town influenza broke out in the first week in January, 1890, introduced, it was believed, by steamers from Europe. The date of the commencement of the epidemic in Natal is quite uncertain, and some medical men even denied that there had been any epidemic there at all. Others again asserted that influenza began to be epidemic as early as October, 1889, or before it began to prevail in Europe outside Russia. This was the date assigned by one observer for the outbreak in Durban, but the evidence of other observers in that town was most conflicting. The disease seems to have reached Basutoland about the end of March and to have spread northwards. British Bechuanaland was attacked at the end of April, the disease being thought to have been imported from Cape Colony. As late as October, 1890, influenza became epidemic in the Shiré Highlands in Eastern Africa.

Of islands off the coast of Africa it is to be noted that in May, 1890, feverish colds " apparently representative of the influenza in a modified form " were reported from Orotava, in the Canary Islands; while as early as January of that year an outbreak of influenza occurred on H.M.S. *Australia* while at St Vincent in the Cape Verde Islands. The Azores were also visited by the

[1] *L'Épidémie d'Influenza en Égypte pendant l'Hiver* 1889–1890, par Dr Engel Bey. Le Caire, 1894.

disease, while St Helena seems to have escaped until the month of August 1890. In Mauritius influenza appeared as an epidemic in July 1890, or according to some observers not until August. Rodriguez and Réunion, Madagascar and the Seychelles, were all visited by influenza in that year, but with varying degrees of severity.

The Western Hemisphere was reached by influenza almost, if not quite as soon as the western shores of Europe. In Jamaica the disease was recognised as epidemic on December 6th, 1889, or some time before it had begun to be epidemic in England. In New York the earliest case was said to have occurred on December 11th. In Boston the disease broke out in the middle of December, and on the 17th of that month 30 prisoners in the Suffolk County gaol were attacked by it within a period of twenty-four hours. On the 19th influenza had appeared at Buffalo, at Detroit, and at Kansas City; next day it was reported as occurring among post-office employés at Rochester, N.Y., and a week later it had broken out at Philadelphia, Baltimore, and many other places in the United States. The Southern States were attacked rather later than the more northerly ones, and it was thought that a low temperature increased the mortality caused by the disease if it did not favour its spread. In St Louis and New Orleans the epidemic first appeared about January 1st, 1890. Its crisis, as shown by the highest point reached by the mortality from all diseases and from respiratory diseases in particular, seems to have occurred in Boston, New York, and Brooklyn in the week ending January 11th; in Cleveland, Providence, and Minneapolis in the following week; in Charleston and Washington in the week ending January 25th, and in St Louis, Milwaukee, and San Francisco in that ending February 1st.

Canada also suffered severely from the first great wave of pandemic influenza in the winter of 1889–90. In Ontario the date of the earliest case was given as December 4th, 1889. In Quebec the disease was even said to have been observed as early as November 20th, but it did not become epidemic until the middle of the following month. In Manitoba the first noted occurrence of influenza was on December 23rd; in New Brunswick it was believed that the epidemic began "about October or November 1889." In Vancouver Island December 22nd was the date assigned for the first cases. The North-Western Territories were attacked apparently some time in December 1889 or January 1890, and both here and in British Columbia the "Indians," who live in separate bands or communities, scattered throughout a wide extent of country, suffered greatly from the epidemic. Newfoundland was apparently invaded in February.

In regard to the West Indies the early occurrence of influenza in Jamaica has been already mentioned. In some other islands of the group the date of the appearance of the epidemic is less certain. In Trinidad it was even denied that the disease had occurred at all, although a disorder very like influenza is said to appear there every year in the months of January and February. Whether an outbreak which occurred in the Trinidad Leper Asylum early in 1890 was of the nature of this local, annually recurring disorder, or whether it was part of the great pandemic wave of influenza has never been determined. At Tobago the malady appeared about the end of January; at Antigua some time in the same month; at St Kitts and Nevis early in February; at St Vincent at the end of January; while at Barbados it was not until the middle of April that the disease broke out. The Bahamas, Grenada, and St Lucia were believed to have escaped, and the Bermudas were equally fortunate, with the exception of Ireland Island, where a very sharp epidemic occurred among the officers and men of a number of Her Majesty's ships lying in the harbour. The infection was imported to the island by a Swedish corvette, the *Saga*, and the outbreak was almost confined to the harbour, only eleven cases occurring on shore.

In Mexico influenza was reported to be present from January 17th, 1890. On the last day of that month it had also appeared in the town of Guatemala, 300 miles from the Atlantic coast. British Honduras is said to have escaped altogether in 1890, though in subsequent years it has been less fortunate.

The course of the epidemic in South America during the first year of its prevalence was very erratic, though as a whole this great continent seems to have suffered from the ravages of the malady no less than the others already briefly considered. In British Guiana the disease first made its appearance in Georgetown in the week ending August 2nd, 1890, yet long before this it had invaded the greater portion of the remainder of the continent. Thus, in Chile, on the west, influenza was prevalent from the month of January, and in Buenos Aires and Monte Video on the east from February 2nd. From the latter it was believed to have spread northwards to Brazil. In Peru, Callao was invaded in April; and in the same month the disease appeared in Guayaquil, Cucuca and Quito, in Ecuador. In that country the course of the epidemic was said to have been from south to north. Finally in September and October 1890, a severe epidemic of what was described as a "hybrid" of influenza and whooping-cough prevailed in the Falkland Islands, particularly in the East Island.

Such, in barest outline, was the course followed by the great pandemic wave of influenza, which began in the autumn of 1889, and in the course of a few months spread over almost the whole globe—to China and Japan in the far east, to California and Chile in the west, to Iceland and northern Canada in the north, and to New Zealand, the Cape, and (perhaps) the Falkland Islands in the south. That every portion of the inhabited lands within this vast area—which practically comprises the entire earth's surface—was visited by the disease, and that every separate community within those lands suffered from its attacks cannot, however, be asserted. There were many islands, and even countries of considerable size, which seem to have escaped its ravages, at any rate during the first year or two of its prevalence. Some of these have been already named. But it is of importance to note that many of those geographical areas which remained untouched for a considerable period after neighbouring areas had been invaded by the disease, were themselves invaded at a later period, and in many instances suffered from recurring epidemics of very great severity. Whether any geographical area of considerable size has remained *entirely* free from influenza from the year 1889 up to the present date is not easy to state. For the present it appears open to doubt, though possibly certain towns or villages, or even communities and inhabited areas of still larger size, may have been so fortunate as to escape its ravages altogether.

The subsequent history of the recent great pandemic of influenza, which perhaps has not yet come to an end, cannot be told in detail here. Few of the countries invaded by the disease when it first swept over the earth's surface in the winter of 1889–90 have not suffered from later recurrences, as severe as, or more severe than the first outbreak. In some countries there has been a tendency to annual recrudescence of the disease, which has in many instances made its reappearance in either the spring or the autumn. But it is impossible to assert that influenza follows any definite rule, either in relation to season or in regularity of recurrence. Epidemics have occurred in the depth of winter in many instances, and recrudescences in the summer months are not wholly unknown. In England the winter seems to have been more especially favourable to the renewed prevalence of the disease. On several occasions there has been observed a tendency towards recurrence of influenza at or about the same time at widely separated parts of the earth's surface; but this may be explained provisionally by supposing that the recurrence of similar seasonal, climatic, or other conditions in these widely separated areas favours the revival of an infection which has still remained from the preceding outbreak of the disease.

LEPROSY.

History. There can be little doubt that leprosy is one of the oldest of diseases. The references to its existence in ancient and medieval writers are often obscured by the uncertainty and confusion which existed between true leprosy and a variety of other disorders, but it is none the less certain that leprosy is of great antiquity.

Throughout Biblical times—at least from the date of the Jewish exodus from Egypt—it was a familiar disorder in the countries that would now be spoken of as the Near East. An almost equal antiquity is claimed for the disease in India and China. It is mentioned by some of the ancient Greek writers. Through the early medieval period it was no doubt a comparatively common disorder; and special laws for its control, and asylums for the reception of lepers, are known to have existed in several European countries as early as the eighth and ninth centuries of the Christian era. A little later the disease appears to have been prevalent in Europe to a truly terrible extent, for between the 11th and 13th centuries leper-houses in vast numbers were established. How far this increase in the number of leper-houses is to be taken as implying a real increase in the prevalence of the disease is uncertain; and equally uncertain is the influence in spreading the disorder which has often been ascribed to the Crusades. In recent centuries leprosy has greatly diminished in Europe. It has, however, remained endemic in large areas of Asia and Africa, and has been newly introduced to regions of the world where it was unknown before, such as many parts of Australasia, South Africa, and North America.

The *bacillus lepræ*, which is associated with the disease, was first observed in the tissues of lepers by Hansen in 1873, some ten years before Koch published his discovery of the closely allied, but probably distinct bacillus of tubercle.

Recent Geographical Distribution. *Europe.* In Europe leprosy is no longer a widely prevalent disease. The countries containing the largest proportion of lepers are Spain, the countries in the Balkan peninsula, and Iceland.

In the British Isles leprosy is extinct as an indigenous disease. A small number of cases are known to exist, either in an active or cured condition, but it is believed that in all the infection has been contracted outside these islands. They probably do not exceed a score.

In France a certain number of lepers are found on the south coast, on the shores of the French Riviera[1]. They are said to be most numerous at Contes, Eza, Peille, and La Turbie, just above Monte Carlo. Sporadic cases are also met with in Brittany. In Paris it is believed that there are a considerable number of persons affected with the disease; one estimate has placed the number at between 160 and 200. All have, so far as is known, contracted the disease abroad. The appreciable risk which these unfortunate people afford to their neighbours has been recently the subject of a protest by M. Hallopeau.

In Holland 17 lepers were known to exist in 1897.

In Germany leprosy is known to exist in Memel, in the extreme north-east corner of the country, close to the Russian border, and on the shores of the Baltic Sea[2]. The disease was introduced here from Russia about the year 1870. In all there have been 27 cases in Memel, of which 17 have ended fatally.

In Hamburg there were ten lepers in 1898, of whom 5 were born in Germany, 2 in Portugal, and 1 each in Sumatra, Mexico, and Brazil. Seven were infected in Brazil, and one each in Mexico, Penang, and the Philippines. No spread of the disease from person to person has been seen in Hamburg. One leper was also reported in Mecklenburg-Schwerin in January 1900, but the other German States were quite free from the disease.

In Belgium only 4 lepers, and in Denmark not more than three were known to exist a few years ago[3].

---

[1] Jeanselme. *Report of the Leprosy Conference in Berlin.*

[2] R. Koch. *Die Lepra-Erkrankungen im Kreise Memel.* Jena, 1897.

[3] Ehlers. *Janus*, 1898, p. 140. "The Geographical Distribution of Leprosy."

In Norway, at one time the most active centre of leprosy in Europe, the disease has steadily declined in recent years, and this change has been attributed to the effects of the strict segregation which has been practised in that country since the isolation law of 1885 came into force. In the year 1856 there were believed to be 2833 lepers in Norway, but in 1896 their number had fallen to 681, and in 1898 two leprosoria were closed and converted into hospitals for the treatment of tuberculosis. The principal centre of the disease is said to be still in the neighbourhood of Bergen, just as it was in the Middle Ages[1].

In Sweden there are believed to be from 70 to 75 cases of leprosy, of which some 30 are isolated. Of this number 36 live in the province of Helsingland and 15 in Dalecarlia[2].

Russia has always been believed to contain a large number of lepers. Recent investigations have shown that, while the infection is widely spread throughout the country, the actual number of known lepers is not large in proportion to the area and population of the country. There is reason, however, to believe that many more cases exist than are known to the authorities. It will be convenient to consider both Asiatic and European Russia together. It is only since the year 1887 that accurate knowledge of the distribution of, and numbers suffering from, the disease in the Russian Empire has been available. Between that year and the year 1895 information of the existence of leprosy has been received from every portion of the empire, with the exception of the following governments : Vilna, Kalisz, Lomza, Lublin, Siedleç, Radom, Novgorod, Penza, Tambof, Tula and Elizavetpol, and the provinces of Semipalatinsk and Turgai in Central Asia. In these governments and provinces leprosy has apparently always been non-existent.

In a large number of other governments and provinces, such as those of Archangel, Volhynia, Kovno, Warsaw, Kielçe, Piotrkow, Ploçk, Suvalki, Olonetz, Simbirsk, and Yaroslavl in European Russia; Kutais and Daghestan in the Caucasus; Akmolinsk and Semirétchinsk in Central Asia; and Tobolsk and Tomsk in Siberia, the number of lepers recorded is small and does not exceed one, two, or three in a few scattered towns or villages. From several of these areas, also, it is to be noted that the disease is apparently disappearing, and during the last five years for which

[1] *Ibid.*   [2] Sederholm (at the Berlin Conference).

returns are available (1891–5), the following governments have been free from recorded cases of leprosy: Archangel, Mogilef, Kielçe, Suvalki, Olonetz, Kutais, Yeniseisk, and Akmolinsk.

There remain 14 governments and 9 provinces in which leprosy is an endemic and more or less widely prevailing malady. These territorial divisions, in which leprosy is most rife in Russia, are to some extent grouped around the shores of various seas or inland lakes, particularly the Baltic Sea, the Gulf of Finland, Lakes Peipus, Pskof and Virtz-Jarvi (in Livonia), the Black Sea, Sea of Azof, the Caspian and Aral Seas, the Sea of Okhotsk, and the Lakes of Tchaldyr (in the government of Kars), of Goktcha (in that of Erivan) and of Baikal (in that of Irkutsk, Siberia). The exceptions to this rule are certain centres of leprosy in the Batalpashinsk division of the Kuban province, the Piatigorsk and Naltchinsk divisions of the Terek province, and the Stavropol government (all in the Caucasus); and the Samarkand and Ferghana provinces (in Central Asia).

As the result of a special inquiry made by the Medical Department of the Ministry of the Interior in 1895—1897, the most important centres of leprosy in Russia have been grouped as follows :—

1. *The Baltic Province Group.* In the provinces bordering on the southern shores of the Baltic Sea 564 lepers were registered in the years named. The ratio they bore to the total population was 12·8 per hundred thousand. This part of the country would appear, from these figures, to be more deeply tainted with the disease than any other, but the returns from Eastern Siberia were extremely imperfect, and it was thought probable that leprosy was in fact more rife in this latter region than on the Baltic. Among the Baltic provinces Livonia occupied the first place with as many as 26 lepers per 100,000 inhabitants ; Courland had 19·9, Esthonia 10·9, and the government of St Petersburg not more than 2·2 in the same unit of population.

2. *The Black Sea and Azof Sea Group.* In this group 235 lepers were registered, or 1·8 per 100,000 inhabitants. The Kuban province returned 6·2, and the Territory of the Don Cossacks 3·2 lepers per 100,000 of the population. While it is true that these areas as a whole border on the seas named and are tainted with leprosy to a considerable extent, it has also to be remembered that the principal centre of the disease in the Kuban province is,

as was pointed out above, in the interior of the country, and that in the governments of Bessarabia, Taurida, and Kutais, all of which border on the Black Sea, the number of lepers recorded was exceedingly small.

3. *The Caspian Sea Group.* The number of lepers registered here was 141, equivalent to 5·1 per 100,000 persons. Astrakhan was the worst government in this group, with a ratio of 7·6 lepers per 100,000, the Terek province coming second with a similar ratio of 5·5. Daghestan, at one time thought to be a centre of leprosy, appears on the contrary to be almost free from the disease. In the recent special inquiry alluded to above no return was received from Daghestan, but in the ordinary annual reports of the Russian Medical Department only single cases were returned from here in 1890, 1894, and 1895, and none in 1891–1893. In the government of Baku, which lies on the western coast of the Caspian Sea, a considerable number of lepers were registered.

4. *The Aral-Caspian Group.* In this group only 54 lepers were recorded. This was equal to 2·2 per 100,000 inhabitants. The Syr-Daria province returned figures equivalent to a ratio of 2·7. It is, however, pointed out that in all probability a large number of lepers exist in these regions who never come under the observation of medical men (from whom the returns here quoted were, in every instance, obtained), and that presumably the number of persons suffering from the disease is very considerably larger than the returns indicate. There is also a great paucity of medical men in Transcaspia, Uralsk, and the other Central Asiatic provinces, and it is impossible for them to have an accurate knowledge of the number of lepers in the very large areas under their care.

5. *The Arctic and Pacific Ocean Group.* In this enormous territorial area, which includes the whole of northern and eastern Siberia, only 60 lepers were recorded in 1895—1897. But so sparse and scattered is the population of these Siberian border provinces that this figure represented a ratio of 12 lepers in every 100,000 inhabitants. The figure, it must be added, is much below the truth. Returns were received from but a few of the larger towns, and from the annual reports of the Russian Medical Department it is quite clear that a very much larger number of lepers exist here than the (imperfect) special return appeared to

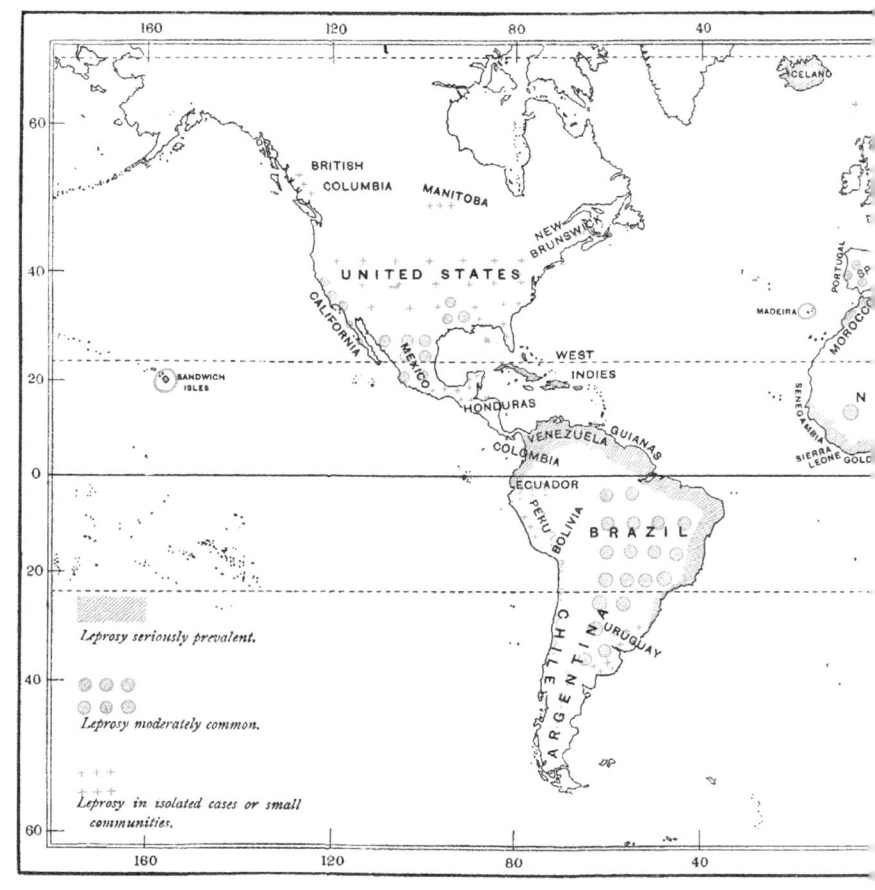

DISTRIBUTION OF LEPROSY

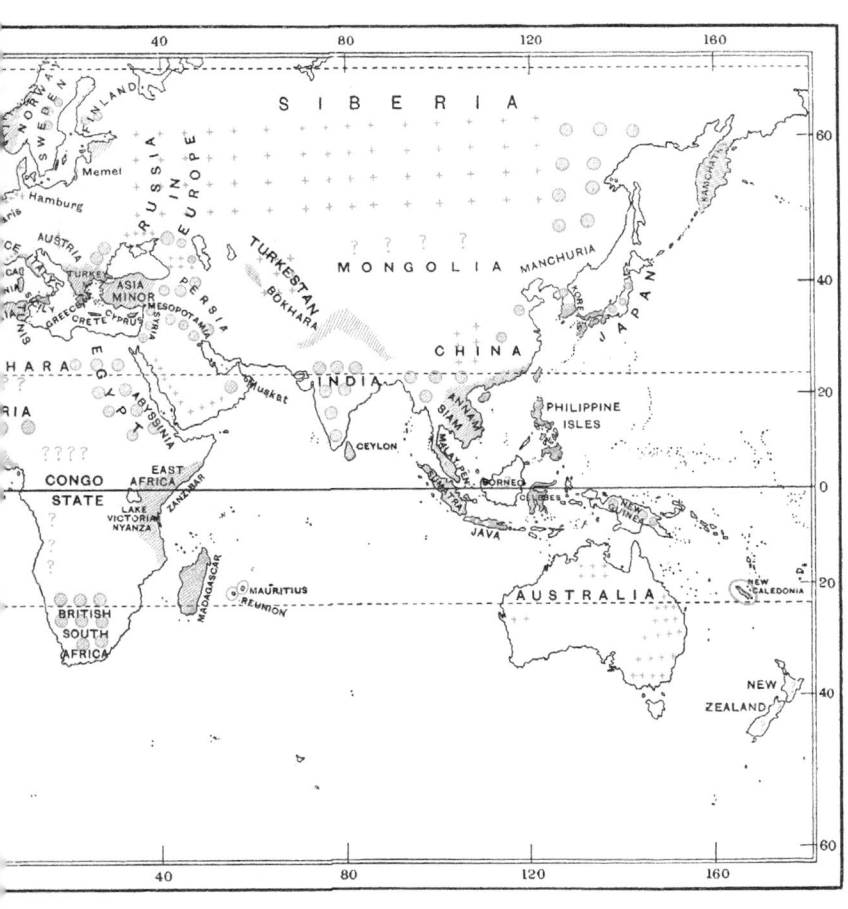

indicate. So little is truly known of the nomadic and other indigenous tribes in these far-away regions that it is impossible to estimate even approximately the extent to which eprosy or any other disease prevails amongst them.

6. *The Lake Baikal Group.* In the Irkutsk government the number of lepers recorded was equivalent to 2·2 per 100,000 inhabitants. In the Transbaikal province, whence no special returns were received, from 1 to 7 lepers have been recorded in each annual report of the Russian Medical Department since the year 1887.

There remain to be mentioned certain centres of leprosy in Russia which show no grouping round the shores of seas or lakes. Thus in the inland government of Kars, in Transcaucasia, the number of lepers recorded in 1895—1897 was equal to 7·4 per 100,000 inhabitants; in that of Erivan the ratio was 3·0 per 100,000 inhabitants; while in that of Tiflis only a few isolated cases were returned. In Central Asia, again, 42 cases of leprosy were registered in the provinces of Samarkand and Ferghana, both far removed from any sea or lake. The ratio to the population was 1·8 per 100,000. In these two provinces lepers are mostly congregated into what are termed *makhaukhans*, or leper colonies.

Finally in the other Russian governments and provinces not yet named only 37 lepers were found to exist in 1895–1897 ; and as the population of these areas is very little short of 30,000,000, it will be seen that the proportion of persons affected by leprosy in them is infinitesimal. Many of these cases of the disease have been shown to have been imported from elsewhere where leprosy is endemic. The most frequent agents in thus transporting this, and indeed many other diseases, are those peasants who annually leave their homes to obtain work in other parts of the country, and, on returning, bring the infection with them.

In the entire Russian Empire it was found by the special return made in 1895—1897 that just 1200 lepers were known to exist, of whom 633 were males and 567 females. But from what has been said above as to the trustworthiness of the returns, it cannot but be held that the actual number of lepers in Russia largely exceeds that figure. Prof. Petersen, at the Berlin Conference on leprosy held in 1897, stated that he estimated the total number of lepers in Russia at about 3000.

In Finland the number of lepers in 1893 was estimated at 51 (33 males and 18 females); in 1896 the number had risen to 67 (42 males and 25 females)[1].

In Austria-Hungary leprosy is believed to exist only in the form of isolated, sporadic cases.

In Spain leprosy is apparently widely spread, though accurate knowledge of its distribution and of the numbers affected appears to be wanting. It is said to be commonest on the south and east coasts, and particularly in the provinces of Valencia and Alicante, both on the east coast. There are villages in these provinces where as many as ten or twelve families of lepers are known to exist. The disease is apparently increasing in Spain as in Italy; for. in the province of Valencia the number of known lepers was returned in 1888 as 69, and these were distributed in 13 parishes, while in 1893 the number had increased to 120, and they were distributed in 20 parishes. Enforced segregation is not attempted in Spain, and the overwhelming majority of lepers in that country live among the general population without restraint. A leprosorium exists at Granada, and another (with accommodation for 40 inmates) at Seville, and it may consequently be assumed that the disease is common in the southern provinces, of which these are the principal towns. The existence of leprosy in Spain has been attributed to importation of the infection some fifty years ago from the Philippine Islands. In Portugal the provinces of Beira, Algarve, and Estremadura all contain lepers. One observer has reported details of 466 cases of the disease in this country[2], but many more are believed to exist.

In Italy there are said to be a considerable number of scattered foci of leprosy. Lepers are not infrequently met with in the Italian Riviera, particularly in the neighbourhood of San Remo; at Turin 13 Piedmontese suffering from the disease have been seen since 1890; in the neighbourhood of Naples 15 leprous families were known to be living some years ago; and in 1886 eighteen lepers were counted by one observer at Albertello in Apulia. There is reason to believe that the disease is increasing in Italy. Up to 1891 the annual recorded deaths from leprosy in that country had diminished, as the following figures show[3].

In 1887 the recorded deaths from leprosy were 22 and the

[1] Fagerlund.   [2] Falcao.
[3] *Annuario Statistico Italiano*, 1898.

ratio per million inhabitants was 0·7; in 1888 the deaths and ratio were respectively 15 and 0·5; in 1889, 19 and 0·3; in 1890, 7 and 0·2; while in 1891 only a single death from leprosy was registered. But in the succeeding years the corresponding figures have been as follows: in 1892, 4 and 0·1; in 1893, 3 and 0·1; in 1894, 5 and 0·2; in 1895, 8 and 0·3; and in 1896, as many as 17 persons died in Italy from leprosy, or 0·6 per million inhabitants. There is said to be an entire absence of precautionary measures against the spread of the disease on the part of the people.

The countries of the Balkan peninsula are also considerably tainted with leprosy. Ehlers, who a few years ago made a personal investigation into its distribution in this part of Europe, has published the following facts[1]. Roumania is the only Balkan State which issues trustworthy statistics, and in that country, according to the most recent reports, there were 208 lepers. In Servia, on the other hand, only three cases have been recorded[2], and the same number are said to exist in Bulgaria. It is almost certain, however, that the disease is much more widely prevalent in these countries than these figures would imply. The same remark also applies to the official return of 133 cases of the disease in Bosnia and Herzegovina. In Montenegro it has been estimated that about 100 persons, out of a population of 227,000, are afflicted with the malady. In Turkey leprosy is common enough; in Constantinople alone von Düring counted 258 lepers, and estimated their total numbers at between 500 and 600. The disease is particularly common among the Spanish Jews living in this city. There is a leprosorium at Scutari, which contained some 35 patients in 1897 and 30 in October, 1902. There are a few lepers also on Mount Athos, the mountain of the monks, in the Ægean Sea.

In Greece the number of persons affected by the disease is known with no greater certainty than in the other countries just mentioned, one recent estimate placing it as low as 99, and another at between 400 and 500. It is probable that the truth lies between the two extremes, but Prof. Hadji Michali of Athens does not think it can exceed 200.

In some of the outlying islands in the neighbourhood of the continent of Europe leprosy has not only been endemic

[1] *Bulletin de la Soc. Franç. de Dermat. et Syphilog.* Juin, 1897.
[2] Lazarewitch (Belgrade).

for a very long period, but it has shown far less tendency to diminish here than it has on the continent itself. Thus Iceland has been an important centre of the disease at least since the thirteenth century, and it would appear that the number of lepers in the island, though less than in medieval times, is increasing rather than decreasing. Ehlers, who visited Iceland in 1889 and again in 1895, found that the number of affected persons was at least three times as great as the official records implied[1]. Thus in 1889, while the official returns mentioned 47 cases only, he himself counted without any special search as many as 144. Six years later 15 additional cases were seen, bringing the known number up to 159; and it was believed that the true number of lepers in the island was probably about 200. Four miserable leper asylums which existed before 1848 were abolished in that year, as the disease was thought to be declining, but the diminution, which was apparently due to the violent epidemic of measles which in 1846 carried off one-third of the whole population, was only temporary, and, as pointed out above, the number of lepers in Iceland is now increasing. In the year 1895 a law enforcing segregation of lepers in their own homes was passed. Recently (1898) a new law has been enacted providing for the compulsory notification of every fresh case, and a new hospital has been constructed at Reykjavik for the segregation of at least a certain number of lepers.

In the Faröe Islands leprosy is extinct, and has been so since the middle of the eighteenth century. The fact is a remarkable one, for the diet of the inhabitants of these islands and of the Icelanders is said to be precisely the same, while in regard to climate and conditions of life the differences, if any, should be rather more favourable to the prevalence of the disease in the Faröes than in the more northerly island. In the Shetland and other islands to the north of Scotland leprosy has also for long been non-existent.

Many islands in the Mediterranean are endemic homes of the disease. It exists in the Balearic Isles. Corsica and Sardinia are both affected by it. In Sicily 153 cases of leprosy were reported in the year 1888, but it was believed that this figure was far below the truth. In Elba, Malta, and Gozo leprosy is far from un-

---

[1] Prize Essays on Leprosy. *New Sydenham Society*, Vol. CLVII, 1895. "On the Conditions under which Leprosy has declined in Iceland."

common. In Malta it was the cause of six deaths in the year 1898. Crete contains a large number of lepers[1], and it is believed that the recent disturbances in that unfortunate island led to an increase of the disease there and a diffusion of it to adjoining countries, particularly Greece. Thus Ehlers himself saw, in Athens and the Piraeus, 26 cases of leprosy in fugitives from Crete immediately before the outbreak of the Greco-Turkish war. In Cyprus lepers are also numerous, and they are apparently increasing, for in 1889 the number segregated on land set apart for the purpose near Nicosia was 63, and this had increased to 65 in 1890, and 74 in 1891.

Asia. In Asiatic Turkey leprosy is a common disease. It is widely spread in Mesopotamia, Syria, and Asia Minor. One estimate of the number of lepers in Asia Minor has placed them at one per thousand of the whole population[2]. In Persia the disease is endemic, but is—or was thirty years ago—confined to certain districts of the hill country of Irak Ajemi, in the provinces of Azerbijan and Khuzistan, and particularly in the localities of Chamsé, Kaswin, Sendjan, and Karadagh[3]. It is said to be absent from the Persian shores of the Caspian, although, as already pointed out, it is certainly met with in the Russian provinces bordering upon that sea. Leprosy exists to some extent in Arabia, and particularly at Muskat.

The prevalence of the disease in Asiatic Russia has been briefly described above, and it is unnecessary to repeat what has been said concerning its distribution in the immense areas of the Asiatic continent which form part of the Empire of the Tsar. It may, however, be added that the peninsula of Kamchatka forms an important centre of leprosy, and that some of the indigenous tribes of Eastern Siberia, such as the Yakuts, are said to be afflicted with the disease to a very serious extent. Bokhara also contains a large number of lepers, particularly in the neighbourhood of Sharshaouz and Kitab. Near the first-named place are two *kishlaks*, or settlements, set aside for these unfortunate persons. Lepers in Bokhara are avoided by other people, who regard the disease as a scourge imposed by the Divinity as a punishment for grievous sins. All persons suffering from it are

---

[1] Ehlers and Cahnheim reckoned their number in 1901 at about 600 in a population of 280,000. The Greeks rather than the Mohammedans suffer.

[2] Von Düring.   [3] Hirsch.

sent to one of the leper *kishlaks*, where apparently their sole or main means of sustenance consists of such alms as they can obtain on bazaar days, when they are allowed to visit the principal towns to beg from the charitable[1].

In Chinese Turkestan leprosy is believed to be rare or absent, neither the Chinese, Kirghiz, Mongols, nor Manchus suffering from it. The Turkomans and others in the plains and desert oases of the north seem to be free from the disease, but south of the Celestial Mountains it is not uncommon[2].

India has from time immemorial been one of the principal centres of leprosy in Asia. At the present day the disease is common throughout the greater part of the peninsula. Its distribution, however, is far from even, and while there are areas almost free from its ravages, there are others where it is among the most prominent of diseases. It is particularly prevalent in what has been called the middle belt of the Himalayas, and the native tribes in the Simla division are said to suffer from the disease to a greater extent than the inhabitants of any other part of India[3]. Kangra, Kulu, and the native States of Mundi, Suket, Bilaspur, Sirmore, Chamba and others also contain a large number of lepers. It is believed to be increasing here ; and it is noteworthy that no caste or religion appears to be more susceptible than any other. In Kashmir, according to the census of 1890, the number of lepers was 202[4]. For some unexplained reason the real Kashmiris are said to be comparatively exempt from the disease, and a considerable portion of the lepers in Kashmir are natives of the Punjab who flock to the cooler country in the hot season, when the disease becomes more troublesome. The Gujurs, a tribe of cowherds living on the Himalayan slopes, are said to suffer greatly from leprosy. In other parts of India the malady is common, but is apparently decreasing, the number of lepers returned in the census of 1891 being below that of 1881. The highest ratios in 1891 were found in certain districts of Bengal (*e.g.* Bankura, and Birbhum), in the Simla district, in Dehra Dun, in Kumaon and in some others. The members of the Indian Leprosy Commission (1890–91) found that the disease prevailed mostly inland, and

[1] Grekof.
[2] Raemdonck. *Report of Leprosy Conference in Berlin*, 1897.
[3] Carleton. *Transactions of the First Indian Medical Congress.*
[4] Mitra. *Ibid.*

away from the great rivers, and that the lowest ratios were those in the Gangetic valley, in the Doabs, and along the west coast from Karachi to Ahmedabad.

In Ceylon leprosy is far from uncommon. The leper asylum at Hendala admitted 108 fresh patients in the year 1897, which with 246 remaining over from the previous year made the total number of leper inmates 354.

In the Malay Peninsula lepers are very frequently met with. No trustworthy statistics as to their numbers are available, but it has been pointed out that certain unwritten social rules and the knowledge of the disease possessed by the natives, even by those in the most secluded spots, indicate a wide prevalence of the disorder. The Malays themselves suffer from it, but not to the same extent as the Chinese immigrants who have settled in the peninsula. Cantlie has strongly urged the view that throughout Indo-China, the Malay Peninsula, and the East Indian islands, leprosy has been introduced and maintained principally, if not solely, by the immigrant Chinese.

"The Chinese," he writes, "mix very little with any but their own people, and though leprosy is undoubtedly endemic in some parts of the Straits its comparative rarity among the Malays is hard to account for except on the theory that the Chinese introduced it, and by their exclusive habits have not spread it as widely as they might otherwise have done.

"To look at the matter from another point of view, the Malays and the aborigines seem comparatively free from the scourge, yet the Sandwich Islanders, their relations in blood, are the most leprous people on earth. How can this anomaly be accounted for? It may be, though it is only a suggestion, because the habits of the Malays and Hawaiians are so different. The Malay is a born rover, he is always coming into contact with other people, and his constitution is inured to change. The Hawaiian on the other hand is an isolated being who till less than one hundred years ago never saw a soul of his own race. Such isolated communities are always prone to new diseases, as we see even on the little island of St Kilda, and hence the Hawaiian constitution was more fitted for the reception of leprosy than that of the Malay[1]."

There can be little question that, whether the disease is mainly

---

[1] Prize Essays on Leprosy. "Report on the Conditions under which Leprosy occurs in China etc." *New Sydenham Society.* 1897.

or solely maintained in these regions by the Chinese or not, it is a comparatively common disease in the Straits Settlements. In the hospitals at Singapore 65 fresh cases of leprosy were admitted in the year 1898, and 26 lepers died. In that at Penang 81 fresh cases were admitted, and in the leper asylum at Pulo Jerejak, in Province Wellesley, the number of.fresh cases in the same year was 180. In Perak leprosy is widely prevalent both among Malays and Chinese, but the aboriginal race of Sakais are not known to suffer from it. The same immunity appears, to be shared by another aboriginal tribe, the Jakuns, in Johore, for, though leprosy is said to occur amongst them, Wilson, who repeatedly visited them, never succeeded in seeing a case. In Johore the Chinese suffer more than the Malays. In Muar leprosy is not uncommon, and here the Chinese also suffer most, the Malays being rarely affected, while concerning the aboriginal races no evidence is forthcoming. At Sungei-Ujong, where the disease is rare, only the Chinese are affected and the Malays escape it entirely.

In Siam leprosy is said to be rife in parts, and even to prevail to an alarming extent. There are no leper establishments, at any rate in southern Siam, and "every hamlet has its leper and every village or town its leper quarter." In the Laos country, too, leprosy is a very common disorder.

Among the Annamites of Cochin China the disease is widely prevalent. Segregation is extensively practised, indicating perhaps a prolonged acquaintance with the malady, though whether the measure is successful in checking its spread does not appear. Leprosy is found equally in the mountains and the plains of Cochin China.

In the islands of the East Indies this disorder appears to behave very much as it does in the adjoining Malay peninsula. In Sumatra it is common, particularly on the north-east coast, where about 1000 lepers are believed to be in existence, the number on the south-west coast being only 156. The native Malays on the coast are said to be little affected, and the principal sufferers here, as throughout the Far East, are the Chinese. There is a leprosorium at Medan, near Dili, on the north-east coast, for the segregation of coolies from the plantations. In the interior of the island there is said to be much leprosy among the Battaks, and also among the Tamil immigrants from Madras, as well as among the Javanese and Chinese. More than one per cent. of the Chinese are said to be lepers[1].

In Java the disease is possibly less common; one observer[2], at least, never saw a case there. On the other hand Albricht, a Dutch physician, is quoted as saying that it is rife in the island, and particularly in the plains[3], and he is supported by Broes van Dort, who states that the central part of the island contains from 30 to 38 lepers, the western portion 42, and the eastern portion the large number of 2703. The latter adds that the disease has assumed dangerous proportions since the Dutch introduced European civilisation, as formerly the natives killed all persons attacked by leprosy or exposed them to die of starvation.

In the Molucca or Spice Islands, lying close to the equator, leprosy is very prevalent, especially at Amboina and Ternate.

Borneo appears to be almost, if not quite free from the disease[4]. In 1888 there were said to be no lepers in the island, but in the next two or three years several cases developed among Chinese immigrants, who were imported in large numbers from the Kwangtung and Fokien provinces of China (the "cradle of leprosy" in the East) to work on the newly-opened tobacco estates. Later, coolies were imported from Singapore, Java, and Sumatra. In 1890 the first case of the disease was reported, and a palm-mat shed was built on an island to receive lepers. It never contained more than ten inmates, all of whom, except one Portuguese convict, were from southern China. In 1893 the lepers were all deported and since then the north of the island has apparently been free. Leprosy is also believed to be absent from the south of the island.

The distribution of leprosy in the Chinese Empire is a matter of great interest. It has been studied fully by Cantlie, from whose essay the following facts are in the main gathered. In Manchuria the disease is said to be unknown as an endemic, the few cases met with occurring in Chinese from the south. It is also said not to exist in Peking[5] although that city is full of Manchus, Tatars and other Mongolian tribes, in addition to Chinese proper; and on these grounds it is believed to be absent

---

[1] Martin. *Archives de Méd. Navale.* June, 1899.
[2] Kohlbrügge.   [3] Quoted by Cantlie.
[4] Skertchley. (Addendum to Cantlie's Prize Essay.)
[5] Dudgeon, quoted by Cantlie.

from the province of Chih-li, in which Peking is situated. It can scarcely escape notice, however, that these are very slender grounds upon which to base a negative conclusion as to its presence in northern China. The author quoted has adduced, in support of this view, the case of Eastern Siberia and Sakhalin, which, he concludes, from the absence of the malady in Vladivostok and at the mouth of the Amur, are also free from it. It has already been pointed out above that, as regards Siberia, leprosy is endemic throughout the whole of the eastern portion of this great territory, and though it appears to be absent from the island of Sakhalin—for many years past a place of exile for Russian criminals—it is not uncommon, as will be shown later, in Corea, which is much nearer these northern Chinese provinces than Sakhalin. It must therefore be taken, at present, as not quite proven that leprosy is entirely absent from these provinces. Raemdonck, however, quotes the statement of missionaries that neither the Chinese nor Mongolians are affected by leprosy in Mongolia.

In the province of Shantung, on the Yellow Sea, one or two lepers are said to be met with in every village, at any rate in the neighbourhood of Wei-hsien. In the coast provinces immediately to the south, however, namely those of Kiang-su and Chekiang, leprosy is believed to be not indigenous, and is only met with very rarely, if at all. It is believed also to be unknown along the course of the great Yangtze-kiang, as otherwise lepers from there would come for treatment to Shanghai, yet, according to Jamieson, the disease is "the rarest of rarities" in this city.

In the inland province of Hu-peh leprosy is known, but is no great scourge; it is at any rate rare in Hankow, and has not of recent years been mentioned in medical reports from Ichang. In the adjoining province to the west, that of Sze-chuen, it is also rare, occurring only in the Chinese and not in the aboriginal Lolos.

Returning to the coast, a change is found south of the province of Chekiang, and in the two south-eastern coast provinces of Fokien and Kwangtung leprosy is exceedingly common. It would seem that for a distance of about one thousand miles— between Chifu and Foochow—the coast of China is free from leprosy; that it exists to the north of this area, and to a limited extent in the inland provinces in the same latitude, and that it suddenly becomes widely prevalent as the coast winds round from east to south. The two provinces named, those of Fokien and Kwangtung, are described as the cradle of the disease in China, and the source whence it has spread to many other countries and islands in the Far East. In both it is rife; in Amoy, in Canton, Fatshan, Swatow, Pakhoi, Hongkong, and Macao—everywhere it is described as prevalent.

The island of Hainan, off the south coast, contains many lepers, but only among the Chinese, who dwell on the shores. The aborigines, of mixed Mongolian and Caucasian race, allied to the Laos of North Siam, are believed to escape. In Yünnan, an inland province west of Kwangtung, leprosy is also said to be not rare[1]. Concerning its existence in the remaining provinces of China little appears to be known.

In Corea the disease is present in the south, but is said to diminish towards the north, and to be no great scourge in any part of the peninsula[2]. In and around Fusan it prevails to a considerable extent, twenty cases having been seen in ten months. At Gensan no cases were seen during fifteen months. In Séoul one observer saw only two cases in eighteen months, while another met with a considerable number of lepers there[3].

Japan must be regarded as one of the principal homes of leprosy in the east. The number of lepers in the Empire has been estimated at from 75,000 to 100,000[4]. The disease is said to be twice as prevalent in the southern as in the northern provinces, for while in the latter the number affected varies between 100 and 600 in each, in the former there are from 500 to 1200, and in the province of Kumamoto alone as many as 2473 lepers were recently said to be in existence. In Japan the mountainous regions are, it is stated, more affected than the coast, and the temples at Narita and Minobu are much visited by lepers. The Japanese believe leprosy to be a punishment for sins, and the unfortunate lepers flock to the holy hills and wander from one shrine to the other as an act of expiation. In Japan generally the disease is said to be very rare among the rich noble class of the population,

---

[1] Michoud. *Medical Reports of the Chinese Imperial Maritime Customs.* No. 48.

[2] Cantlie. [3] *Chinese Customs Medical Reports*, No. 48 and No. 42.

[4] Ashmead. *Janus*, 1899, p. 71. This figure is perhaps an exaggeration; the official statistics of the Sanitary Bureau of the Japanese Home Office returned the total number of lepers in the kingdom at only 23,647, in September, 1897.

who suffer from tuberculosis more than any other malady; to be more frequent among the great middle class, whose principal disease, however, is syphilis; and to be specially rife among the outcast element of the population, who suffer from leprosy more than any other disease[1]. The Ainus of northern Japan and the Kurile Islands are said to be entirely free from the disease.

In Formosa the disease is common, particularly in the cultivated portions of the north of the island[2]. In some villages the number of lepers is said to be very large. Here again, just as in the Malay Peninsula and in China, the aborigines—the Pepohoans—are free from the disease, notwithstanding the fact that they have been brought in contact with the Chinese, the principal sufferers, for over two centuries. In the south of the island leprosy is said to be comparatively rare.

The principal seat of leprosy in the Philippine Islands appears to be in the neighbourhood of Manila, where a hospital of 180 beds, built in 1577 and rebuilt in 1774, is devoted to the treatment of lepers.

Australasia. In New Guinea leprosy occurs sporadically in many parts of the island.

In the Pacific Islands, with the exception of the Hawaii group, leprosy has gained but little hold. This exception is, however, an extremely important one, and the Hawaii or Sandwich Isles have of recent years become one of the principal centres of the disorder. Before 1848 the latter was, it is believed, unknown in the islands, but in that year the discovery of gold in California and consequent demand for coolie labour led to a great migration of Chinese across the Pacific from west to east. By 1856 Hawaii was in constant communication with California and China, and had become a centre of trade, and it was about this time that leprosy first appeared. The spread of the disease was extraordinarily rapid, and by 1882 some 2000 persons, or five per cent. of the native population, were believed to be lepers. In 1865 the leper settlement on Molokai was established, and between that date and 1899 a total of 5125 lepers have been sent thither. Apparently most of the Hawaiian islands are affected, but the disease is worse in some than in others[3].

[1] *Archives of Surgery*, 1898, p. 380.
[2] Rennie. *Chinese Customs Med. Reports*, No. 45.
[3] At the present time the disease appears to be slowly but steadily diminishing in Hawaii.

New Caledonia is reported to be passing through an almost equally severe experience, and leprosy is widely spread among the natives. The numbers are variously estimated at from 1500 to 4000. A few are isolated at Belep. At Tahiti the disease has committed great ravages.

In New Britain and in the New Hebrides, on the other hand, leprosy is said to be unknown. The title of Leper Island attached to one of the latter group of islands is therefore a misnomer. In Fiji the disease is described as being very common.

In Australia[1] leprosy appeared in the latter half of the 19th century; its introduction was probably brought about by the immigration of Chinese in the early days of the discovery of gold. In Victoria the disease has been practically confined to a score or so of cases among the Chinese inhabitants. In New South Wales, however, where some 70 cases have occurred since 1859, many natives who have never left the country have suffered, while rather fewer Chinese have been attacked here than in the other Colonies[2]. In Queensland a considerable number of lepers both of Chinese and other nationality have been recorded in recent years. South Australia appears to have escaped the disease entirely; but in the Northern Territory a small number of cases, mostly of Chinese, have been seen and dealt with. In West Australia only two lepers, both Chinese, had been met with up to 1897. Tasmania has remained free throughout. A table for the whole of Australia shows a general increase of recorded cases since 1877, and an apparent increase, in New South Wales and Queensland, since special legislation for lepers was adopted. It is believed that more perfect registering and the greater attention paid to the disease account for this apparent increase.

In New Zealand leprosy is said to be endemic among the Maories, but the exact extent of it is unknown.

Africa. The malady has apparently been introduced to Algeria from Spain. Between 1884 and 1897 nearly sixty lepers were traced

[1] Prize Essays on Leprosy. " The Epidemiology of Leprosy in Australia." By J. Ashburton Thompson, M.D., *New Sydenham Society.* London, 1897.

[2] On January 1st, 1896, 37 lepers were under treatment in the lazaret at Little Bay; 3 fresh cases were admitted in the year, 5 died, and 19 out of 20 Chinese patients were returned to China. On December 31st only 16 cases remained. On December 31st, 1898, this number had fallen to 14. From 1883 to 1898, 18 whites, born in New South Wales but of European descent, were admitted to hospital, and 33 Chinese.

in Algiers, of whom 24 were Spaniards, 2 French, 2 Italians, and 1 Maltese; while 27 were natives, 8 being Algerian Jews and 19 Mussulmans. The Spaniards came from the provinces of Alicante and Valentia, where, as stated above, the disease is very prevalent. In Tunis and Morocco leprosy is said to be endemic, and in Egypt it is common, especially in Upper Egypt and the Delta. It is also far from rare in Abyssinia, especially in the mountains of Samen and on the shores of Lake Tsana.

In East Africa leprosy is common, extending apparently along the greater part of the coast from north to south. In British East Africa, between Lake Victoria Nyanza and the sea, it is said to be very common, particularly the nervous form of the disease[1]. In German East Africa, somewhat further south, leprosy is more frequently seen round the great lakes than it is on the coast, and it is particularly common on certain islands in the Victoria Nyanza Lake. In the Congo Free State it is apparently rare[2].

On the West Coast of Africa it is not common. It is unknown in the Cameroons[3] and at the mouths of the Niger. It exists, however, on the Gold Coast, in Sierra Leone, and in Senegambia. One observer (Jonkin) did not meet with it on the coast between Sierra Leone and Old Calabar, while he found it was fairly common about the Upper Niger and Benue rivers. It must, however, be far from rare upon the coast, for at Lagos numbers of lepers are seen, and as many as 99 cases of the disease were treated in the contagious diseases hospital there in the year 1898. Jonkin[4] describes an inland endemic area of leprosy as extending some 500 miles westward of Lake Tchad, and beyond the waters of the Upper Niger; the town of Kano is the great centre of the disease. North of this leprosy holds its own to the southern fringes of the Sahara, but there is no reliable information as to its presence in any of the desert towns or oases.

The disorder has long been endemic in South Africa[5]. It

[1] Kobb, *Beiträge zu einer geographischen Pathologie Britisch-Ost-Afrikas.* Giessen, 1897.

[2] *Congrès Nat. d'Hygiène...de la Belgique et du Congo; Comptes Rendus.* Bayet says it is known, but not common, in the Congo Free State, and is mostly seen on the Lower Congo.

[3] A. Plehn. *Janus,* 1896–7, p. 383.

[4] *Report of Conference on Leprosy in Berlin,* 1897.

[5] Prize Essays on Leprosy. "Leprosy in South Africa," S. P. Impey, M.D., *New Sydenham Society.*

existed in Cape Colony before the advent of the white man, and appears to have been brought from the north. It affected to a considerable extent the Hottentots and to a somewhat less extent the Bantu, or Kafir races. It is uncertain whether the Bushmen suffered. In 1845 the well-known leper asylum on Robben Island was established, and between that date and 1894 the number of lepers admitted to it was 1697[1]. These were not all drawn, however, from Cape Colony, as by an arrangement with the other colonies, and with other States, lepers from a wide area in Southern Africa are sent to this place.

In Basutoland leprosy was probably introduced after the great Basuto war of 1835; in 1894 some 256 lepers were believed to exist. In the area recently known as the Orange Free State it has increased *pari passu* with its increase in Cape Colony, and in 1894 the lepers here numbered some 150. Griqualand East and the Transkeian territory are more seriously affected by the disease. The Griquas are half-breeds of Hottentot race; in 1863 they removed for political reasons on a long and tedious trek from (what became) the Orange Free State and Griqualand West. Leprosy spread in the course of their journey, and over 500 lepers were believed to be living among them in 1894.

In the area recently under the rule of the Transvaal Republic many natives were known to be suffering from the disease before the late war. The exact number was uncertain. A small leper hospital for about 20 patients existed in Pretoria. In the colony of Natal the number of lepers increased from two in 1843 to over 100 in 1886, and to over 200 in 1894; the malady has spread from kraal to kraal and from tribe to tribe until no tribe is free from it. It is, however, confined to the blacks in Natal. In British Bechuanaland it is very rare, and this is ascribed to the dryness of the climate and the sparseness of the population[2]. In Pondoland there are no lepers, the chiefs rigorously excluding, under pain of death, all affected persons. In Zululand and Swaziland the disease is known to exist, but accurate information in

[1] On January 1st, 1898, 532 lepers were under treatment at the asylum; on January 1st, 1899, the number had increased to 560. In the leper asylum at Emjanyama on the same dates the corresponding figures were 182 and 255 respectively.

[2] The rarity of leprosy in Bechuanaland may nevertheless be doubted. In the Vryburg district it is said to be very common in the large native locations. (*Cape of Good Hope, Reports on the Public Health for the year* 1898.)

regard to its prevalence is not available; it is probably not increasing, because the natives are believed to put to death or to ostracise all persons suffering from it. In Mashonaland lepers are not frequently seen.

The number of lepers in South Africa was estimated in 1894 at 1750, distributed as follows: Cape Colony, 600; Griqualand East, 250; Transkei, 357; Basutoland, 250; Bechuanaland, 10; Natal, 200; Orange Free State, 150; Transvaal, 30. It is probable that these figures are considerably below the truth.

In many of the islands adjoining the continent of Africa leprosy is known. In the Canary Islands there are a considerable number of lepers, and the asylum at Las Palmas is said to draw its inmates from each of the seven islands in the group. In Madeira leprosy is also met with, and there is a leper hospital of ancient date.

In Madagascar the malady is endemic and is said to be spreading since the old laws of isolation were relaxed. In Mauritius and Rodriguez it is also endemic, attacking all the various races living there, but especially the coloured Creoles. It is also met with in the Seychelles.

North and Central America. Throughout the continent of North America, from north to south, leprosy exists, mostly in the form of scattered cases, but here and there, particularly as the tropics are approached, it is much more common.

In Canada only some 32 lepers were believed to be living in 1894; but they may have increased since[1]. The majority of these (21) were in Tracadie, on the southern shore of the province of New Brunswick, where a leper asylum has been built. The inhabitants of this shore are mostly of French origin, some coming from Acadia and some from the province of Quebec. Leprosy has existed here since the year 1815, and since that year 167 deaths from the disease have been recorded. The number of cases is unknown. On Cape Breton Island it has almost disappeared, only two cases remaining in 1894. In British Columbia 8 lepers, all of them Chinese, were isolated at the Darcy Island asylum in 1897. In Manitoba 3 lepers, all from Iceland, are, or were recently, living in the province of Winnipeg[2].

---

[1] Prof. J. C. Graham. *Transactions of the Congress of American Physicians and Surgeons.* Washington, 1894.

[2] *Annual Report of the Provincial Board of Health of Ontario for* 1897.

In Canada generally the few cases of leprosy which exist can be traced to importation by foreigners, and there seems little tendency for the disease to become domiciled or to increase. It has, however, lingered for a long period on the shores of New Brunswick, and the Icelander lepers in Winnipeg may yet prove to be a danger[1].

In the United States leprosy is also to some extent an exotic; it is comparatively uncommon and shows no great tendency to increase. It is, however, widely spread throughout the various States. The regulations as to the treatment of lepers vary in the different States, but in none does there appear to be an accurate record of the number of affected persons existing at the present time. The following is an estimate made in 1894 of the number of lepers "heretofore recognised" in the States:—Arkansas, 3; California, 158; Dakota, 2; Florida, 6; Georgia, 1; Idaho, 2; Illinois, 13; Indiana, 2; Iowa, 20; Louisiana, 83; Maryland, 4; Massachusetts, 5; Minnesota, 120; Missouri, 2; Mississippi, 2; New York, 100; New Jersey, 1; Oregon, 3; Pennsylvania, 6; Utah, 3; Wisconsin, 20. This gives a total of 560, spread over a considerable number of years[2]. It must be noted, however, that this figure is probably much below the truth. Thus Osler quotes Dyer as stating that on January 12, 1898, he knew of 124 positive living cases in Louisiana alone, including 25 in the Leper Home in Iberville parish. The same authority added that it was justifiable to estimate the number of lepers in the State of Louisiana as between 300 and 500.

There appears to be nothing very characteristic in the geographical distribution of the above cases in the United States. In a group of States, however, in the north of the Republic, the numbers are said to illustrate the tide of settlement of Scandinavians in transit from New York to a country and climate resembling those of Scandinavia. In these (Minnesota, Wisconsin, and Iowa) 170 lepers are said to have immigrated from Norway[3], but only 30 now remain. It would seem that in the United States, as in

---

[1] If they have not already done so. In 1897 four cases of leprosy were discovered in Manitoba (1 in the North-West Territories, 2 in Winnipeg, and 1 in Selkirk). Whether these four cases included the three cases mentioned in the Ontario Report I am unable to say. *Report of the Provincial Board of Health of Manitoba for* 1897.

[2] Nevins Hyde. *Transactions of the Congress of American Physicians, etc.*

[3] Armauer Hansen.

Canada, the disease tends to diminish rather than increase after its importation by immigrants. At present the larger number of lepers in the country appear to be in Louisiana and California.

Leprosy also exists in Mexico, and some thirty cases are usually under treatment at the lazaretto in the City of Mexico. The disease has been known here since the time of Cortes, but the authorities believe that it is not extending. It is most frequent in the State of Sinaloa on the west coast, and in Durango to the east of it; it is also common in Guanajuato and Jalisko and in Guerrero on the northern frontier. In Guatemala and Salvador scattered cases of the disease are met with. In British Honduras the disease is also known: 3 cases were treated at the Public Hospital, Belize, in 1898, and 1 at the Corosal Hospital. In Nicaragua and in the mountain regions of Costa Rica it exists in sparsely scattered cases. Lepers are also seen in Panama, but the numbers are unknown.

In the islands off the continent of North America leprosy is much more common than on the mainland. In the West Indies the number of lepers is large and is apparently increasing. It is endemic in Cuba, and in Havana is a leper asylum for some eighty patients. Many hundreds of leprous persons are, however, believed to be living in that city, or were before the war, and are (or were) under no sanitary control whatever. Nevins Hyde estimated the number of lepers in Cuba in 1894 at between 300 and 500, but now there are probably more. In Haiti the disease is very widely spread, and also in Jamaica, where during the past twenty years the average new admissions to the leper asylum have been thirty per annum, rising in some years to forty, and sinking in others to fifteen. In the Bahamas some 15 or 16 fresh cases come into hospital each year. The Lesser Antilles are also more or less seriously affected by the disease. In Guadeloupe and Martinique one person in 860 is said to be a leper, in St Christopher the disorder is rife, while in Montserrat, Grenada, and Dominique it is rare. In Trinidad it is increasing; 233 lepers were interned at the asylum there at the end of 1897, and 265 at the end of 1898. In the Danish Antilles there were in 1896 22 lepers at St Thomas, 82 (4 per 1000 of the population) at Ste Croix, but none at St Jean. The lepers were all of the negro or mixed races. There is a leper asylum at Ste Croix. In St Lucia, in Antigua, and in St Vincent cases are dealt with every year; in the latter island 19 lepers were in the asylum in 1896. In Barbados the disease must be common and increasing, for in 1898 the daily average number of lepers resident in the lazaretto throughout the year was 114; in 1897 it had been 111, and in 1896 only 107. A few cases of the disease are known to exist in Puerto Rico. In the Virgin Islands leprosy is said to be rare.

In the Bermudas, lying alone in the Atlantic Ocean, a few cases of leprosy have been under observation for a considerable period; at the beginning of 1898 they were twelve in number, at the end of that year they were ten.

South America. Along the north coast of South America the malady is far from uncommon. In Venezuela many cases are seen. In British Guiana leper asylums exist at Mahaica and Gorchum, and in the former, in 1898, 99 fresh cases were admitted in addition to 374 patients in the wards at the beginning of the year. As the corresponding number at the beginning of 1895 was only 359 it seems that the disease is increasing. In French Guiana, where it is thought to have been imported by slave boats, one-tenth of the population are believed to be leprous; the Red Skins here are said to escape. In Surinam (Dutch Guiana) from 500 to 2000 lepers are believed to be living. Colombia is seriously ravaged by the disease; the number of lepers here is said to have increased, in forty years, from 400 to over 27,000. In 1898 only 841 out of this large number of affected persons were segregated, in three asylums, at Agua de Dios, Contrataçion, and Caño de Soro.

Along the west coast of South America leprosy is believed to be rare. Occasional cases have been seen in Peru. Of the prevalence of the disease in Ecuador, Bolivia, and Chile little appears to be known.

The whole coast of the Brazils is said to be affected. The last census of the country gave a return of some 5000 lepers, but there are thought to be about twice this number. The disease is specially prevalent in the states of Parà, Pernambuco, Bahia, Minas Gerães, Matto Grosso, Rio de Janeiro, and São Paulo.

In Uruguay, on the other hand, leprosy appears to be rare. In the Argentine Republic lepers have been admitted to a hospital at Buenos Aires since 1892. In that year 5 cases were admitted, 12 in 1893, and 15 in 1894. The disease is said to be increasing in the interior of the country, and to prevail especially in Corrientes, Entre-Rios, Cordoba, and Buenos Aires.

Factors determining the Geographical Distribution of Leprosy. The present geographical distribution of leprosy is to a great extent a legacy of the past, to a less extent the result of forces and conditions now acting. In the Old World it has remained, almost from time immemorial, in many countries, such for example as India, Syria, the Levant, and probably in China; while in others, as for example the greater part of Europe, it has to a great extent disappeared, leaving behind it a few isolated centres of the disease—such as in Iceland and in Spain, in the Balkan countries and on the coast of Norway.

Whatever may ultimately prove to be the truth as to the contagiousness of the disease there is no question that leprosy can be transported from one country to another. It has been so introduced in recent times into Algeria from Spain; it was, it is believed, imported into Spain from the Philippine Islands; it has been on several occasions imported into the United States from Norway; and it has been carried to Manitoba in Canada from Iceland.

Within the past half-century it has been introduced into some of the Pacific Islands, such as Hawaii and New Caledonia, with the most disastrous results; and in the south-eastern portion of Asia, in the East Indies, and in Australia, it has been mainly, perhaps solely, distributed by the movements of Chinese from the endemic centres of leprosy in China. In the distant past the examples of the carriage of leprosy from one country to another by the movements of affected persons were no less striking.

Climate, soil, elevation, and similar general conditions appear to have but little influence on the distribution of the disease. It is almost as common in semi-arctic Iceland as in tropical Ceylon; and it is as rife in some of the elevated Himalayan tracts as in the plains of India. It shows, however, a certain tendency to prevail along the coasts of seas and lakes and rivers. The tendency is not absolute, and there are many places in India, in Russia, in Central Africa and elsewhere, where leprosy shows no definite relation of this nature, and where, on the contrary, the disease is commoner inland than on the coast. On the whole it would seem that it is rather in local than in general conditions that the disease finds influences favourable or unfavourable to its prevalence.

Dirt, insanitation, poverty, overcrowding, and improper diet are almost constantly found associated with the existence of leprosy, and it is difficult not to ascribe to them a considerable share in determining its prevalence. It is possible that these and other factors are all essential to the multiplication and diffusion of the *bacillus lepræ*, which must be regarded as the essential cause of the disease.

The view which ascribes the production of the malady to the consumption of fish, particularly of badly salted and decomposing fish, finds some slight support in the geographical distribution of the disease, in that, as stated above, lepers are found in many countries along the shores of seas, lakes, and rivers in which fish abound, and that the disease is frequently seen in persons who eat fish of the kind just described. A considerable number of instances have, however, been recorded of leprosy occurring in individuals and in races who have never tasted fish of any sort, while in India not only does the distribution of the disease show no special relation to seas and rivers, but leprosy is found to be common in people whose religion and customs forbid them to touch fish.

The influence of race upon the prevalence of leprosy is not easy to determine. Some races appear to be more liable to become leprous than others, but such differences may as often be ascribed to differences in mode of life and of exposure to the cause of the disease as to true racial differences. All races can become subjects of the malady; but at the present day Europeans and the white races generally seem to escape it much more readily than the natives in countries where it is endemic—as for example, China, India, and South Africa.

Few questions in medicine have given rise to more controversy than those connected with the mode of spread of leprosy. It was formerly almost universally held that the disease was hereditary. It is now much more generally held that heredity has little if any share in its diffusion. The members of the Leprosy Congress held in Berlin in 1897 were of this opinion, and the balance of evidence is certainly against the view that leprosy can be transmitted from parent to child in the same way that syphilis can be transmitted.

The question of the contagiousness of leprosy has given rise to similar differences of opinion. But as it has been definitely proved that the *bacillus lepræ* is given off in the discharges from the sores of lepers, and in many instances in the saliva, the nasal secretions, the urine, and even the milk of patients, it is clear that persons living in the immediate neighbourhood of lepers must run considerable risk of the bacillus entering their tissues. The numerous instances in which lepers have carried their disease from one part of the world to another have been referred to above, and instances are not unknown of persons coming from places entirely free from leprosy, contracting the disease from intercourse with persons suffering from it, and communicating it to others on their return to the leprosy-free country. The spread of the disease in Canada and the northern United States has distinctly followed the track of the Scandinavian and Icelandic immigrants who imported it to the continent. All these facts point strongly to the conclusion that leprosy, like tuberculosis, is an infective malady, and that it spreads from person to person like any other germ-disease.

That the distribution of leprosy can to some extent be controlled by human effort should follow from this conclusion, and the possibility of so controlling it has been shown in more than one instance. The complete isolation, under anything like tolerable conditions of life, of persons suffering from a disease that may last for many years or even decades, is obviously extremely difficult. But the success which has attended even the modified form of isolation practised in Norway is encouraging; and at least justifies the assertion that human effort to keep the disease within limits is a factor, and might be a very much more important factor than it is, in determining its geographical distribution.

LIVER ABSCESS.

General Characters and Etiology. Abscesses of the liver may arise from a variety of causes, such as traumatism, gall-stones, hydatids, general pyæmia, and the like. But the solitary, so-called tropical abscess of the liver, which is met with solely in warm climates, presents so distinct a clinical group of symptoms that it is almost certainly due to some constant and specific cause, and is usually regarded as a definite and separate disease. This form of liver abscess will alone be dealt with here. Its causation is uncertain. . The pus in such an abscess is usually sterile, or contains the ordinary pyogenic organisms or the bacillus coli. In several instances the *amœba coli*, identical with that found in some dysenteric stools, has been discovered in the pus of liver abscesses. The close association of this disease with dysentery will be discussed later.

Geographical Distribution. *Europe.* Abscess of the liver is said to be by no means rare in parts of Spain, Italy, and Turkey, and it is occasionally met with in Greece. In the south of France cases have been seen, but it is rare here, and even in northern and central Italy and northern Spain it is much less frequent than in the southern part of those peninsulas. In England and other countries further north true "tropical" liver abscess is only seen as an importation from the warmer zones of the earth.

Asia. In some parts of the Caucasus this disease must be extraordinarily prevalent. In the Tiflis hospital abscess of the liver is said to be found in 4 per cent. of the autopsies[1]. It is possible that in a certain proportion of these cases the abscess is

---

[1] Pantiukhof, " *K Statistikié Kavkazskoi Patologii*"; Tiflis, 1898.

not of the specific "tropical" character now under consideration, but it seems probable that it is so in the majority of instances.

In Persia and Arabia liver abscess has been frequently seen in Europeans, particularly along the shores of the Red Sea and Persian Gulf. But of all Asiatic countries India is the one where this disease is most common. The returns for the British army in India in the years 1886–95 showed an annual admission-rate from abscess of the liver of 2 per 1000, with a death-rate of 1·13 per thousand strength. In the two following years the admission-rate remained the same, while the death-rate rose to 1·22 and 1·29 respectively. The admission-rate for "hepatitis" in the same period varied between 21 and 23 per 1000 yearly. The Burma Coast and Bengal-Orissa head the list of Indian territorial divisions in regard to frequency of liver abscess, but it seems to occur in all parts of the peninsula. In Rajputana it is apparently rare, although hepatitis is a common disorder. The natives in India suffer far less than the European residents.

Liver abscess is frequent in many of the East Indian Islands, but here also it is mainly the European that is attacked. In 1897 among the men of the Dutch army in the East Indies there were 23 cases in Europeans and only 2 in natives. In the hospitals and gaols of the Straits Settlements liver abscess is no rarity, and hepatitis is exceedingly common. In Cochin China hepatitis must be also very prevalent, as it is said to account for 3·8 per cent. of the mortality from all causes[1], and it may therefore be surmised that liver abscess is also a common disorder here[2]. Both diseases are occasionally mentioned in reports from southern China, but they do not seem to be very prevalent there, and are rare in the northern portions of the empire. In Japan abscess of the liver is also infrequent and only indigenous in the southern islands.

Australasia. There is mention of both hepatitis and liver abscess as occurring in the Philippine Islands, Hawaii, Tahiti, New Caledonia, and other Pacific Islands, but information as to their existence in Australia and New Zealand seems to be lacking.

[1] Reference in *Janus*, 1897–8, p. 96; I have not noted whether the figure quoted referred to deaths among Europeans only or to total deaths.

[2] Hirsch, however, quotes authorities showing that liver abscess is not common in Cochin China.

Africa. Abscess of the liver is almost as common in Egypt, Nubia, and Algeria as in India. In Tunisia cases are not infrequently seen. In Mauritius and Madagascar the disease is very prevalent, and on the West Coast, particularly about Senegambia and the shores of the Gulf of Guinea to the north of the estuary of the Congo, it seems to be equally so. Further south on this coast it becomes rarer. South Africa is, however, not wholly free from it. In the Kimberley hospital it is—or was before the war—frequently seen, but several of the cases were in persons from "up country." Even in the Durban hospital 4 cases were treated in 1898. I have not noted the returns for other years.

America. In Canada liver abscess is not an indigenous disease. In the United States it is most often seen in the southern States, but even as far north as Baltimore it is said to be "not very infrequent" (Osler). In Mexico it is commoner in the west than in the east. It is frequent in Guatemala, San Salvador, Costa Rica, and Panama. On the other hand in the West Indies, though the disease is met with, it is, at least in many of the islands, anything but common, and the same statement holds as to its prevalence in the Guianas. For Brazil the information available is rather contradictory, but there seems to be no doubt that in the north of Chile, in Peru, and in Venezuela liver abscess is extraordinarily common—in striking contrast with its infrequency in Guiana and the West Indies.

Factors concerned in the Geographical Distribution. The term "tropical" as applied to liver abscess, it will be seen, is not a strictly accurate one; for though most often met with in the tropics, the disease is quite common in many subtropical countries, and is, it seems, not wholly rare in the more temperate States of North America and Southern Europe. The term is, however, a useful one, and serves to show that the affection is mainly one of hot climates. So far as the very imperfect statistical information from most countries goes, it may be judged that the disease is not solely dependent upon heat for its prevalence. It is more common, for example, in Mauritius, Algeria, and the north of Chile than in the West Indies or Guiana, although the latter are nearer the equator than any of the former.

For its production in the individual the most favourable conditions for liver abscess seem to be residence in a hot country and exposure to chill. Immoderate indulgence in alcoholic drink also strongly predisposes to it, and in India and elsewhere cases of the disorder have diminished to some extent among European troops with a diminution of drunkenness.

Dysentery is another and probably a very important predisposing factor in the production of liver abscess. The true relation of the two diseases is still a matter of some uncertainty. It is however certain that an attack of dysentery often precedes the development of an abscess of the liver. In the Caucasus—to judge, at least, from the returns of the Tiflis hospital already alluded to—almost all the cases of liver abscess are complicated with dysentery. In India too there is a certain parallelism between the statistics of the two diseases in the different geographical divisions of the country. In 1891–95, for example, the Burma Coast and Bengal-Orissa headed the list for both diseases; and in 8 out of the 12 geographical groups a high dysentery rate coincided with a high hepatitis rate. But it is said that, when more closely examined, the figures do not show a very exact parallelism. It has also been pointed out that the great rise in the dysentery rate after the last Afghan war was not associated with a rise in the hepatitis rate. In summing up the evidence on this point the late Sanitary Commissioner with the Government of India stated that in from 21 to 59 per cent. of fatal cases of abscess of the liver association with dysentery had been found; while on the other hand in 20 to 25 per cent. of fatal cases of dysentery liver abscess had been met with[1]. This certainly shows considerable à priori grounds for believing that there is a close relation between the two; and this view is still further supported by the discovery (by Kartulis, Powell, and others) of the amœba coli, the supposed cause of tropical dysentery, in the contents of liver abscesses.

Racial influences are probably of some importance in the distribution of liver abscess. In India, in Malaysia, on the West Coast of Africa, in Egypt and elsewhere the European is much more often the subject of the disease than the native. In the Caucasus the Russians are far more frequently affected than the people of the country. The difference between the white and the coloured races is most marked in India; and less so in Africa and the

[1] *Annual Report of the Sanitary Commissioner with the Government of India*, 1894.

West Indies, where the negroes suffer to a greater extent than the natives in India. It is doubtful whether this difference is to be attributed to true racial differences, or whether it is not largely due to other causes. The European who goes for the first time to a hot country has to pass through a process of acclimatisation in which not a little strain is thrown on the functions of the liver. Should he carry with him the habit of taking considerable or excessive quantities of alcohol, and should he further not realise the great risk of chill in a hot climate, he will be naturally far more predisposed than the native to develop liver abscess. The fact that among European troops in the tropics the men suffer far more than the officers, and the women scarcely at all, shows that the liability to this disease may be much more a question of individual habits and circumstances than of racial susceptibility.

MALARIA.

General Characters and Etiology. Under the general term malaria are included a number of intermittent and remittent, acute and chronic, malignant and benign disorders, which modern research tends to show may possibly be distinct and separate diseases. For the present, however, it is convenient to deal with them together.

Recent investigations have demonstrated that malarial diseases are associated with the presence in the blood of certain microscopic plasmodia or hæmatozoa, and that these vary in appearance and character with the different types of the disease. For a full description of these bodies and their relations to the varieties of malaria the reader is referred to the writings of Laveran, Celli, Manson, and others, and for an account of the clinical manifestations of malaria to any recently published text-book on Medicine. Here it must suffice to give a brief account of the mode in which malaria is spread, and of the relation of the mosquito to the malarial organism, as these questions have obvious and important bearings upon the geographical distribution of the disease.

The work of English and Italian observers has proved within the last few years that the various forms of malaria are transmitted mainly, and perhaps wholly, by means of the mosquito. Like the filaria sanguinis hominis the malarial parasite passes one portion of its existence in the body of the mosquito. The passage of the organism from the blood of the human being to the stomach of the insect takes place when the latter is in the act of "biting" a malarial patient and sucking his blood. It is believed again to reach the tissues of the human being by like means, that is to say by passing from an infected mosquito during the act of biting.

Whether this is the sole method by which man can be infected is uncertain, and some still hold the view, abandoned by others, that the malarial parasite may escape from the mosquito into water, which on being drunk would convey it into the stomach of a human being.

It is believed that only certain mosquitoes are capable of transmitting the malarial parasite. These mostly, but perhaps not exclusively, belong to the genus *Anopheles*. This genus contains a very large number of species, and several of them have been shown experimentally to be capable of harbouring the malarial organism, but the principal carrier of malaria is believed to be *Anopheles claviger*, or *maculipennis*. Of the other great genus of mosquitoes, the *Culices*, the majority are regarded as incapable of transmitting malaria ; but Grassi believes that *Culex penicillaris* and *Culex malariæ* may carry the disease.

A very large number of phenomena connected with malaria and its epidemiology are still obscure, in spite of the light thrown upon many others by the "mosquito theory." But that the theory is in the main correct can scarcely now be questioned, and further observation will no doubt remove many of the obstacles which now stand in the way of reconciling the theory with *all* the observed facts of the disease.

The highly important stages in the life-history of the malarial parasite which have been briefly summarised above, and about which an immense literature has recently accumulated, have the closest possible bearing upon the geographical distribution of the disease itself. If, as there seems every reason to believe is the case, a certain species, or certain species, of mosquito is or are essential to the life-cycle of the parasite, it is clear that malaria will occur only where it or they are found. A constantly increasing body of evidence has been adduced from many countries tending to show that the distribution of human malaria is in fact identical with that of its mosquito host or hosts, to the extent that it is never found in their absence, though there are certainly places in which the malaria-mosquito is found and the disease is unknown. More accurate knowledge is, however, yet needed as to the exact species of mosquitoes capable of carrying malaria, and as to their presence in all those geographical areas where malaria is found, before precise statements as to the distribution over the earth's surface of the malaria-bearing insects can be made, and in the present chapter the distribution of the disease itself will alone be considered.

History. The early history of malaria is obscure, and Hirsch states that it is not until the sixteenth century that the records of the disease begin to be of value. Certainly from medieval times it has been one of the most universal of all maladies, and in many countries it has caused an amount of suffering and death—to say nothing of incapacitation and consequent economic loss—that can only be characterised as appalling. Until within the last few years the cause of malarial diseases was unknown, the method by which they were spread was uncertain, and they themselves were regarded as to a great extent beyond human means of prevention. They were accepted as one of the permanent, inevitable conditions of life in those countries where they prevailed, and considerable tracts of the earth's surface were either abandoned as uninhabitable, or looked at askance by the colonist and the traveller as presenting such risks to health and life that they were to be avoided at all costs. The recent acquisition of knowledge as to the cause and mode of spread of malaria, and the successful results already obtained by applying this knowledge, have opened up a wholly new prospect for these countries. They have shown that, even if malaria cannot be wholly exterminated, it is possible to live in a malarial country and yet escape its attacks.

The malarial parasite was first demonstrated by Laveran in 1880. His researches were confirmed independently by Marchiafava and Celli. The mosquito had for a considerable period been vaguely suspected of taking some part in the diffusion of malaria, but it is only within the last five or six years that the true relation of the one to the other has been established. The "mosquito theory" has been gradually built up by the labours of Manson in China and elsewhere, of Ross in India, and of the Italian observers Bignami, Celli, and Marchiafava. The most striking incidents in the gradual proof of the theory have perhaps been the demonstration by Ross of the life-history of proteosoma in birds (an organism allied to that of malaria in man); the successful production of malaria by Manson in 1901 as the result of allowing infected mosquitoes brought from Italy to bite healthy persons in London; the direct cultivation of the malaria organism in the bodies of mosquitoes by many observers; the preservation of healthy persons in the most malarious districts of Italy by strict precautions against mosquito bites (certain Italian railway officials, and Drs Sambon and Low in 1900); and similar successful results obtained by the campaign against mosquitoes in Hong-kong, Cuba, the West Coast of Africa, and elsewhere. To these results few have contributed more than the members of successive expeditions sent to notoriously malarial districts by the Royal Society, by the Liverpool School of Tropical Medicine, and by other bodies.

Recent Geographical Distribution. *Europe.* In the British Isles malaria, except in certain parts of the fen country, is almost extinct as an indigenous disease, and in the cases that are met with the patients have usually contracted the disorder abroad. The term "malarial diseases" still forms a heading in the reports of the Registrar-General, but the deaths in this group in England have steadily decreased in the last two decades.

In France[1] malaria is met with on the west coast among certain saline marshes about the mouths of the Seudre and the Loire. It is said that when these marshes are kept in good order for the production of salt, and only sea water admitted to them, they cease to be malarial; but when they are abandoned or spoilt for the production of salt (*marais gâts*) and fresh water mixes with them, they become very favourable to the development of malaria. The neighbourhoods of Rochefort and of Marennes were until recently very unhealthy, but the latter has of late years improved. Fever is common in Vendée and Loire Inférieure, but it is every-where diminishing. A notable lessening of the disease has taken place on the shores of the Lac de Grand-Lieu, near Nantes, at St Philbert de Grand-Lieu. The type here is principally tertian, but other types are met with. At Machecoul an endemo-epidemic of malaria occurs from July to November, at first of tertian and later of quartan type, with occasional pernicious forms; while malarial cachexia is commonly seen. At Nantes and higher up the borders of the Loire malaria is much less frequent than it was formerly, and the same is true of Tours, where it is now unknown in the garrison. Tertians also predominate here, but other types occur. At Saumur severe remittents are sometimes met with. Formerly the whole French coast from the mouth of the Gironde to that of the Adour (Landes) was very malarial, but the fixing of

[1] See a valuable article on the distribution of malaria by Laveran; *Janus*, 1896–7, p. 301.

dunes, the planting of sea-pines, and the cultivation of the soil have succeeded in almost completely extinguishing the fever.

On the French Mediterranean coast there is little malaria east of the Rhone estuary, as the shore-line here is considerably elevated. But to the west of this river the coast becomes low and marshy and fevers are very common. On the Île de la Camargue, which forms practically the delta of the Rhone, and along the Beaucaire canal, at Aigues-mortes, are a series of marshes which are notoriously malarial.

In the interior of France malaria is no longer very common or severe. It still exists in the neighbourhood of La Dombes. Many years ago a large experiment in pisciculture was made here; a series of lakes were constructed, which were used for fish-rearing for two years, and in the third year were drained, dried, and re-stocked. This alternation of wetness and dryness of the soil led to the development of a most intense endemic form of malaria which literally decimated the population. From 1802 to 1842 the mean life of the inhabitants of La Dombes was only 24 years. In the last half-century great improvements have been made in the district, and half of the land has been reclaimed for agriculture, with the result that malaria has greatly diminished. La Sologne (in the department of Loir-et-Cher) and La Brenne (in that of Indre) are also feverish localities, owing to the presence of lakes and marshes. There too the rearing of fish, which has been a principal industry since the year 1450, has led to the development of a severe form of malaria. As late as 1852 the inhabitants were in a pitiable condition of misery and suffering, but since then wide measures of drainage and cultivation have greatly improved the health of the district, and malaria is now far less common than formerly.

In Germany malaria is apparently diminishing rapidly. In the three-year period 1883—1885 as many as 5·2 per 1000 of the cases of all diseases annually treated in hospital were cases of some form of malaria, but in the next period (1886—1888) the figure had fallen to 2·8, and in 1889 to 1891 it was only 1·9. It is severe only on the plains, and is most prevalent among the fens of Oldenburg, along the marshy banks of many rivers, and at certain points on the shores of the German Ocean.

Holland, at one time the home of the most severe and fatal forms of malaria, is now comparatively free from the disease, in spite of its slight elevation and low coasts. This change has been brought about entirely by the energy of its inhabitants, who have carried out extensive measures to prevent the inroads of the sea and to drain the marshes which formerly existed. At the present day malaria is met with on the coasts, in the neighbourhood of Amsterdam and Rotterdam, and about the mouths of the Escaut and Over-Yssel rivers, and in the island of Walcheren, of evil memory. The continued diminution of malaria in Holland is shown by the following figures : the deaths from continued fever have steadily declined from 1·2 per 10,000 inhabitants in 1889 to 0·3 in 1898; in the same period the deaths from intermittent fever fell from 0·2 to 0·1, and those from pernicious forms of fever from 78 to 1 —figures too small to represent in proportion to the popula-tion[1].

A marked fall in deaths from malaria has also taken place in recent years in Belgium[2]. The mean annual deaths from "fièvres paludéennes" in that country was 378 in the decade 1871—1880, and only 214 in the decade 1881—1890, while in the year 1895 it fell to 84. In 1897 the deaths were 101, which is equivalent to about 15 per million inhabitants.

In Denmark malaria must be either almost extinct or of an extremely mild character. In 1897 one death from intermittent fever was recorded for the urban population of the whole country, while in each of the five preceding years not a single death from this cause occurred in the same group of inhabitants[3].

In Norway malarial fever is also said to be unknown, and this is doubtless due to the mountainous nature of the country and its precipitous coasts, with consequent absence of fever-breeding marshes. In Sweden, on the other hand, the disease exists. As an endemic it is absent north of the Angerman river, in 63° N. latitude. It occurs in the neighbourhood of Stockholm and in the district of Nyköping. It has at times (*e.g.* in 1846) become epidemic along the Baltic coast, and inland on the shores of the Mälar Lake, and at some points on the Wenern Lake. It is a remarkable fact that in 1846 it also appeared in Luleå, a marshy district at the head of the Gulf of Bothnia, close to the 66th parallel of northern latitude. This would appear to be the most

[1] *Annuaire Statistique des Pays-Bas*, 1898.
[2] *Annuaire Statistique de la Belgique*, 1898.
[3] *Statistisk Aarbog, Kjöbenhavn*, 1899.

C.   16

northerly point in Europe in which malaria has occurred, as it is even further north than Archangel.

In Russia malaria is a very widely-spread disorder, and there are few portions of the country entirely free from it. There is reason to believe that it is becoming more prevalent, as the number of registered cases of malaria treated in hospitals throughout the empire rose from an average of 1,624,098 in the years 1887—1892 to considerably over 3,000,000 in each of the years 1894 and 1895[1]. Part of this rise, but not all, was probably due to the increase of hospitals and to improved registration. The disease is found as far north as the government of Archangel, and is not unknown in the town of Archangel itself, on the shores of the White Sea. It is also met with in the northerly governments of Olonetz and Vologda. In the Baltic provinces—Courland, Livonia, and Esthonia—which have frequently been regarded as specially malarious, the disease is, on the contrary, far less prevalent than in many other parts of the country.

Malaria is extremely common in the Don Cossack territory in southern Russia, perhaps more so than in any other part of the country; thus, out of a population of about 2¼ millions, over 73,000 cases of this disease were registered in 1895. It is exceedingly common also in the governments of Kherson and Simpheropol (or the Crimea), on the shores of the Black Sea and the Sea of Azov. The group of governments in south-western Russia, known collectively as the Ukraine (Kief, Tchernigof, Poltava, etc.), are also highly malarious. Throughout southern Russia there are large areas of imperfectly drained, marshy country which favour the prevalence of the disease. It is met with along the valleys of the great rivers, such as the Don, the Bug, the Dnieper, the Dniester, and the Volga. The government of Astrakhan at the mouth of the Volga is particularly malarious, the disease occurring especially in the marshy regions in the delta of that river and about the shores of the Caspian Sea. Malaria is very frequently seen also among the Bashkirs, living in villages along the river Ural; the disease makes its appearance here in summer as the lakes and pools round which the villages are built begin to dry up. In the north-eastern governments along the upper reaches of the Volga, and its great feeders, the

---

[1] Annual Reports of the Medical Department of the Russian Ministry of the Interior.

Kama, Viatka, and Ufa, malaria is also widely endemic and the cause of a high annual sickness-rate.

In Austria-Hungary the malady is endemic, and causes a considerable amount of illness, but it is apparently slightly decreasing. Thus in four successive triennial periods, beginning with 1883–85 and ending with 1892–94, the proportion of malarial cases treated in the civil hospitals to cases of all diseases was 11·6, 8·4, 9·4, and 9·4 per 1000 respectively[1]. The banks of the Danube, the Save, the Drave, and the Theiss, are notoriously malarial.

Along the coasts of Spain and Portugal there are many points where malaria is prevalent. It is especially so in the marshy district near the mouth of the Guadalquivir. In Spain malaria was, some twenty years ago, more widespread than in any other country in Europe, with the exception of Italy, but there has been some improvement in recent years.

Italy is now the most important endemic centre of malaria in Europe. The disease here attains the proportions of a great national misfortune. Owing to its ravages about five million acres of land are uncultivated, and each year it attacks some two million of the inhabitants, and causes the death of over 15,000 persons. The lower animals have not escaped it, and whole herds of cattle have, it is said, within comparatively recent times fallen victims to it[2]. In upper Italy the growing of rice is the principal cause of the prevalence of the disease, and it is particularly common in districts where stagnant water and marshes abound. In Tuscany the chief centre is formed by the Maremma, a series of uncultivated plains separated from the sea by dunes, which prevent the escape of water and lead to the formation of many pools and marshes. The Pontine marshes, which adjoin the Maremma, occupy the Mediterranean coast for a distance of some 26 miles; they are so low and flat that as far inland as 11 miles from the sea the land is only 4½ feet above sea-level. These marshes have a turfy, moist soil, and produce a rich vegetation. The Roman Campagna, on the other hand, has a soil almost denuded of vegetation, except in spring. L'Agro Romano is a marshy area of over 1000 square miles. All these localities are notoriously malarial. The city of Rome itself, in spite of the nearness of these dangerous marshes, enjoys a comparative immunity from

---

[1] *Bulletin de l'Institut International de Statistique.* Vol. X. Rome, 1897.
[2] Celli.

the disease. On the coasts of the Adriatic are the malarious marshes of Apulia and at the extreme south those of Basilicata and Calabria, the latter forming the "instep" and "toe" of the boot-shaped peninsula.

On the eastern shore the disease begins to be severe south of Cape Gargano, and in the south some of the worst forms of continued and pernicious fevers, such as are rare in the north, begin to be common. On the western coasts of Italy severe foci of malaria exist at Grosseto in the Agro Pontino, and in the neighbourhoods of Salerno and Pæstum. Malaria is on the whole decreasing in Italy, and there has been a steady and almost unbroken diminution in the number of recorded deaths from the disease, from a total of 713 per million inhabitants in 1887 to one of only 450 in 1896[1]. The last-named figure is however still a very high one, and malaria remains as in former days the most fatal disease in Italy.

In the Balkan peninsula the malady is of very frequent occurrence. In Roumania, Bulgaria, and Servia it is endemic along the course of the Danube. In the Dobrudsha it is very prevalent. In Montenegro it is rife in the fertile plains watered by the Tshernitza, and other rivers flowing into the Lake of Skutari. This lake is situated on the border between Montenegro and Albania, and as it is now much smaller than it was in former times it has come to be surrounded by extensive swampy tracts, in which the conditions for the development of malaria are highly favourable[2]. In many parts of Turkey in Europe malaria is extremely prevalent. The province of Salonika has always been a *foyer* of paludic fevers, and the disease is frequently reported from various parts of the province. In Dedeagatch, in Kavala, and in other parts of European Turkey the reports received by the Ottoman Board of Health from its sanitary officers show clearly that malaria is the leading disease.

In Greece malaria, fortunately of a mild type, is exceedingly common, and is said to account for one-third of the total sickness in the country. The great majority of the cases are intermittent in character, the continued and pernicious types being rare.

Many islands in the Mediterranean are malarious. The coasts of Corsica and Sardinia are unhealthy from this cause. In Corsica the east coast, which for a distance of some 60 miles in length

[1] *Annuario Statistico Italiano*, 1898.
[2] Levy. *British Medical Journal.* April 1st, 1899, p. 801.

and two or three miles in width is low and marshy, is notoriously fever-stricken, and in July, after the harvest is in, the inhabitants are obliged to fly to the mountains to escape the disease. In Sardinia malaria is very prevalent, and in the south-east of this island, and also in parts of Sicily, the mortality from this cause is excessively high, rivalling that from the same disease in the Basilicata and Pontine marshes. In these malarious districts of Italy and the adjoining islands the deaths from this class of fevers alone amount to 8 per 1000 inhabitants per annum.

In the island of Crete paludism also prevails to a considerable extent, and during the recent military occupation by the Powers many of the European troops stationed there suffered considerably from various forms of malarial fever.

Asia. In the Caucasus paludic fevers are excessively rife in the province of Kuban, on the eastern shores of the Black Sea, where in 1895 very nearly 100,000 cases were registered among a population of only 1,693,000. Malaria is not confined to the coast, but occurs with equal intensity throughout every district of this flat and marshy country. In the province of Terek on the east it is only a slightly less fatal scourge. In the government of Stavropol, just to the north of these, malaria is considerably less common, but is still a very serious factor in the pathology of the country. These three provinces, which constitute the Caucasus proper, are all flat, low-lying, and more or less marshy. The governments of Transcaucasia, on the other hand, are for the greater part mountainous, and here malaria is very much less common. It is, however, present to no inconsiderable extent in every government lying between the Black and the Caspian Seas. It is especially prevalent along the valleys of the Kura and Rion rivers and on the Black Sea coast.

Throughout Central Asiatic Russia malaria is rife. It is impossible to judge of its extent here from the official figures of registered cases of sickness or death, as the great majority of cases in the native population do not come to the cognizance of the Russian medical authorities. On the eastern coasts of the Caspian Sea, around the shores of the Sea of Aral, along the banks of the Zeriafshan, Amu Daria and Syr Daria rivers, and in the neighbourhood of the irrigation channels which intersect the country, malaria is always present. At times it has taken on the characters of an epidemic of intense severity. This was the case in 1897, when in the provinces of Transcaspia and Samarkand, and especially in the neighbourhoods of Merv and Tashkent, malaria of most virulent character prevailed to an alarming extent. In Merv another very severe outbreak of the disease occurred in 1900.

In Bokhara these fevers are very widely spread, and often of very malignant character[1]. They prevail principally in the summer, but in the valleys they occur all the year round, only lessening in the winter. The localities particularly notorious for fevers are Hissar, Kuliab and Patta-Hissar, and generally the towns on the banks of the Amu Daria and its branches. So unhealthy is Hissar in the summer that its inhabitants annually evacuate it at the end of February for a more salubrious spot. The Russian settlers also suffer considerably.

In Siberia malaria is also widely spread. Though there are no doubt large areas of country more or less free from it, the disease is nevertheless found in every province, from north to south and from east to west. It is common along the banks of the great rivers, as the Yenisei, the Obi, the Lena, and the Amur.

It is probable that a great part of the Arabian coast is more or less malarious. In both Mecca and Medina all forms of malarial fevers are seen among the Moslem pilgrims, and "pernicious fevers" and malarial cachexia are among the commonest causes of death among them. To what extent these affections are of local origin, or how far they are contracted by the pilgrims elsewhere it is difficult to say. In the higher table-lands of Assyr malaria is entirely unknown.

In many parts of Asia Minor malaria is common. In the neighbourhood of Trebizond there are many marshes and all forms of the disease are seen. At Adalia, and on the islands of Mitylene and Samos it is a source of much illness. It is particularly prevalent along the shores of both the Black and Mediterranean Seas, and at many places on the Turco-Persian frontier (Khanikin Penjvin, etc.). Mesopotamia, the marshy district between the Tigris and Euphrates, is exceedingly malarious. In Persia malaria is found principally along the shores of the Caspian Sea and the Persian Gulf, and along the Turco-Persian frontier. It is also found at considerable elevations in the central Persian table-lands, and is not uncommon in Teheran itself.

[1] *Bolnitchnaia Gazeta Botkina*, 22—24, 1899, "Bokhara and its Sanitary Condition," by Dr I. I. Grekof.

The coast of Palestine has long had an unenviable reputation for "fevers"; they are very prevalent at Tripoli, Jaffa, and elsewhere. Afghanistan and Baluchistan are both spoken of as malarious countries.

In India malaria or allied fevers annually cause the death of between four and five million persons. These deaths are classified under the general term "fever," and while it is possible that a certain number of them are due to other pathological conditions it is probable that the overwhelming majority are caused by malaria in some of its many forms[1].

The disease is met with in almost all parts of the Indian peninsula. It is commonest on the low-lying coasts of the Bay of Bengal, and along the banks and especially at the mouths of great rivers, such as the Indus, the Ganges, and the Brahmaputra. A form of fever of intense severity occurs in the belt of waterlogged forest at the foot of the great Himalayan range to which the name of the Tarai has been given. The most malarious of the territorial divisions of India are those of the Central Provinces, Assam, Bengal, Bombay, Coorg, and Berar. The provinces with the lowest mortality from malaria are those of Lower Burma, the Madras Presidency, and Mysore[2]. The deltas of the Indus and Ganges, already mentioned, consist of low, marshy lands overgrown with jungle, and liable to frequent inundations, and they are consequently notoriously unhealthy. Similar conditions, but less pronounced, are found throughout the greater part of the coast line; the western shore of the peninsula is said to be more malarious than the eastern. Bombay itself is built on a marshy island in which malaria is rife. In Bengal, in the neighbourhood of Calcutta, is a triangular tract of land which is believed to be the most unhealthy area in India, with the single exception of the Tarai. This tract is bounded by the high banks of the Ganges and the Madhamati on the north and east, by the districts of the Southern Parganas, Birbhum, Bankura and Midnapur on the west, and by the tidal swamps of the Sundarbans on the south. It contains some fifteen or sixteen thousand square miles, and includes the districts of Murshidabad, Nadia,

---

[1] Crombie (*Indian Medical Gazette*, April, 1899) has made a useful provisional classification of the "Unclassified Fevers of the Tropics."

[2] *Annual Reports of the Sanitary Commissioner with the Government of India.*

Burdwan, Hooghly, Jessore, Khulna, Howrah and the 24 Parganas.
The population is over eleven millions. No part of this area is
more than 60 or 70 feet above sea-level, and in the 24 Parganas,
Howrah, and Khulna the average elevation is only from 10 to
20 feet. The intensity with which malaria prevails here may be
gathered from the fact that the average mortality from this disease
alone is in many parts 50 per thousand per annum, that considerably over one-half of the inhabitants have enlarged spleens, and
that in some villages this condition is found in as many as
81·2 per cent. of the population. The disease is worst in low
and undrained situations, but less frequent on the elevated banks
of rivers and in well-drained sites[1].

In India generally malaria has been particularly active in recent
years of famine. In the Central Provinces, after the disastrous
and sudden cessation of the monsoon rains in September 1896,
the disease became unusually virulent and widespread, and caused
a mortality for the year of no less than 29·5 per 1000 of the
inhabitants. "There was an increased production of the poison,
owing to the abrupt cessation of a heavy rainfall, and diminished
resistance due to the famine[2]."

Many parts of Ceylon are highly malarious, and occasionally
epidemics of great malignity prevail in some districts of the island.
In the hospitals and in the gaols the enormous majority of cases
of sickness are due to malaria in one of its forms. It is a cause
of a very high annual mortality in the island generally; the deaths
from "remittent fever" and "malarial cachexia" more than doubled
between the years 1891 and 1897. In the same period those
registered as from "ague" diminished considerably, and those
from "simple and ill-defined fever," a term which appears to
include the greater number of malarial cases, also somewhat
declined. But in 1898 the deaths reported under all these
headings, and particularly those ascribed to "malarial cachexia,"
increased enormously[3].

Paludism is endemic throughout Farther India and the Malay
Peninsula. Many parts of Siam are highly malarious, and certain

---

[1] Gregg. *Transactions of the First Indian Medical Congress.*

[2] *Annual Report of the Sanitary Commissioner with the Government of the
Central Provinces*, for 1896.

[3] *Ceylon Administration Reports. Report of the Principal Civil Medical
Officer and Inspector-General of Hospitals*, for 1898.

forests and valleys possess a deadly reputation for the prevalence of the most severe and fatal forms of fever. The valley of the Mekwang, particularly in the neighbourhood of Luang Prabang, is notoriously malarial, and "untold numbers" of natives are said to perish of the disease each year. In the forest belt behind Chantabun jungle fever is rife, and throughout the plains of Cambodia malaria, mostly of the intermittent form, is commonly met with[1].

In the Malay Peninsula proper malaria is also an important factor in the pathology of the country. At Singapore, in Penang, and in Province Wellesley large numbers of persons annually come under treatment for some form of remittent or intermittent fever. In the hospitals and prisons throughout the Straits Settlements more cases of malarial fever or malarial cachexia come under treatment than of any other disorder, though as a cause of death in these institutions malaria is exceeded by dysentery, beri-beri, tubercle, and anæmia. Throughout the peninsula generally the more severe forms of the disease are rare. It is noteworthy that among the pilgrims annually visiting the Holy Places in the Hedjaz the Malayans seem to be particularly liable to attacks of intermittent fever.

In French Cochin China, Annam, and Tongking malaria is widely prevalent. In the French settlements it is the cause of three-fifths—and sometimes even of two-thirds—of the sickness-rate, and one-fourth of the mortality. The disease is mostly met with on the coasts, and particularly at the delta of the great Mekong river, which in its upper course forms the boundary between Siam and Annam. This area of low-lying land is mainly devoted to the growing of rice, and is subject to annual inundations,—conditions which have universally proved favourable to the development of malaria.

Throughout China malaria is exceedingly prevalent, though as a rule it occurs in the more benign forms of intermittents. Severe remittents and attacks of the most pernicious types of the disease are, however, occasionally met with in some parts. It is said to prevail more or less throughout the valley of the great Yangtze-kiang, and is especially active after the floods to which that river is annually liable. It is rife at many places along its banks. At Chung-king intermittent forms of fever are very common. At

[1] H. Warington Smyth. *Five Years in Siam*. London, 1898.

Hankow tertians are the most frequent, then quartans and quotidians; the most severe cases are met with in foreign residents and
are said to be imported from southern China or from Formosa.
In the natives of the surrounding districts fever prevails from the
beginning of spring to the end of autumn. At Kiu-kiang the quo-
tidian type is the form that most often comes under observation.
Malaria is also more or less rife along the whole of the south coast
of China. All types of the disease are seen in the south-west province
of Yünnan. At Lung-chou, Pakhoi, and other places in the
Kwang-si province the disease is endemic, and it is equally so
in the adjoining province of Kwang-tung (Canton). In the Lappa
Customs district, and especially at Macao, malaria prevails in the
hot and damp summer months; at Wuchow on the Canton river
it is also common. The east coast of China is malarious in many
places. Hongkong is notorious for the prevalence of fevers.
In the neighbourhood of Swatow tertian intermittents are very
common, quotidians and quartans somewhat less so, while remittent and continued fevers are much less frequent. At Ningpo
malaria is the most prominent disease. Around Shanghai it is
also commonly met with; it was unusually prevalent here in the
summer of 1889.

Further north, at Chifu, on the Straits of Pe-chili, malaria
became widely epidemic in the year 1895 as the result of turning
up long-undisturbed ground for road-making and building. Severe
forms of remittent fever are not rare here.

In Peking intermittent forms of fever are common enough, but
they usually assume a benign character, and pernicious attacks are
rare. At Tientsin remittent forms are not infrequently observed[1].

In Corea malaria is the predominating disease. At Séoul,
Chemulpo, and elsewhere in the peninsula it prevails principally
in the spring and autumn.

In Japan malarial fevers are exceedingly common, particularly
on the plains where rice is extensively cultivated. An area of
malaria "of quite extraordinary intensity" is said to have developed
in recent times in the Suruga province on the Oigawa river, where
the rice fields are lower than the level of the river. This district
contained in 1882 about 51,000 inhabitants, and it was said that

---

[1] See the successive *Reports of the Medical Officers in the Chinese Customs
Service.*

the annual number of cases of malaria amongst them was not less than 16,000 to 20,000[1].

Many other islands off the Asiatic continent are homes of malaria. In Formosa all forms of the disease are met with, but tertians are the most frequent, while severe bilious remittent forms, which are said to have been common in foreigners after the opening of the port of Tamsui, are now extremely rare. About one-fourth of the patients treated in the hospital at Tamsui are victims of malaria. The worst regions in the island are certain valleys shut in by hills, and flat lands lying between the mountains and the sea, such as the Kapsulan plain and in the neighbourhood of Tamsui. The aboriginal inhabitants of the mountainous districts are said to enjoy relative immunity; the greatest sufferers are the Chinese, particularly those not born in the island, and the men of the Japanese garrisons[2].

Throughout many of the East Indian islands malaria is widely endemic. In Borneo it is the chief disease. Dr Nieuwenhuis, the first European to cross the island from east to west (in 1894) states that certain regions are free from it; thus the two capitals on the south and west coasts, Bandjermasin and Pontianak, are not malarious, although they lie in the midst of a marshy country and water and mud abound in the neighbourhood. The same was found to be the case with Palembang, in Sumatra, and at Sambas, in western Borneo. In the more elevated regions of the Bornean interior, malaria is rife; it is said to cause a high annual mortality among the Dyaks, the principal inhabitants in these regions, while the Malays in the plains suffer but little. It is not a question of racial immunity, as the latter easily contract malaria if they go to the hills, and, as already stated, the Malays, both in their own country, and when on pilgrimage to Mecca and Medina, are frequently subject to its attacks. In the north-eastern portion of the island malaria is by far the most prevalent disease, and annually causes a vast amount of sickness and death. In Labuan, off the north-west coast, it figures more prominently than any other malady in the annual hospital reports. In Java malaria is very common among the hill tribes; it is however said not to be endemic among dwellers in the hills, but to attack persons who

[1] *Sei-i-Kwai*, 1898.
[2] A. Rennie in *Reports of the Medical Officers of the Chinese Customs Service*. No. 45.

from time to time are compelled to labour on the coffee plantations there[1]. At Batavia and other Dutch stations a heavy
mortality from malaria occurs among the Dutch residents and
soldiers stationed there. Very severe or pernicious forms of the
disease are common, and Batavia at one time had earned the
unenviable name of "the Dutchman's Grave." Of recent years
marked improvement has taken place in the health of the town,
mainly owing, it is believed, to improved sanitation, to a gratuitous
supply of quinine to the people, and possibly in part also to an
improved water-supply. Koch has stated that the malaria sickness-rate in the Dutch Colonial army in Batavia had diminished
by 50 per cent. in the fifteen years ending in 1899[2].

Throughout the Philippine Islands malaria is also extremely
prevalent. The Spanish troops in this archipelago suffered greatly
in 1895 and the following years, when the disturbances brought
about by the war led to a greatly increased prevalence of the
disease.

Australasia. Many of the islands of Oceania are, on the other
hand, remarkably free from malaria, although the presence of
marshy, low-lying coasts seems to invite its prevalence. It would
appear to be rare in Fiji, only 2 or 3 cases coming under treatment each year in the colonial hospital at Suva. In New
Caledonia marsh fever is said to be unknown, in spite of the
abundance of marsh-land.

In the adjoining group of the New Hebrides, on the contrary,
malaria is, it is stated, very common[3]. Severe forms of malarial
cachexia are met with here among the Papuans. The Sandwich
group and the Marquesas Islands are reported to be free from
the disorder.

In British New Guinea, however, it is the disease most
frequently met with, attacking both natives and Europeans.
It is commonest in the low-lying districts, and appears to be
absent in the hills. Its distribution is very irregular, and some
of the most swampy spots are exempt while other places far
removed from swamps are notoriously malarious[4].

---

[1] Kohlbrügge. *Janus*, 1897–8, p. 221.
[2] *Deutsche Med. Wochenschrift.* Feb., 1900.
[3] Bernal, *Archives de Médecine Navale.* Aug., 1899.
[4] See Sir W. MacGregor's *Annual Report* for the year 1897–8. He states
that it is very remarkable that again and again parties of twenty to thirty men,

Malaria is a rare disease in Australia. In West Australia it was the cause of only 12, 6, 13, and 4 deaths respectively in the years 1889—1892 inclusive; in 1893 and 1894 it caused not a single death; in 1895 only 2 deaths from remittent fever were registered, and in each of the three following years only a single death. Intermittents were the cause of scarcely more deaths. South Australia is equally free from malaria, and only single deaths from remittent fever and none from intermittents were registered here in the years 1893 to 1898. In New South Wales less than a score of deaths per annum are attributed to malaria. In Victoria it appears to be unknown, at least as a cause of death, and the same is true of Tasmania. The immunity of Australia, as contrasted with the prevalence of the disease in New Guinea, is remarkable.

New Zealand is also free from paludism, in spite, it is said, of the presence of marshes of considerable extent.

Africa. The north coast of Africa is not especially malarious. In Tunisia some foci of the disease are met with in scattered parts of the Regency, but cultivation is beginning to improve them, and it is hoped this class of fever will disappear in time. In Tripoli and in Morocco malaria is endemic and more or less widely spread, but it occurs in a comparatively mild form. In Egypt the disease is met with, especially at the mouths of the Nile, and forms of fever are common along the whole valley of that river, prevailing especially after the annual inundations begin to subside.

In Abyssinia malaria is met with on the plains, but is said to be unknown in the hills. At the northern foot of the great central plateau of Abyssinia is a series of vast marshes, clothed with luxurious vegetation and known as the Kollas, and here cases of fever are exceedingly numerous. In Algeria the malady is prevalent on the coast and along the muddy banks of rivers. The worst regions are in the neighbourhood of Lake Alloula and on the coasts of Chiffa, about the marshes of Ferguen, Chaïba, and

natives and Europeans, have spent several consecutive weeks in low-lying mangrove and swampy country and have left it without a single sick man; while, on the other hand, of the same number of men landed in Cloudy Bay and camped on dry ground and at a distance from swamps, all save one man became ill of fever. Some of the worst cases have been contracted in a dry, limestone district.

Mazafran, all in the province of Algiers. In the province of Oran the plains of Sig and of Habra are feverish, and in that of Constantine the plains of La Seybouse and around Lake Fezara have the same reputation. At the time of the French conquest of Algeria the troops suffered very severely from the ravages of malaria, but of recent years cultivation and drainage have done much to improve the health of the colony.

In the great Sahara desert malaria is said to be unknown, except in the oases, but at some of these, as for example at Biskra, it is rife.

The West Coast of Africa has, from time immemorial, been ravaged by malaria. Few places along the entire coast are free from it, and in many the disease occurs in its most malignant and fatal forms. In Senegal malaria becomes violently epidemic each year at the end of the rainy season ; severe types of remittent fever are very common, and not infrequently bilious complications are present. In the hospital at Dakar malaria accounts for 57·5 per cent. of the total cases treated. From January to July the disease quite disappears, but only to break out with renewed virulence in the following months[1]. In the British colony of the Gambia the disease is rife, and the same periodic prevalence is observed. From July to October the climate of Bathurst is extremely bad ; fevers abound, the natives suffering from them as well as the European residents. The rest of the year is said to be fairly healthy[2]. In Sierra Leone, long notorious for the deadly character of its climate, malaria is no less prevalent.

The whole coast of Guinea is intensely malarious. The Gold Coast, the Ivory Coast, and the Slave Coast, the mouths of the Niger and the shores of the Cameroons, have an unenviable reputation for their extreme unhealthiness. At Cape Coast Castle malaria has long been the cause of a very high mortality. At Kumasi the types of fever met with are less severe than on the coast. In the Cameroons the prevalence of the disease is said to vary directly with the rainfall. The ravages it commits here may be estimated from the fact that of thirty missionaries sent out by the Bâle mission between 1886 and 1893, ten died, and five had to return to Europe on account of their health. In

[1] *Annales d'Hygiène et de Médecine Coloniales*, 1898, I. Dr Clerac, "Le Paludisme au Sénégal."
[2] *Colonial Report* for the Gambia for 1898.

a group of ninety men stationed there 438 attacks of malaria, or nearly five per head, occurred in the course of one and a half years. And out of one group of 44 deaths, 34 or 77 per cent. were caused by malaria[1].

In the French Congo malaria is the most prevalent of all diseases.

In Central Africa malaria is seen along the course of the great rivers and around the larger lakes, but it is of less severity and prevails to a much less extent than on the coasts. In some parts, however, a severe type of this disease and a high degree of prevalence are met with. In the Congo Free State malaria attacks both the black and the white races, and in the latter pernicious forms of it are not rare. The valleys of the Congo and the Zambesi are notoriously feverish. In British East Africa and in German East Africa the disease is not uncommon. The prevalence of the fevers varies to some extent with the height above sea-level; they are however met with at considerable heights, in Uganda prevailing up to an elevation of 5000 feet. They are severe around Lake Nyassa (at Bandawe, for example), and also on the shores of Lake Tanganyika. In the Masai highlands on the other hand there is said to be no malaria; and about the Victoria Nyanza and in the Shiré Highlands and about Zomba (Portuguese East Africa) the disease is not severe. In British Central Africa fevers of tertian type are not uncommon. Along the east coast of Africa generally malaria is far from rare, and in places, as at Zanzibar and about the mouths of the Zambesi, it prevails to a very grave extent. About Mombasa it is very common; but the coast of British East Africa is said to be less feverish than that of the German territory immediately to the south.

In many parts of South Africa malaria is observed with great frequency; but nowhere in this part of the continent does it attain the extent or severity which characterise it on the West Coast and in certain regions in the interior of Central Africa. Its prevalence varies considerably in different years. In the region of the Transvaal[2] malaria is endemic in many places; in 1893 the disease was prevalent throughout the greater part of the Republic, but in 1894 on the other hand very few cases were recorded. In the Barberton Hospital 449 cases of the disease

[1] F. Plehn. Ref. in *Arch. de Méd. Navale.* Sept. 1899.
[2] *Journ. Trop. Med.* October, 1899.

were treated in the five years 1893—1897. The climate of the Transvaal is sub-tropical, and in the north and east are numerous low-lying swampy districts with a reputation for fever. Mashona-land is a very malarial country, and many of the labourers recently employed in railway construction there have suffered from the fever in some form.

In Natal malaria is certainly not common. The coast of south-eastern and southern Africa is incomparably more healthy than either the eastern or western coasts further north. At Delagoa Bay, however, severe malarial fever prevails at times (*e.g.* in 1890). At Port Elizabeth and at East London fevers of distinctly malarial type are recognised. At Port Elizabeth 23 cases of the disease were treated at the Provincial Hospital in the year 1898. At Durban doubt has been expressed whether true malaria exists as an endemic, but 134 cases returned as cases of malaria were treated in the Durban Hospital in 1898[1].

The floods to which some of the South African rivers give rise after heavy rains lead at times to the prevalence of malarial fever. This was the case at Gordonia in 1898, when ague became epidemic after severe floods caused by the Orange River in the month of January. In the same year it also prevailed at Victoria West. At Graaff Reinet a considerable number of cases come under observation annually.

Further inland, on the other hand, malaria would seem to be common in many places. In Rhodesia, within the tropics, malaria is more frequent than in other parts of South Africa, except in the north-eastern Transvaal and Zululand. Among the British troops stationed in South Africa in the year 1897, malaria attacked principally a regiment garrisoned in Rhodesia, and many of the cases of the disease treated in the hospitals at Capetown are drawn from this part of the colonies. At Buluwayo malaria is the predominating disease; the mortality from it is, however, not high among the Europeans, though among the natives it is con-siderable[2]. In Zululand malaria is commonly met with, especially in the districts of Ndwandwi, Lower Umvolosi, and Lebombo[3].

Many of the islands off the continent of Africa are highly malarious. In Madagascar the disease is rife along the entire

[1] *Report of the Durban Hospital* for 1898.
[2] *Report of the Memorial Hospital at Buluwayo* for 1897–8.
[3] *Colonial Report* for 1891.

coast and in the river valleys. It is found especially on the west coast, where a series of feverish marshes extends between the mountains and the sea. As the mountains are approached malaria becomes rarer, but it is met with even on the high plateaux of Antsihanaka. On the east side of the island the malarial zone is narrower, because the hills are nearer the coast. In the French expedition of 1895 the route chosen to gain the interior of the island was from the western side, with the result that the troops suffered to a disastrous extent. In ten months 6000 men died, mostly from some form of malarial fever, and the survivors were reduced to extremities. Some points on the coast are relatively immune, as for example Nossi Komba in the north-west. In Antananarivo malaria is not common, but in certain places not far from the capital it prevails to a considerable extent. In the little island of Nossi-Bé, off Madagascar, malaria is excessively prevalent. In 1880 it was the principal factor in a general mortality of 89 per thousand in the general population, and of 75 per thousand in the French troops stationed there.

At Mayotta, in the Comoro Islands, "paludism absorbs the entire pathology" of the island; no one escapes it, and the garrison has to be changed every year.

In the Mascarene Islands malaria is now endemic. The history of the disease here is one of great interest. Before the year 1857 malaria was unknown in Mauritius or Réunion. During the next few years a limited number of cases only occurred; but in 1866–9 a violent epidemic of the disease swept through both islands, and since that date malaria has been the most important and prevalent of disorders met with there. Malaria is said now to "dominate the pathology of the island [of Mauritius], not only in the sense that it alone gives rise to more than half the total mortality, but also in respect to its influence in modifying the character and determining the seasonal prevalence of other diseases[1]." The same appears to be the case in Réunion, but Rodriguez, which adjoins and has apparently the same climatic and other characteristics as Mauritius or Réunion, has hitherto escaped entirely. The fact is one of the most remarkable in the distribution of malaria. At the present time malarial fever and cachexia are the direct cause of one-third of the entire mortality in Mauritius.

[1] Davidson.

C. 17

In St Helena malarial fever is also met with, but to no great extent.

North and Central America. Malaria is rare in the western hemisphere to the north of about the 45th degree of north latitude. Its northern limit is therefore further south here than in the eastern hemisphere. It has not been seen in Greenland. In the marshy tracts of the Hudson Bay territories it is also said to be unknown. In Canada it must be exceedingly rare, if it exists ; it is scarcely mentioned in recent annual health reports of any of the provinces in the colony. The disease begins to appear about the shores of the great lakes of Huron, Erie, and Ontario.

In the United States it is common on the coasts of North Carolina, Pennsylvania, and New Jersey. In the prairie ravines west of the Missouri and on the Kansas plains it is common, and it occurs in many places in the great central plain between the Mississippi and the Alleghany Mountains. In Massachusetts malaria appeared, it was believed for the first time as a serious epidemic, in 1885[1]. It broke out in a village in the eastern portion of the State, and later spread to many other towns in the neighbourhood. Serious outbreaks occurred at Uxbridge, N. Saugus and Woburn. In a small district in Woburn, covering an area of about half a square mile, it was found by a house to house inquiry that 1900 cases of malaria had occurred in the course of the years 1894, 1895, and 1896. In Delaware malaria is also met with.

Along the States bordering the shores of the Gulf of Mexico —Florida, Georgia, Alabama, Louisiana, and Texas—malaria is a very common disorder. The delta of the Mississippi and the neighbourhood of New Orleans are markedly feverish. On the Pacific coast of the States, the disease is most prevalent in California, where the Sacramento and San Joaquin valleys are said to be specially unhealthy. In these valleys malaria is believed to have existed only since 1846, and to have become commoner with the introduction and development of hydraulic mining[2].

---

[1] The disease was known in Massachusetts prior to 1885, but was never severe.

[2] P. King Brown, M.D. *Journal of the American Medical Association.* Vol. xxxii. Jan.—June, 1899.

Both the Pacific and Atlantic shores of Mexico are malarious. In the Bay of Vera Cruz, on the east, a series of marshes exists behind the barrier of dunes which edges the sea, and here fevers are rife. The high plateaux more inland are free, and in the town of Mexico itself, although surrounded by marshes, the disease is not met with. It is found on the coasts of Yucatan and is endemic throughout the greater part of Guatemala, especially in the departments of Retalhuleu, Suchitepequez, Escuintla, and others. In British Honduras malaria "dominates the pathology and augments the death-rate[1]"; as a rule it is of a mild type, but at times severe outbreaks occur, as for example in 1871–72, when it became widely epidemic. In Stann Creek district is a flat alluvial swampy tract of black, pultaceous offensive soil, covered with a sandy marl left by the receding sea, and here malaria prevails, but not of a severe type. It is endemic, but occurs in a mild form, in Nicaragua, Costa Rica, and San Salvador, even at considerable altitudes.

In Panama malaria is excessively prevalent, and some of the worst forms of the disease are met with. Colon is notorious for its feverishness, and it will be recalled that the labourers on the ill-fated Panama Canal suffered from malaria to a disastrous extent.

The majority of the West Indian Islands are malarious. Cuba is excessively so, and the sufferings of the American and Spanish troops in the recent military operations there from both malarial and yellow fevers were very great. It was estimated that fully 7000 Spanish troops died from these causes alone. In the American camps malaria was the most formidable disease; 90 per cent. of the cases assumed the æstivo-autumnal type; tertians were rare, and quartans almost unknown. Havana is a well-known centre of the disease. In Jamaica[2] malaria is an important factor in the sickness-rate, but less so than in Cuba; a severe, remittent form of the disease was epidemic in the island in 1897. In the Lesser Antilles the malady is very intense in Guadeloupe and Martinique, while in Anguilla and Barbados it is said to be unknown. In St Lucia all types, from the mildest to the most severe, are met with, remittents being, however, the commonest. In St Vincent remittents and

[1] *Report of the Colonial Surgeon for* 1898.
[2] *Reports of the Island Medical Department.*

intermittents are not uncommon, and the same is true of Grenada and of San Domingo. In Trinidad malaria is endemic in many places; in the Mayaro district it is always very prevalent; in the Manzanilla and Toco districts severe forms are often seen, and there are few parts of the island where it is not an all-prevailing disease[1]. In the Virgin Islands a severe form of it is endemic; in 1897 an epidemic of a pernicious malarial fever visited the islands of Anegada, Virgin Gorda, and Tortola. In the Bahamas malaria is not uncommon.

South America. Along the northern shores of the South American continent the inhabitants suffer considerably from paludic fevers. In British Guiana all forms of malaria are exceedingly prevalent, and the disease causes a higher mortality than any other[2]. In Dutch Guiana it is no less rife. In French Guiana three-fourths of the total amount of sickness is due to paludism in its many forms. Four zones have been described here, viz. the maritime zone, consisting of the islands off the shore (Îles du Salut, etc.), which are comparatively healthy, the littoral zone, which is less healthy, but not so feverish as the third zone, consisting of vast marshy tracts known as the *savanes noyées* or *savanes tremblantes*, which are so deadly malarious that all efforts at colonising this region have failed. The fourth zone is that of the mountains, which are sparsely inhabited and of which little is known. In Venezuela malaria is rife throughout the basin of the Orinoco river and along the sea-coast.

The shores of Brazil are said to be malarious more or less throughout their entire extent, while in the interior of the country the disease is rarer; the valley of the Amazon is, however, feverish. In the Argentine Republic malaria is met with, especially in the provinces of Tucuman, Salta, Jujuy, Corrientes, and in certain districts in the north, where marshes are numerous, water abundant, and the temperature high. In Paraguay, Uruguay, and La Plata there are many marshes in the midst of the great plains, and inundations by the rivers which flow from the slopes of the Andes are frequent, yet in spite of these conditions the climate is healthy and paludic fevers are said to be rare.

---

[1] *Annual Report of the Surgeon-General for* 1898.
[2] *Report of the Surgeon-General for* 1898–9.

On the western side of the continent malaria is certainly met with on the coasts, but becomes rare or extinct as the great range of the Andes is approached. In Colombia the coasts are malarious, and the same is the case in Peru and Bolivia. Chile has on the whole a healthy climate, and south of latitude 35° malaria is said to cease altogether in the western hemisphere.

Factors which govern the Geographical Distribution of Malarial Diseases. If, as was pointed out at the beginning of this chapter, certain mosquitoes are essential to the transmission of malaria, one of the most salient factors in the geographical distribution of the disease must necessarily be the geographical distribution of these mosquitoes; and, further, the well-known influence of external physical conditions upon malaria-prevalence must now be regarded as acting not simply upon the malarial parasite itself, nor even upon the parasite and its human host, but also upon the insect in which it passes part of its existence. The geographical distribution of the specific malaria-bearing mosquitoes has not yet been fully worked out. The influence of many external conditions on the distribution of the disease has, on the other hand, long been recognised, and will be briefly discussed here, without attempting to determine in what proportions this influence acts directly on the parasite, the human subject, or the malaria-bearing mosquitoes.

Malarial fever may occur in almost all latitudes, including the subarctic, temperate, and tropical zones. It is however commoner and more severe in hot climates than in warm, and in warm than in cold, and from the equator to the poles there is a fairly regular decrease in both the prevalence and intensity of the disease. In Europe the most northerly point at which it has been seen is at the head of the Gulf of Bothnia in the 66th degree of N. latitude, and in the Russian province of Archangel. In Asia it has been met with in Siberia, in the northerly government of Yakutsk. In America it has occurred as far north as Southern Labrador, but as an endemic its present limit here is about the 45th parallel. Warmth, therefore, has clearly a powerful influence over malarial prevalence. This is further shown by the comparative rareness of malaria at high altitudes. Low, marshy countries have from time immemorial been regarded as the great homes of malaria, and though the disease is exceptionally found in elevated and dry regions, it is infinitely more common and severe at lower altitudes and in moister climates. Flat, hot, marshy coasts, as those of Western Africa; the deltas of great rivers, as of the Mississippi, the Ganges, the Rhône, the Loire, the Volga, the Mekong, the Indus, the Brahmaputra, and the Nile; moist plains at the foot of great mountains, as the Himalayan Tarai or the *savanes noyées* in French Guiana; low-lying lands subject to inundations by rivers or seas; swampy rice or paddy fields; and land that has gone out of cultivation all offer ideal conditions for the prevalence of malaria.

Moisture is essential to the existence of malaria, mainly no doubt because the malaria-bearing mosquito, like all other mosquitoes, passes one portion of its life-cycle in water. The presence of marshes, stagnant lakes, and pools is highly favourable to malaria prevalence, and one of the most powerful measures in ridding a country of these fevers has been the draining of the soil and the filling in of notoriously unhealthy collections of standing water. A measure of this kind has been long practised empirically in Holland, in France, and elsewhere, with most successful results in the lessening or disappearance of malaria. It is now being extensively practised in other countries with a more conscious and certain aim—the removal, that is to say, of the great breeding-grounds for the malaria-bearing mosquito—and it is being supplemented by still more energetic attempts to do away with even the smallest collections of standing water—such as holes and irregularities in the surface of land about dwellings, and all hollows or receptacles in which water might possibly stagnate and become a breeding-place for these insects.

It was formerly believed that in addition to warmth and moisture a certain amount of decomposing organic matter in the soil was necessary for the prevalence of malaria. No doubt decomposing vegetable matter is one of the commonest concomitants of a warm and moist soil, but it does not seem to be essential to the production of malaria, for the disease can exist in its absence.

To what extent malaria can be carried by the air is uncertain, but it is usually believed that the infection cannot be diffused far by air currents either in a horizontal or in a vertical direction.

The crews of ships lying some distance from a highly "feverish" coast have often escaped attack, and there seems to be no authenticated instance of malaria being wafted by the wind to ships passing a malarial coast. Outbreaks of malaria on shipboard have occurred, but they have been apparently, and in some instances certainly, due to the drinking of contaminated water by the crews. Manson speaks of at least one authenticated instance of this kind, but he also expresses the belief that the malaria germ soon dies in water, so that after a few days' voyage a ship's water-supply would cease to be dangerous. The tendency of malaria to cling, as it were, to the earth and to show but little power of rising high above it has long been recognised, and in Italy and many other malarious countries the peasants secure a relative or certain immunity from attack by building their dwellings on supports at a considerable height from the ground.

Race is perhaps a factor of some importance in the malaria problem. All the great divisions of the human family can and do suffer from the disease, but some appear to be much less susceptible to it than others. The negro races are said to be to some extent immune to it, as well as the aborigines of the Formosan hills. It was formerly thought that the Chinese and Malays were also relatively immune to malaria, but recent evidence lends no support to the view. How far any apparent immunity observed in certain native races is congenital and how far acquired it is difficult to say. Acclimatisation seems to be possible, though in the case of malaria, as in that of many other diseases, what is called acclimatisation is, as Manson puts it, less "an unconscious adaptation of the physiology of the individual" than "an intelligent adaptation of his habits." It has now been shown that malaria is largely a preventable disorder, and that it is possible to escape it by a careful avoidance on the one hand of chill and other conditions tending to increase the susceptibility of the individual, and on the other of the introduction of the germ by the bite of the mosquito.

While malaria may be regarded as the type of a permanently endemic disease, it has shown great fluctuations in the countries where it is endemic, and has also the power of invading others where it was before unknown. The latter phenomenon is of the greatest interest and requires close study in connection with the now generally accepted view that the disease is mosquito-borne. Should it be proved that the appearance of malaria in any district has coincided with the appearance there of the mosquito which is believed to transmit the germ, no more conclusive evidence of the specific relation of the insect to the disease could be found. Several striking examples of recent introduction of malaria into regions hitherto almost, if not quite, free from it are known, and some have been already mentioned in the course of this chapter. One of the most remarkable was the sudden appearance of the disease in a severe form in the island of Mauritius in 1866. Before that date the island was practically free from malaria, but in that and the two following years a very severe form of the disease became widely epidemic there, and ever since, as has already been noted, the whole pathology of the island has been dominated by the existence of this endemic malaria. Davidson[1], who has made a most careful study of the diseases of Mauritius and particularly of this malarial invasion, states that he has been informed that in or about the year 1866 a malaria-nursing mosquito was for the first time introduced into the island. If this should be substantiated it will furnish most important evidence of the truth of the mosquito theory.

Another instance of the appearance of malaria in a healthy district is that already cited of the outbreak in Massachusetts. I am indebted to an American friend, himself one of the victims, for some interesting information in regard to the Woburn outbreak in 1894-5. Almost every household was attacked, and the disease was locally attributed to the presence of certain highly insanitary pools in the neighbourhood. Whether there was in this case an invasion of the district by a particular kind of mosquito before the outbreak my informant was unable to say.

In these instances the exact means by which the disease was introduced anew can only be a matter of conjecture. In other cases, however, a malarial outbreak has appeared to be clearly attributable to the act of man in artificially bringing about a set of conditions which are well known to favour the existence of malaria. On several occasions the construction of ponds for fish culture has been followed by the appearance of the disease, and historical examples of this kind in the disastrous malarial endemics of La Dombes, La Sologne, and La Brenne in France have been mentioned above. In some ways analogous to such outbreaks, though not quite so easily explained on the mosquito theory,

[1] *Janus*, 1898, p. 149.

are those epidemics of malaria which have followed on earth disturbance on a considerable scale, such as for the construction of railways, canals, or other public works. Of these a very large number have now been observed in all parts of the world. Among recent examples may be mentioned the following. During the construction of the railway in the Transvaal in 1893 the workmen suffered severely from malaria. At Chifu, in China, an unusually intense malaria epidemic in 1895–6 was attributed to the turning up of long undisturbed ground for road-making and building[1]. A severe outbreak of fever in the Galle and Matara districts in Ceylon from July 1895 to June 1896 was said to have been "clearly traced" to soil disturbance, the result of cutting the railway from Galle to Matara[2]. Similar observations have been made in Formosa, and nearer home the fact may be recalled that at the time of the construction of the underground railway in London, and during the excavation of the St Martin Canal in Paris (in 1811), and also while the fortifications were being raised round the latter city in 1840, malarial fever became for the time quite prevalent in the neighbourhood of the works.

Whether these outbreaks can be wholly explained by supposing that operations of this kind lead to the formation of surface irregularities in which water collects and stagnates, offering ready breeding-grounds for mosquitoes, or whether the problem is not a more complex one than this simple explanation would make it, must be regarded at present as a matter of speculation.

On several occasions and in many countries where malaria has for long been endemic it has at times taken on the most severe epidemic characters without any of the obvious reasons just discussed, such as the construction of marshes or lakes, or disturbance of the earth by man. A very striking example of this kind was the intense outbreak of malaria in Merv and Tashkent and the neighbouring country in the year 1897. The disease is always more or less present there, but in that year it caused a most violent outbreak, resulting in very heavy mortality. Less marked examples of epidemic fluctuations of an endemic malaria might be cited from many other parts of the world.

Finally the diminution or disappearance of malaria in countries or districts where it had formerly been endemic are phenomena

[1] *Chinese Customs Medical Reports.* No. 52.
[2] *Ceylon Administration Reports (Medical),* 1896.

which offer many points of interest. It has long been recognised
that the draining of marshes and swamps, the cultivation of
uncultivated land, the construction of dykes and dams to prevent
the flooding of low-lying areas by rivers or by the sea, all tend to
the disappearance of malaria. Nowhere have these measures been
more successful than in Holland, where malaria was once disastrously common, and where now it is a comparative rarity. In the
near future even more striking examples of the power of man to
control, if not extinguish malaria may be available. The "malarial
campaign" has begun in earnest in many parts of the globe. The
"mosquito theory" if it has not quite succeeded in explaining all
that is obscure in regard to this disease, has already led to practical
results of such value that the world can at present afford to wait
for a complete theoretical solution of the malaria problem. In
Italy direct attacks on the malaria-bearing mosquito combined
with the most careful measures to protect the individual from its
bites have already shown that even in the most malarial regions it
is possible to escape the disease, and a wider adoption of these
and similar measures may perhaps ere long lead to the diminution,
if not to the total extinction of malaria in the peninsula. In the
British possessions on the West Coast of Africa a determined war
is being waged against malaria, and the effect has been most
encouraging. The hope indeed is already expressed of obtaining
there a result that at any early period would have seemed incredible
—the removal, that is, of the eternal reproach against those shores
of being "the white man's grave." It has at least been shown
that, although it may be impossible to eradicate malaria from
West Africa, the European may now take up his residence there
with enormously increased powers of protecting himself against
the attacks of the disease. In Cuba and elsewhere the energetic
application of similar measures is said to have already diminished
the deaths from malaria in a most striking way.

KALA-AZAR.

General Characters and Etiology. Kala-Azar, or black
fever, is the name given to a disease met with in Assam and elsewhere, the true nature of which has given rise to some controversy.
The descriptions of the disease by different authors vary considerably, but in general terms the leading symptoms appear to

be debility, wasting, recurrent attacks of high fever in the first stage and of low fever in the second stage, rapid enlargement of the spleen and generally also of the liver, darkening of the skin (whence the name), cachexia, dropsy, and sometimes intercurrent attacks of pneumonia or dysentery.

Giles, who first investigated the disease for the Indian Government, formed the opinion that it was nothing more than ankylostomiasis in anæmic subjects. Rogers, on the other hand, reported to the Government in favour of the view that kala-azar is a form of malaria of a peculiarly infectious character. Ross, in a third report (1899), endorsed the latter view. Bentley has suggested that it is identical with Malta fever[1].

Geographical Distribution. Rogers states that this disease first appeared about the year 1887 in the Rungpore district of Bengal; and that it crossed the Brahmaputra and so entered Assam. At first it was seen mostly in the low-lying malarious country between the Brahmaputra and the Garo Hills, in the Goalpara district of Assam. It seems to have steadily spread over a considerable area of Assam and to have caused a very serious increase in the mortality of that country. The districts most severely affected have been those of Kamrup, Darrang, and Nowgong. The Nagar Hills have escaped completely, and some of the other hill districts have only furnished a very few cases. The Nowgong district has suffered most severely; in 1897 as many as 12,102 deaths from kala-azar occurred in this district alone[2].

In the last five years there has been a tendency to decrease in the number of deaths from this disease, and the hope is expressed that it may be slowly disappearing from the country. The following are the returns of deaths from kala-azar in the whole of Assam from 1893 to 1899:—1893, 10,247; 1894, 13,164; 1895, 15,894; 1896, 15,637; 1897, 18,612; 1898, 16,472; 1899, 14,246.

---

[1] *Brit. Med. Journ.* Sept. 20, 1902.

[2] *Report on Sanitary Measures in India in* 1899–1900.

MEASLES.

General Characters and Etiology. The clinical characters of measles are too well known to require description here. The cause of the disease, however, is not known, for though it behaves in all respects like other infective fevers of microbial origin, no micro-organism clearly specific to it has yet been demonstrated.

History. " Morbilli" or " Rosalia" was first recognised as a separate disease distinct from smallpox in the sixteenth century. But under these terms both measles and scarlet fever seem to have been dealt with, and it was not until later that Sydenham and others clearly showed the distinction between the two. Measles has probably had a wide distribution, at least since medieval times. It is a typical epidemic disorder, and is subject to great fluctuations in all countries.

From time to time these epidemics have extended (or, perhaps, synchronised) over wide areas, including even two or three adjacent countries. Such were the epidemics of 1796—1801 in France, Germany, and England; of 1834-6 over the greater part of Northern and Central Europe; and of 1846-7 in Northern and Western Europe and North America. Measles nevertheless can scarcely be said to have ever become truly pandemic, in the sense that influenza has from time to time.

Recent Geographical Distribution. *Europe.* In England and Wales measles is among the commonest of all diseases. It is constantly present, and shows remarkably little fluctuation from year to year. In the decade 1881-90 the annual mortality from this cause was 440 per million inhabitants, and in the decade 1861-70 it was exactly the same; in the intermediate decade it was somewhat lower, 378 per million.

In some recent years it has been unusually prevalent, in 1896 causing a mortality of no less than 572 per million, a figure which had only once been exceeded before, when, in the year 1887, it rose to 602.

In Scotland the measles mortality has varied, much as it has in England, in recent times. In the decade 1881–90 the mortality ratio from this cause was 385 per million. It is noteworthy that a high measles death-rate (500 per million) occurred in Scotland in 1895, in the year preceding an unusually high death-rate from the same cause in England. In 1896 the mortality figure for measles in Scotland was 360 and in 1897 it was 490. The disease is essentially one of urban populations in Scotland as in England.

In Ireland the mortality from measles has varied very widely in recent years; in 1888 there was a severe epidemic of the disease in the island.

In France the urban mortality from measles (*i.e.* in towns with over 10,000 inhabitants) has varied much in recent years; in 1886 –90 it was equivalent to a ratio of 460 per million, and has since varied between 170 and 380. In Paris the malady now causes more deaths than any other infectious disease[1].

In Belgium the deaths from measles remained remarkably constant in the years 1870 to 1895, the annual mean mortality from this cause scarcely varying from (approximately) 500 or so per million inhabitants. In 1897, however, it fell to about 300 per million.

Holland, on the other hand, has shown a very fluctuating mortality from measles, varying between 110 per million in 1895 and 360 per million in 1889, the minima and maxima respectively in the decade ending in 1898.

In the German Empire as a whole the death-rate from measles was returned in 1894 as 332 per million, which was a slight increase on the mean of 297 per million returned for the preceding seven years. In Prussia alone the rate was 311 in 1894 against a mean of 319 in the preceding septennium and one of 450 in the seven-year period before that. Saxony and Würtemburg have

---

[1] The deaths from measles in Paris in the five years 1895 to 1899 were respectively 679, 658, 821, 876, and 904. These figures indicate a considerable increase, in either the prevalence, or mortality, or both, of measles in the French capital.

had comparatively low measles rates, the former recording means of 294 and 271 in the two septennia named and a rate of 133 in 1894, while in the latter the three corresponding figures were 154, 221, and 272 respectively. In Bavaria and Swabia measles is, next to croupous pneumonia, the most common of all diseases.

No less marked fluctuations in the recorded deaths from measles are seen from year to year in Denmark. In the urban population of that kingdom the ratio rose from a mean of 279 in the septennium 1880—1886 to one of 403 in the next septennium, but fell to 297 in 1894. The rates for individual years show wide variations. The same is true for Norway, where the rates are on the whole low; in 1893 the recorded deaths from measles were scarcely more than 60 per million, but in 1894 they rose to about 230 in the same unit of population. In Sweden the ratio was 62 per million in 1894 as compared with annual means of about three times that number in the preceding fourteen years. Judging from these figures it would appear that measles is, on the whole, a considerably less prevalent disease in Scandinavian countries than in the British Isles.

Measles appears to be more prevalent in Russia than in any other European country. In the three years 1893–95 the annual mean death-rate from this cause was 870 per million per annum in European Russia. No part of the country is free from it. It appears to be somewhat less prevalent in the northern government of Archangel, but is exceedingly rife in the adjoining governments. The Polish governments have on the whole shown lower mortality rates than other parts of the country, and occasionally it has been low in the Don Cossack Territory. But everywhere great fluctuations are shown from year to year.

In Austria measles caused a mean mortality of 479 per million per annum between 1880 and 1886, and approximately the same in the next septennium. More recently it seems to have caused fewer deaths.

In Hungary the deaths from measles were somewhat higher, and in Croatia and Slavonia very much higher, at any rate in the period 1887 to 1893, when they were equivalent to a mortality of 879 per million per annum.

Switzerland shows, like so many other countries, wide variations in the annual recorded measles mortality; for example in the fifteen principal Swiss towns the deaths were in 1896 equal to a ratio of about 260 per million, and in 1897 to one of about 180 only.

In Italy the mortality rate for measles has varied between 292 per million in 1894 and 806 per million in 1887, the minimum and maximum rates respectively in the ten years ending in 1896, with a tendency to diminution towards the end of that period. In Spain the disease appears to be exceedingly common.

In the Balkan Peninsula measles is also well known. It is practically never absent from Constantinople, and records of it from the Salonika province and from other parts of European Turkey are by no means rare. In Greece also measles is a frequent disorder; it was seriously epidemic in Athens in 1898.

In Iceland measles has on four occasions caused severe epidemics, but it is not endemic in the island.

In the Faröes the disease has also caused four epidemics in the past, but if it exists there now, it certainly cannot be common. It is not mentioned by a recent traveller to those islands, who discusses at some length the principal diseases met with there[1].

In Malta measles is frequently seen (150 cases of the disease were treated in 1898), and in Cyprus a mild type of the disorder occasionally becomes epidemic.

Asia. Measles is constantly present in the Caucasus and in Transcaucasia. The number of registered cases of the disease here fluctuates considerably, but is usually somewhat lower than in European Russia generally, and the case-ratio per million of the population is about equal to that of the Polish governments. The mortality ratio in 1895 was about 250 per million for the whole of the two regions, and something under 200 per million in each of the preceding eight years. The imperfections of registration of both cases and deaths in this part of Russia has to be borne in mind, and similar caution is required in regard to the returns from Central Asia and Siberia. Of the numerous races in the Caucasus the Imeritians are said to be peculiarly susceptible to measles.

The returns from Russian Central Asia have always indicated a remarkably low prevalence of measles, and could the figures be regarded as trustworthy it would be permissible to state that in the years 1887—1895 measles prevailed in these regions to only one-twentieth, or sometimes to only one-fiftieth, of the extent to

[1] *The Faröe Islands*, by J. Russell-Jeaffreson, F.R.G.S. 1898.

which it prevailed in European Russia in the same period. Untrustworthy as the figures are, it is probable that the great difference between them and the corresponding figures for European Russia is only in part accounted for by the greater imperfections in registration, and it is almost certain that the disease is very considerably less prevalent in Central Asiatic than in European Russia. This conclusion is the more permissible in that in Siberia, where the imperfections of registration are probably almost as great as in Central Asia, the number of registered cases of measles have, in the period named, borne a ratio to the population only slightly lower than in European Russia. It is noteworthy that among the Bashkirs, living on the European-Asiatic frontiers, about the Ural and Volga rivers, measles is rare. In the northerly government of Yakutsk it has been quite common in recent years, and in 1900 a most violent epidemic of the disease prevailed in the extreme north-eastern portion of Siberia, Kamtchatka, and the Commander Islands. In this epidemic both adults and children suffered, and in some places as many as one-fourth of the population are said to have died from it. It was believed that the infection was carried from Vladivostok.

In Asia Minor measles is not unknown. It was epidemic at Trebizond in June 1898 ; at Bayazid in 1900 ; at Erzerum in the same year, and again after the severe earthquakes there in 1901. In the island of Samos a mild epidemic prevailed in the autumn of 1901. In Syria, Mesopotamia, and Arabia it appears from time to time, and cases of the disease are by no means rare among the Mussulman pilgrims at Mecca and Medina in the Hedjaz.

In India measles is quite a frequent disease, though probably less so than in Europe. It attacks both Europeans and natives, and at times becomes epidemic. It would seem to be comparatively prevalent among the Gurkhas.

In Ceylon it causes more annual deaths than any other of the zymotic diseases. Slight epidemics are frequently recorded in many parts of the island. It varies greatly in its prevalence, and in 1891 caused a mortality of 190 per million inhabitants ; in 1894 the mortality rate was only 60 ; and in 1897 it was again about 190 per million.

In Burma measles is also very common, and in the Southern Shan States it is one of the principal causes of mortality in children.

In the Dutch East Indies it prevails to some extent. Among the Netherlands troops stationed in this part of the world a few mild cases occurred in the year 1897.

In China measles is an exceedingly common disease, attacking both the native Chinese and European residents. At Hankow many cases were recorded as occurring in native children in 1890–91, and again in 1894. At Kiukiang a wide-spread epidemic occurred in 1895, attacking especially the children of a Roman Catholic mission school and orphanage; as many as 140 children were down with it at the same time in this institution. At Chungking it commonly occurs in native children in the spring and autumn. At Swatow it was epidemic in the summer of 1891. It is seldom absent from Shanghai, and at times becomes epidemic there. In the Pakhoi district it prevailed widely in December 1897 and January 1898, attacking "nearly every house in the district." In Hongkong it is a common disease.

In Japan measles occasionally causes widespread epidemics. A recent writer in the Sei-i-Kwai medical journal[1] states that the great epidemics recur at intervals of about twenty years, while in the intervals "small endemics" and isolated cases are frequent.

Australasia. In British New Guinea measles seems to be unknown among the native inhabitants.

In Australia, where measles is said to have been introduced in 1854, it now occurs at times in the form of epidemics, and the annual mortality from the disease varies very greatly. In Victoria it has not been very active in recent years; in 1897 it caused only 3 deaths in Melbourne. In South Australia it was epidemic in 1891, causing as many as 261 deaths in a population of about 320,000 (a ratio of approximately 800 per million), but in the following years the disease became quiescent again until 1898, when it was somewhat prevalent, causing 54 deaths in a population which had considerably increased in the interval. In New South Wales measles has been the cause of little mortality in recent years.

In West Australia measles caused no deaths in the years 1889 to 1892; 21 deaths and 6 deaths respectively in the two following years; then again caused no deaths for two years, and in 1897 and 1898 was responsible for 1 death and 34 deaths

[1] October, 1898.

C.   18

respectively. It will be noted that a recrudescence of the disease occurred here, as in South Australia and New Zealand, in 1898.

In Tasmania the disease at times becomes widely epidemic, and then for several years together almost disappears. A severe epidemic occurred in 1885, and another still more general one in 1893 and 1894. In 1893, when the disease was also epidemic in New Zealand, as many as 2773 cases were notified and 36 deaths were recorded, and in 1894 there were 547 notifications with 15 deaths. In 1895 the disease still prevailed to some extent in Tasmania, but was very mild in character and no death from it was recorded. The 36 deaths in 1893 were equivalent to a mortality ratio of about 225 per million—not a high figure for an "epidemic" year. The mildness in type of the disorder is clear from the figures quoted above.

In New Zealand measles was widely epidemic in 1893, causing 525 deaths. This outbreak had been preceded and was followed by periods of quiescence; but in 1898 there was a recrudescence, causing 57 deaths.

Measles is common in many of the Pacific Islands. In Fiji, where it was introduced from Australia in 1875 and caused a terribly fatal epidemic, the disease is now endemic. It attacks both whites and natives; in 1897 34 cases of measles (29 of which were in Fijians) were treated in the colonial hospital at Suva. It is known to exist also in New Caledonia, in the Society Islands and elsewhere. In the Marquesas group, however, it is said to be unknown.

Africa. In the countries forming the northern shores of Africa measles is not a very common disease. In Morocco no certain mention of it appears to be on record. In Algeria, on the other hand, it occasionally becomes epidemic and is generally a cause of high mortality among children. In Tunisia it is said to be very rare; an epidemic, however, occurred in 1892 in Tunis, causing 139 deaths. Its rarity here may be judged by the statement that in the French army of occupation the average annual sickness-rate from measles is only 0·89 per 1000 strength, as contrasted with a rate of 5·59 for the rest of the French army in other parts of the world.

In Egypt epidemics of measles occur with very considerable frequency, and the disease is spoken of as the most fatal of all infectious diseases among Egyptian children. Dr Engel Bey has stated that more than one-third of the total deaths in children under five years of age which occurred in the towns of Lower Egypt in the five years 1886–1890 were due to this cause.

On the West Coast of Africa measles is undoubtedly met with; and while it appears to be far from common in some parts, in others it at times causes severe epidemics. Davidson states that in Senegal it is said to be more severe than in France, but whether this points to a wider prevalence or to a greater case-fatality in individuals is not clear. In the colonial hospitals of Sierra Leone and Lagos no cases of measles were treated in the year 1897, and I have found no certain mention of the disease in either colony. On the Gold Coast, however, the disease exists, and 3 cases were admitted to the colonial hospital in 1897. In the British possessions on the Gambia measles is at times prevalent; in the latter half of 1891 it gave rise to an epidemic of severe type. It is known also to exist and to attack the natives in the Belgian Congo. In the German Cameroon territory, on the other hand, it is said to be unknown.

Whether the disease exists to any extent among the negro races in Central Africa appears to be doubtful. Pruen states that he never succeeded in recognising it in the natives of Central Africa, but he adds that he is far from sure that it does not exist among them. Cook states with regard to Uganda that "measles are very common, and many infants die of subsequent broncho-pneumonia, their unclothed bodies being but poorly fitted to withstand the evil effects of chill[1]."

On the eastern side of the African continent the degree of prevalence of measles is a matter of some uncertainty. Epidemics of the disease are said to be far from rare in Zanzibar. In Abyssinia it is known. In British East Africa it is probably rare.

In Madagascar measles is said at times to become epidemic.

In the Mascarene Islands, Mauritius, Réunion, and Rodriguez, measles at times becomes epidemic; in the intervals it remains sporadic, and, though the individual cases are often grave, the disease seems for a time to have lost its infectiousness. A similar character is common to this malady in many other parts of the world. In the Comoro Islands measles is occasionally epidemic. In Madeira and the Canary Islands the disease is known, though apparently not in a severe form. In St Helena it has not been

[1] *Journ. Trop. Med.*, June 1, 1901.

mentioned as prevailing in the recent Health Reports on that island, and the belief has been expressed that it has probably not occurred there since an epidemic in 1807.

In South Africa measles is in many parts a very prevalent disease and its behaviour appears to differ in no way from that with which we are familiar at home. From the Annual Reports on the Public Health of the various colonies in South Africa it may be gathered that the disease at times causes more or less severe epidemics in many places throughout the colonies. In the year 1897, for example, it was epidemic at Albany, at Aliwal North, in the Jamestown district and in many other places; in 1898 outbreaks occurred in Strydenberg sub-district, in Phillipstown town and district, and elsewhere. All the places just mentioned are in Cape Colony. In regard to the native territories, Tembuland, the Transkei, and Pondoland, measles has not been mentioned in some recent reports from these regions, but it may be doubted if this indicates that the disease is not known there. In the Transvaal and Orange Free State measles has from time to time caused severe epidemics. In Bechuanaland the disease is certainly known, and is said to have usually accompanied or followed smallpox, the natives calling it *sekoripanye se senyè*, or the little-pox. In Natal it is at times epidemic, and also in Zululand.

In South Africa generally measles appears to have been unknown before the advent of the white colonist, but it is now a common disorder. The disastrous effects of measles in the "concentration camps" necessitated by the recent war will be fresh in the memory of all. In these camps children in large numbers and at the most susceptible age were brought together, and it is not surprising that when measles was once introduced among them it spread rapidly. The high death-rate is easily accounted for by the terrible ignorance displayed by the children's mothers in the management of the disease, as revealed in recent blue-books published on the subject.

North and Central America. In Greenland measles is said to be unknown, but, somewhat further south, it appeared in the Hudson Bay Territory in 1864, and ten years later was the cause of a very deadly epidemic among the natives in Kadiak.

It is a comparatively common disease throughout the greater portion of Canada. In the province of Quebec it is permanently present; there is no season in which it is not seen in some town or district, and it is often prevalent in many places at the same time. Its behaviour here is as apparently capricious and irregular as in many other parts of the world. In the province of Ontario the disease finds frequent mention in successive annual Reports of the Board of Health as prevailing in many parts of the province. In Manitoba it is not infrequently epidemic, and on several occasions outbreaks of the disease have been clearly due to an importation by immigrants from elsewhere. In Nova Scotia sporadic outbreaks of measles not infrequently occur. In New Brunswick it is one of the most conspicuous diseases; it is said to occur in "epidemic waves" at varying intervals; in the city and county of St John 108 cases of the disease were registered in 1897. In British Columbia measles is at times epidemic in many parts of the colony.

In the United States measles is met with throughout, occurring especially in the spring months of March, April, and May. It is said to be mainly a rural disease in the States, and it is noteworthy that in 1897—to take the returns of a single year—no death from it was recorded in the towns and cities of the States of Arizona, Arkansas, Delaware, Idaho, Mississippi, Nevada, and N. Carolina, while in many other States the numbers of deaths from measles were small. In Massachusetts the disease has been diminishing in prevalence in recent years. In Michigan the annual mortality in the years 1881–1895 has varied, very widely and very irregularly, between 19·3 (in 1895) and 206 (in 1888) per million living. The disorder is at times extremely fatal among the indigenous Indians in the United States.

In Central America measles is neither frequent nor severe. In Mexico it is met with at times. In British Honduras it is not common, but at times causes widespread epidemics.

In the West Indian Islands measles at times prevails to a considerable extent. In Cuba it is said to be rare; but in Jamaica it has recently been very frequently seen. It was epidemic in 1897 in the eastern district of the latter island, and in 1898 outbreaks of the disease were recorded in many places in the island. In Barbados a wide epidemic occurred in 1896 after the arrival of a British regiment from Halifax in December 1895; no parish in the island escaped. In St Lucia the malady was also prevalent in 1896, and in December of that year it became widely epidemic, but in a mild form, in St Vincent. It was also slightly epidemic in Trinidad and Tobago in 1896. Throughout the Leeward Islands it appears to have been somewhat rare in recent years.

South America. In British Guiana measles occasionally becomes epidemic; it did so in 1898, and in 1890 it was widely prevalent at the same time as influenza. In Dutch and French Guiana the disease is seen occasionally, but it is usually milder than in Europe. It is met with throughout Brazil in the form of recurring epidemics, and it is said to be more severe in type here than in Central America and the West Indies. One of the worst epidemics on record occurred among the natives on the banks of the Amazon in the year 1749. Measles is also frequently epidemic in Paraguay, Uruguay, and the Argentine Republic, and it is no less familiar in Bolivia and Chile on the western side of the continent. For Ecuador and Peru, and for Venezuela, further north, there appears to be no trustworthy evidence as to the occurrence or behaviour of the malady[1]; but it can certainly exist in Colombia, an epidemic having broken out at Palmira in that country in December 1900[2].

Factors determining the Geographical Distribution. Measles is one of the most widely prevalent of all diseases. It is found in cold, temperate, sub-tropical, and tropical countries. It is rarer in some countries than in others, but its degree of rarity seems to show little definite relation to the distance of the country from the equator, save that on the whole measles appears to be commoner in temperate countries than in tropical, and that in such northerly places as Iceland, Greenland, and the Hudson Bay territory it is either unknown or only occurs as an imported disease. It is probable that the last phenomenon is due rather to the isolated position of these countries or islands than to their northerly latitude; as when measles has been imported to Iceland and to the Hudson Bay territory it has raged with terrible violence. In these epidemics the disease appears to prevail until it exhausts all the available material, and then to disappear and remain absent until it is imported again.

Measles is far more prevalent in urban than in rural populations, and is particularly rife in densely populated large towns.

---

[1] Neither Hirsch nor Davidson makes any mention of measles in these countries. [2] *Journ. Trop. Med.*, April, 1901.

London and Liverpool, for example, almost always return very high death-rates from measles. Otherwise the general distribution of the disease in England and Wales varies very considerably in different years. This is to a great extent accounted for by the varying periods which elapse between successive epidemic prevalences of the disease in different districts. Measles is essentially a disease of childhood, and when it prevails in a district for a year or two it seems to exhaust all the available patients of susceptible age, and then time is needed for the appearance of a fresh supply of subjects. An exception to this rule may perhaps be found in very populous towns, such as London or Liverpool, where the disease finds a constant supply of susceptible material. The Registrar-General has pointed out that "when there has been a severe outbreak in any year in a given area, it will be found that the returns of the next year show, as a rule, a subsidence in the area itself, but an extension to the adjoining districts; another year, and these districts are also comparatively free, while a wider circle of surrounding districts has become infected[1]." Numerous examples of this behaviour of the disease have been observed in recent years. It has further been pointed out by the Registrar-General that high minimum rates of measles mortality are usually found in large populous areas without breaks in the population, while the occurrence of high maximum rates in these same areas points either to gross carelessness or to a bad sanitary condition, or probably to both. On the other hand, low minimum rates are apparently associated with populations broken up into detached groups, "since the infection may under such circumstances fail to be conveyed from one group to another, in which case it may die out for a season."

It is very probable that the disease behaves in a similar manner in other countries, and that this is, in part at any rate, the explanation of its rarity in remote islands, such as Iceland and the Faröes, and of its extreme severity when it is newly imported to a community long free from it.

While, as already pointed out, measles is found in all latitudes, there is reason to believe that it is often of a somewhat milder and more benign character in tropical and sub-tropical climates than elsewhere, but even this rule is not without marked exceptions. In relation to season, measles varies greatly in different parts of

[1] 1891 Report.

the world, and it has shown itself capable of prevailing as a severe epidemic in all seasons. It is, however, in most countries more frequent in the colder than in the warmer months.

Neither the character of the soil nor the degree of elevation above sea-level appears to have any great influence upon this disease. In its relation to race, measles appears to be as indifferent as in its relation to most other external conditions. All races are susceptible to it, and it has been shown that it is just as capable of attacking the Chinaman, the Hindu, and the Negro as the European. It appears on the whole, however, to be decidedly less common in the African negro than in most other races.

Measles is one of the most infectious or "contagious" of all the infective fevers. It is spread by the movements of infected persons, and also probably by means of fomites. Its intensity in different epidemics varies widely; some of the most severe and fatal outbreaks, as already stated, have occurred in communities long free from it, to which it has been newly imported. The instance of the intense and fatal epidemic of measles in Fiji has been mentioned above. In this outbreak the infection was believed to have been brought to the islands from Sydney in the retinue of king Kakobau, and as many as 20,000 of the natives, or between one-fifth and one-fourth of the population, died from it. Other equally deadly outbreaks of measles under similar circumstances were those of 1749, among the natives on the river Amazon; of 1829, in Astoria; of 1846, among the Indians of Hudson Bay territory; of 1852, among the Hottentots at the Cape; of 1854 and 1861, among the natives in Tasmania; and of 1874, in Mauritius. Epidemics of such high fatality as these are rare, and can perhaps only occur under the special conditions just mentioned—that is to say, the introduction of measles infection into an isolated community long free from it. The disease, however, may be almost equally fatal under certain other conditions, of which perhaps over-crowding and imperfect attention to nursing and medical treatment are the most important. Measles is, indeed, a disorder the lightness or severity of which depends, to an extent that can scarcely be exaggerated, upon the care and attention bestowed upon the individual patient or patients. Hence when it has broken out under circumstances where such care and attention were well-nigh impossible, or where through ignorance and culpable neglect they were not applied, measles has often become a very alarming and deadly disease. Examples of such outbreaks during war-time are well-known, and would usually fall into the first of the two classes of outbreaks just named—that is to say, outbreaks under conditions where proper care and attention were well-nigh impossible. Such were the fatal measles epidemics among the Confederate troops in America in 1866; among the Garde Mobile during the siege of Paris in 1871; and in the National Army of Paraguay, in the war with Brazil. In the latter instance, one-fifth of the army were said to have been swept away by measles in three months, "not from the severity of the disease," but "from want of shelter and proper food." In the second class of outbreaks just mentioned—that is to say, those where proper care and attention were not applied from sheer lack of knowledge and culpable neglect—must be placed the recent measles epidemics in the concentration camps in South Africa. If it is harrowing to read of the terribly high mortality among the children brought together largely for reasons of humanity—for they would otherwise have been exposed to death on the veld— it is still more harrowing to read of the astounding ignorance and shocking treatment, or rather the lack of all proper treatment, care, and attention, to which those attacked were exposed by their mothers and the women of the camp generally.

Measles, as stated, is essentially a disorder of childhood. The majority of deaths attributed to it fall in the first two years of life. It is rather less fatal to females than to males. So far as is known at present, measles is neither a milk- nor a water-borne disease, and it is not known to affect any of the lower animals.

MEDITERRANEAN FEVER.
(MALTA FEVER, ROCK FEVER, NEAPOLITAN FEVER, UNDULANT FEVER.)

General Characters and Etiology. It is only within comparatively recent years that the specific character of Mediterranean or undulant fever has been recognised, and for the present its geographical distribution is a matter of some uncertainty. While the disease in its typical form presents characteristic features by which it can without much difficulty be distinguished from other diseases, there is a group of fevers the symptoms of which resemble those of Mediterranean fever sufficiently to have caused no little confusion and uncertainty. The most important member of this group is that to which the name typho-malarial fever has been given. Considerable discussion has taken place as to the exact nature, or even as to the existence of this fever as a specific disease, and recently the suggestion has been made that typho-malarial fever and Mediterranean fever may be one and the same malady. A similar suggestion has been made with regard to kala-azar. For the present, however, it seems better to consider separately what is known of the affections described under these names.

Mediterranean fever has been defined as follows :—An endemic, pyrexial disease, occasionally prevailing as an epidemic, having a long and indefinite duration, an irregular course, with an almost invariable tendency to undulatory pyrexial relapses. It is usually characterized by constipation, profuse perspirations, and accompanied or followed by symptoms of a neuralgic character, by swelling of the joints with effusion, and other rheumatoid symptoms. After death the spleen is found to be enlarged and often softened, many of the organs are congested, but Peyer's glands are neither enlarged nor ulcerated, nor is ulceration present in other parts of the small intestine. There is a constant occurrence in certain tissues of a definite species of micro-organism[1]. This micro-organism is in the form of a coccus, and is usually known as *micrococcus melitensis*; it was first described by Bruce in 1887.

The disease thus described appears to have a somewhat wide distribution over the earth's surface, and evidence is constantly being brought forward of its endemicity in places where it was not hitherto known to exist, and occasionally it would seem to become epidemic in places where it is unknown as an endemic disease. The specific cause of the fever, however, appears to remain for long periods, if not permanently, in certain geographical areas, and "endemicity" may be regarded as an important feature of the disease.

Recent Geographical Distribution. *Europe.* In the British Isles this disease is never seen, excepting in patients who have contracted it abroad.' In France, according to Jaccoud, Mediterranean fever is occasionally met with in Paris, but only as an imported disease from Italy. Whether it exists along the Mediterranean shores of France is uncertain, but the late Dr Hughes, who made a close study of the disease, expressed the belief that it is not common there. Throughout Northern and Central Europe generally there is an absence of evidence of its existence, either as an endemic or àn imported disorder, and it is essentially in the south of Europe and the islands of the Mediterranean Sea that the fever is commonly met with. It is seen in the south of Spain, and its comparative frequency at Gibraltar has added the name of Rock Fever to the long list of names by which the disease has been known at different times and places. Italy is another important endemic home of the disease. In Naples and many places near, such as Caserta, Benevento, Campobasso, etc., it is met with. It became epidemic in Naples in 1872. In Middle and Upper Italy it is seen at Rome, Ariccia, Terano, Fermo, and Padua, the frequency of the disease diminishing from south to north. It was described in Southern Italy in 1879. It is of common occurrence in Sicily,

[1] *Mediterranean Fever or Undulant Fever*, by M. Louis Hughes, Surg.-Capt., Army Medical Staff. London, 1897.

where it is endemic at Catania and Palermo, and is also met with at Taormina and other villages on the southern side of Mount Etna. Epidemics of the disease occurred at Catania in 1872, 1878, and 1884. In the Balearic Isles the fever is believed to be endemic in Minorca. In Sardinia it is well known at Cagliari and in the interior of the island, while there is also good reason to believe that it exists in Corsica[1]. Malta is, perhaps, the worst of all the endemic centres of this disease, and Malta fever is one of the names by which it is most familiarly described. It is one of the most serious of all maladies in the island, and is a graver risk to the troops and others living there than is, for example, enteric fever. The great frequency of this fever in Malta may be seen from the following recent figures :—in 1898, while only 12 cases of scarlet fever, 14 of diphtheria, 150 of measles, and 294 of enteric fever were registered, as many as 724 cases of Mediterranean fever were reported[2]. It is possible that the disease also exists on the shores of the Adriatic; Brunner of Trieste has recorded the case of a workman employed in constructing a railway in Southern Dalmatia who suffered from it[3].

Further east, Mediterranean fever is met with in some parts of the Balkan peninsula. It is said to occur in Constantinople, but bacteriological proof of its existence there is lacking, and I have not succeeded in hearing of a single case which could with absolute certainty be regarded as of this nature. A form of unclassified fever is, however, very common in this city, which in some cases bears considerable resemblance to Malta fever in its duration and general symptoms.

A similar disease, believed by some to be the same disease, is also seen at Athens (Typhaldos). In Cyprus the fever is spoken of as endemic in many of the towns (Karageorgiades)[4]; and in Crete it is said to have become so since the year 1890 (Capetanakis).

[1] Kretz records the case of a physician who suffered, in Ajaccio, from a fever of six months' duration, and whose blood after recovery gave positive results with the agglutination test. (*Wiener Klin. Woch.*, No. 49. 1897.)

[2] Malta: *Public Health Report for* 1898. The figures quoted appear to include those of Gozo as well as of Malta; but there are discrepancies in the report which do not seem easy to explain.

[3] *Wiener Klin. Woch.*, Feb. 1900.

[4] Williamson (*Brit. Med. Journ.*, Sept. 14, 1901), on the other hand, says the fever is not common in Cyprus.

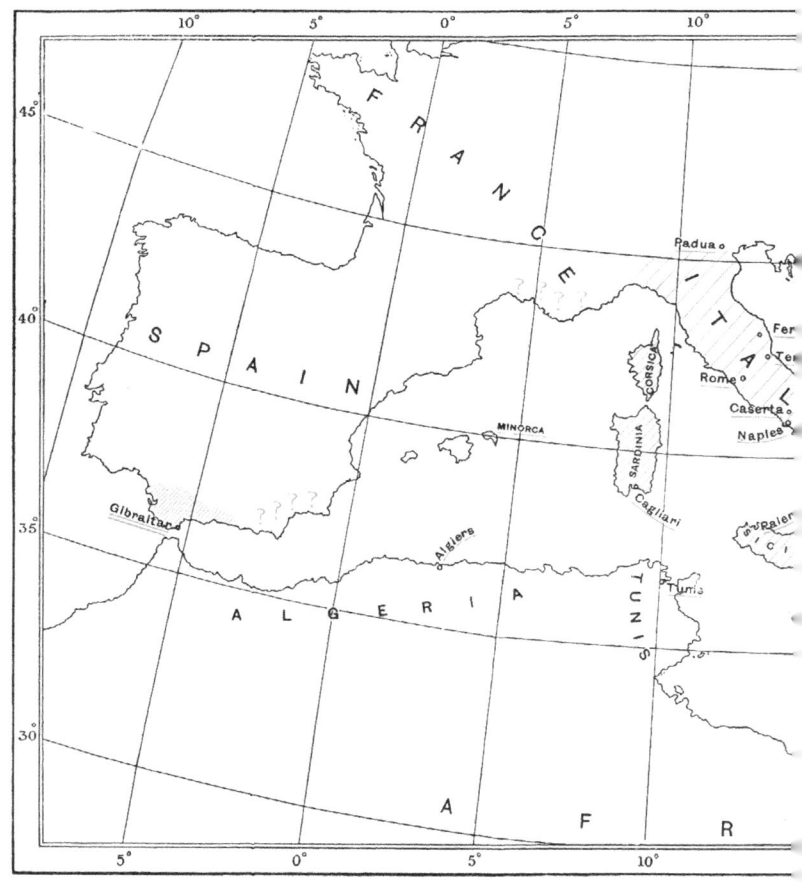

DISTRIBUTION OF MEDITERRANEAN OR MA

MALTA FEVER IN THE MEDITERRANEAN BASIN

Africa. The southern shores of the Mediterranean appear to favour the prevalence of the disease to which this sea has given its most convenient name no less than the northern shores. Thus Mediterranean fever is met with in Tunis, in Algiers, and perhaps in Egypt. In Tunis, Perini stated that he saw the disease first in the year 1879; and he added that, when the French occupation took place, an epidemic of it broke out at Goletta, the muddy sea-port town near Tunis, and that it had since then been endemic there.

Regarding Algeria the evidence is less clear. Brault states that he has seen cases there which clinically resembled Mediterranean fever, but no bacteriological inquiry was made to confirm the diagnosis. Recently the disease has been recognised in the Canary Islands, but it is here apparently of a milder character than in the Mediterranean[1].

In regard to the existence of the fever in Egypt it appears that cases are not infrequently seen at Alexandria, but Hughes states that reliable information was difficult to obtain, and that most of the cases there proved to be instances of relapse in men transferred from Malta, where they had already suffered from the disease. Morgan, on the other hand, is quoted as mentioning (in 1892) the occurrence of a few cases of Mediterranean fever in men stationed in Alexandria who had never been in other parts of the Mediterranean. Bruce and others state that it is unknown in Tripoli, and Hughes, who visited Benghazi and Derna in 1893, failed to find evidence of the disease there, though at the same time he dwells on the unreliable nature of the evidence available. On the African shores of the Red Sea cases of this fever have been seen. At Suakim and Massowa Milnes has described cases indistinguishable from those seen in Malta, Naples, and Crete, and Pasquale, Rho, and others have furnished similar evidence of the existence of the disease there.

On the eastern shores of Africa, in Zanzibar and on the neighbouring coasts, a fever closely allied to, if not identical with Mediterranean fever is also said to be present[2]. Similar cases, about which the like doubt exists, have also been reported from Mashonaland[3].

[1] Melland, Brit. Med. Assoc. Annual Meeting, 1902.
[2] *Journ. Trop. Med.*, September, 1898.
[3] *Ibid.* Nov. 1900.

Asia. Several authorities mention the existence of Mediterranean fever at Smyrna in Asia Minor. Cases of a very similar disease have also been described as occurring at Jerusalem, but here, as elsewhere, the confusion existing between Mediterranean fever and so-called typho-malarial fever renders it impossible to make any more definite statement. The same is true in regard to the doubtful existence of Mediterranean fever at Aden.

The question of the presence or absence of this disease in the Indian peninsula is one of some difficulty. Here, as in Asia Minor and at Aden, cases described by some as typho-malaria, and by others as resembling true Mediterranean fever, have been met with, and these will be referred to subsequently. In 1879 an outbreak of a disease occurred at Chorbattia, of which the clinical description and the temperature charts are said to have been practically indistinguishable from those of Mediterranean fever. Squire, again, mentions a similar occurrence at Dinapur, in Lower Bengal, in the 70th regiment, in the year 1883. Hughes also describes the case of a man landed at Malta from a troopship coming from India. He was said to be in the second week of an attack of enteric fever, which had developed seven days after he left Bombay. The symptoms resembled those of Mediterranean fever rather than of enteric or malaria, quinine had no beneficial effect, and after death (which occurred on the 14th day of the disease), while no enteric lesions were present, the post-mortem appearances were just those met with in acute and fatal cases of Mediterranean fever. Unfortunately it proved impossible to confirm this opinion by bacteriological examination. The same author quotes the evidence of many Indian medical officers in support of the view that this disease does exist in India ; and adds that Professor A. E. Wright and others at Netley have obtained a serum reaction with the blood of so-called enteric fever cases invalided from Sabathu (near Simla, in the western Himalayas) exactly similar to that obtained in the case of Mediterranean fever. Many of these patients had had no previous Mediterranean service, and had suffered at Netley from symptoms indistinguishable from those of the disease in question. More recently observations have been published pointing to the existence of the malady in Bombay, in several places in the Punjab, and in

Simla itself[1]. To sum up briefly it may be said that there is constantly increasing evidence that a disease closely resembling, if not identical with Mediterranean fever exists rather widely in India[2].

The evidence for the presence of Mediterranean fever in China is of the same nature as that just quoted for its presence in India. Thus Durand and others have described cases of an analogous disease as occurring in Hongkong. Manson states that he has probably seen cases in China, and Wright has described a case of continued fever in a person from Hongkong, in which the blood gave a serum reaction with the *micrococcus melitensis* similar to that of patients who had suffered from the fever now under consideration. Quite recently too it has been stated that this fever exists endemically in the Philippine Islands. It would seem to be quite common in Manila : one observer saw three cases there in a single week, each of which gave a marked agglutination reaction with the micrococcus even in high dilutions[3].

America. Very great doubt existed until recently concerning the prevalence of Mediterranean fever on the continent of North America, but recent investigations have shown that it probably does exist there. One case apparently of this nature was mentioned at a meeting of the Pathological Society of Philadelphia (February 1st, 1899). The patient was an army officer, and it was thought that he had contracted the affection in Puerto Rico, in the West Indies ; his blood gave a reaction with the *micrococcus melitensis* similar to that with blood of a Mediterranean fever patient. This was believed to have been the first case of the disease that had ever occurred in the States, and the supposed source of infection was, as just stated, not in the States themselves but in a West Indian island. Eight cases have been observed also in the Army and Navy Hospital at Hot Springs, Arkansas, but all the patients seem to have contracted the disease elsewhere —five apparently in the Philippines, one in Cuba, one was a sailor from a man-of-war that had been cruising along the coasts

---

[1] *Indian Medical Gazette*, Sept. and Nov. 1900, and Jan. 1901.

[2] Three cases of the fever, in each of which the agglutination test gave positive results, were observed in the Swat Valley in 1900. (Greig, *Indian Medical Gazette*, March, 1901.) The disease is also believed to exist in Assam. (Hislop, *Indian Medical Record*, May, 1902.)

[3] *Philadelphia Medical Record*, Nov. 1900.

of South and Central America, and the eighth was a sailor from a vessel stationed in West Indian waters.

Finally it must be added that at the General Medical Congress held in Washington in 1893 a class of fever was described as met with in Venezuela, and this was later compared with the fever of Malta and Gibraltar, which it appeared to resemble closely.

Factors concerned in the Geographical Distribution. Until fuller and more certain data are available to show the exact distribution of this disease over the earth's surface, it is difficult to draw conclusions as to the influence of various conditions in determining it. Year by year, almost month by month, fresh observations are published from many parts of the world tending to show that this fever is diffused much more widely than had been hitherto believed to be the case. So that any conclusions that may be drawn as to its relation to external circumstances can only be put forward tentatively, and as liable to be upset by later knowledge.

But so far as present knowledge goes, it allows us to say that the great home of this fever is the basin of the Mediterranean Sea; that it is not equally common on all the shores of that sea, but that where it is common here it is very much more so than in any other part of the earth's surface. In the Mediterranean basin it is most common and most severe at Malta, at Gibraltar, in the Balearic Islands, in Cyprus, in Crete, and on the south Italian coast. But it is seen more or less in many other parts of both the European and African shores of the sea, and in the Levant. Elsewhere the disease has occurred with more or less certainty in scattered parts of the continents of Asia, Africa, and America, but not hitherto in Australia or the Pacific Islands. In many of the places it has been reported from, there is not only great uncertainty as to the diagnosis, but there is also evidence that, even if existent, it is a very rare disorder.

Should so-called typho-malarial fever prove to be identical with Mediterranean fever, then very important modifications would have to be made in the above statements, and the Mediterranean basin could no longer be regarded as the great centre of this disease.

A certain degree of warmth would seem to be favourable to the prevalence of Malta fever. It is not, however, a tropical disease, and seems to prefer a warm, temperate climate to extremes either of heat or cold. It has never been seen in really cool climates; in Italy it becomes less frequent from south to north, and in Europe generally the most northerly records of its (even doubtful) appearance have been the occurrence of imported cases in France and the British Isles. In America, too, the suspected cases already referred to seem all to have been contracted in the warmer zones of that continent.

The relation of the disease to temperature is further shown in its seasonal variations. It is most common in the summer, the largest number of cases occurring in July; it begins to diminish in the autumn, and is rare in winter and spring. Though commonest at low altitudes, Mediterranean fever has been seen, at least in India, in hill stations at a height of 7000 ft. above sea-level.

It is not an infectious disease. In those countries where it is known to exist it is essentially an endemic disorder, but it has at times taken on epidemic characters. Hughes believed that it was largely associated with insanitary conditions and particularly with a sewage-soaked soil. As with so many other diseases, however, the true relations of insanitary conditions to this fever are very difficult to establish. It certainly attacks persons of the well-to-do classes living in large and well-ventilated dwellings to almost, if not quite as great an extent as the poorer classes of the population living in conditions of overcrowding and dirt.

Malta fever may perhaps be an air-borne disease. There seems to be no positive proof that it is water-borne. It is mainly a disease of young adult life, being commonest in the age-groups between ten and thirty, and very rare after fifty. Men seem to suffer more than women.

MUMPS.

History. The disease known as mumps is very clearly described in the writings of Hippocrates, and is mentioned by all the ancient Greek and Roman medical authors. It is not, however, until the beginning of the 18th century that the records are concise enough to enable the distribution of the disease to be traced. Since then it has been a widely diffused disorder, occurring as an endemic or as successive epidemics in almost all parts of the world.

Recent Geographical Distribution. *Europe.* Mumps is a common disorder in all European countries. Statistics are to a great extent wanting to indicate the relative frequency with which it occurs in different parts of the continent. But no country is free from it. It is frequently seen in the British Isles and in all countries of Western Europe. Throughout European Russia it is no rarity, and is found as far north as the shores of the White Sea, and in the south on those of the Black Sea. It is a very common disorder in many of the Balkan States, and constantly comes under observation in Constantinople. In Bavaria and Swabia mumps appears to occur with remarkable frequency. In the far north epidemics of the disease have been observed in the past in Iceland, Lapland and the Faröes.

Asia. The distribution of mumps throughout Asia is not easy to follow. The disease, from its comparative harmlessness, is frequently omitted by writers from the list of maladies prevailing in a given country; while, owing to the rarity with which it causes death, it is often excluded from tables of the principal causes of mortality in official statistics, when such exist.

In the Russian possessions in Asia—Siberia, the Caucasus, and Russian Central Asia—mumps is no rarity. In the three

years 1893–1895 cases of the disease were reported from every
one of the provinces of the Caucasus proper and of Transcaucasia;
in the last-named year the total number of cases recorded in both
territories was 3238, with 142 deaths.   Similarly throughout all
parts of Siberia mumps is met with, including the Amur province,
the territory of Chita on the Russo-Mongolian frontier, the far
northerly province of Yakutsk, and the island of Sakhalin.   In
Central Asiatic Russia it seems to be just as common as in Europe,
and cases of the disease are yearly recorded from Transcaspia, from
the Turkestan provinces, and from the Semipalatinsk and Semirétchinsk territories between the latter and Siberia proper.

Little is known of the prevalence of mumps in Mesopotamia,
Arabia, and Asia Minor.   It occurs from time to time in Persia,
and was epidemic on the Turco-Persian frontier in the winter of
1901–1902.   It is also seen on some of the islands off the coast
of Asia Minor, and was epidemic in Mitylene at the latter end of
1901.

In India mumps is very frequently seen.   The natives of
India are perhaps rather specially susceptible to the infection.
Native children are constantly brought under observation suffering
from it, and adults do not escape it.   Buchanan speaks of it as
"exceedingly common" among the Indian natives[1].   In the
native army in India considerable numbers of the men are yearly
treated for it.   Thus in 1896 as many as 1081 cases of mumps
were recorded in this section of the Indian army, and in 1897 the
cases numbered 742.   Many cases are also observed in the Indian
gaols; the figures fluctuate considerably, but in 1896 as many as
945 prisoners were attacked by the disease, while in 1897 this
figure rose to 2264.   In the latter year the largest proportion of
cases were reported from the North-West Provinces and Oudh.
European residents and troops stationed in India are also affected
by the disease.   Mumps is met with not only in the plains of
India, but also at considerable heights.   An epidemic of the
disease was reported from the Hindu Khush in 1892, affecting
principally the Sepoys of two regiments of the Kashmir Imperial
Reserve.   The infection was, however, believed to have been
brought up from the plains.   In the following year (1893) the
men of another newly-arrived regiment were attacked by the
disease.   In neither instance, apparently, did it spread to the

[1] *Journ. Trop. Med.*   Sept. 1899.

inhabitants of the villages in or near which the troops were quartered.

In Ceylon mumps prevails to some extent from time to time.

The Chinese are not immune to mumps, though whether the disease is common in China it is difficult to say. In Chung-king an epidemic of mumps prevailed among the native inhabitants in the winter of 1897–8[1], and it has occurred elsewhere in China, as for example in the Chung-poo valley, in the province of Fokien. Recent medical writers upon the pathology of that country are, however, for the most part silent in regard to this disease.

In the East Indian Islands mumps is probably rare. The disease is said to be unknown among the hill tribes of Java. Kohlbrügge has mentioned an epidemic of mumps affecting foreigners in the island, but adds that the infection did not spread to the natives[2].

Australasia. Mumps is not unknown in Australia, but appears to be a comparatively rare disorder there. In West Australia only one death from the disease was reported in the mortality statistics for the ten years 1889–1898; that death occurred in 1895. In South Australia in the same ten years a single death from mumps was reported in 1891, and another in 1895. In Queensland similarly this disease makes a very rare appearance in the annual health reports as a cause of death. In New Guinea it appears to be unknown.

Mumps occasionally appears in some of the Pacific Islands. In the Solomon group epidemics of the disease are not unknown, and sometimes it is severe enough to be a cause of death. In Fiji an occasional case is admitted to hospital.

Africa. In many parts of the African continent mumps is met with, and is indeed by no means a rare disorder.

Along the west coast it is frequently seen. In most recent years the Colonial Report for the Gambia has recorded the occurrence of several cases of mumps in the colony and of deaths from the disease. It is mentioned among the maladies to which the native inhabitants of the Belgian Congo are liable[3]. In the Cameroons there was a small epidemic of mumps in 1895; and in Uganda it is said to be very common.

---

[1] *Med. Reports of Imperial Chinese Maritime Customs.* No. 55.

[2] *Janus,* 1897–8, p. 221.

[3] *Cong. Nat. d'Hygiène et de Climatologie Méd. de la Belgique et du Congo,* 1897.

Throughout South Africa mumps is not a rare disease. It occurs in the form of epidemics, and these occasionally assume an intense character, affecting a large proportion of the population. An outbreak of this kind occurred in Namaqualand in 1898, after many years' complete freedom from the disease, and in the large centres of population as many as three-fourths of the inhabitants were attacked. The malady is not absent from the pathology of Cape Colony, and it has been the cause of epidemics in several parts of it in the last few years.

The disease is also known in Madagascar and in Mauritius.

North America. Recent reports from the Canadian provinces show that mumps is a well-known disease in most, if not all of them. In the Ontario reports it finds frequent mention as occurring in many parts of the province. In Manitoba a considerable number of cases are yearly seen. In New Brunswick the disease is also not unknown, and even as far north as Alaska it has occurred epidemically.

Throughout the United States mumps is an exceedingly common disorder; it was the cause of a most violent epidemic among the Confederate troops during the War of Secession. It is also met with in Mexico.

In the West Indies epidemics of it are far from rare. Recent reports from Jamaica make mention of outbreaks in several parts of the island.

South America. With the exception of mention of an epidemic of mumps in Peru, medical literature appears to be silent as to the existence of this disease on the South American continent.

Factors governing the Geographical Distribution of Mumps. This disease has a practically universal distribution over the earth's surface. It can apparently exist independently of latitude, climate, altitude above sea-level, and conditions of soil. It is a highly "contagious" disorder, and frequently occurs in the form of epidemics, sometimes of great severity, though not of long duration. Institutions, such as schools, orphanages, or prisons, are the more usual scenes of such epidemics, but larger communities may be affected. Mumps may even be epidemic over whole countries, but has never become truly pandemic. It prefers cold and wet weather to warm and dry, and the majority of epidemics occur in the winter and spring.

The disorder is most often seen in children and adolescents, but adults are by no means exempt from it, and it has often attacked soldiers in barracks or in the field, witness the very severe outbreak during the American War of Secession already mentioned. Many other epidemics affecting troops in India, France, Germany and elsewhere might be quoted.

Sanitary conditions, or the reverse, seem to be wholly without influence upon the prevalence of this disease.

Its relation to race is somewhat uncertain. All the great divisions of the human family have at some time suffered from it ; but in some epidemics among mixed communities the disease has shown a certain selective power, attacking only persons of one race and sparing others. This was the case in the outbreak in New Archangel, Alaska, in 1843–4, when only the natives were attacked while the Europeans escaped. In a recent epidemic in Java the natives escaped while the foreign residents were attacked.

MYCETOMA (MADURA FOOT).

General Characters and Etiology. Mycetoma is a peculiar affection of the foot, sometimes of the hand, and very rarely of the knee, thigh, jaw, or neck, which is met with most often in India. The foot is the part of the body usually affected, and hence the name Madura foot so commonly applied to the disease. In its most characteristic form it begins with the appearance of nodules in the subcutaneous tissues. The nodules enlarge and coalesce ; they may soften and discharge an offensive, oily fluid, containing small granular bodies, of a yellow, pink, or black colour. The foot becomes swollen to twice or thrice its original size ; all traces of the constituent tissues of bone, muscle, and tendon disappear and become replaced by a more or less uniform gelatinous mass, in which are a number of cysts and sinuses containing either a black, truffle-like substance (the black variety) or a yellowish fish-roe-like substance (the white variety).

The disease never ends in a spontaneous cure. The limb becomes useless and the muscles wasted, and unless the affected member be amputated the general health may suffer considerably, and death may ultimately occur from exhaustion or some inter-current disease.

In the white variety of mycetoma a ray-shaped fungus resembling that of actinomycosis is found in the tissues of the affected limb, to which Vincent has given the name of *Nocardia* or *Streptothrix maduræ*. The fish-roe-like bodies are made up of aggregations of the fungus. In the black variety a somewhat similar, but less distinctly ray-formed mycelial growth is present. The ray-fungus of the white variety has been successfully grown in artificial cultures, but does not appear to be inoculable in the lower animals. While most observers regard the fungus as the cause of the disease, Berkeley, Cunningham, and others have expressed the belief that it is merely an accidental infection of a part already the subject of mycetoma, the true cause of which must, according to this view, be looked for elsewhere.

History. The earliest reference to this disease is apparently to be found in the writings of Kämpfer, published early in the 18th century. Probably the malady is of considerable antiquity in India. In recent times special attention has been directed to it by Vandyke Carter (1859 and subsequent years), who described its clinical and anatomical characters, while the parasitology of the disease has been the subject of careful studies by Vincent, Boyce and Surveyor, and Kanthack.

Recent Geographical Distribution. Madura foot was believed at one time to be met with only in India. Later observations have shown that it does occur elsewhere, but India is still the principal home of the disease. Hirsch states that it is found in many parts of the Karnatic (Malabar Coast and inland places near); at Pondicherry and Karikal; at Bellary, Tanjore, Guntur, Madura, Cuddapah, Trichinopoli and Combaconam; on the slope of the western Ghâts; in Ratnagiri, Poona and many other districts of the Bombay Presidency; in Kattiwar, Gujerat and Kutch; at Karachi and other places in Sindh; in Bhawalpur, Bikanir, and other parts of Rajputana; in Jhelum in the Punjab and in Sirsa and Hissar in the North-West Provinces. It appears to be absent from the eastern half of the peninsula, or if it occurs there it is only in the form of imported cases. In many parts of the areas in which it exists it is a disease of very considerable frequency.

The first notice of the occurrence of mycetoma outside India seems to have been in 1883, when Collas recorded its existence in the island of Réunion. Cases have also been seen in French Guiana, and at Saigon in Cochin-China. Libouroux has stated that one instance has been observed in Constantinople, and Bassini has operated on a single case at Padua in Italy. Gémy and Vincent have described a case in a native of Morocco, and Bérenger-Férand, who saw the disease among the negroes of Senegal, ventured to express the opinion that mycetoma exists in all parts of Africa, in the latitude of Senegal, from the Atlantic to the Red Sea. Positive evidence, however, in confirmation of this view seems to be lacking. Rho has stated that it is unknown at Massowa on the Red Sea coast[1]. A few cases have been seen in Algeria.

Quite recently evidence has been published of the existence of mycetoma in Central America and the West Indies[2]. Two unmistakable cases have been seen in Cuba, in which *Streptothrix maduræ* was present. In Costa Rica Madura foot is said to be quite common, though it has often been confused with elephantiasis. In Nicaragua also it seems to be by no means rare. There appears, then, to be good reason to believe that there is a mycetoma area in the tropical portions of the New World comparable to the Indian and African areas of this disease.

Factors Determining the Distribution. The exact distribution of this disease over the earth's surface is so imperfectly known that it is difficult to discuss to any useful purpose the various conditions and causes which have led to its presence in one country and its absence in another. Of recent years it has been shown to exist in places where it had never before been heard of, and fresh observations may at any time show that it has a wider distribution than is even now believed.

But, so far as may be gathered from the records hitherto published, mycetoma is found only in hot countries. It appears to be confined to the tropical or subtropical regions of Africa, Asia, and Central America; and has only been seen in quite limited parts of those regions. It occurs on all varieties of soil. The influence of race upon its distribution is uncertain. The fact that it is much commoner in India than elsewhere can scarcely be attributed solely to a special susceptibility of the Indian native to the disease. In those parts of India where it occurs the Hindu population suffers more than the Mohammedan, while the Europeans and Eurasians entirely escape it. The European and Eurasian may perhaps be immune to the disease, or the fact that they do not suffer from it may be merely due to their custom of wearing boots, while the native goes bare-footed. The African negro and the natives of Morocco can become subjects of the affection.

Nothing is known as to the manner in which the disease is contracted, or spread from person to person. The known facts concerning mycetoma favour the view that it is a "place disease"; that is to say that the specific cause of it remains for long periods

[1] *Janus*, 1896–7, p. 598.
[2] *Pan-American Congress*, *Med. Record*, Feb. 25, 1901.

together in certain defined areas. It seems probable that the specific cause of it, whether this be the ray-fungus or something else, exists in the soil or water of these infected areas, and gains access to the tissues by penetrating the whole or abraded skin of the foot. No other explanation so readily accounts for the ease with which the barefooted native contracts mycetoma, while others living in the same regions and covering the feet escape. But up to the present neither the ray-fungus nor any other suggested cause of the disease has been detected in the soil or water of the affected areas. Nor is there any evidence to show that mycetoma can be spread by direct infection from person to person.

OPHTHALMIA, TRACHOMA, AND OTHER EYE DISEASES.

Exception may well be taken, on scientific grounds, to a discussion of the geographical distribution of eye diseases as a whole, and not of each particular affection separately. Iritis, glaucoma, cataract, corneal ulcers, purulent ophthalmia, granular lids and the like are each due to more or less different causes, and are pathologically so distinct that it may seem to serve but little purpose to deal with them together. But, on the other hand, it has to be pointed out that, at present, materials do not exist for a comparative study of the distribution of each eye affection separately; while at the same time there are a number of known facts of much interest, relating to the distribution of some of these affections, which it may be useful to group together. In the present chapter, while a large number of eye affections will be incidentally mentioned, the greatest prominence will be given to ophthalmia and trachoma.

Geographical Distribution. *Europe.* In the British Isles all forms of eye affection are met with from time to time. Purulent ophthalmia and trachoma are practically confined to the poorer classes, living in conditions of dirt and overcrowding. They are probably rarer in England than in any other country in the world. In Ireland the chronic form of trachoma is said to be relatively common.

In the continental states of Europe these diseases are found in greatest intensity among peasant populations, in barracks, in schools, and particularly among that class of the general population whose personal habits, and neglect of treatment and of precautionary measures to prevent their spread, greatly favour the prevalence of these affections. The late Dr Van Millingen published the following table, showing the percentage of trachomatous cases to all other cases of eye disease in different countries[1].

England 0·07	Holland 7·05	Greece 25
Scotland 0·7	Spain 11·09	Italy 25
Ireland 3	Hungary 14	Portugal 25
America 3	Turkey 18	Central Asia 45
France 4	Bulgaria 20	Africa 50
Belgium 4	Russia 25	Roumania 52

Germany does not appear in this table. But in Eastern and Western Prussia granular ophthalmia seems to be exceedingly common. In 1897 Hirschberg examined some 7000 school-children in these provinces, with the result that in no school were less than 5 per cent. found to be trachomatous, and in one school the proportion rose to as high as 47 per cent[2]. In 1898, again, Rimpler found in a gymnasium at Göttingen that 26·4 per cent. of the students were trachomatous[3]. In Russia purulent and granular ophthalmia are remarkably common among the peasants, as well as many other forms of eye disease. As the result of neglect of these conditions blindness is sadly frequent. In 1893 a well-known benevolent institution in St Petersburg inaugurated a system of "flying squads" of surgeons and nurses to visit all parts of the empire and try to relieve some of the vast mass of suffering from eye disease. Seven such "squads" were sent in the first year. In 1898 the number had risen to 36; they visited many parts of European Russia, the Caucasus, Central Asia, and Siberia, and treated 50,222 patients, of whom 3432 were found to be irreparably blind. In many parts of Russia, as also in Central Asia, Montenegro and elsewhere, the frequency of eye diseases is attributed to the absence of chimneys to the peasants' huts, and the irritating effect of the smoke which consequently fills the rooms. In the Faröe Islands on the other hand, where eye diseases are common, they are ascribed to exposure to the weather in that cold and damp climate; and the same cause may perhaps explain the cases of ophthalmia that are occasionally seen in French fishermen off the Newfoundland coasts.

Asia. The frequency of these diseases in the Caucasus, Russian Central Asia, and Siberia has been pointed out above. The

---

[1] *Annales d'Oculistique.* Sept. 1895.
[2] *Klinisches Jahrbuch,* Bd 6.
[3] *Deutsche Med. Woch.*

Bashkirs suffer greatly from them[1], and in Bokhara they are extremely prevalent[2]. In Arabia, Syria, Asia Minor, and Mesopotamia few diseases are commoner than conjunctivitis. In Persia ophthalmia is disastrously prevalent, not only in the cities, but also in the mountain chains between Shiraz and Ispahan. At Khanikin and Suleimaniyé, on the Turko-Persian frontier, trachoma and ophthalmia abound, and at Kermanshah the majority of the inhabitants suffer from these diseases.

In many parts of India conjunctivitis occurs with great frequency. In Ceylon it is common in some parts, but the inhabitants of the high central area are said to be relatively immune (Hirschberg). In the Straits Settlements ophthalmia must be prevalent, to judge from the returns of the Singapore prisons and the Penang hospital. In the Dutch East Indies both natives and Europeans suffer to a considerable extent from simple and granular ophthalmia. In Java, among the hill tribes, conjunctivitis is common, while trachoma on the other hand is little seen[3].

In China all forms of eye disease are extraordinarily prevalent; with the exception of alimentary troubles no group of diseases is more so. Conjunctivitis, entropion, cataract, keratitis, pannus, granular lids, and trichiasis abound. Ulcers of the cornea, with perforation and staphyloma are also very common; pterygium is said to be frequent, but glaucoma infrequent[4]. Granular eye disease seems to be particularly rife in China. In Yünnan it abounds, and in Southern China generally it is said to constitute 78 per cent. of all forms of eye disease. It is scarcely less common in Japan, where the proportion of such cases is said to be 75 per cent.[5] In Corea ophthalmia is the second most frequent of all diseases, and blindness is very common.

Australasia. In Australia ophthalmia is said to be of very frequent occurrence among the aborigines. It also occurs among the white population; in 1895, for example, it was almost epidemic at Hampden, and in 1897 at Lowan, Victoria; in the former it was so prevalent that a school had to be closed in consequence. In British New Guinea, on the other hand, there is apparently no contagious ophthalmia[6]. In some of the Pacific Islands, as for

[1] *Meditzina.* 1896, No. 2   [2] Grekof.
[3] Kohlbrügge.   [4] Coltman.
[5] Mujashita of Tokio.
[6] *Brit. New Guinea. Annual Report,* 1898.

example Fiji, conjunctivitis and other eye diseases come under treatment, but with what degree of frequency I have no means of saying.

Africa. All forms of eye-disease seem to be particularly frequent in Egypt. As the name Egyptian ophthalmia, often applied to acute granular ophthalmia, implies, this disease is rife in that country. Trachoma is probably more prevalent here than in any other country in the world. Van Millingen estimated the proportion of trachomatous cases to all other cases of eye disease in Egypt at 80 per cent. In this as in many other hot countries the frequency of ophthalmia is largely favoured by the glare of the sun, by dust, and by swarms of flies.

In Algeria trachoma is exceedingly common, and often attacks whole families[1]. An intermittent form of ophthalmia has also been described as occurring here which is thought to accompany or replace paroxysms of acute intermittent fevers. It is said to be always unilateral and nearly always to affect the left eye. This form of eye disease has also been seen in the United States, but never in China[2].

On the West Coast of Africa many forms of eye affection are seen, some of them in large numbers, but trachoma seems to be remarkably rare. The negro races of Senegambia and the Guinea Coast are said to be immune to it. Plehn never saw a single case in the German Cameroon Territory. In Central Africa and on the East Coast other forms of eye disease abound; they are spoken of as the curse of the country, and nowhere in the world, it is said, is blindness so common as in this continent. In Bechuanaland and generally throughout South Africa ophthalmia is also exceedingly prevalent.

America. To what extent the majority of eye diseases occur in Canada and the United States I am unable to say. From the table already quoted from Van Millingen, it would seem that trachoma is a comparatively rare disease in America. It is said that none of the indigenous tribes of Canada suffer from it. Some of the Indian tribes of the United States, however, are subject to it, and the Chinese found in the big American cities are said to be particularly liable to it. In Kentucky a peculiar form of the affection is said to be met with, to which the name "Kentucky

[1] Gros. *Janus*, 1898, p. 181.
[2] Yarr. *Journ. Trop. Med.* Nov. 1899.

trachoma" is locally given.   The negroes in the United States are said to escape the disease.

For South America, ophthalmia is spoken of as an "endemo-epidemic" disease in British Guiana; and in Peru, Chile, and Bolivia cases of ophthalmia are not only very frequent, but are often of a very obstinate character.   In Brazil it is no less frequent and severe, and usually takes the form of so-called "Egyptian ophthalmia" (Lombard).

Factors concerned in the Distribution of Ophthalmia and Trachoma.   All forms of conjunctival inflammation, whether simple, catarrhal, purulent, follicular, granular, acute or chronic, appear to be more or less contagious.   Some, particularly those of gonorrhœal origin, and the acute and chronic granular forms, are highly contagious.   They are consequently found to be most common among overcrowded communities, where the chance of the infective discharge being carried to the eyes of others is greatest.   They are rare among the well-to-do classes in all countries, and most frequent among those classes who use towels or handkerchiefs in common, and who are ignorant or careless as to the necessity of treatment and of taking precautions to prevent their spread.   Dirt and mal-hygiene greatly favour their prevalence.   In many hot and sunny countries the glare of the sun, clouds of hot, irritating and foul dust, and the carriage of infection by means of flies, all contribute to diffuse the disease widely.

High, pure mountain air seems to be really inimical to the spread of these troubles, and they are certainly found at their worst in low-lying regions.

True epidemics of ophthalmia are by no means rare, particularly occurring in institutions, such as schools and barracks.

In regard to trachoma, or the granular form of ophthalmia, it seems probable that racial susceptibility has a good deal to do with its distribution. The Mongolian races appear to be particularly liable to it.   The high proportion of trachomatous to other cases of eye disease in China and Japan has already been pointed out, and the same tendency is observed among the Chinese inhabitants of American cities, the East Indies, and the Malay Peninsula. Manson has suggested that this marked liability to trachoma in these races may be due to the peculiar elongated almond-shape of the Mongolian eye.

At the other extreme there are certain races that possess a relative or almost positive immunity from this form of eye disease. Thus it is said that the African negro rarely or never suffers from it. The almost or quite complete absence of trachoma from the West Coast of Africa, already pointed out, supports this statement; and it is further said that in the United States negroes have remained free from trachoma even when the disease was epidemic among whites in their immediate neighbourhood. Thus the "Kentucky trachoma," already alluded to as common among the white population of Kentucky, never attacks the blacks. Some years ago, it is said, in Tennessee, a number of white labourers, mostly Irish, and a number of negroes were employed together in railway construction, and trachoma broke out with great severity among the former, yet there was not a single case among the latter. It is probable however that the immunity of the negro race to trachoma is not absolute. The late Prof. Van Millingen of Constantinople pointed out that, while the negro eunuchs in the harems of that city, who live a lazy, comfortable life, escape the disease, other negroes living in a state of overcrowding and dirt suffer considerably from it, trachoma accounting for 24 per cent. of all their ocular affections[1].

The indigenous tribes of Canada are also said to be immune to trachoma. The Crees and Santeux tribes of Manitoba, living side by side with the Russian Mennonites, appear to escape its attacks, although the latter suffer from it severely. The Indian tribes of the United States, on the other hand, are liable to become subjects of trachoma[2].

---

[1] *Annales d'Oculistique.* Sept. 1895.

[2] For a discussion of this question see papers by M. T. Yarr, F.R.C.S.I., in the *Journal of Tropical Med.*, May, 1899, and in the *Brit. Med. Journal*, May, 1899, from which the facts in the text are quoted.

PELLAGRA.

General Characters and Etiology. Pellagra is a nutritional disease, occurring most often in the spring, either as an epidemic or endemic disorder. The clinical features are more or less fully described in all text-books of medicine, to which the reader is referred for details. Here it will suffice to state that the disease is characterised at first by general weakness, headache, giddiness, spinal pain, gastric disturbance, not infrequently diarrhœa, and a peculiar skin eruption. This eruption appears on the back of the hands, on the face, on the feet (of bare-footed people), and more rarely on the back and chest. There is a general erythematous condition, with drying and exfoliation of the epidermis; the skin becomes very rough and dry, and sometimes suppuration and crust formation are observed. In severe cases the cerebro-spinal and sympathetic nervous systems appear to be deeply involved; there may be delirium and various spasmodic and paralytic phenomena, and a condition allied to scurvy on the one hand and to typhoid on the other may supervene. Severe mental trouble, particularly melancholia, imbecility, and suicidal mania, may follow prolonged or repeated attacks. The disease has been known to last from ten to fifteen years, and even longer.

Pellagra is generally believed to be caused by the consumption of damaged grain, and there is very strong evidence that the disease is an intoxication produced by eating badly kept and decomposing maize or maize-flour. This question will be discussed more fully in a later part of the chapter.

History. This malady was first recognised and described in Spain, where it is said to have been endemic, at least in the Asturian district of Oviedo, since 1735. Since that date it has become very widely prevalent in other parts of the peninsula.

A little later it began to extend in Italy, where isolated cases are known to have occurred even before the date just mentioned. The earliest record of pellagra in France is nearly a century later, and it was not until 1829 that the first accounts of it in that country appeared, though the disease itself was observed there as early as 1818. In 1833 the malady was recognised in Roumania, in 1839 it appeared in Corfu, and in quite recent years it has been frequently seen in some parts of southern Russia.

Recent Geographical Distribution. The distribution of pellagra over the earth's surface is a peculiarly limited one. Up to the present it has only been met with in Europe and in Africa, and in both continents it has only been seen in certain countries. These countries are Spain, Italy, France, Roumania, Bosnia, Croatia, Russia, and the island of Corfu in Europe, and Egypt in Africa.

In Spain the disease is known as the *mal de la rosa*. It is —or was up to 1879—particularly prevalent in Asturias, and especially in the communes of Regueras, Llanera, Corbera, and Careño. In addition to the above places, Hirsch states that pellagra affects many of the inhabitants of Lower Arragon and Burgos, of the provinces of Guadalajara and Cuenca, of the district on the frontiers of Navarra and Arragon, of the Ebro valley in the province of Zaragoza, of the level banks of the Douro and Tormes in the province of Zamora, and of Galicia. In some villages as many as 2 per cent. of the population were found to suffer from it.

At the present day Italy is one of the principal centres of pellagra. The history of the disease in that country since its appearance there early in the 18th century has been one of constant extension, both in the number of persons attacked and in the area of country affected. But it has never extended to the southern provinces—the Abruzzi, Campania, Apulia, Basilicata, and Calabria—and has remained confined to the northern and central provinces. Lombardy, Venetia, and Emilia are the areas in which it is most prevalent. In Lombardy it is most common in the provinces of Brescia, Bergamo, Cremona, and Milan; in Venetia, in the provinces of Padua and Rovigo; in Emilia in those of Ferrara, Piacenza, and Parma. In Piedmont, Liguria, Tuscany, Rome, Umbria, and the Marches the disease is met with to a less extent. Pellagra is a cause of no inconsiderable mortality in Italy, as shewn by the following figures. In 1887 the deaths from this disease for the whole country were equal to 125 per million of the population; in 1888 the ratio was 117; in 1889, 104; in 1890, 123; in 1891, 142; in 1892, 141; in 1893, 106; in 1894, 98; in 1895, 105; and in 1896, 99 per million. The deaths from pellagra in these years were very much more numerous in Italy than the deaths from acute and chronic rheumatism together; they were about the same as the deaths from croup, and were not very much less numerous than those from scarlet fever. Lombroso has stated that as many as five per cent. of the village populations in northern Italy have been known to suffer from the disease.

In France pellagra is more or less endemic in the coast-region of the Gironde, in the Landes, in the Hautes-Pyrénées and Basses-Pyrénées, in the Haute-Garonne and Aude. Sporadic cases of the disease have also been reported from a number of other parts of France, including Paris.

The disease is believed to have existed in Roumania since the year 1833 at least, although cases were not treated in hospital until 1846. Since then there has been a considerable increase in its degree of frequency. Of the two divisions of the country, Moldavia and Wallachia, the former is more affected than the latter. This area of pellagra prevalence is contiguous with the Russian centre of the malady. Up to the present the disease has only been reported in Russia from the governments of Bessarabia and Kherson. In Bessarabia it has been for some years endemic, and it has recently become so in the Tiraspol *uyézd*, or district, of Kherson[1]. In both governments many Moldavian inhabitants are found, and these appear to be especially liable to the disorder. The *volost* (or collection of villages) most affected is that of Lungovo, but it must also be very prevalent in Dubosary, for one medical man, in the month of April, 1901, alone, registered as many as 30 cases of the disease there.

In Bosnia and Croatia pellagra is certainly met with and is perhaps common[2]. In Turkey, on the other hand, it is unknown.

In Corfu pellagra first appeared in 1839, and after 1856 it became truly endemic.

Until recently it was believed that the disease was confined to countries between the 42nd and 46th degree of north latitude.

[1] *Bolnitchnaia Gazeta Botkina*, May 23, 1901.
[2] *Wien. Med. Woch.* 1901. No. 10.

But it is now known that it prevails in Egypt, and even as far south as Assouan, which is almost within the tropics (latitude 24° N.). It is much more frequent in Lower than in Upper Egypt, where it is in fact a rarity. The disease is said to be unknown in the Fayoum, and to be rare in the neighbourhood of Luxor. It is rare also in Alexandria, Port Saïd, and Suez, but is common in the provinces nearest to Cairo. Some cases have been seen at Tokar by the Red Sea, among starving Arabs, after a famine in 1891[1].

Factors concerned in the Geographical Distribution of Pellagra. From the date of the earliest observations on pellagra in Spain down to the present day the disease has been ascribed to the use of maize as food, and this view of its mode of production is now almost universally held. It is based on the facts that the disease did not appear in Europe until after the introduction of maize as an article of food, that it disappears when maize is excluded from the dietary, and that it is endemic only in those areas where maize is the chief article of diet. Its relative frequency, moreover, in the different parts of those countries where it is met with seems to depend on the extent to which maize or maize products are eaten by the people. In Italy this is especially the case, and pellagra is there most common where polenta, or maize porridge is the staple food of the peasants, and least so where this is largely combined with, or replaced by other articles of food, such as wheaten bread, potatoes, and chestnuts. In Egypt, again, pellagra is most common in Lower Egypt, where maize is much grown and eaten, and rare in Upper Egypt, where maize is little eaten or not at all, and where millet is the chief grain food of the peasants. In the Fayoum, too, and near Luxor, where as we have seen the disease is very rare, the natives scarcely eat any maize.

The large majority of writers upon pellagra agree that good sound maize, well cultivated and well harvested and properly stored, will not give the disease, but that when the reverse of these conditions obtains pellagra is likely to follow. Thus the affection is unknown in those tropical or sub-tropical countries where maize is indigenous, and where it can be grown and harvested under the most favourable conditions, and it is only found in countries where this cereal is an exotic, and where care is not exercised in

[1] Sandwith, *Brit. Med. Assoc.* 1898. Paper on " Pellagra in Egypt."

supervising all the steps between the sowing of the grain and its ultimate consumption in the food of the peasantry. The grain may be cut before it is ripe, or it may be cut and stored in a damp state, or it may be stored in pits in the ground where it becomes damp and rotten, and all these conditions favour the development of pellagra. In Roumania and in Italy the disease is commonest where the climate or soil are such as to render the production of a good crop of maize most difficult. Hence, also, the disease is most often seen, and indeed may become epidemic, in years when there has been a bad maize harvest, and "when the maize corn has been malformed, gathered half ripe owing to the peasantry being short of food, and stored or used in the wet state." In Bessarabia and the Kherson government in Southern Russia it is the Moldavian inhabitants who suffer most, and it is precisely they who live on maize (*kukuruz*) for the most part badly kept. At all times the Moldavians are said to prefer maize to other flours, but in famine years they are compelled to subsist mainly upon it. This flour is badly preserved, and becomes sour and more or less offensive to most tastes. But the Moldavians make a variety of *yastva* or food-stuffs out of it, and hence the great extent to which they have suffered from pellagra in the recent years of bad harvest and famine.

The exact nature of the pellagra poison is not known. It has been surmised that the disease may be due to some chemical substance produced in decomposing or fermenting maize, or to some micro-organism growing in the grain, or to a parasitic mould growing on or in the grain.

Lombroso, after careful observations and experiments, was of opinion that the disease was not brought about by any parasitic mould, but that it was probably due to a toxic chemical substance developed in the decomposing grain. This view is now most generally held, and it does not exclude the possibility that the toxic substance, whatever it may be, is developed in the maize by the action of certain micro-organisms which are constantly found in such grain.

Such factors as the climate, soil, and geological and physical characters of the country affect the distribution of pellagra only in so far as they make the growth and harvesting of maize a difficult or an easy proceeding. In like manner race, in itself, is probably of no importance in determining the prevalence of the disease. In some instances, it is true, one people—as, for example, the Moldavians in southern Russia—has appeared to suffer more than another in the same district. But this has been due to no racial susceptibility, and simply to the fact that the affected persons have eaten maize in the condition already described, while the non-affected have not done so.

PLAGUE.

General Characters and Etiology. Clinically plague occurs under three principal forms :—the bubonic, the pneumonic, and the septicæmic. Some observers have spoken of a fourth form,—the abdominal or enteric, but it is doubtful whether this form exists. The clinical varieties of the disease have been thought to depend upon the different modes of penetration of the bacillus into the tissues—the bubonic by its passage through the skin and its multiplication in a lymphatic gland ; the pneumonic by its being inhaled and multiplying in the lungs ; the abdominal (if such variety exist) by its entering the alimentary passages. There is, however, evidence which seems to point to the possibility that each variety of the disease may have a distinct and specific character, bubonic cases giving rise to fresh bubonic cases, and pneumonic cases to fresh pneumonic cases. The Indian Plague Commission is of opinion that this, rather than the accident of the mode of penetration of the bacilli into the tissues, determines the character of the attack. But the evidence on the point is conflicting.

Each variety of the disease may become septicæmic before death by the passage of the bacilli into the blood-stream. In some instances the attack seems to be septicæmic from the beginning.

A mild form of plague or bubonic fever, or even a condition of enlarged glands without fever, is often seen before and during a plague epidemic, and is known as *pestis ambulans* or *pestis minor*.

History. The oldest known historical references to the occurrence of plague are probably contained in the early books of the Old Testament. The Biblical use of the word plague is, like its medieval use, more often than not, in the general sense of a pestilence or epidemic disease, rather than in the specific sense of the disease now known as bubonic plague. The expression "a plague of leprosy" is a common one, and there is no reason to suppose that any of the ten "plagues" of Egypt were of the character of the malady now under discussion. But in a few instances there are some grounds for believing that the disease mentioned was in truth bubonic plague and no other. The most remarkable reference of this nature is found in the First Book of Samuel[1]. The whole history of the outbreak there described

---

[1] 1 Samuel iv, v, vi. The Philistines defeated the Israelites at the battle of Eben-ezer, and took the Ark. After arriving at Ashdod the pestilence appeared. In the ordinary version the reading is "and (He) smote them with emerods." In fear that the presence of the Ark in their midst was the cause of their misfortunes they sent it first to Gath and then to Ekron. But in both places the disease appeared. In the former "He smote the men of the city, both small and great, and they had emerods in their secret parts"; in the latter "there was a deadly destruction throughout all the city," and "the men that died not were smitten with the emerods." This is not the place to discuss the meaning of the word "emerods," which occurs rather frequently in the Old Testament, and which some have explained as meaning hæmorrhoids. But it may be noted that in the Revised Version the word "tumour" is substituted in the text for the word "emerod," and in the marginal notes an alternative reading is in each case given, "or plague boils." It may therefore be accepted, at least, that this pestilence was one in which "tumours" were a prominent symptom; and "tumours in their secret parts" might well be taken as a description of inguinal buboes. In the meantime there seems to have been a veritable plague of mice. In the ordinary version there is only incidental reference to "the mice that mar the land" (1 Samuel vi, 5); but in the Septuagint version the references are more distinct. Thus in the latter version it is stated (chap. v, 6) that "in the midst of the land thereof mice were brought forth, and there was a great and deadly destruction in the city"; and further (chap. vi, 1) that "their land swarmed with mice." So that at the time of this pestilence there was clearly an unusual number of mice, and perhaps some unusual migrations of these animals, bringing them more to the notice of the people.

After retaining the Ark for seven months, during which period the plague apparently did not abate, the Philistines, in alarm, determined to return the Ark to the Israelites. After consulting their priests and diviners, they placed on the Ark images in gold of their "tumours" and of the "mice that marred the land" besides a coffer of jewels as a guilt-offering, and sent back the Ark, drawn by two milch-kine and followed by some of the lords of the Philistines. The milch-kine brought the holy vessel to Beth-shemesh, and the lords of the Philistines, having seen the Beth-shemites obtain possession of it, returned the

is so full of interest that I have discussed it at some length in the footnote below, and it need only be briefly summarised here. The Philistines had conquered the Israelites and captured the Ark, and retired with it to Ashdod. There a whole series of misfortunes befell them : the idol Dagon fell and was broken in pieces, a plague of mice ravaged the land, and at the same time the people were smitten with "tumours." This pestilence followed the Philistines in their subsequent movements, and seems to have been carried by them to the Israelites, who suffered from a most deadly epidemic. The exact nature of the disease here referred to is, of course, uncertain, but it is not perhaps stretching the limits of possibility too wide to suppose that a plague epidemic smote the Philistines, that it spread with them in their movements from city to city, and was finally carried by them when they returned the Ark to the Israelites. Geographically the whole epidemic was confined to the country lying in the extreme south-east corner of the Mediterranean Sea ; commencing in "Ashdod and the coast thereof" (that is to say the region south of Jaffa) and spreading inland. The mention of "the mice that marred the land" has a more than passing interest, in view of the well-known susceptibility of these animals to plague.

The many other epidemics mentioned in the Old Testament may or may not have been examples of plague, and it would serve but little purpose to dwell upon them in detail. As an example of a somewhat remarkable one may be mentioned the "plague" which visited the Israelites after the punishment of Korah, Dathan, and Abiram and caused the death of 14,700 people[1]. Another pestilence mentioned in Biblical writings is that which was sent as a punishment for David's disobedience in numbering the people, in direct contravention of the divine injunctions[2].

same day. But another fearful epidemic smote the Beth-shemites, as a punishment for having looked into the Ark, and no less than 50,070 persons perished.

For the curious it may be mentioned that in Deuteronomy xxviii, 27, the disobedient are threatened with "the botch of Egypt and with the emerods," to which, as before, the Revised Version adds a marginal note "*or* tumours, *or* plague boils." In support of the view that the plague was in fact referred to, it may further be pointed out that throughout medieval times and down to the reign of Elizabeth the plague was constantly spoken of as the "botch," the word signifying the bubo, or the most prominent and characteristic feature of the disease.

[1] Numbers xvi. [2] 2 Samuel xxiv.

David was given the choice of a famine, a war, or "a pestilence and a distemper upon the Hebrews for three days." He chose the last, and seventy thousand persons perished of the pestilence. The story of it is told in some detail by Josephus, but there is no clear evidence as to what was the nature of the disease.

Vague and uncertain as are the references to plague in the Bible, they are equally so in the secular literature of the pre-Christian era. The historians of the time make not infrequent mention of pestilences, particularly as attacking armies and besieged cities, but, with scarcely an exception, there is nothing to show positively that these were outbreaks of bubonic plague. The famous Plague of Athens (430–428 B.C.) was almost certainly not of this character. Thucydides, who himself suffered from it, has left a sufficiently detailed description of the symptoms to justify this conclusion, and to indicate that the mortality, which was excessively high, was probably due to many diseases (of which small-pox and typhoid fever would seem to have been the most important), aggravated by the horrors of overcrowding in a besieged city.

The sole certain mention of a plague epidemic in the pre-Christian period is contained in the writings of Rufus of Ephesus, a physician of the time of Trajan, who refers to the records of certain Greek physicians concerning an epidemic in the third century before Christ. This epidemic ravaged Syria, Egypt, and Libya, and of its nature there can be little doubt, for the contemporary physicians quoted by the author just named described the disease as "accompanied by an acute fever, by terrible pain, by a trouble of the whole body, by delirium, and by the appearance of large buboes, hard and without suppuration, not only in the usual positions, but also behind the knee and at the elbow."

In the second century of our era (164–180), during the reign of Marcus Aurelius, a widespread epidemic seems to have ravaged the greater part of Europe. But of far greater intensity was the historical outbreak in the reign of Justinian. It began in Egypt in 542, and spread thence to Syria, Byzantium, and Europe. This great pestilence lasted, with intermissions, for at least 52 years. The evidence is perhaps not quite conclusive whether the various epidemics occurring in different parts of the Roman world during this long period are rightly grouped as constituting one great pandemic outbreak of plague, or whether a number of other diseases, such as small-pox, have not been included with them. But in any case it appears safe to say that from the year 542, for the next half century and possibly longer, plague was generally very prevalent[1].

Between the plague of Justinian's time and the Black Death of the fourteenth century many outbreaks of pestilence occurred in different parts of Europe, some of which there is good reason to believe were of the nature of bubonic plague.

Of all great historical outbreaks the Black Death is the most important, as it was the most deadly and appalling in its consequences of any either before or since. Its origin is obscure. Plague had been present in certain parts of Europe during the early half of the fourteenth century, but in these places the disease had apparently shown little tendency to become widely spread, and nearly all writers agree that the intensely fatal form of plague to which the name of Black Death has been given was a fresh importation from the East. It was in Genoa in the spring of 1347; and in the same year it was in Constantinople, Cyprus, Greece, and (later) in Marseilles.

The manner in which it was imported to these places has been the subject of much controversy, but certain facts may be regarded as now fairly well-established and as tending to show the source of this great pandemic. The contemporary imperial historian, John Kantakuzin, who himself saw the disease in Constantinople, stated that it had come from the country of the Hyperborean Scythians, that is to say, from the steppes of southern Russia and the Crimea. Another Byzantine writer, Nicephorus Gregoras, ascribed it to the same source. There is evidence in the Russian

---

[1] The principal authorities for this great historical event are the writings of two illustrious lawyers, Procopius of Constantinople and Evagrius of Antioch, each of whom became prefect of his own city, and both of whom were eye-witnesses of the principal events of the plague ; while one (Evagrius) not only himself suffered from it, but lost his wife, many children, and a large number of relations and servants. To these witnesses may be added Bishop Gregory of Tours, who makes brief reference to the plague in France. This pestilence began at Pelusium in Egypt and spread thence to Alexandria and to Syria. It took two years to reach Constantinople (April 544), and was gradually carried to nearly all parts of the then civilized world. It was in Gaul in 546. In most countries it began at the sea-coast and spread inland. Procopius has left a description of the disease seen by him in Constantinople which is so full of detail as to leave no doubt that it was the plague. Evagrius describes at least four recurrences of the disease at Antioch, the last in the year 592.

chronicles to show that plague was prevalent in 1346 in the country round the river Don, at Surai and Bezdej on the Volga, as well as in the Caucasus, Armenia, and Khiva, and it may be regarded as almost certain that southern Russia was the part of Europe first invaded by the Black Death[1].

How the Tatar inhabitants of those broad steppes first contracted the disease is far less certain. They had constant relations with China, and hence it is thought the infection may have been derived from that country. The supposition is *à priori* a plausible one. The Tatar khans, at least until the time of their first serious defeat by the Russians under Dmitri of the Don at the battle of Kulikovo in 1380, kept open the route from China by the northern roads through what are now called Semipalatinsk, Semirétchinsk, and the Kirghiz steppes of Turgai and Uralsk, to the north of the Aral and Caspian Seas. An infectious disease existing in China might possibly have been carried by the Tatars to Europe. Unfortunately, however, there is no positive proof of the existence of plague in China at that time. On the contrary, the Chinese annalists make no mention of the disease for the half century before this period, and the contemporary Arabian traveller Ibn-Batuta, who spent the four years 1342–46 in China, does not seem to have seen or heard of the disease there.

An alternative suggestion has been made that the Black Death came from India, and there are many considerations in favour of this view. Plague is known to have been prevalent in India in the early part of the fourteenth century. Between the years 1325 and 1351 Muhammad Tighlak's army in Malabar mostly perished of the disease. The Arabian traveller just quoted, Ibn-Batuta, himself saw and was attacked by the disease at Muttra, Muhammad Tighlak's capital—apparently in the year 1332[2]. Further it may

[1] The *Troitzki Chronicle* states that there was pestilence then " in the East, in the Horde and in Ornatch " (at the mouth of the Don), and "in Sarai and in Bezdej" both on the Volga ; and " in other countries," and a great plague "among the Bessermans" (or Khivans), "and among the Tatars, and the Armenians," and "among the Abazians" and among the Genoese and Venetians in Taurida and in Azof, and among the Tcherkessi. The *Pskof Chronicle* states boldly that the plague came from India. (See also Karamzin's *Istoria Gosudarstva Rossiiskavo*, Vol. IV and Notes.)

[2] *Travels of Ibn-Batuta*, translated from the abridged Arabic MS. copies by the Rev. Samuel Lee, B.D., London, printed for the Oriental Trans. Com., 1829.

be noted that the Russian contemporary chroniclers and other later writers (*e.g.* Fracastori in the 16th century) ascribed the origin of the Black Death to India. The fact that plague was epidemic in Khiva, Armenia, and the Caucasus, about the period of its appearance on the Volga and the Don, seems to show that the Black Death was imported to Europe by way of southern Central Asia, and this lends still greater probability to the view that it may have come from India[1].

Whatever may have been the source of this outbreak, it spread slowly over the whole of Europe, including Scandinavia and Iceland, and was even carried to Greenland. The northern part of Russia was not invaded until the year 1851, and then the disease came from the west, and not from the south-eastern steppes, where it had prevailed five years previously. It has been computed that from first to last the Black Death caused the death of some twenty-five million people.

Throughout the remainder of the fourteenth, and in the two succeeding centuries, plague was practically endemic throughout the greater part of Europe, and from time to time caused severe epidemics affecting larger or smaller areas. It was not until the seventeenth century that it began to recede, as it were, eastwards; that is to say, the western portions of Europe gradually became free from the plague, and while the disease remained in the East and still from time to time spread westwards, each succeeding epidemic was more limited than the one before it. Denmark seems to have been the first European country to be freed from the disease; the last epidemic occurred there in 1654. The dates of the last epidemics in other countries appear to have been as follows: Sweden, 1657; England, 1666 or 1667; Switzerland, 1667–8; France (with the exception of the great outbreak in Marseilles and its neighbourhood in 1720) also in 1667–8; western Germany in the same years; and Spain, 1677–81.

---

[1] The additional argument in favour of India as the source of the Black Death, advanced by Hirsch and others, that the pulmonary form of the disease was specially prominent in the Black Death and has always been a distinguishing feature of plague in India, can scarcely now be held as valid. Pneumonic plague has been shown to be no rarity in modern epidemics outside India; and it is open to question whether lung symptoms were not just as prominent in other great epidemics of plague as in the Black Death of the 14th century.

At this time and in the succeeding century plague seems to have been confined as an endemic disease in Europe to the Balkan peninsula, whence from time to time it was still able to spread to the north and west and cause epidemics of no little severity. Russia, then very frequently at war with Turkey, suffered more from these outbreaks than any other country. A serious epidemic occurred there in 1709; and in 1710, when Peter the Great was attacking the fortress of Riga, 60,000 persons perished from plague. In 1718 there was another epidemic, and in 1770-2 there occurred in Moscow one of the most fatal pestilences ever seen in Russia[1]. Throughout that century a number of outbreaks occurred in Austria, in Hungary, and on the shores of the Black Sea and the Adriatic.

In the first half of the nineteenth century plague showed but little activity in Europe. In 1807 it spread from the Caucasus to Astrakhan and Saratof on the Volga. In 1813 it caused an epidemic in Malta, and in 1815 another at Noja in Italy. In 1820 it appeared in the Balearic Isles. In 1828-9 the Russian troops in Wallachia were attacked by it, and in the same year a rather severe epidemic occurred at the taking of Kars by the Russians.

In 1837 plague broke out in the Greek island of Poros, in 1838 there was a small outbreak in Odessa, and between 1839 and 1843 a few isolated epidemics occurred in Transcaucasia. In Constantinople plague seems to have prevailed rather frequently up to the year 1839, but after a small outbreak there in 1841 the disease completely died out, and with its disappearance from the Turkish capital the whole continent of Europe became exempt from plague and remained so, with one slight exception, until the last few years of the century.

In the latter half of the nineteenth century the centre of interest in the story of plague shifts from Europe to certain remote portions of Asia and Africa. As this period is of special importance, as preceding the present wide extension of the disease, it deserves to be dealt with in some detail. But before doing so it is necessary to recount, very briefly, the early history of plague outside Europe.

---

[1] See papers by the author upon Plague Epidemics in Russia, in the *Practitioner*, November, 1894; and the *Indian Med. Gazette*, September and October, 1898.

In Asia, plague is no doubt a disease of some antiquity, but its early history here is even more obscure than in Europe. For China there is no certain reference to any great epidemic before the 18th century, in the latter half of which there is reason to believe that plague prevailed in Yünnan, the province where it is still endemic[1]. A contemporary author wrote of this as follows :—

"Then in Chau-Chau [in Yünnan] it happened that in the daytime strange rats appeared in the houses, and lying down on the ground perished with blood-spitting. There was not a man escaped the instantaneous death after being infected with the miasma. Tau-Nan composed thereon a poem entitled 'Death of Rats'—the master-piece of his ; and a few days after he himself died of this queer rat epidemic[2]."

There can be little doubt that this was the plague as we now know it.

In India plague was certainly epidemic in the early fourteenth century, in the time of Muhammad Tighlak. At the end of that century (1399), after Timur's ravaging hordes had left the country, there was another severe pestilence, perhaps of plague, in the districts he had traversed. In 1443, again, Sultan Ahmed I. lost so many of his men from plague that he was compelled to return to Gujerat ; this disease is spoken of by historians as "ta'un," the name still used in the East.

The famine in India of 1590-4 was followed by a pestilence which depopulated whole cities, villages, and hamlets, and which may have been the plague. In 1611 plague appeared in the Punjab ; it was called "waba" or "waba-o-ta'un." "Waba" or "veba" and "ta'un" are still the names for plague in Turkey and Arabia, in Persia and Hindustan. This epidemic spread through Lahore and the Doab to Delhi and northwards to Kashmir. No place in the peninsula, it is said, escaped it. It lasted eight years. In Kandahar it was accompanied by a swarming and unusual

[1] Lowson (*Ind. Med. Gazette*, January, 1897) states that he caused a search to be made into Chinese literature, but no record was found of any historical epidemic which made a mark in the annals of the country. On the other hand the Chinese annals contain innumerable records of pestilences—as well as of other great national disasters, such as earthquakes, floods, and supernatural visitations—and it may well be that some of these pestilences were epidemics of plague.

[2] See note in *Nature*, Feb. 16, 1899, by Mr Kumagusu Minakata.

behaviour of mice. Later, in 1684, the army of Aurungzebe suffered from it, and from that year to 1691 plague raged severely over the greater part of western India, particularly in Bombay, Surat, Goa, Ahmedabad, Bulsar, Daman, Thana, and Sindh.

During the 18th century the Indian annals are silent as to the behaviour of plague, but early in the 19th the disease again became active. In 1812 it broke out in Cutch, and spread thence to Kathiawar, Ahmedabad, Radhanpur, the southern part of Sindh and other places. This outbreak died out about 1821. In 1823 occurs the first account of plague in the hills of Garhwal and Kumaon, where it has ever since been endemic. In 1828–9 a plague-like disease was present at Hansi, and in 1836–8 occurred the well-known outbreak at Pali in Jodhpur State, Rajputana.

In Persia the earliest epidemic of the plague is believed to have been that of 1571–75, and between that date and 1835 that country was visited several times by the disease. In Mesopotamia there was an outbreak of plague in 1596 in Baghdad, and then follows a period of two centuries without any record of its existence there. But in 1773 it reappeared in Mesopotamia, and again in 1800–1, 1830–34, and 1867. Syria, Armenia, and Asia Minor have been the scenes of frequent outbreaks from the earliest historical times, and together with Egypt and Turkey were regarded as the permanent homes of plague at the end of the 18th and beginning of the 19th centuries. The Red Sea coast of Arabia has more than once been visited by plague, and even in Mecca isolated cases have been seen. No serious epidemic has however occurred at any time in the Moslem Holy Places, and this comparative immunity is ascribed by the faithful to the special protection of the Prophet[1].

In Africa, the earliest known scene of a plague outbreak was in Egypt and the adjoining country of Libya on the west. The epidemic of the third century B.C. already referred to mostly prevailed here; the Great Plague of Justinian's reign also began here. For a time, perhaps, plague was endemic in Egypt, and for

---

[1] As is pointed out later in the text, Medina has always escaped, but Mecca has been less fortunate. Thus in 1815, when plague was raging in Jedda and Yanbo, great numbers of the inhabitants of the former place fled to Mecca, "thinking to be safe in that sacred asylum, but they carried the disease with them, and a number of Mekkans died, although much less in proportion than at Djidda [Jedda]." Burckhardt, *Travels in Arabia*, London, 1829.

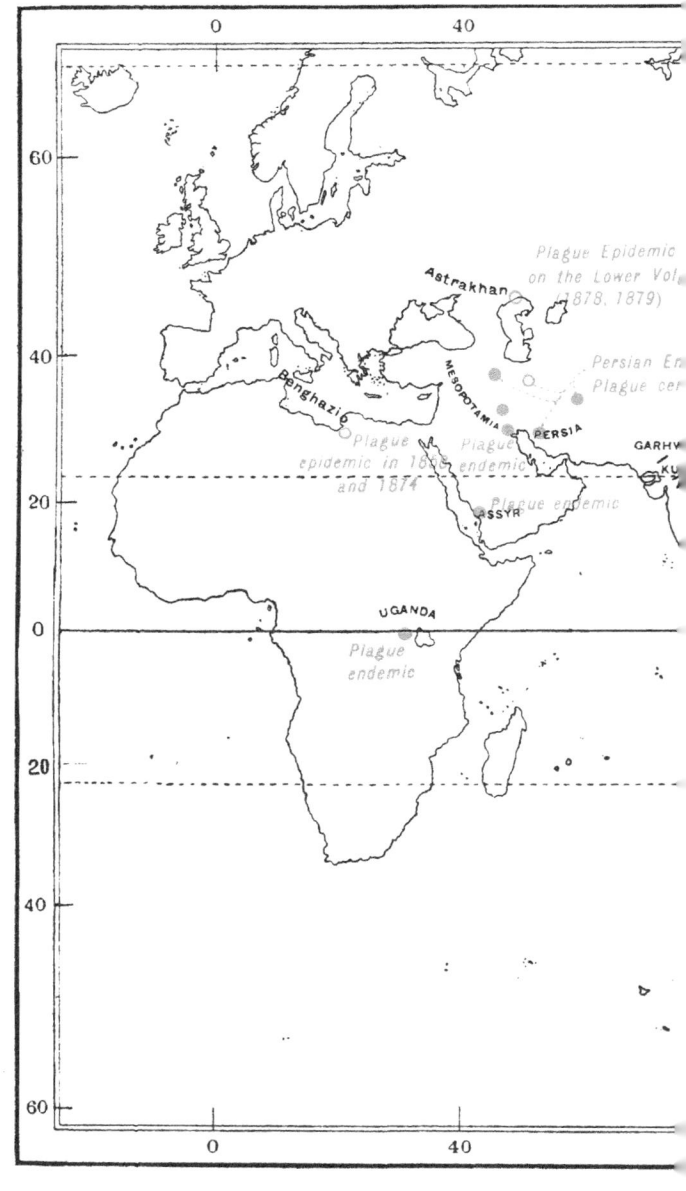

REGIONS IN WHICH PLAGUE

BETWEEN 185

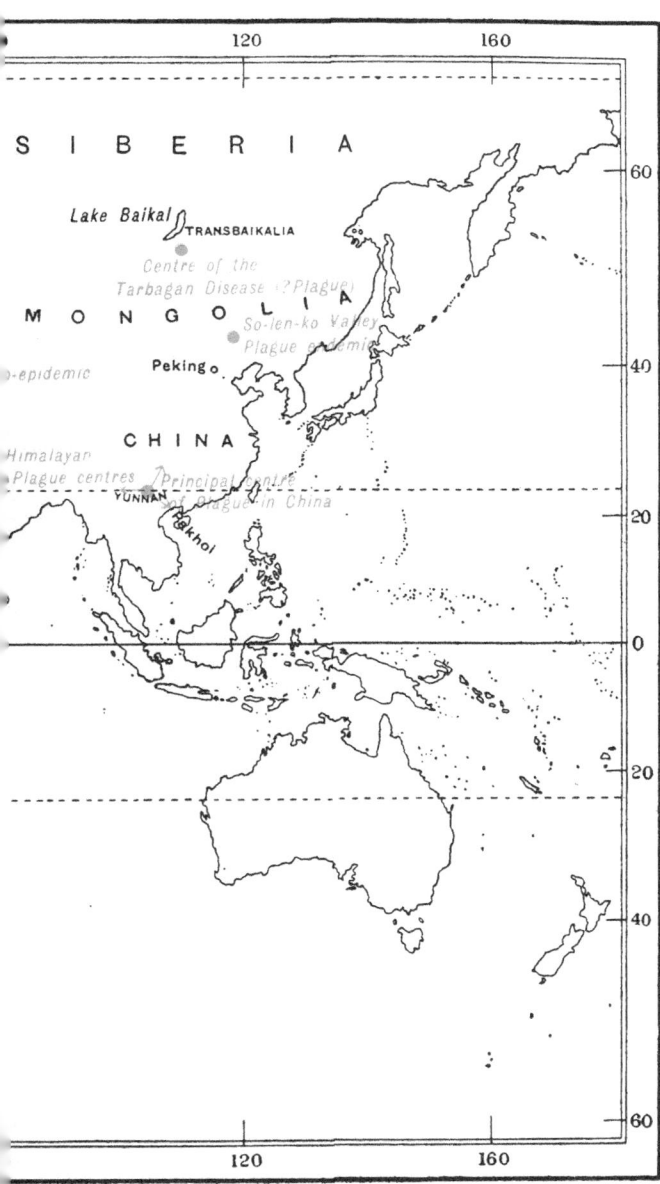

S I B E R I A

60

Lake Baikal ⌡ TRANSBAIKALIA
        Centre of the
    Tarbagan Disease (?Plague)

M O N G O L I A
                        So-len-ko Valley
                        Plague epidemic

    Peking o   40

-epidemic

        C H I N A

Himalayan
Plague centres /Principal centre
        YUNNAN   of Plague in China   20

Pakhoi

                                                0

                                            20

                                            40

                                            60

120   160

WAS ENDEMIC OR EPIDEMIC
AND 1894

long it was customary to regard that country as the permanent home and source of the disease. How far this was actually the case it is difficult to say. It is more than possible that some confusion has long existed in the popular mind between the Biblical Plagues of Egypt and the specific disease to which we give the name of plague. Throughout medieval times almost nothing seems to be positively known of what the disease was doing in that country, but from the 16th century onwards there are more certain records, and Egypt then seems to have justly shared with Eastern Europe and the Levant the reputation of being the source of most of the European epidemics.

Towards the close of the 18th century there were frequent outbreaks in Egypt, but even then there appear to have been intervals of from two to five years' duration when the country was more or less free from it. Lastly in 1844, three years after its disappearance from Constantinople—its last foothold in Europe —plague died out in Egypt, and did not appear there again until the year 1899.

Algeria was the scene of several epidemics of plague between the years 1552 and 1837 ; Tripoli and Tunis probably suffered in each of these outbreaks, as the infection travelled from Egypt westward.

Morocco was visited by the disease on six occasions between the years 1678 and 1819. So far as is known plague had occurred in no other part of the African continent before the middle of the century just closed.

Endemic Centres and Epidemic Outbreaks of Plague in the last half of the 19th Century. For many years succeeding the year 1850 plague was confined to a limited number of comparatively small and somewhat remote areas, in which the disease was either truly endemic or in which it recurred at some-what long and irregular intervals. Europe had been free of it for a considerable period. America and Australasia had never, so far as is known, been visited by plague in historical times. The southern hemisphere was also believed to have entirely escaped its ravages, and all the known centres of plague were confined to the continent of Asia and the northern part of Africa[1].

---

[1] See "The Endemic Centres of Plague," a series of papers by the author in the *Journal of Tropical Medicine*, March—May, 1900.

In Asia the most important, if not the only regions in which plague had existed as an endemic disease since 1850 and before 1894, were certain areas in China, India, Mesopotamia, Arabia, Persia, Siberia, Mongolia, and possibly Russian Central Asia. In Africa similar foci of plague have existed or developed in the same period in Uganda and in Tripoli.

(a) In *China*[1] plague has been endemic in the south-western province of Yünnan, bordering on Tibet and Burma, possibly for a century or more, and certainly since the great Mohammedan rebellion in 1860. Within an area bounded approximately by the 100th and 104th degrees of east longitude, and the 23rd and 28th degrees of north latitude, the disease has prevailed more or less severely every year. Outside this area and outside this province other centres also appear to have existed for some time, as for example at or near An-pu in the Kwangtung province, while in many other places plague has broken out frequently though at somewhat irregular intervals, and it is uncertain whether the disease is truly endemic or is on each occasion imported from the centre in Yünnan. Among these places may be mentioned Pakhoi, and Lungchou in the province of Kwang-si. It has at times spread to many other places in this province and in that of Kwangtung, and less often to the province of Kwei-chou, to the north-east of Yünnan, and to Laos, between Yünnan and Siam. It never spread to Tongking or Annam, or to northern China. The outbreaks in the later years of the period under consideration were much less severe than those in the earlier years. But in 1893 and 1894 a marked accession of activity was observed in the endemic area of Yünnan, and in many of the other places mentioned, and in 1894 the disease appeared for the first time in history in Canton, whence it spread to Hongkong. If, as there is good reason to believe, the later appearance of plague in India and elsewhere was due to an extension of the infection from one of these places, then the Yünnan centre of endemic plague must be regarded as the cradle of origin of the recent wide extension of the disease to so many parts of the earth's surface.

(b) In *India* plague has been endemic in the districts of Kumaon and Garhwal, on the southern slopes of the Himalayas,

---

[1] For detailed references concerning plague in China see the author's paper in the *Journal of Tropical Medicine*, March, 1900, p. 200.

since the year 1823. The disease, known locally as *mahamari*[1], has appeared in one or both provinces at very irregular intervals since that date. Outbreaks occurred in 1834–5, in 1837, and in 1846–7. Between 1847 and 1854 the disease became epidemic in each successive year. In 1853–4 it spread far beyond its usual limits and caused a serious degree of mortality in the adjoining plains. It was epidemic again in 1859–60, in 1869–70, and in each of the years from 1875 to 1878. In 1884 and each of the following four years it was present in some part of the districts, and again in 1891, 1892, and 1893, a few cases of *mahamari* were observed. In 1894 it is doubtful if any cases occurred, in 1895 there is no mention of it, but in 1896 and 1897 limited outbreaks were dealt with by the authorities.

The Himalayan centres of plague have been much less active of recent years than formerly, in the sense that the later outbreaks, though possibly more frequent, are the cause of a much smaller mortality than the earlier ones. The quiescence of the disease here in the years 1892 to 1896 was in marked contrast with the unusual activity of the Chinese centres at the same period, and it is almost certain that the recent epidemic prevalence in India is due to an extension of plague from China, and not to an extension of mahamari from the Himalayas.

(*c*) In *Mesopotamia* plague appeared, after an absence of 33 years, in 1867[2]. The scene of that year's outbreak was the Hindieh marshes on the Lower Euphrates. In 1873 it was epidemic in the same region, but on both sides of the river, and in 1875 it broke out in a district rather more to the south. In 1876 Baghdad and other places suffered severely, and 20,000 persons were believed to have died of it in that year. In 1877 it was epidemic again in Baghdad and elsewhere, but not to so fatal an extent as in the previous year. In 1880 plague broke out in the El-Zayad tribe about the end of September; thence it spread to

[1] It appears possible that some of the epidemics of *mahamari* here enumerated were epidemics, not of plague, but of some other disorder, as the term is said to be used by the natives for an outbreak of any very severe disease. It is probable, however, that the large majority were epidemics of plague.

[2] The history of plague in Mesopotamia, Persia, and Arabia is based upon reports contained in *Parliamentary Paper* C. 2262, "Papers relating to the Modern History and Recent Progress of Levantine Plague" (1879), and upon numerous unpublished reports and papers contained in the Archives of the International Board of Health at Constantinople.

Chenafié in November, and reached Djaara in January 1881. From Djaara it spread to Nedjef, where in the course of some four months it caused 3000 deaths. In 1884 it was again epidemic, attacking especially the populations of Bedra, Zorbatia, Djizan, and Mendelli. In 1892 plague was prevalent among the Arab tribes on either side of the Shott-el-Ghraf, in the Muntefik district. Since that date the only known occurrence of plague in Mesopotamia has been a small outbreak of the disease in Baghdad in 1901, followed by a number of mild cases of a bubonic disorder during several months, and by a second small outbreak officially recognised as plague in January 1902.

(d) In *Persia*, where plague had appeared for the last time in 1835, a small epidemic of the disease occurred in 1863 in the extreme north-west corner of the country, in and around Maku. In 1870 it again broke out on an elevated tableland to the south-east of Lake Urumia and in 1876 it was epidemic in the province of Khusistan in the extreme south-west of the country. In December 1876 it was present in two villages 25 leagues distant from the shores of the Caspian Sea, and in March 1877 it appeared in Resht, quite close to the coasts of that sea. In Resht and its neighbourhood 4000 persons were believed to have died from the disease. In September it was reported from the Khalkal district of Azerbijan, and in the same month it was (possibly) present in Astrabad and (doubtfully) in Herat. In January 1878 plague was again epidemic in villages in Persian Kurdistan and remained there until April. In 1881 it re-appeared in the northern part of Kurdistan, and in two villages in Khorassan, near the Afghan frontier. In 1882 it was epidemic at Soöutch-Boulak and Uzundéré; in 1883 at Djivanrao and in its neighbourhood; in 1885 in and near Hamadan, the place of assembly for Persian pilgrims *en route* to the Shiah burying-grounds of Nedjef and Kerbela; in 1889 in villages in the neighbourhood of Merivan; and in 1890 in the district of Mahideste, near the Turco-Persian frontier. In 1896 a small outbreak of plague is said to have occurred among a Turkish regiment of Hamidié cavalry stationed in the *nahyé* of Aughnot, in the sandjak of Guendj, close to the Persian frontier. The appearance of the disease at Bushire and at Bender-Dilem, both on the shores of the Persian Gulf, in 1899, will be briefly mentioned later. It is unknown whether it was due to an imported infection from India, or to a manifestation of the disease already endemic in the country. Finally in 1900 a severe outbreak of plague again occurred in the Djivanrao district, and was the cause of considerable mortality.

(e) A centre of plague, or of a disease very closely allied to it, has existed for some years past in the *Transbaikal* province of Siberia[1]. The exact nature of the disease here is uncertain, but as its principal features are fever and the appearance of glandular swellings or buboes it is probable that it is in truth the plague itself. It is associated with, and apparently dependent upon, an epizoötic disease which attacks a species of marmot known locally as the tarbagan (*Arctomys-bobac*). These animals abound in the plains of Transbaikalia, and they are also found over a wide area in Europe and Asia. In Europe they extend as far west as Russia and eastern Germany, and in Asia they are found in Siberia, Mongolia, and Tibet. So far as is known they suffer from the disease in question only in the plains beyond Lake Baikal, and possibly in a certain part of eastern Mongolia, where, as will be shown later, human plague is also endemic. The outbreaks of tarbagan plague occur at short but irregular intervals, in the autumn of certain years. The disease is intensely infectious; it attacks persons who touch the bodies of animals that have died from it, and from them spreads to others in their immediate neighbourhood.

(f) In the valley of So-len-ko in eastern Mongolia, at a distance of twelve days' ride from Peking, plague has been endemic at least since the year 1888. It breaks out each year in the late summer and sometimes causes a very high mortality in a number of villages, the inhabitants of which live in a state of filth excessive even for Chinese villages. The marmot referred to in the last paragraph, *Arctomys bobac*, is found in the neighbourhood, and it is said to suffer from an analogous disease, but in this region the dependence of the disease in man upon that in the animals is much less direct and obvious than in the Siberian centre just described.

(g) The possibility of the existence of plague as an endemic disease in *Russian Central Asia* has recently been pointed out. The evidence for its existence is of the following character. In

---

[1] See paper by the author, "Plague in Siberia and Mongolia and the *Tarbagan*," in the *Journ. Trop. Med.*, February, 1900.

October 1898 a severe epidemic of plague broke out in the village of Anzob, in a remote mountainous region near the borders of the Samarkand province and Bokhara. Upon inquiry it was found that many of the inhabitants of this and an adjoining village had suffered during the past twenty years from attacks of a disease the principal symptoms of which were fever and glandular swellings. For the present it remains uncertain whether this endemic disease is in truth plague. Dr Levin, who conducted the inquiry on behalf of the Russian Government, thought that this disease and the outbreaks of plague on the Lower Volga and on the Caspian (which will be mentioned again later) might be due to carriage of infection from Mongolia by means of Kalmuck and Kirghiz pilgrims, who might have visited certain sacred Moslem shrines in Mongolia and brought the disease back with them. Against this view, however, must be set the enormous distances between the areas just named and the Mongolian centres of plague. Moreover the Mussulman inhabitants of the Kirghiz steppes east of the Volga are said to have no "holy places" of any kind in Mongolia; while the Kalmucks, who live to the west of the Volga, are despised and looked upon as heathens by both Mongolians and Kirghiz and are not allowed to pass through the Kirghiz steppes[1].

(*h*) Plague has for the greater part of a century been endemic in the hilly country of *Assyr*, in south-western Arabia. The disease is believed to have been imported here from Egypt, as the result of the invasion of the country by Egyptian troops in the second decade of last century. The first known outbreak of plague here was in 1816. A second serious epidemic was that of 1853. Since 1874 the disease has been more active, and the cause of several epidemic outbreaks. From 1874–79 it was very prevalent among the Beni-Sheir tribe of Arabs; in 1889, in 1893, in 1894, in 1895, in 1896, in 1900, in 1901, and in 1902 more or less serious extensions of the disease took place; but it is remarkable that in no instance has the infection left the hills and spread to the plains. The Assyr *foyer* of plague is practically confined to a strip of mountainous country, mainly occupied by the Beni-Sheir and Beni-Amr tribes of Arabs. The climate is temperate; in winter it is cool but not excessively cold. The inhabitants are extremely poor, and live in a state of much filth and misery.

[1] Goldberg, *Bolnitchnaia Gazeta Botkina*, 1900, No. 25.

Although the centre is not infrequently visited by Bedouins from other parts of Arabia, it is only on very rare occasions that any of them contract the disease, and even when this occurs they never carry it to other places[1].

(*i*) In Africa one of the most important centres of plague is situated in *Uganda* on the north-western shores of the Victoria Nyanza. How long the disease has existed in this locality is unknown, but it has certainly been there for the last twelve or fifteen years, and probably considerably longer. *Kaumpali*, as it is called locally, has been described by many writers upon Central Africa, and its true nature has been known for a considerable period. Complete scientific proof that it is indeed the plague was obtained by Koch and Zupitza in 1898. It is most severe in Buddu, the most southerly province of Uganda, and around Bukoba and Kitangule on adjoining German territory. It at times causes an excessively high mortality, as for example in 1898–9, when the Sultan of that region reported that no less than 467 deaths from plague had occurred in the course of a year in Bukoba out of a population of only 715. It was also very active in and around the capital after the wars of 1889 and 1890. Another sharp epidemic occurred in Buddu about 1896, and it has recently been carried by native traders to some of the islands on the Victoria Nyanza[2].

(*j*) In the province of *Benghazi* in the regency of Tripoli, North Africa, plague has become epidemic on two occasions in the period under review. In 1858 it prevailed rather widely in four out of the five divisions of the province, and in 1874 an epidemic of great intensity (attacking 533 persons in a population of 734) almost decimated a few tribes inhabiting the Cyrenaic plateau. Since that date there has been no word of plague in Tripoli, and it appears probable that the disease is no longer endemic in the country.

---

[1] Doughty (*Travels in Arabia*, Vol. I. p. 618) has described a number of ruined villages in the north-east of Arabia, in a direct line between the plague-centres in Mesopotamia and Assyr, the inhabitants of which had been carried off by a disease described as plague, or *waba*, apparently about the year 1870. Had this malady truly been the plague, the occurrence might have established an important link between the two plague-centres described in the text. But from the brief account given of the symptoms it seems most probable that they were due to some other affection rather than plague.

[2] Cook, *Journ. Trop. Med.*, June 1, 1901.

(*k*) Once and once only during the period now under review has plague appeared in Europe. It will be recalled that in March, 1877, the disease was severely epidemic in Resht near the Persian shores of the Caspian Sea. There is reason to believe that it may have been carried some months previously to the Russian shores of that sea. In Baku three deaths occurred in November 1896, all in the same house and all after a short attack of fever accompanied by buboes. In the autumn of 1877 a very similar disease appeared among the Russian troops in the Caucasus, engaged in operations connected with the Russo-Turkish war[1]. At the same time some cases of a like character were seen among the Russian troops in European Turkey. In the summer of 1877 a disease, of which fever and buboes were the principal features, was prevalent in the town of Astrakhan and in a number of villages around. These cases have been regarded as of the nature of a mild form of plague and they form an important link connecting the outbreaks in Persia, and particularly that in Resht, with the epidemic on the Volga. The outbreak on the Volga was confined to six villages, of which Vetlianka was the most important. It lasted from October 1878 to the early months of 1879, and was self-limited, coming to an end before any organised measures for its extinction were put into force[2].

Epidemic Extension of Plague since 1894. The above are the principal, and so far as is known the only places on the earth's surface in which plague prevailed in the latter half of the nineteenth century before the year 1894. That year was an epoch in the history of the disease, as it marked the commencement of the widest, though far from the most severe, pandemic

---

[1] The late Count Loris Melikof was in command of these troops, and a year and a half later he was appointed to take entire charge of the plague measures in the government of Astrakhan after the outbreak of plague at Vetlianka. It was when holding the latter appointment that he recalled that a disease which was described at the time as "typhus fever with buboes," had in 1877 caused a high mortality among his men in the Caucasus. The medical authorities in St Petersburg came to the conclusion that in all probability this disease had been plague, and apparently plague of a far from mild form, since the case mortality was from 75 to 80 per cent. (*Protocols of the Society of Russian Physicians in St Petersburg*, 1879.) Galanin (*Bubonnaia Tchuma*, St Petersburg, 1897), who served as surgeon in the Russo-Turkish war, states that analogous cases occurred among the troops in European Turkey.

[2] For the details of this epidemic see the author's papers upon plague in Russia already referred to.

extension of plague in historical times. The course of the disease in the last eight years has been as follows.

Early in 1894 plague broke out in Canton, where it had never before been seen. In the preceding year the disease had been unusually active in the endemic centres already described in southern China, and it had extended to places on the direct route between those centres and Canton. The epidemic in Canton, a city of a million and a half inhabitants, living in the midst of filth and insanitary surroundings, lasted several months and was a severe one. In May 1894, if not earlier, plague appeared in Hongkong, and there is little doubt that the infection was brought there from Canton, which is only 90 miles distant, and is in constant communication with the former. On May 10th Hongkong was officially declared to be an infected port. Throughout the hot and rainy season the disease was widely prevalent. In July it began to decline, and by August the epidemic was practically over. In the following year no epidemic recurred; but in 1896 there was a severe recrudescence of the disease, lasting from January to June. In 1897, plague, though still present in Hongkong, caused no widespread epidemic, but in 1898 a third outbreak of great severity took place. The alternation of epidemic and non-epidemic years which characterised the period 1894 to 1898 was broken in 1899, when plague was prevalent to as serious an extent as it had been in the preceding year. Through the winter of 1899–1900 the disease was quiescent, but in 1900, in 1901, and in 1902 it was again epidemic.

During subsequent years plague has continued with varying degrees of activity in Canton. For 1895 there is an absence of positive information, but in 1896 and in each succeeding year there have been recrudescences of the disease in the city.

Many other places in and near China were invaded by plague in 1894 or subsequent years. In 1894 a few cases were imported to Swatow, and in the following year a small epidemic occurred there. In 1897, 1899, and 1900 this port was the scene of renewed outbreaks. From Swatow the disease spread to Chao-Yang. In 1895 it appeared at Pochin and Hoihow in the island of Hainan (opposite Pakhoi). Amoy also suffered considerably in 1895, 1896, 1897, and 1899, the recrudescences occurring each year in the months from May to October. The two Portuguese customs ports of Macao and Lappa, both near

Hongkong, were scenes of severe epidemics of plague in 1895 and 1896, and of less severe ones in 1897, 1898, and 1900. In 1898 a small, localised outbreak occurred at Nha-Trang on the shores of Annam. The outbreak here, at first attributed to the presence in the port of a scientific laboratory in which important experiments with plague material had been carried out, was later shown to be probably due to an importation from some Chinese port, such as Pakhoi.

Finally plague appeared in 1899 in the town of Newchwang on the shores of the Gulf of Liao-tong in northern China. The town is now largely under Russian influence, and the infection was thought to have been imported by coolies who were brought in large numbers from southern China to assist in the construction of the Manchurian branch of the great Siberian railway. The outbreak was never severe here, and with the exception of a few cases in Hai-chou, Hai-chen, and Port Arthur the disease was confined to the town of Newchwang. In 1900 plague was again epidemic here.

From China plague spread to Formosa and Japan. In Japan, Kobe, Nagasaki, and Osaka have been the scenes of recurrent outbreaks of considerable severity.

The unfortunate experience which the Indian peninsula is now passing through from the ravages of plague began in the year 1896. In September of that year the disease was recognised in Bombay, but as it had by that time taken a serious hold of the city it is certain that it had already been there for some considerable time, and possibly since the early part of the year. The origin of the first outbreak in the city has never been determined, but the balance of probability is in favour of the view that the infection was imported from China. The evidence in favour of the alternative view that it was due to an extension of *mahamari* from the Himalayan centres already described is so slight as to be practically negligible. Since 1896 the disease has been truly epidemic in Bombay. Each year, approximately at the same period (December to March or April), plague has raged with, if anything, increasing intensity, and the mortality caused by it in the past winter (1901-2) was as high as that in the first year of its prevalence.

In October 1896 plague had already passed beyond the limits of Bombay and the island upon which that city is built. In that and the following three or four months imported cases of plague were detected in various places in the Bombay Presidency, and at longer or shorter intervals these cases were followed by others among the resident population of those places. This process was continued throughout the winter, until there were few portions of the Presidency, from Shikarpur in the north to Kolhapur and Goa in the south, that had not been invaded by the disease. In some places the epidemic was more severe than in Bombay itself. Thus in Mandvi, on the island of Kutch, 11 per cent. of the population were attacked in the first year's epidemic, while in Goa the proportion was no less than 33 per cent. Karachi also suffered to some extent, having an attack rate of about 4 per cent. In Poona the first year's outbreak was also of considerable severity.

During the summer of 1897 there was a marked decline in the intensity of the epidemic throughout the Bombay Presidency. But in the late summer months a recrudescence of the disease began, and in the following winter plague was as widely prevalent as in the preceding year. Bombay and Karachi suffered a second severe visitation. In Hubli, in the extreme south of the Presidency, a violent outbreak occurred in July, August, and September, that is to say at the height of the monsoon, when in the previous year plague had been everywhere relatively quiescent. An autumn epidemic occurred this year (1898) throughout the southern Mahratta country, Dharwar, Satara, Belgaum, and Kolhapur suffering especially. From November to the following March (1899) there was a general decrease in the prevalence of plague throughout the Presidency, excepting in Bombay and Karachi, in each of which a third epidemic outburst prevailed in the early months of the year. In the summer of 1899 the disease again died down, but it never wholly disappeared, and in July a serious increase took place in the Poona district and in the Belgaum state. In Poona the most violent and deadly outbreak yet recorded raged throughout the late summer months, the mortality for three weeks together remaining at the terrible figure of over 500 per 1000 per annum. (The highest in Bombay in any week has been 142 per 1000.) In the states of Kolhapur, Nasik, and Belgaum the disease was active at this time, and also at Hyderabad in Sindh. In the winters of 1899-1900, of 1900-1, and of 1901-2 Bombay suffered from its fourth, fifth, and sixth serious epidemics, and in many other parts of the Presidency plague was prevalent.

Outside this Presidency the spread of plague was at first remarkably slow. In the North-West Provinces small outbreaks have occurred at Hardwar, Kankhal, Jawalapur, and Khandraoni. In the Punjab plague appeared in October 1897, and in the following winter a large total number of deaths occurred in nearly 100 villages scattered throughout the Hoshiarpur and Jullundur districts. In the winter and spring of 1901-2 the Punjab suffered from a violent outbreak of plague, which was for several weeks the cause of an enormously high mortality. Hyderabad in the Nizam's territory has also been visited by plague on several occasions since the autumn of 1897, and the mortality throughout a great portion of the territory has been at times considerable. The Central Provinces have on the whole suffered but little from plague, a few cases only having been reported from Hinghenghat, Nagpur, and Bhundara in 1898 and early 1899. In the summer of 1899, however, a more severe outbreak occurred in Hinghenghat, and in October the disease was spreading in Nagpur. Rajputana has at no time suffered severely. In Bengal the infection of plague was extremely slow in gaining a footing. Calcutta itself was not invaded, with the exception of some controversial cases in the autumn of 1896, until the month of March or April 1898. A very mild outbreak prevailed there for the following three or four months ; a second and more severe epidemic ensued in the early months of 1899, and in February to April 1900 and again in the early months of 1901 and 1902 still more serious epidemics, causing a mortality comparable to that in Bombay, prevailed in Calcutta. In other parts of Bengal the course of the disease has been similar to that in Calcutta, and after the occurrence of scattered cases in a very large number of towns and villages an epidemic of great gravity occurred in many places in the first quarter of 1899. The towns and districts most seriously affected then and since have been those of Patna, Saran, Gaya, and Monghir.

Southern India has also not escaped the ravages of plague. Already before the end of the year 1897 the infection had spread as far south as Travancore. But it was not until the following year that the disease gave rise to a serious epidemic south of the Bombay Presidency. In August of that year a violent outbreak occurred in the province of Mysore, the city and cantonment of Bangalore suffering severely. In the Madras Presidency the malady has, since the year 1898, given rise to epidemics in several places, but it has nowhere raged with great intensity in this part of India.

Throughout the Indian peninsula plague has caused an amount of sickness and death in the course of the past six years which it is difficult to estimate. The returns from many places are incomplete, and it is probable that a considerable number of cases and deaths have escaped registration, but a moderate computation must place the mortality from the disease at little less than 600,000, a figure which, though high in itself, is incomparably lower than the total mortality from cholera in India in the same period. Farther India has hitherto suffered much less than India proper from the plague. A few imported cases have been dealt with in Rangoon, and an outbreak of some intensity occurred at Akyab in 1901. In Penang a small outbreak occurred in July 1899. In Singapore several imported cases have from time to time been dealt with, but the disease has never become epidemic here or elsewhere in the Malay Peninsula.

Plague, it has been pointed out, remained epidemic in China outside its endemic foci for two or more years before it appeared in India. Not long after its appearance in India it began to be heard of in other parts of the world. Jedda, the great pilgrim port of Arabia, was one of the earliest places to be invaded by the disease; at the end of May or early in June, 1897, cases of plague began to occur there, and an epidemic fortunately of small proportions, prevailed during the next four or five weeks. In 1898 a second outbreak occurred between the months of March (or, possibly, January) and April, scattered cases continuing to occur until July. A third limited outbreak developed in February 1899. It is a matter for congratulation that in spite of the large number of pilgrims to Mecca and Medina who passed through Jedda in each of these years no widespread epidemic of plague occurred among them. Only two known cases of plague occurred in Mecca in 1899, and Medina is still able to boast, as it did in the time of Burton's memorable pilgrimage, that the *Ta'un* or plague has never entered its gates.

The origin of the first outbreak in Jedda has never been explained. It was believed not to have been caused by an imported infection from the Assyr centre already described. More probably it came from India, but it is noteworthy that the departure of Indian pilgrims to the Hedjaz had been prohibited by the Indian Government as early as February 20th, 1897, or three months before the disease appeared in Jedda. Possibly it was brought to Jedda indirectly from Bombay by traders from the coasts of Hadramut and Muskat, who have frequent relations both with Bombay and with the Red Sea coast of Arabia. In 1900 a few cases of plague occurred in Yanbo, to the north of Jedda.

At the end of August, 1898, a very fatal epidemic of plague occurred in the village of Anzob, and a few deaths from the disease in an adjoining village, in the remote mountainous district of southern Turkestan, close to the borders of Bokhara. This has been already referred to, and the possibility has been pointed out of the endemic existence of plague in this region. The source of the outbreak remains unknown; it was a very severe one locally, causing the death of 237 persons out of a population of 387, or 61 per cent. of the total inhabitants.

In the month of November of the same year plague appeared in Madagascar and in Mauritius. The former was probably invaded first. The occurrence was of historical importance as the first known appearance of plague in the southern hemisphere[1]. Its subsequent spread to South America, the Pacific Islands, and Australia has further shown that the view previously held that plague could not exist south of a certain latitude is untenable.

In Madagascar the disease prevailed with some severity from November 1898 to February 1899, in which period it caused some 200 deaths. After an interval of quiescence it became active again in July, and lingered on in Tamatave—the locality most, if not solely affected—until towards the end of the year, when the island was officially declared plague-free. A third small outbreak occurred in October 1900, and a fourth in 1902. The original source of infection in Madagascar is uncertain; it has been suggested that it was imported by Arab sailing-boats, or *boutres*, which make frequent calls at ports on the Indian coasts of Madagascar and the Mascarene Islands. It has also been ascribed to a direct importation by a boat from Bombay: the first victims are said to

---

[1] Strictly speaking the centre in Uganda is mainly in the southern hemisphere, but it is practically an equatorial region.

have been natives employed in unloading the cargo of rice brought by this boat.

In Mauritius plague was recognised in January 1899, but the infection had probably gained the shores of the island some two months earlier. The epidemic here soon became one of considerable severity, from 50 to 100 cases of the disease occurring weekly through the months of August, September, October, and November, 1899. In the following winter months the mortality decreased, but the disease never completely disappeared. It again became epidemic in the autumn of 1900, and has ever since been more or less active in the island.

In February 1899 two deaths from plague occurred in the Transvaal; one at Middelburg, on the line between Lorenzo Marquez and Pretoria, and the other at Kaapmuiden, near Barberton, some hundred miles distant from Middelburg. Both victims were natives of India, and in both the infection had probably been imported from that country. Whether the two patients had had any relations with each other is not clear; if not, the occurrence of the two cases at about the same period is at least remarkable. There was no spread of the disease in their neighbourhood. But in, or before August, 1899, several deaths from a disease believed to be plague were reported from Magude, two days' journey from Lorenzo Marquez. On September 14th it was officially announced in the House of Assembly at Cape Town that the disease at Magude was plague, and that 42 cases, all fatal, had occurred up to that date. Later information in regard to this outbreak has been scanty and uncertain. In October or early November cases of plague occurred in Lorenzo Marquez itself, and it was feared that a grave extension of the disease might ensue on the outbreak of the war in South Africa. Fortunately this fear, which was fully justified by numerous precedents in past history, has not been substantiated by events. But late in 1900 the disease was epidemic at King William's Town, and during 1901 outbreaks of some little severity occurred in Cape Town and Port Elizabeth, while isolated cases were reported from Mossel Bay, Ladysmith, Mafeking, Hermon, Imvani, Somerset West, and Stellenbosch.

In May, 1899, the presence of plague in Alexandria was officially announced. Egypt had not been the scene of a plague epidemic for some fifty-five years, and the announcement gave rise to no little alarm. The exact date of the outbreak here is uncertain, but there is reason to believe that the infection was already in the city in the early months of the year. The earliest recognised cases occurred among the Greek community. The mode of introduction of the infection into Alexandria could not be explained with certainty. The city is in frequent communication with Bombay, and it is possible, though not proven, that the disease was imported by infected goods, or by rats, from the Indian port. The outbreak was never severe in Alexandria, but it lingered throughout the summer and autumn, reappearing on three separate occasions after the disease was believed to have been stamped out and the measures for its control had been relaxed. Throughout the winter Alexandria remained free from plague, but in the spring and summer of 1900 the disease again appeared and caused a certain amount of sporadic sickness and death. A small number of cases also occurred in that year at Port Said, Damietta, and Cairo. In 1901 limited outbreaks of plague were recorded in Zagazig, Minieh, Mansoura, Alexandria, Mit-Gamr, Tanta, Zifteh and a few other places, but in 1902 the disease was confined to Alexandria.

In the month of July, 1899, plague was reported from Oporto on the shores of Portugal, and from Kolobovka, a village at the mouth of the Volga. These outbreaks derived importance from the fact that they were the first reappearance of plague in Europe since the epidemic on the Volga in 1878–1879. The epidemic in Oporto was slightly the earlier of the two. It probably began in June, but the first recognised case occurred early in July. The epidemic reached its highest point in October, in which month 95 persons were attacked by the disease. It then declined, and by the end of the year was almost, if not quite extinct. The origin of the Oporto outbreak could not be ascertained. It could scarcely have been due to a revival of infection remaining from a previous epidemic, as Oporto and the whole of Portugal had been free from plague for some two centuries. The earliest cases were among dock-labourers or persons living near the docks, and it is probable that the infection was in some way imported by sea. No recognised importation could, however, be traced. It is interesting to note, as probably not without bearing on this question, that, during the last six years' prevalence of plague, Portuguese possessions have suffered considerably throughout the world. In India Goa was

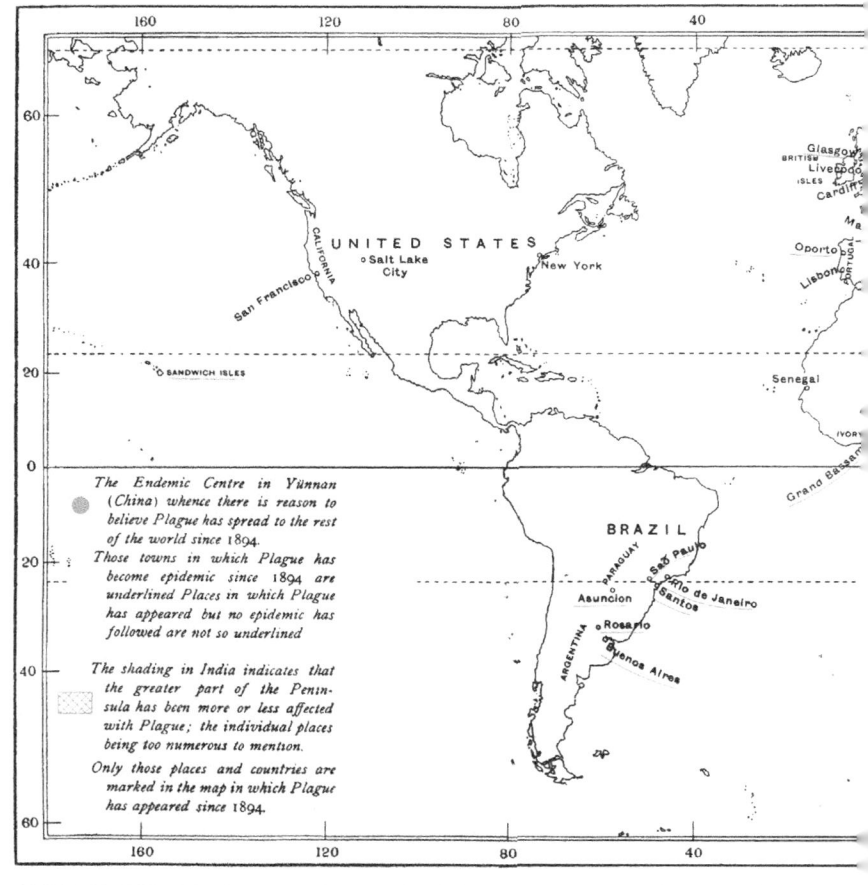

The Endemic Centre in Yünnan (China) whence there is reason to believe Plague has spread to the rest of the world since 1894.

Those towns in which Plague has become epidemic since 1894 are underlined Places in which Plague has appeared but no epidemic has followed are not so underlined

The shading in India indicates that the greater part of the Peninsula has been more or less affected with Plague; the individual places being too numerous to mention.

Only those places and countries are marked in the map in which Plague has appeared since 1894.

DISTRIBUTION OF PLAGUE BETWEEN THE YEARS 1894 AND 1902

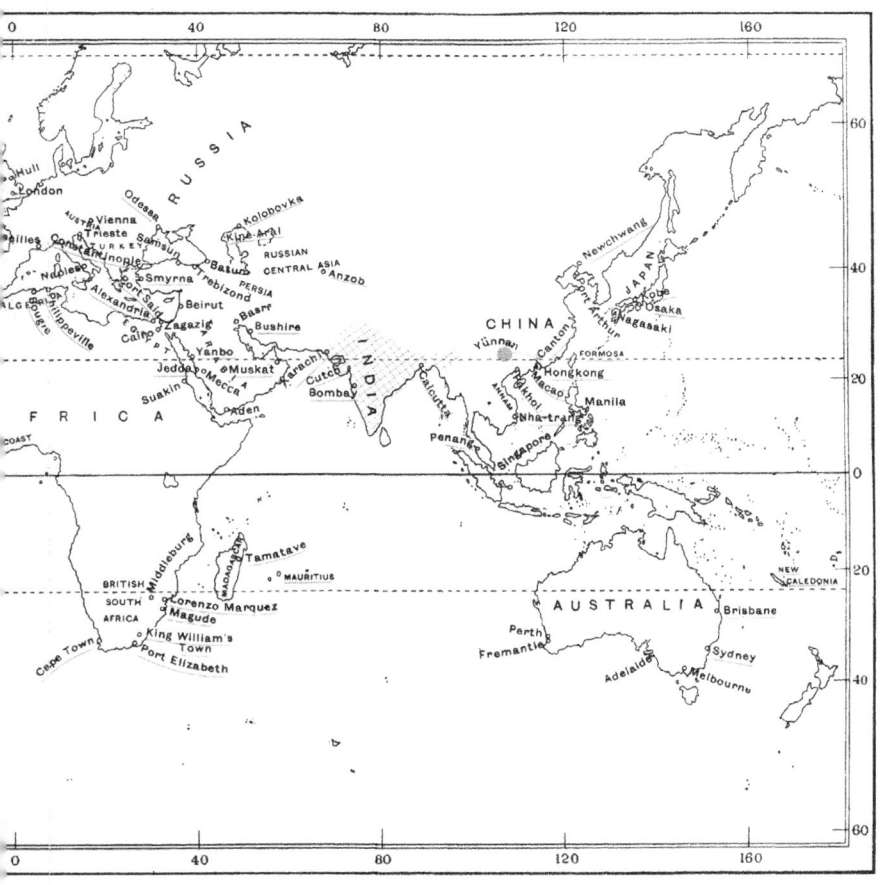

the scene of a very severe epidemic, and Daman lost one-third of its inhabitants in 1897 ; in China, the Portuguese ports of Macao and Lappa have had the infection of plague in their midst—at intervals if not constantly—since 1895 ; in South Africa it was present in Portuguese territory (Magude and Lorenzo Marquez) in 1899; and in the autumn of that year epidemics of some severity prevailed in Brazil and Paraguay, countries having constant relations with Portugal. To this may be added the fact that numbers of Portuguese are found in or near many other places and countries where plague has recently prevailed.

Oporto was almost the only place in Portugal in which plague appeared. In Baguin, a village some twelve miles from Oporto, nine cases with five deaths occurred in October. At Braga, a manufacturing town forty miles to the north of Oporto, two imported cases occurred. In Lisbon plague caused the death of Dr Pestana, a distinguished physician and bacteriologist, but no other cases, imported or indigenous, occurred in the capital. A slight recrudescence of plague occurred in Oporto in 1900, since which date Portugal seems to have been free from the disease.

The outbreak at Kolobovka, at the mouth of the Volga, on the shores of the Caspian Sea, began, it is believed, on July 16th (28th), 1899. Between that date and August 9th (21st) twenty-four persons were attacked by the disease, and in all but one the attack proved fatal. Special interest attaches to this epidemic, as in no less than 20 out of the total 24 cases the disease developed in its pneumonic form. There was no spread of the infection outside the village named. Exactly the same difficulty was met with here as elsewhere in attempting to account for the origin of the epidemic. It is difficult to believe, as suggested by one Russian authority, that the infection was imported here from the centre in Mongolia, some 4000 miles away. It is just conceivable that it may have come from the neighbourhood of Anzob, in the Samarkand province, but of this no proof was forthcoming.

A second outbreak of plague occurred in this neighbourhood at the end of November, or early in December, 1899. In this instance the disease appeared in two Kirghiz *aouls*, or settlements, in the maritime division of the Krasnoyarsk *uyézd*, or district, of the government of Astrakhan. The names of these two settlements were Kishkiné-Aral, and Kiné-Aral; they are situated on islands, at some distance from the mainland. Some 59 cases, every one of which proved fatal, occurred here, and all are said to have been of the pneumonic form of the disease. Other doubtful outbreaks occurred in 1900˙ at Vladimirovka in the government of Astrakhan, and at Tolovka, in that of Samara.

In 1899 an outbreak of plague of some severity occurred on the Ivory Coast of the Gulf of Guinea. It lasted from March to May and is said to have decimated the black population.

In the latter part of September, 1899, telegraphic news was received that plague had appeared in Asunçion, the capital of Paraguay. Paraguay is an inland country in the heart of South America ; plague had never before in history appeared in the western hemisphere, and it seemed almost incredible that, having travelled so far, it should appear not, as in so many other places, on the coast, but some 600 miles up country. Later all doubt was removed as to the specific nature of the outbreak, and it subsequently appeared in Buenos Aires, Rosario, Rio de Janeiro, Santos, São Paulo, and elsewhere, indicating that a considerable area of this part of the South American continent was affected by the disease.

In the same year plague broke out for the first time in history in Australasia. In Australia itself outbreaks of varying intensity, but none of very great severity, have been reported from Sydney (New South Wales), from Melbourne (Victoria), from Brisbane and Rockhampton (Queensland), from Perth and Fremantle (West Australia), and from Adelaide (South Australia). A few isolated cases have also occurred in the interior, but practically plague in Australia has been limited to the seaports. In the Pacific Islands some cases occurred in Honolulu at the end of the year 1900, and in New Caledonia about the same period.

In 1900 and 1901 several small outbreaks of plague occurred in different parts of the Turkish Empire. In Smyrna the disease was present from May to July 1900; in December a small but very fatal epidemic of the pneumonic form of the disorder was reported from Tomasso, a neighbouring village, and a case of bubonic plague in the town itself followed in January 1901. In May and September single cases occurred here, and again one case in December, one in January, and one in October, 1902. A small group of cases were discovered at Beirut on the Syrian coast in July, 1900, and more than once plague has occurred in that port upon ships coming from elsewhere. In January, 1902, a single case occurred there in a patient who had left Baghdad some time before, and may have imported the disease from that city.   In June, 1900, there was a single case at Trebizond, and in 1901 a small explosion of the disease occurred at Samsoun. In Constantinople plague first appeared indigenously in January, 1901, and again in May, and it was fitfully present in the city and its neighbourhood throughout the greater part of the rest of the year.   In 1902 it again gave rise to a few sporadic cases.

Batum and Odessa were the scenes of limited outbreaks in the latter half of 1901, and in Odessa a considerable number of cases of the disease were dealt with in the summer and autumn of 1902.   In the autumn of 1900 a limited epidemic appeared in Glasgow, and in the following year the late summer was marked by another small explosion there.   About the same time as this second outbreak Liverpool was visited by the disease, but it attacked only a very few people.

In Cardiff and Hull single cases of plague have also occurred, but, with the exception of imported cases of the disease on board ships, no other parts of the British Isles have as yet been visited by the malady.

Finally mention must be made of the occurrence of three cases of plague in Vienna in 1898, traced to infection from a laboratory in which observations on the plague bacillus were being conducted; of an (uncertain) outbreak of the disease in Algeria in 1899; of the appearance of plague in Aden in the spring and summer of 1900; of small epidemics of the disease at Bushire and Kishm on the Persian Gulf; of its reported outbreak on more than one occasion at Muskat; of a certain number of cases of the disease in the Chinese part of San Francisco in 1900 and 1901, and of a few cases in the Philippine Islands in the same years; of a single case on the island of Chios, and of a concentrated but brief outbreak at Naples in 1901; of a single death at Salt Lake City, Utah, in the same year; of small epidemics in Baghdad in May 1901, and January 1902, with possibly continued presence of the disease there in the mild form of bubonic fever between the two recognised outbreaks; and of a large number of cases of the disease at Nairobi in British East Africa early in 1902.

Plague on Board Ship.   During the past eight years of plague prevalence a very large number of instances of the appearance of the disease on ship-board have been recorded. These have occurred in a great many parts of the world. In this manner plague has actually been imported to innumerable ports, but it has been exceptional to find such imported cases followed by a local epidemic; and even when this course of events has been observed, there has usually been no apparent connection between the imported case and the subsequent indigenous cases. On the contrary it would seem that imported cases on ship-board, probably from the fact that they are recognised and dealt with, do not offer great risk, and that when plague has become epidemic in a port or country hitherto exempt, it has gained access there in some much more insidious way—either by the importation of infected rats or infected merchandise.

Cases of plague on ship-board have usually occurred upon ships that have left an infected port, and most commonly within a comparatively short period after sailing. It must be confessed, however, that there have been some notable exceptions. For example, in the autumn of 1896 some cases occurred on ships in the Thames that had left India nearly six weeks previously. A case of plague developed on the s.s. "Golconda" at Marseilles on December 17, 1898, though the ship had left Calcutta—then, technically, a plague-free port—some three weeks earlier. An exactly parallel incident occurred on the s.s. "Peninsular" in September, 1899, and many other instances of the kind might be mentioned.

It is noteworthy that in the majority of instances of plague on ships the patient has been one of the crew, and more often than not a fireman, or one of the ship's servants whose occupation would bring him into relation with that part of the ship where the stores are kept and where rats would most likely abound. In very many, but in far from all of the published instances, dead or sick rats have been seen upon ships where plague has occurred on board. As a rule a ship outbreak of plague is limited to a very few cases, the disease being much more easily controllable than yellow fever, beri-beri, or influenza under the like conditions. Sometimes, however, the plague infection seems to cling to the fabric of the ship itself. Perhaps the most striking example of this was the s.s. "Mahallah," a Red Sea mail boat, on which (in 1898) cases of plague persistently recurred after each of three successive and very thorough disinfections.

Factors concerned in the Geographical Distribution of Plague. The geographical distribution of the malady is frequently varying. In the past it has tended to remain for long intervals in certain countries or districts, which have at one time been of the nature of low-lying alluvial plains, or deltas at the mouths of rivers, at another (as in the latter half of the nineteenth century) of remote mountainous regions, away from the main routes of trade or travel, and sparsely inhabited by highlanders of various races, who appear to have nothing in common save that they live in the midst of filth and insanitary surroundings.

On at least four known occasions in the world's history plague has become more or less truly pandemic. The first occasion was in the reign of Marcus Aurelius in the second century. The second was in the sixth century, in the time of Justinian, when plague ravaged the Roman world for some fifty or more years. The third occasion was the terrible outbreak of the Black Death in 1347 and the following years, in which some 25,000,000 persons in Europe alone are believed to have lost their lives. That pandemic spread to the furthest known limits of the eastern hemisphere north of the equator. The fourth wide extension of plague is now running its course and has been briefly described above.

The present pandemic has shown that plague can exist and become epidemic both to the south of the equator and in the western hemisphere. It has shown that its limits are not of necessity determined by lines of latitude or longitude nor by isotherms, and that when once the infection is imported to any portion of the earth's surface it will tend to become epidemic there, provided the local conditions are favourable. At present the most southerly place in which plague has appeared, though not in epidemic form, has been Melbourne, in (about) 38° south latitude. In the north, the disease became epidemic in 1349 as far north as Iceland and Greenland, and it is not beyond the bounds of possibility that the present pandemic may extend considerably further towards the pole than it has hitherto.

In general terms, however, plague is mainly a disease of temperate and subtropical countries, prevailing most frequently, and as a rule most intensely, when the temperature is neither excessively hot nor remarkably cold. Its seasonal variations in countries where it has been endemic or epidemic have been more constant than in the case of some other diseases. In the endemic centres in Mongolia and in Yünnan ; in the recent epidemic recurrences in Alexandria, Oporto, Hongkong, Constantinople, Madagascar and Mauritius; in the historical plague prevalences in Europe and in Northern Africa, plague has usually been most active in the warmer part of the year. In some Indian cities, on the other hand, where the cool season more closely resembles the warm season of more temperate climates, plague has in each of the last six years declined in the summer and increased in the winter months. For India taken as a whole the curve of plague in recent years has shown two maxima and two minima. The first maximum has fallen in the warm and dry month of March ; the second in the warm and moist months of October and November. The first minimum has coincided with the commencement of the rains in the hot and moist month of June ; the second, with the cool and dry months of December and January. Upon these data and a large number of others examined by the Indian Plague Commission the conclusion has been arrived at that neither the temperature nor the humidity of the air has any influence in India upon the amount of plague prevalence. In individual places in India the epidemic outbursts of plague have occurred at very different seasons of the year. Some, as was pointed out above, have been observed in the coldest months. Others, as for example those in Hubli and Bangalore in 1898 and in Poona in 1899, have occurred in the hot season.

Neither extremes of heat nor extremes of cold suffice to prevent the appearance of plague. To the examples of outbreaks in hot weather just named may be added the small epidemic at Jedda in 1897, when the thermometer stood for a whole week at a time (at the very middle of the epidemic period) at about 111° Fahr. in the shade, and was as high as 174·2° in the sun. At the other extreme may be mentioned the outbreak on the Lower Volga during intensely cold weather between October 1878 and January 1879, and that on the Persian mountains near Lake Urumia in 1870, when plague broke out while the villagers were snowed up in their dwellings, 4000–6000 feet above sea-level.

It must be doubted whether elevation above sea-level is of any great importance in plague distribution. It was for long believed that the disease lingered in low-lying alluvial plains and river deltas. But in recent decades it has been in high table-lands or hilly districts that plague has been for the most part endemic. Of this character are the endemic centres in the Himalayas, in Arabia, and in Mongolia, and among examples of plague epidemics at great heights may be mentioned several in Persia at 4000–6000 feet above sea-level and the outbreak at Anzob in Russian Central Asia (1898), at an altitude of from 6000 to 7000 feet.

There is little to show that racial susceptibilities greatly influence the distribution of plague. No race is certainly immune to it. In China and India, and elsewhere, Europeans have to a certain extent escaped it, but not a few have been attacked, and all have probably been not only much less exposed to infection than the natives, but also better able to guard against it by taking judicious precautions. In India the Eurasians suffered much more than the Europeans but much less than the natives. At Cape Town the proportion of Europeans to natives attacked has been remarkably high. It is noteworthy that in spite of its geographical position between China and India, in both of which plague has recently been so widely active, the Malay Peninsula has practically escaped the disease; and to this may be added the statement that in Cape Town the Malay inhabitants, who are rather numerous, have not suffered. Possibly this race has a certain immunity to plague, but more extended observations on the point are needed.

Plague is one of the infective fevers and may therefore be spread by the movements of infected persons. There is, however, good reason for believing that its distribution over the earth's surface is brought about less by the movements of infected individuals than by other means. The plague virus can exist for long periods outside the human body, not only in inanimate objects (*fomites*) but also in the lower animals, many of which are capable of suffering from plague, while others may harbour the bacillus and carry it from one place to another without actually suffering from the disease. Among the lower animals that are capable of actually contracting plague under natural conditions the most important are rats, mice, monkeys, bandicoots, squirrels, cats, and marmots[1]. Of these, rats suffer far

[1] See the author's papers upon "Plague in the Lower Animals," *Brit. Med. Journal*, May 12th and 19th, 1900.

the most frequently, and are of all animals—not, perhaps, excluding man—the most important agents in the spread of the disease. Mice are less susceptible, but appear in some instances to have suffered considerably, and may have been agents in spreading the infection. Monkeys, bandicoots, squirrels, and cats have on rare occasions been attacked by plague, but there is nothing to show that they have aided to any considerable extent in diffusing the disease in man. Marmots undoubtedly spread a disease allied to plague in Transbaikalia and possibly in Mongolia. The evidence in regard to dogs, pigs, sheep, goats, and jackals is less conclusive, but there is no reason to believe that these animals play any great part in spreading plague. Horses and cows do not appear to have ever contracted the disease under natural conditions, and may be disregarded as active agents in spreading it. Birds are equally immune, and there is no evidence to show that either they or reptiles or fishes have ever aided in multiplying and diffusing the infection. Insects, on the other hand, are probably agents of considerable importance in its spread.

The close association of rat epizoötics with human plague is perhaps one of the most important phenomena in the epidemiology of plague. The appearance of the disease in rats is sometimes accompanied by migrations and other unusual behaviour among these animals. When plague is present rats probably become infected with ease, either by living in infected soil, by eating contaminated grain or other food-stuffs, by feeding off the flesh of other animals (including human beings) dead of plague, by sniffing at infected rags, articles of clothing, or dressings from plague patients, and perhaps by means of the bite of an infected insect. It would be out of place here to describe the symptoms produced in these animals, or to discuss the bacteriological and other evidence to prove that the disease in them is in truth the same disease as in man[1]. Their identity has now been placed beyond the possibility of doubt. In modern times few outbreaks of plague have occurred without fresh evidence being adduced of the presence of the disease in rats. In earlier times, it is true, authors rarely made mention of any unusual phenomena among rats during a plague epidemic. For the two great pan-

[1] I have discussed the evidence at some length in the *Brit. Med. Journal*, May 19, 1900.

demics of the disease, those of the 6th and 14th centuries, no record can apparently be found in the writings of any of the principal historical authorities that would show that there was then any unusual disease or mortality among rats[1]; and not less remarkable is the absence of reference to rat-mortality in connection with a large number of other epidemics, as, for example the great pestilence in Marseilles in 1720, that in Moscow in 1770–71, and apparently the greater number of outbreaks in the nineteenth century in Europe, Egypt, Arabia, Mesopotamia, Tripoli, and elsewhere. In some of these instances there was, perhaps, no rat sickness, but it is more likely that in many of them it was present but overlooked. For unless special attention were directed to it, or unless it were present to a striking extent, a phenomenon of this kind would easily escape notice, or if observed might easily escape record by an observer or historian.

On the other hand sickness and mortality among rats or other rodents have been recorded in connection with not a few of the historical epidemics. The Biblical mention of the "mice that marred the land" at the time of a plague (?) epidemic among the Philistines and Israelites has already been alluded to. Probably one of the oldest references to plague in rats is that contained in the Bhâgavata Purana, one of the most important and perhaps one of the most ancient of the sacred Hindu writings. The people are there instructed to quit their houses and go to the plains as soon as they observe that "rats fall from the roof above, jump about and die." Clearly there was risk of disease appearing in man if he remained in his house, and it is a fact of the highest interest that the instructions contained in this ancient writing are followed to this day in many parts of India when an unusual sickness or mortality is observed among rats. Avicena (A.D. 1000) states that, when plague is approaching, mice and other animals that usually live underground leave their holes and stagger as though drunk[2]. Lodge, one of the best known of English writers

---

[1] Probably all kinds of rats were less common in the civilised world at those periods of history than they are now. The brown rat, for example, is thought to have reached England only as late as the beginning of the 18th century. A complete knowledge of the distribution and comparative frequency of the different varieties of rats in all countries and at all periods might throw much light upon this interesting question.

[2] Reference in *Indian Plague Commission Report*, p. 132, footnote; where the statement of Avicena is quoted from an article by Dr Rudolf Abel in the

on plague, makes mention of unusual phenomena among rats in connection with the disease[1]. When plague was prevalent in India in 1611–18 rats and mice seem to have been affected by the disease. The Emperor Jehangir has left a detailed account of the epidemic as he observed it in Agra, and in one passage he succinctly states that a sick rat was seen in a certain house, and that the rat was killed by a cat, which belonged to the daughter of the owner of the house. A slave, who was ordered to remove the body of the rat, was subsequently attacked with plague; and later the girl who owned the cat, her mother, and finally all the members of the family were seized with the disease. This record, so far as the sequence of facts is concerned, might have been taken from a Plague Report of yesterday, instead of from an Indian author who lived nearly three centuries ago.

During the Great Plague of London in 1665 it was apparently feared that rats and mice might be the means of spreading the infection, and there was a general crusade against these animals. "All possible endeavours," writes Defoe, "were used also to destroy the mice and rats, especially the latter, by laying rats-bane and other poisons for them, and a prodigious multitude of them were also destroyed[2]." Almost equally remarkable is the reference to rat disease in Yünnan in the 16th century, to which detailed allusion has already been made. In the last half-century the endemic prevalence of plague in this Chinese province has,

---

Zeitschrift für Hygiene, 1901. I have not had access to this article, and can therefore only observe with surprise that the Commissioners remark:—"it is noteworthy in this connection that the minute bibliographical researches of Dr Rudolf Abel, extending over the whole field of classical, mediæval and modern literature, bring out clearly that there are no records of epidemics among rats occurring concurrently with plague among men."

[1] The passage runs: "And when as rats, moules, and other creatures (accustomed to live underground) forsake their holes and habitations it is a token of conception in the same, by reason that such sorts of creatures forsake their wonted place of abode." No mention is made here of actual disease in the rats, only of unusual migrations.

Note also the death of "dogs, cats, mice, and rats" in Leeds in 1645 during a plague epidemic.

[2] In Hongkong, in innumerable places in India, in Alexandria, Odessa, Oporto, and elsewhere "a prodigious multitude" of rats have been destroyed in the last few years, under precisely the same conditions and with precisely the same object. The trite saying that "History repeats itself" has perhaps seldom been more strikingly exemplified.

in like manner, been constantly associated with sickness in rats; and the same remark applies to the endemic centres of plague in the Himalayas. Finally, during the recent wide extension of the disease, from almost all quarters of the world where plague has broken out there has been a steady stream of evidence, establishing beyond question a close correlation between plague epizoötics in rats and plague epidemics in man.

To what extent rats are the means of carrying plague over the earth's surface is still a matter of doubt. It is, however, certain that in recent years the epidemic appearance of plague in a country far removed from any other country where the disease was known to exist, has very rarely been traceable to the passage of an infected person from one to the other. In this respect plague differs essentially from a disease such as small-pox, the spread of which can almost invariably be traced to the movements of human beings. It is therefore almost certain that the infection of plague is carried from country to country by other than human agency, and the most probable explanation is that it is carried either by rats or by infected merchandise. As to the carriage of plague over shorter distances—as from town to town in an infected country—it may be said that the appearance of plague as an epidemic has often followed the arrival of a sick person from an infected area, but even then a considerable interval has elapsed between the arrival of the latter and the occurrence of subsequent cases. This interval has very often been characterised by the development of a plague epizoötic among rats. While this has, in India at least, been the most common course of events, it must be added that in many outbreaks no trace of rat-infection has been discovered, and that in others a plague epizoötic among rats has not been followed by a plague epidemic in man.

The part played by insects in the spread of plague is not yet fully determined, but the experiments of Simond, Yersin, and Hankin have shown that the disease may be transmitted, either actively or passively, from rat to rat and from rat to man by means of certain insects, of which probably the flea is the most important. Considerable doubt, however, has been thrown by Nuttall, Galli-Valerio, and others, upon the suggestion that insects are the principal, or even an important means of spreading plague.

While rats and some other lower animals form important media for the persistence of plague infection outside the human body, it is probable also that the infection can remain for considerable periods in inanimate objects without losing its activity. There are many historical instances of plague appearing in a hitherto plague-free district after the arrival of infected goods from a district where plague was present; and though it is easy to find flaws in the records of many of these instances, it nevertheless remains almost certain that the disease can be spread over the earth's surface by means of infected goods or fomites. What particular class of goods is most likely to carry the infection it is difficult to say. The European nations have agreed at successive international conferences to regard the following as "susceptible" articles, the importation of which may be prohibited in time of plague:—body-linen, personal clothing that has been worn, bedding that has been used, rags (including those compressed by hydraulic pressure), used sacks, carpets, and embroideries, raw hides, untanned and fresh skins, fresh animal refuse, claws, hoofs, horse-hair, hair of other animals, silk and raw wool, and human hair.

Possibly the infection may cling to other objects than these. Grain and food-stuffs generally have been regarded with much suspicion, and it is certainly noteworthy that in an immense number of recent outbreaks of plague (e.g. to name a few only, those in Bombay, Calcutta, and many other Indian cities, in Mauritius, in Oporto, in Alexandria, and in Constantinople) the disease has first appeared or has been most active among persons whose employment or place of residence brought them in relation to grain stores, or to bakeries, bread-shops, or groceries, where grain or other food articles were handled. The close association, however, of such places with rats, which are naturally most found where grain and other food are most abundant, must not be forgotten; and it is probable that the phenomenon just noted is often to be explained in this way rather than by supposing that in each instance the grain or other food-stuff was the primary agent in carrying the infection.

The relation of plague to insanitary surroundings has been much debated. The facts appear to be briefly as follows. In all those geographical areas where plague has been endemic during the last half-century the inhabitants have been persons living in much filth, poverty, and unhygienic surroundings. It is impossible to read the accounts of plague prevalence in Yünnan, in Mongolia, in the Himalayas, in Persia, in Mesopotamia, and in Arabia, and not be struck by the extraordinary resemblance between the conditions of life in all these regions. They may be briefly summarised by the words filth and misery. During the recent years of plague activity the disease has caused very serious epidemics only in certain Chinese and Indian towns. The published descriptions of Canton and Hongkong leave no room for doubt as to the filth and insanitation under which the Chinese live; and no one who has explored a Bombay *chawl* or a Calcutta *bustee*, would hesitate to declare that the native part of a great Indian city abounds in filth and in sanitary deficiencies. Outside India and China, in those places where plague has succeeded in establishing itself to any extent—as in Alexandria, Mauritius, and certain South American towns—dirt, and a low level of general hygiene are very prominent features in all the published accounts of the way in which the inhabitants of those places live.

On the other hand, plague has appeared and even caused small, temporary epidemics, in houses or institutions that were irreproachable as regards cleanliness and general sanitary arrangements. Such outbreaks, however, have been exceptional; they have usually occurred where plague was already seriously epidemic among dirty and insanitary conditions in the near neighbourhood, and they have been as a rule controlled with considerable ease.

Of the many conditions implied under the general term insanitation, it seems probable that the aggregation of numbers of susceptible persons together in dark, dirty, ill-ventilated rooms is the most favourable to the epidemic extension of plague. To this may be added as contributory conditions, at least in India, the absence of flooring to huts, enabling rats to gain free access to them, the practice of sleeping on the bare ground, and of going about bare-footed, and a lack of knowledge or willingness to recognise the first cases of the disease, or premonitory rat-sickness, and to take early measures in consequence. Defects in water-supplies cannot in any way be charged with favouring plague extension. Defects in the removal of excreta and of refuse are more liable to suspicion. That dirt and a sewage-soaked soil tend to encourage plague prevalence, if not—as was at one time thought—to generate the disease *de novo*, is an opinion now less generally held than formerly. Fæcal pollution of soil, imperfect removal of refuse, and filth-accumulation have, however, been such frequent accompaniments of severe plague prevalences that it is difficult to deny them all share in the sum of conditions favouring the spread of the disease.

It is not easy to assert exactly how various sanitary defects act in favouring the spread of plague. The plague bacillus under artificial conditions soon dies out, or at least soon ceases to be recognisable, if grown among other saprophytic organisms such as are found in filth, refuse, and sewage-polluted soil. Yet under natural conditions the infection lingers for months together in a saprophytic state between successive outbreaks in human beings (as, *e.g.* in the endemic centres above described); and it is clearly capable of thriving under those very conditions of soil-pollution and filth-accumulation which *à priori* seem to be unfavourable to the life of the bacillus. On the other hand, even in artificial growths, the bacillus likes best the very conditions of moisture, darkness, lack of oxygen, and excess of carbonic acid which are found in close, overcrowded, and ill-ventilated dwellings[1].

While it is probable that insanitation is a very powerful factor in plague distribution, it does not explain all the phenomena of the disease. It fails to explain why plague remains many decades or even centuries together confined to quite limited areas, and then, without apparent reason, spreads widely over the earth's surface. It fails to explain the gradual eastward recession of plague in Europe at the end of the seventeenth century, or its complete disappearance from that continent and from Egypt in the fifth decade of the nineteenth century. No doubt the sanitary condition of most European countries is very much better now than it was in the middle ages ; and this may possibly be in part the reason why the infection of plague has not in the past few years, in spite of repeated importation, been able to sweep over the continent as it did in the fourteenth century, or indeed to obtain a serious foothold anywhere. But it would be rash to assert that it is the whole explanation. No great or startling improvement took place in the sanitation of Europe at the end of

[1] See the President's Minority Report, *App.* IV., *Report of the Indian Plague Commission*, 1901 ; where these and other considerations are developed at great length.

the 17th century, to account for the disappearance there of plague; nor was anything of the kind observed in Egypt or Constantinople to explain why plague died out in those localities in the middle of the nineteenth century.

In the wide epidemic extensions of plague over the earth's surface, just as in the case of influenza and cholera, there seems to be a more or less sudden increase in the activity of the disease in one of its endemic centres, and a carriage of infection thence in all directions along the lines of trade and travel. On no occasion has there been anything to show that a simultaneous increase in activity took place in all or even in many of the endemic centres. The Black Death shewed a clear progression from some eastern country, probably India, to Central Asia and Southern and Eastern Europe, and so to the west. The present extension of plague has also come from the East. A most careful study of the records of all the known epidemic centres of the disease in the last half of the nineteenth century shows that in one alone was there any unusual activity just before plague began its world travels in 1894. That centre, or group of centres, was the one already described in Southern China, and there can be little doubt that the subsequent extension of the disease over the earth's surface is to be traced from it.

PNEUMONIA.

General Characters and Etiology. In the present chapter it is intended to discuss the geographical distribution of croupous or lobar pneumonia only, so far as this is possible. The data for a study of this kind are, however, imperfect, in that under the heading "pneumonia" cases of broncho-pneumonia are very often included, while in not a few reports and statistical returns "pneumonia" is not treated at all as a separate disease but is grouped together with pleurisy, or even absorbed in the more general group of "diseases of the respiratory system." In recent years, however, the specific character of croupous pneumonia has been more and more recognised, and the practice of grouping the disease among the acute infective disorders instead of with respiratory diseases is now very generally followed. This practice appears to be fully justified by the many characters which pneumonia has in common with these disorders, by its tendency in some instances at least to become highly infectious from person to person, and by its occurrence at times in marked epidemic form.

There are indeed, some indications that under the general term "pneumonia" more than one form of specific infective disease may be included. It is certain that some bacilli, as for example the plague bacillus, can primarily infect the lung and cause a definite (lobular) pneumonic condition. It is no less certain that an attack of influenza is constantly associated with the development of pneumonia, though in this case it is still uncertain whether the condition is brought about by a primary invasion of the lung by the micro-organism of influenza or whether it is not a double infection. The frequent co-existence of pneumonia with enteric fever has also given rise to the suggestion that the typhoid bacillus may in these instances have selected the lung instead of the intestines for its primary attack. But for the present the exact explanation of these cases is uncertain, although some authors speak of a "typhoid pneumonia" as a recognised pathological entity.

The mode of production of an acute attack of typical croupous pneumonia is also as yet not wholly explained. A large number of micro-organisms are found in the sputa in such cases, the most constant being the diplococcus of Fränkel and Weichselbaum and the micrococcus of Friedländer, and in some epidemics the *bacillus pneumoniæ* of Klein.

It is possible that the epidemic and highly contagious forms of pneumonia are manifestations of a different disease from the ordinary idiopathic cases of pneumonia, which usually show a low degree or apparent absence of infectivity. A few widespread epidemics, and many localised outbreaks of this nature have been recorded in the past.

History. The antiquity of ordinary croupous pneumonia is unknown. It is probable that the pneumonia of influenza and the pneumonia of plague are as old as the history of those diseases themselves, for there are few ancient or modern writers on either the one or the other who in their descriptions of the disease do not refer to lung symptoms in terms which more or less closely indicate the occurrence of pneumonia.

True epidemics of inflammation of the lungs or pneumonia have occurred in many parts of the world at least since the sixteenth century. Hirsch has summarised most fully the facts, so far as they are known, of these successive outbreaks[1]. In the sixteenth century the records are most numerous for epidemics in Germany, Switzerland, Italy, the Netherlands, and France; in the seventeenth for Germany, Switzerland, and Italy; and in the eighteenth for Italy, Switzerland, France, Spain, England, Denmark, and North America. In the century that has just come to a close a very large number of epidemics occurred in the countries already named. While the overwhelming majority of these in the past have been reported from either Europe or America, it is certain that they can and do occur in Asia, Australasia, and Africa, and mention will be made later of such

[1] Vol. III. p. 125.

epidemic prevalences of the disease in India, New Guinea, and South Africa.

Recent Geographical Distribution. *Europe.* In the British Isles pneumonia is a very common disorder and a cause of considerable mortality, especially in the winter and spring months. It sometimes takes the form of an epidemic; a very severe one of the kind broke out at Middlesborough in 1888, and caused as many as 369 deaths.

In France pneumonia and broncho-pneumonia are responsible for a heavy annual mortality. Thus in the years 1887–1898 in French towns with over 10,000 inhabitants, the deaths from these causes varied between 1650 and 2320 per million of the population. The pneumonia death-rate in French towns is considerably higher than that in many English towns, though lower than that in Christiania or Copenhagen.

In the urban population of Denmark "pneumonia crouposa" is a very frequent cause of death, though of quite recent years the mortality from this cause has tended to diminish. The deaths in Danish towns from this cause in the six years 1892 to 1897 were respectively 611, 588, 431, 405, 425, 444[1]. With the exceptions of "cholerine and acute intestinal catarrh," "phthisis pulmonalis" and "cancer," no other disease or group of diseases causes so many deaths in the inhabitants of Danish towns as the one now under consideration. So also in Norway croupous pneumonia is a cause of more deaths than any other disease save "tubercle of the lung"—the groups of intestinal disorders and cancers proving much less fatal here than the lung affection in question[2].

Throughout the whole of European Russia pneumonia is one of the commonest of affections. It is seen in the government of Archangel on the shores of the White Sea, in the Baltic Provinces, in Poland, in the Ukraine, on the Volga, and on the shores of the Caspian and Black Seas. It seems to be rather more frequent on the whole in the southerly than in the northerly governments.

In Bavaria and Swabia pneumonia appears to be more frequently seen than any of the acute infectious disorders.

Of the prevalence of pneumonia in Spain it is difficult to give

[1] *Statistik Aarbog*, Kjöbenhavn, 1899. The estimated population of these Danish towns was 784,000 in 1895.

[2] *Statistik Aarbog för Kongeriget Norge*, 1897.

an accurate account, but it cannot be rare, for in the Spanish army it is a common cause of invaliding. In the report for that army for the year 1895 it is stated that this disease is frequently observed among the troops in the peninsula or stationed on the Spanish possessions off the African coast. In that year few other diseases accounted for more cases of sickness or of death than pneumonia.

It is a no less common disorder in Italy. The health statistics of that country for the ten years from 1887 to 1896 show that pneumonia caused a higher mortality-rate than any other disease— enteric disorders (including diarrhœa, dysentery etc.) alone excepted. It is in most years more than twice as fatal as pulmonary tubercle, the latter causing an annual mortality of about 1000 per million, while the deaths from pneumonia average from 2000 to over 2500 per million[1].

In European Turkey pneumonia is very frequently seen at the present day. At Preveza, for example, in the first half of 1901 more than one-third of the total deaths were attributed to pneumonia; at Kavala it is far from rare as a cause of mortality, and from many parts of the Salonika province it is frequently reported.

In the Faröe Islands pneumonia is classed together with pleurisy and bronchitis as among the most frequent causes of death.

The evidence for the existence of croupous pneumonia as an indigenous disease in the island of Malta is conflicting. In the Colonial Reports for the possession numerous deaths from this cause are yearly returned. But the late Capt. Louis Hughes has left it on record that during six years' residence in the island he never saw a case of "acute croupous lobar pneumonia." Many cases of catarrhal (lobular) pneumonia had come under treatment in that period, in persons suffering from Malta fever or from enteric, but a case of true lobar pneumonia of indigenous origin had not been observed. Many cases of the latter form of affection had however been seen in men removed from troopships, eight to twelve days out from England[2].

Asia. Pneumonia is not rare in Asia Minor. In reports from Trebizond, Samsun, and other towns it often occupies a

---

[1] *Annuario Statistico Italiano*, 1898.
[2] *Journ. Trop. Med.*, Sept. 1899.

conspicuous position among the diseases named, and cases are frequently reported from Rhodes and other islands off the coast.

Pneumonia is a very frequent cause of death among the Moslem pilgrims to Mecca, and apparently also among the permanent inhabitants of that city. Generally speaking the disease is said to be rare on the coasts of Arabia and common in the interior.

In the Caucasus the disease is no rarity, particularly towards the end of autumn and in the spring. It is the most frequent cause of death among malarial subjects in this part of the world, and at times it becomes truly epidemic[1].

The returns from the Central Asiatic provinces of Russia and of Siberia are very imperfect for this as for most other diseases. As they stand they seem to indicate that pneumonia, while known throughout Asiatic Russia, is of much less frequent occurrence there than in European Russia, but this must be regarded as doubtful. None of the Siberian provinces are free from it, but the largest number of cases are reported from those of Irkutsk, Tobolsk, Tomsk, the Amur, and Tchita, while the returns from the convict island of Sakhalin also indicate a considerable degree of frequency of this disease among its inhabitants.

In Mesopotamia pneumonia is seen, and in the Persian capital many cases occur every winter. At Basra, at the head of the Persian Gulf, the disease is also by no means rare.

In the Indian peninsula pneumonia is a serious affection and a cause of a high degree of mortality. Its distribution here presents many points of the greatest interest. Certain parts of the country appear to be peculiarly favourable to the prevalence of the disease, and from year to year larger proportions of invalidings and of deaths from this cause are recorded from these districts than from any others. The highest degree of prevalence is almost, if not quite invariably, found to occur in the Indus valley, whether the returns of the Indian army or of the Indian gaols be taken as the index. In Peshawar the disease is particularly common. As a cause of death, however, pneumonia would appear to be more frequent in the Punjab than even in the Indus valley. From the returns of the gaols in India it may be gathered

---

[1] Pantiukhof.

that the Upper sub-Himalayan districts are, next to the Indus valley, the most affected by this disease, and then follow, in order of frequency, Central India and the Hills.

On the North-West Frontier epidemics of pneumonia are far from rare. Outbreaks of the kind occurred in the Tirah field force and the Malakand force, and occasionally epidemics of an infectious form of the disease are observed in Umballa in troops returning from field service. In the Hindu-Khush, on the other hand, pneumonia is rare, and in the Hunza expedition there was not a single case, although the men bivouacked without tents at a height of 7000 ft. In Kashmir also the disease is not common. Whether true croupous pneumonia is a frequent disorder among the general population in India it is difficult to say. Buchanan states that, whereas formerly many cases of "text-book pneumonia" were seen there, such cases are now rare, and that, when met with, pneumonia is usually of the catarrhal or lobular form. Possibly this observation may be explained by the frequency with which the latter form of the disease follows on, or is associated with, influenza—a disorder which has been present in India, with varying degrees of severity, since 1890.

Pneumonia is very prevalent in Ceylon, where it is the third most fatal of all diseases.

In some of the East Indian Islands, as for example Borneo, pneumonia is fairly common, but in others it appears to be seen only exceptionally. In the Malay Peninsula and in Indo-China the disease is perhaps rare; in Singapore it is said to be exceedingly rare. The evidence in regard to China is imperfect and conflicting, but it is probable that it is at the present day by no means rare in many parts of that country. In Shanghai it is said to have been almost unknown before 1890, but to have been of not infrequent occurrence since; and if this is so, it may most plausibly be explained by the fact that the influenza pandemic dates from the year named. The disease certainly exists among the native Chinese, and cases are not uncommonly seen at Kiukiang, Swatow, Lung-chow and other places. At Chungking it is said to be rather rare.

Throughout Japan pneumonia is met with frequently, from north to south.

Australasia. In certain parts of British New Guinea an epidemic form of pneumonia is common among the natives at the beginning of the cold season, and apparently the disease is here of a contagious character (MacGregor).

In Australia pneumonia is very widely prevalent and a frequent cause of death, and there is reason to believe that of recent years it has shown a tendency to increase as a cause of mortality. Thus in Queensland the deaths from it increased considerably between the years 1893 and 1897, and it stands second only to pulmonary tubercle as the most fatal of all diseases. It is particularly severe here in winter. A still more striking increase in the pneumonia mortality is shown in the returns from West Australia for the years 1889-98, the figures having risen steadily in that time from 12 deaths in 1889 to 169 deaths in 1898. It is probable that the advent of the influenza pandemic to Australia in the early part of the period in question has been largely responsible for this striking rise in pneumonia mortality.

Pneumonia occurs with varying frequency in the Pacific Islands; in some, as for example New Caledonia, it is said to be particularly fatal to the native inhabitants.

Africa. The malady appears to be decidedly rare in Egypt, in Tunis, and perhaps in Morocco, while in Algeria, on the other hand, it is said to be of very common occurrence and to cause a higher mortality than in either England or France. Certainly in the years 1896, 1897, and 1898, the disease caused a larger number of deaths in Algerian towns with over 30,000 inhabitants than any other group of diseases, with the exception of phthisis and diarrhœal disorders. On the West Coast of Africa pneumonia is said frequently to attack the natives in Senegambia, the Congo Free State, and the Cameroons. Europeans in these portions of the African continent appear to be less liable to its attacks. In Lagos the disease is very common and is particularly fatal to Kroomen.

Throughout South Africa pneumonia seems to be a very common disorder. At the Kimberley hospital many natives are annually treated for it; at Grey's hospital, King William's Town, and in the Durban hospital, it is one of the most prominent of diseases; and from Grahamstown and other places in Cape Colony similar reports are published. Epidemics of pneumonia are said to prevail from time to time in some of the coast towns of South Africa, and the occurrence of dust-storms may, it is thought, stand in some causal relation to these epidemics[1]. The natives in Mount Frere district (Native Territories) are very frequently attacked by it.

It is found that pneumonia is exceedingly common in Uganda and that it is met with throughout the greater part of Central and East Africa. Accurate statistics as to its prevalence are wanting, and such information as is available is largely vitiated by confusion between pneumonia and other respiratory disorders. But it may fairly be stated that all the evidence points to the conclusion that the black races of the African continent are peculiarly liable to be attacked by pneumonia, and that it is often very fatal amongst them.

In Madagascar and Mauritius pneumonia is by no means rare.

North and Central America. In the far north of the American continent pneumonia is widely prevalent. In the Hudson Bay territories and in Greenland it attacks both Europeans and natives. In all parts of Canada the disease is exceedingly common; and throughout the United States it is said to be the most fatal of all maladies, next to pulmonary tubercle[2]. It is certain that pneumonia is either commoner or more fatal, or both, on the continent of North America than in many European countries. Davidson has made a careful study of the distribution of the disease in the United States, and finds that, while of more or less frequent occurrence almost throughout, it is less common on the coasts than in the interior, that it is rare on the South Atlantic coasts, in the Appalachian highlands, in the southern part of the Pacific coast and a part of the Gulf coast, but that generally speaking the principal areas in which it is found to be in excess lie south of lat. 41°. Contrary to what might have been anticipated pneumonia is somewhat less fatal to the Indian races in North America than to the white and black races. The black races succumb most easily to it. There has been a very great increase in the pneumonia mortality in the last half-century in the State of Massachusetts.

In Mexico pneumonia is both frequent and fatal on the central plateau; it is a frequent cause of death in Guatemala; and the natives of British Honduras are said to be particularly

---

[1] Hillier. Lecture at the London Polyclinic, Feb. 1900.
[2] *Census Report*, 1890.

liable to fall victims to its attacks. In Nicaragua and Costa Rica, on the other hand, it seems to be rare.

The disease is frequently seen in many of the West Indian islands. In reports from Cuba, Barbados, St Lucia, and many others it finds frequent mention. In Jamaica it is particularly prevalent in the months of December, January, and February, although there is not more rain in these months than in the summer, nor is it especially cold at that season.

South America. Among the inhabitants of British Guiana pneumonia is a very fatal disease. The Indian races are particularly liable to it, and in some recent years it has been the cause of not less than two-thirds of the total mortality from all causes. Here, as in so many other countries, the occurrence of the influenza pandemic has caused a great rise in the pneumonia mortality. The disease is by no means rare in French Guiana, it is met with in all parts of Brazil, and appears to be particularly common in Uruguay and Paraguay. In the Brazilian army pneumonia and pleurisy together are said to cause nearly one-third of the total mortality. Martin de Moussey[1] speaks of an adynamic form of pneumonia met with in the Andes, which, he states, is common and fatal in all the Andean provinces proper, and in Santiago del Estero, Cordoba, and San Luis in the winter season.

On the western side of the continent it is found that pneumonia is very common both on the coasts of Peru and in the Sierra. The same statement is true in regard to Chile, or, at least, for its capital, Valparaiso. In the Falkland Islands, on the other hand, the disease is almost unknown, in spite of a climate which would, *à priori*, appear to be most favourable to its prevalence.

Factors concerned in the Geographical Distribution of Pneumonia. From the above rapid sketch of the present day prevalence of pneumonia it will be seen that the affection exists to a greater or less degree in almost all parts of the earth's surface. Not only does it occur in all the four continents and in Australasia, but it is met with more or less in every great geographical division of them. It is doubtful whether there is any area of considerable size from which pneumonia is entirely absent, while in a very large number of countries it is extremely common and a cause of an annual mortality which is exceeded by that of few other diseases.

[1] Cited by Davidson.

As already pointed out, however, there are marked differences in the frequency of pneumonia in different countries, and an endeavour has been made—which up to the present has not been wholly successful—to discover what are the influences at work that produce these differences.

"Chill" has from time immemorial been regarded as one of the most important and immediate factors in the production of an attack of pneumonia in the individual, and at no very distant period the disease was regarded as an ordinary inflammation of the lung, the result of exposure to cold. This view, now largely modified, also gave rise to the expectation that pneumonia would be found to be most common in cold climates, and to be least common or absent in the warm zones on either side of the equator. A study of its distribution over the earth's surface, however, soon dispelled this anticipation, and showed that pneumonia can and does exist in all latitudes, and that it is probably quite as common in the tropical belts of Asia, Africa, and South America as in the temperate climes of central Europe or the cold regions of Canada and Greenland. Some observers have even gone further and tried to show that the warmer the country the commoner will pneumonia be. Sanders endeavoured to prove, by means of tables, that the incidence of the disease increases as the equator is approached and the mean temperature increases[1]. But this view appears to err in the opposite direction, and our present knowledge of the disease rather warrants the statement that mean temperature has no effect on its prevalence, but that it can and does prevail in all latitudes, quite independently of their nearness to or remoteness from the equator.

The same is to a great extent true of the relation of pneumonia to altitude. Thus the disease is just as common in the Himalayan districts of India as in the plains; and it is found at considerable elevations in the Mexican plateaux and in the highlands of South America, Arabia, Persia, and the Caucasus. There is, in fact, nothing to show that elevation has by itself any influence on the distribution of pneumonia.

Changes of temperature, on the other hand, are probably important factors in determining its distribution, at least in regard to time. In its seasonal relations pneumonia, in most countries, no matter what their geographical position, shows a tendency to

[1] *American Journal of Medical Science*, 1882.

prevail in the winter and spring, when changes of temperature are most marked and the individual is most exposed to the risk of chill. It has been extremely rarely that a pneumonia epidemic or a maximum prevalence of pneumonia has fallen in the summer months.

It is probable that insanitary conditions aid to a considerable extent in the production of the malady, and that overcrowding and insufficient ventilation are particularly active in this respect. Dirt and escape of sewer-gas into houses may also be not without influence. These factors have usually been more or less at work in the well-known epidemics or prevalences of pneumonia in such institutions as gaols, barracks, or schools. In some epidemics, too, over larger areas the most insanitary dwellings in the area seem to have suffered the most.

In all considerations of this kind, however, in regard to this malady it is important to remember that, as already pointed out, the term "pneumonia" probably includes more than one form of disease, and that as our knowledge of the various kinds of pneumonia becomes greater it will probably be found that they are differently influenced by the different factors just discussed. It is almost certain, for example, that simple idiopathic croupous pneumonia is much more closely associated in its causation with changes of temperature and "chill" than is the epidemic infectious form of the disease, while the latter is almost certainly more intimately connected with the existence of insanitary conditions than the former.

It is certain, too, that in the individual any influences which depress the general vitality predispose to pneumonia, and among these may be mentioned exposure to cold and wet, fatigue, poverty, mental trouble, and exhaustion from other diseases. The influence of race upon the predisposition to pneumonia is probably not very great. All races appear to be capable of suffering from it, but it is probable that the negro is the most susceptible. The tendency of negroes to develope the disease when transported from their native country to other and particularly colder climes has long been recognised ; but here the racial question is obscured by the possible presence of some of the other depressing factors already alluded to.

Varieties of soil appear to have no direct and universal influence upon pneumonia prevalence, nor has the configuration or general geological character of the land. No doubt in individual outbreaks or prevalences these factors are not wholly without their influence, but no general law can as yet be deduced in regard to them. Some have believed that a low level of subsoil water was favourable to the development of pneumonia, but this view has been combated by others. The effect of rainfall is also uncertain, but in some occurrences of the epidemic form of the disease a heavy rainfall has seemed to aid in extinguishing the outbreak. This was notably the case in the well-known epidemic of pneumonia at Middlesborough in 1888.

In regard to the influence of sex it is generally admitted that males are more liable to pneumonia than females, though the disease is usually more severe and fatal in the latter. Persons of all ages are liable to it, but particularly those in the more advanced age-groups. It is peculiarly fatal to old persons and very young children.

RELAPSING FEVER.

General Characters and Etiology. Relapsing or famine fever is, as its name implies, a fever characterised by the occurrence of relapses; usually at intervals of about a week. There is no eruption. The disease is most frequently seen as an epidemic, is often associated with conditions of ill-nutrition and overcrowding, and is always accompanied by the presence in the blood of a specific organism, the *spirillum* or *spirochæte obermeieri.*

This disease presents many points in common with typhus fever, and in France, Germany, and Russia is usually known as "typhus recurrens." It is, however, a distinct and separate disease, and its geographical distribution differs from that of typhus in this respect, that though often found in countries and places where typhus is endemic, and often occurring in epidemics simultaneously with epidemics of typhus, its distribution over the earth's surface is much more limited than is that of the latter disease.

History. The antiquity of relapsing fever is very uncertain. The oldest known references to it appear to be those relating to certain epidemics in Scotland and Ireland in the first half of the eighteenth century. A large number of outbreaks were recorded there, and later in England, in the eighteenth and earlier portions of the nineteenth centuries. On the continent of Europe it is known to have occurred in parts of Russia since the year 1857; and in other countries its history only dates from about the same period, or even later. Relapsing fever was for a long time imperfectly recognised as a separate disease, and it was often confused with other diseases, particularly typhoid fever, and the so-called bilious typhoid, under which name- Griesinger first described an analogous fever met with in Egypt. It was not until the discovery in 1873, by Obermeier, of the *spirillum* which bears his name, that the pathology of the disease was placed on a satisfactory basis.

Recent Geographical Distribution. *Europe.* The British Isles were formerly regarded as essentially the home of relapsing fever. Now, however, the disease is but rarely seen here. In England and Wales a few deaths from this cause are annually recorded by the Registrar-General, but the fever is even rarer than its congener, typhus. The same is true of Scotland; and in Ireland, once the scene of most extensive epidemics, relapsing fever is now seen only exceptionally.

On the continent of Europe the disease has been met with principally in the north, the south remaining almost, if not quite free from its ravages. France is said to have escaped altogether, in spite of its nearness to the British Isles, where the fever was formerly at times epidemic. A few cases only have been recorded in Belgium. In Germany, on the other hand, many epidemics have been observed, in some of which, if not in all, the infection appears to have been imported from Poland or Russia, where the disease is truly endemic. In Denmark relapsing fever is said to be unknown, and in Norway and Sweden it has only been seen on rare occasions. In Russia, however, it is far from uncommon. It has been well recognised there at least since the year 1857, after which date the disease spread widely through the country, but there is good reason to believe that it was prevalent at an earlier period. Reitlinger, whose profound study of this disease as it appears in Russia remains the most authoritative work on the subject, states that relapsing fever has prevailed as an epidemic as far north as the government of Archangel, though apparently as an importation from St Petersburg, the infection being carried by labourers returning to their homes from the capital[1]. In the same way the northerly government of Olonetz has been invaded by the fever, though that of Vologda, which adjoins and is even more closely populated than that of Olonetz, had up to 1874 escaped[2]. In Finland relapsing fever has appeared independently, generally as the result of famines. St Petersburg

---

[1] *Investigations into the History, Geography, and Statistics of Relapsing Fever in Russia* (in Russian). By L. Reitlinger, St Petersburg. 1874.

[2] This government now yearly returns some cases of relapsing fever.

and its neighbourhood, however, at the time Reitlinger wrote, constituted the great centre for relapsing fever, whence the infection was frequently carried in all directions to other parts of the country. The northerly governments always suffered the most severely, and without exception the disease always appeared in the winter—facts which are doubtless to be explained by the greater overcrowding and confinement to close ill-ventilated rooms which especially prevail in the winter in Russia. Moscow has at times suffered severely. Many of the Polish governments, and also those of the Ukraine in the south-west of the country, have been the scene of epidemics. Relapsing fever has been severely epidemic in the easterly governments of Perm and Viatka, and less so in those of Orenburg and Ufa. Odessa has in like manner been visited by frequently recurring outbreaks[1].

At the present day this disease appears to be very widely endemic throughout European Russia. In the three years 1893–5 cases were reported from every government throughout the country: the largest numbers being returned from the governments on the Volga, and the smallest from the Baltic provinces. After the great famine of 1891 the disease was very widely epidemic through a great part of Russia. In St Petersburg an enormous number of cases were treated, occurring especially among the shifting inmates of the "night-refuges," representing the poorest and most wretched portion of the population. The disease is occasionally seen among the Bashkirs inhabiting the steppes in the neighbourhood of the Ural River, but is not very common among them[2].

Throughout southern Europe generally, Austria would seem to be the only country which has at all extensively suffered from relapsing fever. Spain, Switzerland, and Italy have all remained free from it. In the Balkan peninsula the disorder does not appear to be endemic, though in the Russo-Turkish war of 1877–8 several epidemic outbreaks of the fever occurred here among the Russian troops. A few cases were seen at that time in Constantinople, but in each instance they could be traced to importation from Odessa. In recent years the disease seems to have been absent from the Turkish capital. In some of the Mediterranean islands relapsing fever is said to have occurred somewhat frequently in the past.

[1] In the Crimean war both French and English troops suffered severely from relapsing fever.   [2] Nikolski.

Asia. The malady has never been very prevalent in the Caucasus. In former years it affected to some extent the men employed in the fishing industry of Saliany on the river Kura, and in the city of Tiflis. In the ten years 1888–98 only a few isolated cases of the disease have been seen in the latter town[1]. In Batum and other places on the Transcaucasian Railway sporadic cases of the fever have occurred from time to time since 1894[2]. In Turkestan it formerly had a somewhat wide prevalence, the infection being carried from fortress to fortress by the movements of the Russian troops. Of recent years, however, it would seem to have become exceedingly rare throughout Russian Central Asia, the cases returned from these provinces in the years 1893–5 being reckoned by units, while from many parts not a single case was recorded. In Siberia the disease appears to be much commoner. It was originally introduced from Europe by means of exiles and emigrants, and spread over the whole of both western and eastern Siberia, as far as the Primorskaia, or Maritime Province, and the region of the Ussuri river in Manchuria. In recent years cases have been annually reported from most parts of the country; the government of Tobolsk usually returning the largest figures, and that of Yakutsk, in the far north, the smallest.

Of the existence of relapsing fever in the near East, in Asia Minor, Syria, Mesopotamia, and Persia, comparatively little is known. The fevers formerly known as "Levant fever," "Bukowina fever," and "Smyrna fever," seem to have been very probably of the nature of relapsing fever.

In India the disease has been known to exist since the middle of last century. Epidemics have since appeared in all parts of the peninsula. During the past three years it has been severely epidemic in Bombay concurrently with plague, and it is interesting to note that the two diseases have been observed to occur simultaneously in the same individual. To what extent relapsing fever has prevailed in India in connection with the recent unprecedented famine, later publications than those I have as yet had access to will probably show. But it is noteworthy that in the Central Provinces the fever which accompanied the famine of 1896 was described as being *not* of "a

[1] *Statistics of Caucasian Pathology.* By Dr I. I. Pantiukhof. Tiflis, 1898.
[2] *Bolnitchnaia Gazeta Botkina*, 1900, No. 22.

contagious, relapsing, or typhus type[1]." Among the hill tribes of the Himalayas relapsing fever is at times epidemic. There is an absence of records of this disease in Farther India, in the Malay Peninsula, and in the East Indian Islands, but quite recently a single case was observed in Sumatra, the patient being a Chinese coolie immigrant from Swatow, who evidently brought the infection with him[2].

In China relapsing fever is found constantly associated with typhus[3]. It is mentioned as the most common form of fever at Teng-chow-fu, and, this being so, it is permissible to suppose that it is not a rare disease in other parts of the country. In 1877 it was epidemic at Tientsin, and in former years has raged in Hongkong, Peking, and other places in Northern China.

Much uncertainty exists as to whether relapsing fever has ever appeared in Japan. Baelz suggests that an epidemic which ravaged the town of Kochi in the island of Shikoku in 1881 may have been of this nature, but is unable to prove or disprove the truth of his conjecture[4].

Australasia. The Polynesian islands seem hitherto to have escaped the ravages of this fever, and the same statement may apparently be made in regard to Australia and New Zealand. The sole mention of the disease which I find in recent health reports from these countries is the occurrence of one death from it in West Australia in the year 1895, the only death of the kind in the ten years 1889–98.

Africa. Information upon the existence of relapsing fever on the African continent is extremely scanty. If absence of mention of the disease is to be accepted as proving the absence of the disease itself in any given country, then it would appear that this fever is unknown in most parts of Africa. It seems to have been formerly endemic in Egypt, and perhaps is still so, but in the statistics of that country the returns for relapsing fever are included under the general heading of "fièvres typhiques," together with enteric and typhus. In Nubia and Abyssinia it was also believed to have been endemic in former years, and in Algeria several cases were seen during the typhus

---

[1] *Annual Sanitary Report of the Central Provinces* for 1896.
[2] Graham, *Journ. Trop. Med.* June, 1901.
[3] Coltman.
[4] *Sei-i-Kwai.* Oct. 1898.

fever epidemic of 1867. The fever has also been observed in Morocco in connection with famines following on the ravages committed by locusts.

From India the disease appears to have been carried as "coolie fever" to Mauritius and Réunion, but for a long period the former island at least has been free from it.

America. Relapsing fever has on several occasions in the past been introduced into various parts of the United States, principally by means of emigrants from Ireland, where the disease, as was shown above, was constantly epidemic in former years. At the present day I am aware of no evidence to show that this fever exists at all, either as an endemic or epidemic disease, in any part either of the North or of the South American continent. Osler, indeed, states definitely in regard to the United States that, since the epidemic of 1869 in New York and Philadelphia, relapsing fever has not again appeared.

There seems to be a complete absence of mention of the disease in Mexico, the Central American States, and the West Indies.

In South America relapsing fever has never been seen in the Guianas, nor in Uruguay, Patagonia, nor apparently on the Pacific Coasts. But in the Andes it was severely epidemic some thirty or forty years ago. It was described as occurring nowhere at a lower level than 1500 metres, and as attacking communities living as high as 4000 metres above sea-level[1]. This seems to be the only certain mention of the disease in the whole South American continent.

Factors determining the Geographical Distribution. The principal factors in the geographical distribution of relapsing fever appear to be of a rather less complex character than in the case of some other diseases. The fever is invariably associated with the presence in the blood of an organism called, after its discoverer, the *spirillum obermeieri*. For the development and transmission of this organism certain conditions require to be present, the most important of which are overcrowding and want of sufficient food. Its constant association with dearth and famine and the name popularly applied to it of "famine fever" (a name which it shares with typhus), show how close is the relation between

---

[1] Baldow, *Gaz. Méd. de Paris*, 1865. Quoted by Lombard.

the two phenomena. The disease is practically confined to the poorest classes of the community, who are the first to suffer when a bad harvest or a succession of bad harvests sends up the price of food. Overcrowding seems to favour the epidemic spread of the fever, for there is no question that the infection can be communicated from person to person; and in some epidemics it has seemed that overcrowding might be a more important element in the production of the disease than scarcity of food.

Relapsing fever is also spread from place to place by the movements of infected persons, and perhaps of infected articles such as clothing. Its spread through Siberia was shown to be clearly due to the movements of emigrants and exiles; and in European Russia it was in the past, and perhaps is still, spread by the migratory habits of the peasants, who seek work in the capital or elsewhere, and returning to their homes carry the seeds of the disease with them. In America immigrants from the British Isles were the agents bringing the disease into the country, and it should be added that it has never spread far from the immediate surroundings of these immigrants, or tended to become endemic in the States into which it was introduced.

Whether the infection can also be spread by the medium of contaminated articles such as clothes may perhaps be regarded as somewhat less certain than its dissemination by human beings; but important evidence of its diffusion in this manner is forthcoming from Edinburgh, St Petersburg, Breslau, Philadelphia, and New York, and such a mode of spread must be regarded as at least possible.

Neither race, climate, season, elevation above sea-level, the geological nature of the soil, nor the conformation of the country seems to have any influence on the distribution of relapsing fever.

Finally it is of interest to note that while dearth and famine are the most important elements in its prevalence, there is a considerable amount of evidence showing that these conditions may be present without the development of relapsing fever. This fact may be reasonably explained by supposing that the spirillar organism which appears to be essential to the production of the disease is not universally present, and that in its absence no amount of malnutrition or innutrition will suffice to bring about the particular group of symptoms so well recognised under the name of relapsing fever.

RHEUMATISM.

General Characters and Etiology. The term rheumatism includes a considerable variety of conditions—such as acute articular rheumatism or rheumatic fever, chronic articular rheumatism, acute and chronic muscular rheumatism, rheumatic affections of the heart, and a variety of neuralgic and other painful phenomena attributed to the so-called rheumatic diathesis. While some uncertainty exists as to the etiology of these various clinical conditions, and while it is possible that in time some of them will be shown to be specifically distinct and due to separate and different causes, they are for the present conveniently regarded as forming a fairly clear and recognisable disease-group, and they will therefore be dealt with here together.

The geographical distribution of this group of diseases is an exceedingly wide one; and there are few inhabited countries on the earth's surface, from which medical observations are forthcoming, in which some mention is not made of the occasional or frequent occurrence of rheumatic affections.

Their frequency, however, varies considerably in different countries, and it will be the object of the present chapter to show approximately the relative degree of frequency or rarity with which these affections are found in various portions of the globe. Unfortunately, published records are of comparatively little service in this respect; statistics of the prevalence of the disease being for the most part wanting, while mortality statistics are misleading, inasmuch as the number of deaths directly attributed to rheumatism bears but a small proportion to the number of people who have at some time or other suffered from rheumatic affections, or even to the number of those who are distinctly of the so-called rheumatic diathesis.

The pathology of rheumatic affections is uncertain. It is probable that rheumatic fever is not one and the same disease with chronic articular or muscular rheumatism, and modern observations have shown that in its epidemiological and clinical characters it has much in common with the acute infective fevers. Rheumatic fever, in fact, now stands with pneumonia on the border line between the group of inflammatory affections of special tissues due to "chill" and the group of infective fevers. Newsholme has shown that it often occurs in epidemics of some regularity, at intervals of from three to six years.

The bacteriology of the malady has been largely worked at, but with no positive results as yet, although some have suggested that the disease may be due to a primary growth of specific micro-organisms in the tonsils. It was at one time generally held to be a disease of chemical origin, due to an excess of lactic acid in the tissues.

History. It is certain that rheumatic affections are of very great antiquity. Tombs have been quite recently opened in Egypt, and skeletons found inside, the bones of which showed distinctly recognisable rheumatic changes. The recognition and description of these affections, however, is of much later date. Throughout the medieval period great confusion arose between the terms "rheuma" and "catarrh" which were used indifferently in the humoral pathology of the time, and were constantly called upon to explain a variety of diseases. It was only in the 17th century that the word rheumatism came to be restricted more or less to the group of diseases to which it at present applies. Joint affections were long recognised and described under the general term "arthritis," which included not only rheumatism but also gout and other forms of disease. An English physician, Pitcairn, first showed the intimate relation of heart disease and rheumatism in 1778.

Recent Geographical Distribution. *Europe.* Both rheumatic fever and the more chronic forms of the disease are exceedingly common in the British Isles. In the year 1888 a Collective Investigation Committee of the British Medical Association made an inquiry into their prevalence, and found that there was no part of Great Britain or Ireland where both acute and sub-acute rheumatism did not prevail extensively.

In France the official mortality-returns contain no column

for these diseases; but Lombard states that they are frequent throughout the whole of France, and principally so in the centre and north.

Chronic rheumatic affections are said to be very common in the country districts of Belgium. In Germany acute rheumatism and heart disease appear to be particularly frequent in Bavaria and in Alsace-Lorraine, a fact which has been attributed (with what justice I do not know) to the excessive amount of beer-drinking in those countries. All forms of rheumatism are very prevalent in Bavaria and Swabia. In Denmark rheumatic fever is fairly common; it varies slightly in amount from year to year, and in recent times years of maximum prevalence have been 1871, 1873, and 1883–1888 (Newsholme). The deaths attributed to this cause in Danish towns between 1892 and 1897 varied from 37 in the first to 55 in the last year named. The periods of greater prevalence of rheumatic fever in Denmark corresponded to a certain extent with similar periods in Norway, where the years of maximum prevalence were respectively 1871, 1876, and 1888–1889. The returns from Sweden are less complete, but rheumatic fever is common throughout the country, and particularly so in the northern districts.

It is difficult to judge of the frequency of these affections in Russia, but I can assert, from personal observation, that in the capital and along the shores of the Gulf of Finland rheumatic fever, the more chronic forms of articular rheumatism, and muscular rheumatism all occur with considerable frequency. The returns from all the administrative divisions of Russia show also that rheumatism is widely prevalent throughout the country, but as rheumatic fever is not treated separately from the other fevers in the returns it is impossible to judge of its relative frequency in this as compared with other countries.

On the behaviour of these diseases in Austria-Hungary I have no information. But in Italy the acute form of rheumatism caused an annual mortality in the years 1887 to 1896 which varied between 25 and 36 per million inhabitants; while the deaths attributed to the chronic varieties varied between 30 and 39 per million in the same period. These figures probably indicate a considerable degree of prevalence of the disease or diseases. In Spain and in Portugal rheumatic fever is said to be very common. Lombard states that rheumatism is endemic on the plateau of

Castile and in the Asturias, and that it is frequently seen in Seville, Granada, Valencia, and Malaga.

Rheumatism is of frequent occurrence in many parts of the Balkan peninsula. From the Salonika province of Turkey and elsewhere reports of it are frequent. It is probably commoner in the mountains than in the plains. In the hilly country of Montenegro all forms of rheumatism, both articular and muscular, are said to be very prevalent, and this is attributed to the rigour and humidity of the climate. In Constantinople and on the shores of the Bosporus and Sea of Marmora no class of illness is more frequently seen than the various forms of rheumatism, more particularly the sub-acute and chronic forms, both muscular and articular. Generally throughout European Turkey rheumatism and malaria are the two most prevalent of all diseases.

In Iceland and in the Faröes not only the less acute forms of rheumatism but also true rheumatic fever are of very common occurrence.

Asia. In some parts of Asia Minor rheumatism is frequently seen. In Trebizond rheumatic affections are very common, on account, it is said, of the excessive moisture in the air in and around this Black Sea port.

Acute rheumatism is a very frequent disorder in Mitylene and some other islands off the coast, and at Tripoli, on the Syrian littoral, this group of diseases is no rarity.

In the Caucasus rheumatism, both muscular and articular, is one of the commonest of disorders. In central Asiatic Russia (Turkestan, Transcaspia, etc.) it is perhaps much less prevalent than in European Russia, but statistics from these regions are too imperfect to allow of any definite conclusions on this point. Through all parts of Siberia, from east to west and from north to south, rheumatic affections are very common, though to what extent rheumatic fever shares in this frequency I am unable to say. It is said to be of frequent occurrence, however, at Tomsk and in Kamchatka.

Among the pilgrims to the holy cities of the Hedjaz rheumatic affections of all kinds are commonly seen, including sciatica, neuralgia, and other conditions which may plausibly be attributed to rheumatism. In Afghanistan it is very prevalent, and in Persia probably by no means rare. At Khanikin on the Turco-Persian frontier, after the heavy rains of early winter, forms of this malady are extremely common, particularly among the poorer classes of the population exposed to the weather.

Rheumatism is apparently far from rare in the Indian peninsula. Some writers have stated the contrary, and dwelt on the extreme infrequency of all forms of rheumatic affection, including organic heart lesions, in that country[1]. But the official returns show that both Anglo-Indian residents and natives suffer not a little from these diseases. In the English army in India the admission and death-rates from these causes are actually higher than in England. Bengal is said to be the Presidency in which rheumatic diseases are most prevalent, Bombay that in which they are most fatal, while Madras has the lowest admission and death-rates[2]. Chronic forms of the disease are common in the cold season and rains in Rajputana, but the acute forms are rare. They are equally rare in Assam, and in the Malay Peninsula. At Penang and Singapore indeed they are almost unknown. McClosky, during a residence of eight years in the Straits Settlements and the Malay Peninsula, never saw a single case of rheumatic fever in a native, and only one in a European, who had suffered previously at home. Valvular cardiac lesions were equally rare, and a history of rheumatic fever could never be elicited. Only in the post-mortem room had this observer seen chronic valvular lesions in the bodies of natives which were undoubtedly of rheumatic origin[3]. At Perak, similarly, Clarke never saw a single case of true acute rheumatism during a residence of five years[4].

That rheumatic affections occur in many of the East Indian Islands is certain, though their comparative frequency in the different islands is less so. They are far from rare in Java and in British North Borneo; in Sumatra and the Lampongs they are rife, particularly in the latter, and in the Rhio archipelago they are very prevalent. In the Dutch East Indian possessions generally, however, they appear to occur with no great frequency, as in the Indo-Dutch army rheumatism, at least in its acute articular form, is not a common disease, and when it occurs it is usually in a mild form.

---

[1] Buchanan (*Journ. Trop. Med.* Nov. 1899) says that not only acute rheumatism but also chronic rheumatic arthritis and organic heart-murmurs are exceedingly rare in India. Cantlie (*ibid.* Oct. 1899) when visiting several cities in India did not see a single case of rheumatic fever in hospital.

[2] Davidson.   [3] *Journ. Trop. Med.* Jan. 1900.   [4] *Ibid.* Nov. 1899.

The evidence as to the frequency of these diseases in China is rather conflicting. Cantlie has stated, after many years' residence in China, that he never saw a case of rheumatic fever in hospital there[1]. But mention of both acute and chronic rheumatism is by no means rare in the reports of the medical officers in the Chinese Customs service. Thus, for example, 15 cases of articular rheumatism, acute or chronic, were recorded at Lung-chow in 1895–6[2]. Rheumatic arthritis is spoken of as a common disease in the natives of Wuchow[3], where gonorrheal rheumatism is also of frequent occurrence. At Ichang rheumatic pains and swellings of extremities are very common complaints[4]. Finally Coltman roundly states that rheumatic fever is prevalent in Foochow, Shanghai, Soochow, Chifu, Lao Ling and Hang-chow; while chronic muscular rheumatism is common all over China[5]. To this may be added the statement that the Chinese in Formosa suffer much from rheumatism[6]. In Mongolia rheumatism is said to be the most prevalent disorder. In Japan the more chronic forms of the disease appear to be common, while acute articular rheumatism is very rare, at least in Yokohama and Tokio.

Australasia. The sub-acute forms of rheumatism are occasionally seen in both natives and Europeans in British New Guinea. Throughout Australia rheumatic affections appear for the most part to be rare, and are probably a good deal rarer than in England, for example. Only occasional mention of them is found in the health reports from the different colonies, although in South Australia and New South Wales the number of deaths from this cause is by no means inconsiderable, while heart disease does not seem to be particularly infrequent in any of them. Rheumatic disorders are probably equally rare in New Zealand, but in many of the islands of the Pacific—as, for example, New Caledonia, the Society Islands, and the Marquesas group—they are of very frequent occurrence, and in the Sandwich Islands they are, or were some years ago, said to be amongst the most common of disorders. In Samoa, too, many cases of rheumatism are seen.

[1] *Journ. Trop. Med.* Oct. 1899.
[2] *Chinese Customs Med. Reports*, No. 52.
[3] *Ibid.* No. 56. [4] *Ibid.*
[5] *The Chinese, their present* etc. 1891.
[6] *Chinese Customs Med. Reports*, No. 45.

Africa. In Egypt rheumatic affections seem to be remarkably prevalent.

In Tunis both acute and chronic rheumatism are exceedingly frequent among Europeans and natives alike, forming about a third of the diseases treated (Davidson). In Algeria, on the other hand, acute articular rheumatism is said to be of infrequent occurrence, while in Morocco the acute forms of the disease are rare and the chronic forms are common.

Turning to the western regions of the African continent it is found that acute articular rheumatism is very prevalent in Senegal; that in the Gambia a few cases of rheumatism are annually treated in the colonial hospital; that in Sierra Leone and Lagos such disorders form a remarkably high proportion of the total cases admitted to hospital; and that on the Gold Coast rheumatism is among the commonest of diseases to which the natives are liable. Sporadic cases are seen in the Cameroons. In the Congo Free State rheumatism is seen with some frequency, and generally throughout Central Africa rheumatism is prevalent, though most often seen in its more chronic forms, the acute forms being rare (Pruen). Heart disease is also no rarity in Central Africa.

In East Africa rheumatic affections are certainly known. In Uganda they occur with some frequency; in Manyuema they are both common and fatal among the natives, and in Nubia and Abyssinia they are frequently met with.

Finally in South Africa this group of diseases is found to prevail in many parts to a considerable extent. In the native territories of the Transkei, Tembuland, and Pondoland, rheumatism is among the commonest of disorders. In Bechuanaland it is also of frequent occurrence. In Mashonaland, on the other hand, it is perhaps rare; one observer never saw a single case of acute rheumatism during two years' residence there[1]. At most of the large hospitals in Cape Colony and Natal rheumatic diseases contribute a considerable proportion of the cases annually treated.

In Mauritius, Rodriguez, and the Seychelles, this group of diseases appears to be rather prevalent[2]. In Madagascar acute rheumatism is apparently not common among the natives, but the muscular and syphilitic forms are very frequently met with.

[1] Todd, *Journ. Trop. Med.* Nov. 1900.
[2] *Colonial Reports.* Davidson on the other hand says that rheumatism is comparatively rare in the Seychelles.

North America. In Greenland rheumatism is said to be one of the commonest of maladies, but "it is not so certain if acute articular rheumatism prevails there to a corresponding extent." Generally throughout Canada rheumatic affections are met with, but whether with great or little frequency I am unable to say. The same statement has to be made in regard to their prevalence in the United States, where, however, these diseases are said to be commoner in the south than in the north (Lombard). In Massachusetts the mortality ascribed to rheumatism has rather increased in recent decades. Osler is of opinion, from personal observation, that rheumatism is more prevalent in the Canadian city of Montreal than in the United States cities of Philadelphia and Baltimore. In Mexico it is found that rheumatism is of frequent occurrence on the Anahuac plateau, while it is extremely rare in the low-lying coast town of Vera Cruz.

In the West Indies the disease is common enough. Both the acute and chronic forms are often seen in Jamaica, St Vincent, Grenada, Antigua, and St Lucia. In the Virgin Islands rheumatism and malarial fever are the two most prominent of all diseases, and both in the Leeward Islands and the Bahamas rheumatic affections must be particularly prevalent. In the latter "rheumatism" accounted for more admissions to hospital than any other disease in the years 1897 and 1898.

Central and South America. Little appears to be known of the comparative prevalence of these affections in Central America, save that rheumatic fever is rare in Guatemala, while in British Honduras acute rheumatic fever, as well as chorea and valvular heart lesions, which are so constantly associated with it, are said to be entirely unknown.

In British Guiana true rheumatic fever appears to be a rarity, while other forms of rheumatic disease are very widely prevalent. This would seem to be no less the case in French Guiana. In Brazil these affections are moderately prevalent. They are probably more so in the Argentine Republic, where all forms of the disease including rheumatic fever are very common, both in the capital and in the interior, along the Parana, Uruguay, and Paraguay rivers.

In Ecuador, Chile, and Peru rheumatic disorders are frequently seen, and there is perhaps no part of the world where they are more common than along the western shores of the South American continent.

It is not surprising to find that in the Falkland Islands, with their damp, cold climate, chronic and muscular rheumatism are both common and intractable.

Factors affecting the Geographical Distribution of Rheumatic Diseases. For the present our knowledge on this matter is very incomplete. It is certain, as already pointed out, that under the general term rheumatism a considerable number of pathological entities are included. It is only for the sake of convenience and on account of a certain similarity of clinical features that they are grouped together, but there is no true evidence that, for example, muscular and articular rheumatism are the manifestations of one and the same disease, while, as previously stated, there is considerable evidence that the chronic forms of both are pathologically distinct from true rheumatic fever. It is therefore not surprising that the geographical distribution of these various forms of the disease is by no means uniform. Materials for a complete study of their distribution are, for reasons pointed out at the beginning of this chapter, not available, and all that can be stated at present is that in very many countries rheumatic fever is rare or quite unknown, while the chronic articular form and the muscular form of the disease are very common. Among countries where this is the case may be mentioned Morocco, Central Africa, British Guiana, Japan, and some parts of India. On the other hand there are many countries where a frequency of acute rheumatic arthritis is found to co-exist with a frequency of the other forms of rheumatism; and others where all alike are equally rare. It is easier to name the countries where rheumatic affections of all kinds are rare than where they are common, and among such countries may be mentioned the Malay Peninsula and Straits Settlements, some parts of Australia and New Zealand, Guatemala, and British Honduras.

No one will question that, in the individual, "chill" is the most frequent and immediate antecedent of an attack of rheumatism, whether of the acute or of the other varieties; and in general terms the disorders included in the list of rheumatic affections are more common in cold countries than in warm, and particularly common in some regions with a bleak, damp climate, such as that of the Faröes or the Falkland Islands.

But there are very striking exceptions to this statement, and rheumatic affections seem to be nowhere commoner than along the dry and by no means cold coasts of Chile. Their comparative frequency also in the warm regions of New Caledonia, the Sandwich, Society, and Marquesas Islands, many of the West Indian Islands, Central Africa, and India, show conclusively that a low degree of temperature is not necessary for their prevalence. Some, indeed, have believed that rheumatic affections are more prevalent in warm than in cold latitudes, but this is hardly borne out by the facts. In regard to rheumatic fever it is certain that the reverse is the case.

It is probable that sudden changes of temperature, particularly in a moist climate, afford very favourable conditions for the prevalence of all forms of rheumatism, but particularly for the more chronic articular and muscular forms. Neither the geological nor physical characters of a district seem to have any direct influence, except in so far as they affect the climate. These diseases are certainly very prevalent in many countries at high altitudes such as the plateaux of Mexico, the highlands of Norway, Montenegro, and Arabia, and the Swiss Alps, but they are also extremely prevalent at such low altitudes as the coast of Chile, the West Coast of Africa, and generally throughout the British Isles.

RICKETS.

General Characters and Etiology. Rickets is a nutritional disorder common among children. Its main features are slight fever, general sensitiveness of the whole body, delayed ossification of the bones, retarded dentition, and imperfect nutrition of all the tissues. Wasting is marked, and very often the soft bones yield to pressure with consequent deformity. The liver, spleen, and lymphatic glands are frequently enlarged, and there is great diminution in the calcareous salts in the bones. Rickets is usually associated with, and probably caused by, deficient or improper dietary, together with want of sunlight and bad ventilation. Prolonged lactation also favours its development.

History. The disease was first described accurately by Glisson in the 17th century. It is probably of as great antiquity as any of the other diseases to which ill-nourished children are liable, but the materials are wanting for an accurate knowledge of its history.

Geographical Distribution. *Europe.* Rickets is very common among the poorer classes in the British Isles. It is probably equally so in France, Germany, Holland, and Belgium. In many European countries the returns for scrofula and rickets are taken together, and it is impossible to say what proportion of the mortality is due to each of the two diseases. Thus in Holland the mortality from the two disorders together has diminished from 330 per million inhabitants in 1889 to 130 in 1898. In the urban population of Denmark the mortality from rickets alone remained stationary at about 140 per million in the years 1892–97. In Norway *svek*, or rickets, is by no means rare[1]. In Sweden it is called *Engelska sjukan* or "the English disease." In Lapland

---

[1] Axel Johannessen, *Janus*, 1897–8, p. 46.

the disease is said to be almost unknown. Throughout the greater part of European Russia it is common. It is also frequent in Northern Italy, while in Southern Italy and Spain it is rare. In Turkey it is comparatively commonly seen.

In Iceland and the Faröe Islands rickets is said to be a very infrequent disorder.

Asia. The returns from the Caucasus, from Siberia, and from Russian Central Asia, show that the disease occurs in all these regions, but with considerably less frequency than in European Russia. Among the Bashkirs it is said to be quite exceptional to see a rickety child[1]. It is also rarer in Syria, Persia, and Arabia, than it is in most European countries. In India it is so exceedingly rare that a case in a native is considered worthy of record in a medical journal[2]. In Ceylon, on the other hand, it seems to be very much commoner, and it has recently been the cause of from one to three hundred deaths annually in that island (say, from 30 to 100 per million inhabitants). So also in Burma rickets is no rarity[3], at least in the Southern Shan States. In Siam, however, it is said to be unknown (Rasch), and from its very rare mention in reports from the Straits Settlements it must be supposed to be very infrequent there. In many of the East Indian Islands rickets is also very rare or unknown. Nieuwenhuis examined over 5000 children in Borneo without meeting a single case, and in Java it is said to be unknown.

In China and Cochin China rickets is rarer than in Europe, and in Japan it is very little seen.

Australasia. Rickets is probably one of the rarest of disorders in Australia. The only recent mention of it that I have come across is the record of occasional deaths from the disease in West Australia in the last decade. In some of the Pacific Islands, however, it seems to be common, and in the New Hebrides the greater part of the children are said to be either rickety or scrofulous[4]. In Fiji only an occasional case seems to be met with.

Africa. This disease is by no means common in Egypt, in Algeria, in Senegal, and generally in the tropical zones of the African continent. I have no information as to its prevalence in

---

[1] Nikolski.
[2] Maynard, *Indian Med. Gazette,* 1899, p. 167.
[3] Henderson, *Indian Med. Gazette,* July, 1899.
[4] Bernal, *Archives de Méd. Navale,* August, 1899.

South Africa, save that an occasional case is admitted to the Durban hospital. In Madagascar and the Comoro Islands it is almost unknown.

America. In North Greenland rickets rarely occurs.

It is comparatively uncommon also in Canada; and in the United States it is certainly less frequent than in Europe. In the cities, however, it is very prevalent, "particularly among the children of the negro and of the Italian races." In Mexico, on the other hand, the disease is said to be almost unknown. In Guatemala it exists, but causes a low mortality only[1]. In Honduras it is said to be practically unknown.

The disease is also, perhaps, very rare in the West Indies and generally throughout South America, but exact information of recent date seems to be lacking.

Factors concerned in the Distribution. It must be admitted that materials do not exist for an exact comparative study of the distribution of rickets. It is one of the diseases that is often omitted from reports on the pathology of any given country; and in statistical tables it is very often grouped with other diseases. But it may be gathered from the material available that rickets is mainly a disorder of countries with temperate or cool climates. It is rare, or absent in countries with tropical or sub-tropical climates. It is most frequent in the central and northern parts of Europe. In the far north of both the eastern and western hemispheres it again becomes rare.

Wet plains and valleys are thought to be favourable to the prevalence of rickets, while on dry soils and at high altitudes the disease is less common and less severe.

The influence of race is not very clear. It occurs in children of very diverse nationality, but it is noteworthy that in some tropical countries, as for example Java and India, where rickets is almost unknown in natives, it is occasionally seen in Europeans. Negroes are certainly not immune to it.

Improper dietary and impure air are, as already stated, the principal predisposing causes of rickets. Hence its frequency in the poor inhabitants of European and American cities, who live under conditions of overcrowding and general bad hygiene. Hence also, in all probability, its rarity among the Bashkirs and

[1] In 1894 rickets was the cause of 33 deaths in Guatemala; equivalent to about 23 per million inhabitants.

Kirghiz on the Russian steppes, and among the Kabyles of Algeria. It is more than probable also that the infrequency of the disease in warm countries is largely due to the open air life led there, and the beneficent influence on young children of sunlight and fresh air.

It would almost seem that a plentiful supply of pure air is of even greater importance in preventing (and in curing) this malady than the nature of the diet. But the exact share taken by each of these factors in the production of the disease is not always easy to determine. It has been shown experimentally that the condition of rickets in the lower animals can be brought about and controlled by variations in their dietary. Bland Sutton's experiments with lion cubs at the Zoological Gardens have proved this conclusively.

SCARLET FEVER.

History. It was pointed out in the chapter on measles that the early history of that disease was obscured by the confusion that long prevailed between it and scarlet fever. A second source of confusion in the history of the latter is found in its resemblance to diphtheria, miliary fever, and other diseases. So that it is not until the sixteenth or seventeenth century that anything definite is known as to the prevalence of scarlet fever. Probably it existed long before that date, at least on the continent of Europe. Sydenham first clearly differentiated the disease. Nothing is known as to its first appearance in Asia or Africa. It is thought to have been introduced into North America in 1735, into South America in 1829–31, and into Australia and Polynesia in 1847–8.

Recent Geographical Distribution. *Europe.* The mortality from scarlet fever in England and Wales has greatly decreased in recent times. In the three decades commencing in 1861 and ending in 1890, the death-rates from this disease were respectively 972, 716, and 334 per million; and since the year 1890 the fall has been almost unbroken, until in 1900 the ratio was only 119 per million.

This reduction in recorded scarlet fever deaths has been in part ascribed to a lowered birth-rate, with a consequent smaller proportion of children at susceptible ages, in part to improved sanitary conditions and the great increase of isolation hospitals, and in part to the type of disease having become milder in recent years. How far the decline is due to these causes and how far it is due to a really diminished prevalence of the malady is uncertain.

A considerable fall in scarlet fever mortality has also been observed in Scotland; the mean death-rate per million living in the decade 1881–90 was 286, and in 1896 it had fallen to 160, slightly rising again in 1897 to 170 per million, and in 1900 to 190 per million. An even more striking fall has been observed in Ireland, where the average annual mortality in the years 1881–90 was 210 per million, and in 1900 only 55.

In France, in towns with over 10,000 inhabitants, the scarlet fever mortality has fallen from 80 per million per annum in the years 1886–90 to 30 and 40 respectively in the years 1897 and 1898. As these figures refer to an urban population only, they must be held to indicate a very low degree of prevalence of scarlet fever in France.

In Belgium also a parallel fall has been observed; the mean annual recorded deaths in the years 1871–80 were 1963, in 1881–90 they fell to 1303, and in 1897 were only 1019, in spite of an increasing population. The last figure was equal to about 160 per million living. Scarlet fever mortality in Holland is remarkably low, the ratios in the ten years ending 1898 varying between 20 and 50 per million inhabitants per annum; the lowest ratios occurred at the beginning of this period, and the highest about its middle, a tendency to fall again being noticed in the later years.

Throughout the German Empire the scarlet fever mortality, as in England and Belgium, has very markedly declined in recent years. The mean annual death-rate per million living, for the whole empire, fell from 253 in the septennium 1887–93 to 184 in 1894. The decline was no less marked in each part of the empire taken separately; and the returns from Prussia, Saxony, Bavaria, and Würtemburg all indicate a great reduction between the years 1880 and 1894, the figures for the latest year being in some instances only one-third or one-fourth of those for the earlier years of the period.

In Denmark the disease was somewhat unusually prevalent in the years 1893–95, but was very much less so in 1896 and 1897. In the principal Danish towns the deaths in 1897 were equal to a ratio of about 40 per million inhabitants. In Norway a reduction in the deaths from the disease comparable with that already pointed out for other countries was very noticeable in the years 1881–1893. In the five-year period 1881–86 the annual recorded deaths were over 800; in the next quinquennium they fell to about 500, and in 1893 they had steadily fallen to only

195. In the following year, however, they rose again to 275, or approximately 135 per million living. In Sweden the recorded deaths from this disease steadily fell from 1658 in 1891 to 654 in 1896; the last figure was equal to a ratio of about 130 per million.

Scarlet fever, like most other infectious diseases, is much more prevalent in European Russia than in any other continental country. The mean death-rate from this cause in 1893–95 was no less than 1140 per million living. No part of the country is free from the disease; it is quite common in the northerly government of Archangel and in the southerly governments of Astrakhan and the Crimea. The lowest ratios of registered cases in recent years have usually been met with in the Polish governments.

In Austria the deaths from this zymotic in 1896 were equal to a ratio of about 360 per million; in the years 1880–86 the mean annual ratio had been 592, and in 1887–1893 it had been 561. These figures indicate a distinct reduction in recent years, though a much less striking one than in some other countries. In Hungary in the year 1897 the deaths from the disease were about 550 per million; the mortality had considerably declined between 1887 and 1894 in both Hungary with Fiume, and in Croatia with Slavonia, which together make up the Hungarian kingdom, and particularly in the latter, but the returns for 1897 showed a marked rise over those of 1894.

In Switzerland a steady fall in scarlet fever mortality has also been observed; in the 15 principal towns the deaths from this cause fell from 88 in 1889 to 14 in 1897; the last figure was equal to about 23 per million. In Italy the fall has been no less striking; the death-ratio per million having steadily declined from 496 in 1887 to 104 in 1896. For Spain and Portugal recent figures are lacking, but Davidson, in 1892, gave the ratio in towns in the peninsula at only 146 per million. For the Balkan countries trustworthy figures are also for the most part absent; but in Servia the recorded deaths from scarlet fever in 1894 were 1304, which was approximately at the rate of 565 per million. In some other countries in the Balkan peninsula the disease is also of frequent occurrence. In Constantinople it is more or less constantly present, but usually of a mild type. In Montenegro and Albania it is said to be seldom seen, and in

Greece, though often epidemic, it is reported to be perhaps less fatal than in the rest of Europe.

In Iceland the malady is not endemic but is occasionally introduced from without. In the Faröe Islands, though not unknown, it is very rare and usually of a mild type[1].

In the islands of the Mediterranean scarlet fever is met with, though apparently somewhat less commonly than on the continent. In Malta in 1898 only 12 cases of the disease were recorded, a figure much lower than that of most of the other zymotics.

Asia. The recorded cases of scarlet fever in the Caucasus and Transcaucasia have, of recent years, been very much lower in proportion to the population than in European Russia. In 1894 and 1895 the recorded deaths (probably below the truth) were at the rate of about 300 per million living. In Tiflis the disease is said to be almost confined to the Russian inhabitants. In Central Asiatic Russia the recorded cases are still lower, and have usually been scarcely one-tenth of even those in the Caucasus. This may be partly owing to imperfect registration, but the very great difference both in recorded cases and deaths in these regions must be in great part due to a really diminished prevalence of the disease in the Central Asiatic provinces. In 1895 the recorded deaths here were at the rate of only 10 per million inhabitants. In Siberia, on the other hand, the returns of cases have indicated a prevalence of scarlet fever about equal to that in the Caucasus, and therefore much lower than that in European Russia but higher than that in Central Asia. The figures for the remote province of Yakutsk are, perhaps, even more untrustworthy than others, but it may be noted that in recent years either no cases at all or only 2 or 3 per annum have been returned from that part of the country. The disease appears to be far from rare in the Amur province, in that of Tchita on the Mongolian frontier, and in most other Siberian provinces. Among the Bashkirs on the Kirghiz steppes it is said to be of infrequent occurrence.

In Arabia scarlatina is met with; among the pilgrims at Medina several cases are observed in every pilgrim season.

In Syria scarlet fever is said to be seldom seen: in Mesopotamia it is at times epidemic. In Persia it is probably common,

[1] Russell-Jeaffreson, *op. cit.*

in Teheran, at least, several cases occur from time to time. In Asia Minor it is perhaps not uncommon, for I have notes of its epidemic occurrence in Samsun in 1900; in Mitylene in 1901; and in Erzerum with some severity in December 1901, when owing to a series of earthquakes the people were living in a sad condition of overcrowding and mal-hygiene.

In India scarlet fever is an extremely rare disease. It occurs occasionally in the form of a few isolated cases; sometimes in persons newly arrived from England, who may be supposed to have brought the infection with them, but sometimes also in residents to whom no such explanation would apply. The few recorded cases have been mostly, but not entirely, in Europeans. They have been observed in Calcutta, Simla, Saharanpur, Bangalore, Rawal-Pindi, Darjiling, Assam, Cachar, Sitapur, Ranikhet, Hyderabad, Ceylon, and elsewhere[1]. In 1893 no cases were recorded in the British or native army in India; in 1894 there were 8 in the British, and 11 in the native army; in 1895 a single case; in 1896, eight cases in the British troops; and in 1897 as many as 27 in the British troops, but none in the native troops. In a large proportion of the isolated cases recorded in the various parts of India named the disorder could not be traced to any imported infection, either from other parts of India or from abroad. But in 1899, when as many as 83 cases of the disease occurred in the British army in India, it was believed that most of them could be traced to importation by the s.s. "Simla"[2]. In Kashmir the disease is quite unknown.

In Farther India the disease appears to be equally rare. Recent writers on the pathology of Siam make no mention of it. It was not seen by one observer[3] during an eight years' residence in the Straits Settlements (Penang and Singapore). Again, no accounts of it in Annam, Tongking, and Cochin China are forthcoming, and it is indeed said to be unknown there.

In China and Corea scarlatina, to judge by the rare mention of the disease in the reports of the Customs Medical Officers, is not a common disorder, though it is certainly not unknown. At Shanghai 9 or 10 known cases occurred among the foreign residents in 1890–91, and the remark is made that this disease

[1] Caddy and Cook, *Indian Medical Gazette*, August, 1899.

[2] *Report on Sanitary Measures in India*, 1899–1900.

[3] McClosky, *Journ. Trop. Med.* 1900, p. 156.

as well as diphtheria is now losing its former character of an interesting rarity[1]. In 1893 it was widely prevalent in Shanghai, though not regarded as epidemic[2]. Further south on the other hand it appears to be more rare. Cantlie, in a long period of residence, saw only one case in Hongkong, and that was in a patient who had landed from England two days before[3]. In 1898 four cases of the disease were seen by the Medical Officer of Health for Hongkong, two on H.M.S. "Barfleur" and two on H.M.S. "Powerful"; in all it was thought that the infection was imported from England. The disease can certainly exist in the tropics, and several cases have been recorded as occurring on ships in the Pacific and Indian Oceans.

In Japan scarlet fever is "excessively rare" (Davidson)[4]. On the more northerly island of Sakhalin, however, it is far from uncommon, and in 1895 as many as 86 cases with 17 deaths were recorded in the 26,421 inhabitants of this Russian convict-island.

In the Malay Islands the disease appears to be very rare or almost unknown. This is so in Sumatra; while in Java, Borneo, Celebes, and the Philippine Islands it is said to be quite unknown. In New Guinea it is not seen, at all events in the aborigines.

Australasia. In Australia this disorder is at times epidemic. It was so in a mild form in Melbourne in 1896 and 1897. In New South Wales it appears to have been very prevalent in 1897. In South Australia from 1 to 35 deaths were annually reported between 1889 and 1898, the highest figure occurring in the year 1894. In Western Australia it is either very rare or very mild, as with the exceptions of 1 fatal case in 1895 and 2 in 1897 no deaths from this cause were registered in the colony in the ten years 1889–98. The epidemic of 1897 noted in Victoria and New South Wales seems to have spread to Tasmania, where it caused 203 cases and 5 deaths in the year; in the preceding year it was very prevalent at Zeehan and Hobart. Six years before the disease had also been widely prevalent in the island.

[1] *Imperial Chinese Customs Medical Reports*, No. 41.
[2] *Ibid.* No. 46.
[3] *Journal of Tropical Medicine*, July 1899, p. 332.
[4] A writer in the *Sei-i-kwai* (October 1898) states that he saw 9 cases of the disease in Yokohama in 1877, and a few more cases at the end of 1880 and beginning of 1881, when "there were pretty numerous cases in the city."

In New Zealand the recorded deaths from scarlet fever were respectively 19, 31, and 24 in the three years 1889–91, but they have since not exceeded 6 in any year, at least up to 1898. On the whole the recent history of the disease in Australia and New Zealand has shown that it is a far less significant cause of death here than in most European countries. This is in part explained by the disease occurring in a milder form, but is also due to a much lower degree of prevalence.

In the Polynesian Islands scarlet fever is very rare or altogether absent. It is unknown in Fiji, and in the Marquesas Islands, and is only occasionally seen, mostly in the white population, and in mild form, in New Caledonia and the Society Islands.

Africa. The disease is not common in northern Africa. In Egypt it is said to be almost unknown. The infection is not infrequently imported into the country, but it has never spread to any extent, and the disease, when it does attack Egyptian children, is of a mild character. In Algeria it is almost equally rare. In Tunis also it is only exceptionally seen, and the average sickness-rate from this cause in the French army of occupation is only 0·57 per 1000 strength, as contrasted with a rate of 3·98 for the rest of the French army. Throughout the greater part of tropical Africa the malady appears to be unknown or very rare. Pruen, during a long residence in Eastern Central Africa, never succeeded in recognising it, though he was not prepared to say it did not exist there. Cook, during four years' residence in Uganda, never saw a case. It is not met with in Senegal, and no mention of it is to be found in (recent) health reports from Gambia, Lagos, or the Gold Coast. It is absent from the Cameroon district, and is not named by writers on the diseases of the Congo Free State, of Central Africa, or of British East Africa.

In Abyssinia the disease occurs at long intervals; but further south on the east coast it is said to be unknown. It is not seen at Zanzibar.

In South Africa scarlet fever is somewhat more common, though still a rare disease. It is but little mentioned in recent colonial reports, but occasional cases occurred in scattered parts of Cape Colony in 1898, and in 1891 it was epidemic at Hope Town. It is also rare in Natal and in Bechuanaland, and it is doubtful whether it exists in Mashonaland.

In Madagascar and the Seychelles it is unknown. In Mauritius it is seen occasionally, but never spreads.

North and Central America. Scarlet fever is said to be unknown in Greenland or in the northern portions of the Hudson Bay territory, but in Alaska it has appeared on at least one occasion[1]. In the more temperate portions of Canada it is not uncommon. In Quebec it was epidemic in 1889 and more severely so in 1893. The latter epidemic spread all over the province from the west, where it was introduced from the United States. In Ontario it is occasionally epidemic; it was unusually prevalent in Toronto early in 1897, and in that year there was a serious outbreak of the disease in a new settlement of Galician immigrants in Manitoba, where the malady often prevails. In New Brunswick it was epidemic in several places in 1895, and in Nova Scotia and British Columbia it occurs in epidemic form from time to time. The disease is also not unknown in Newfoundland.

In the United States scarlatina is not infrequently observed in the north, but is much less common in the south. It appeared among the causes of death in every State in 1897 with few exceptions; but in Arizona, Arkansas, Florida, Idaho, Mississippi, Montana, Tennessee and North Carolina no deaths from it were recorded in the urban population in that year, while in each of the States of Louisiana, New Mexico, and Virginia only a single death was recorded.

The decrease in the mortality in this disorder which has been so very striking in many European countries in recent years appears to have been also observed in the United States, at least in some of them. In Massachusetts the deaths per million living in the four quinquennial periods contained between the years 1871 and 1890 were respectively 860, 410, 270 and 170 per annum. There was a slight rise in the years 1891–94, when the mean ratio was 250 per million. In Michigan the decrease has only been observed

---

[1] It was severely epidemic in the Yukon district of Alaska in 1862. Dall, who travelled there in 1867, wrote that he "passed by several deserted houses formerly inhabited by some Indians of the Kutchin tribes, who all died five years ago of scarlet fever. This fever was introduced by a trading vessel at the mouth of the Chilkaht River. From the Chilkaht Indians it spread to those of the Upper Yukon and down the river to this point, where all died and the disease spent itself." *The Yukon Territory*, London, 1898.

within more recent years, the most marked decline occurring between the years 1891 and 1894, when the mortality fell from 202 to 51 per million living. Whether a similar fall has been observed in other States I am not in a position to say.

In Mexico scarlet fever is said to be very rare in the warm plains on the coasts, but perhaps more frequent on the cooler central plateau. In Central America generally the disease is probably exceedingly rare. In British Honduras 3 cases resembling scarlatina were observed at Belize in 1898, but they were regarded as very exceptional and no clue could be found to their origin.

The West Indian Islands appear to enjoy a considerable degree of immunity from scarlet fever. Information on this point is imperfect but the disease is said to be unknown in many of the islands and very rare in others. There is some doubt about its existence in Cuba; in Jamaica a single case occurred at Port Royal in January 1898, and the disease was reported from the Western St Andrew district in April of the same year, but in neither instance was there any spread[1]. It is said to be known in Trinidad, but during 1897–1899 it has not been mentioned in the Annual Reports on the health of that island, nor in those of St Vincent, St Lucia, Barbados, the Bahamas, or the Leeward Islands. In the Bermudas 3 cases were notified in 1898. In Puerto Rico the malady is unknown.

South America. The existence of scarlet fever along the northern shores of this continent is doubtful. I can find no mention of the disease in several recent reports of the Surgeon-General for British Guiana. Concerning Dutch and French Guiana, Venezuela, and Colombia information appears to be lacking. It is known in Brazil, and Davidson wrote in 1892 that "scarlet fever, which was rare during the first three decades of this century, has of late years been frequently epidemic, and has made many victims." Whether these later epidemics reached the northern, equatorial regions of this vast country, or were confined to parts further from the equator is not stated, but the earlier outbreaks were apparently limited to the southern and central provinces —Rio Grande, Santa Catarina, São Paulo, Minas Geraes, and to Rio de Janeiro[2].

[1] A severe epidemic of scarlet fever spread over the whole island of Jamaica in 1841. [2] Brunel, quoted by Hirsch.

In Uruguay and Paraguay the disease is said to become frequently epidemic, and often to assume a very severe type; in the former Republic it was the cause of only two recorded deaths in 1897. In the Argentine Republic the disease became epidemic in 1895 and 1896, and caused in the province of Buenos Aires 368 and 375 deaths respectively in the two years. In 1897 the deaths from this cause were only 123, and of these 46 occurred in the northern region, 70 in the central, 7 in the southern and none in the Patagonian region of the province.

For the western side of the South American continent information is scanty. In Chile the disease is said to have been frequently epidemic since 1829, the year of its first appearance. Outbreaks have also been observed in Peru.

Factors concerned in the Geographical Distribution. Scarlet fever is essentially a disease of temperate climates. In the tropics it is almost unknown—perhaps quite unknown as an indigenous disease. In India it is seen very rarely, and of the small number of cases there recorded a considerable proportion have been clearly due to importation. In Farther India it is at least equally rare, or—so far as the absence of positive evidence would indicate— perhaps unknown, and the same is true of the Malay Islands, New Guinea, Southern China, and Japan. In tropical Africa, the tropical portions of North America, Central America, the West Indies, and the tropical areas of South America, scarlet fever is either never met with or is excessively rare.

In sub-tropical countries the disease begins to be rather more common, and there appears to be an almost regular gradation in its degree of prevalence as the distance from the equator increases, until in countries with a temperate or even moderately cool climate it becomes quite common. This gradation can be easily followed in many parts of the world, both in the northern and southern hemispheres. For example along the northern shores of Africa the disease is but little seen; in the southern European countries it is probably somewhat commoner, but still less common than in the countries of Central and Northern Europe. In like manner, while it is probably unknown in tropical Africa, it is met with, but still comparatively rarely, in South Africa, which is, of course, a sub-tropical region, bearing about the same relation to the equator as Algeria or Tunis in the north. In China it would seem to be more common in the north than in the south; it is very rare in

Japan, and comparatively common in Sakhalin, further north. We have seen how in Central Asiatic Russia it causes a very low mortality, a much higher one in the Caucasus, and an extremely high one in European Russia. Similarly, in the western hemisphere, in sub-tropical Mexico it is rare, while in the southern United States it is more prevalent, and in the northern States and Canada it is quite a common disease.

Extreme cold, on the other hand, seems to be almost as inimical to scarlet fever prevalence as extreme heat. In Iceland the disease, as already stated, is an exotic and in the Faröes it is very rare. In Greenland and northern Canada it is said to be unknown, though it has been epidemic at least once in Alaska. In the north of Siberia it is also perhaps non-existent. It must, however, be borne in mind that these remote northerly regions are perhaps too sparsely inhabited to keep the disease going continuously on their own account. Scarlet fever is mainly a disease of childhood and seems to require a large susceptible population to remain permanently endemic in any area. Like measles, as already pointed out, it can occur and cause very severe epidemics in such northerly and inhospitable climes as that of the Yukon territory. But, like measles also, it appears to die out in these regions after exhausting all the available material, and a re-importation of infection and a fresh generation of susceptible persons are then needed before the disease can recur.

While distance from the equator on the one hand and the poles on the other is undoubtedly one of the most important influences in determining the prevalence of scarlet fever, other influences are also at work. Too little is as yet known with certainty as to the seasonal and meteorological relations of the disease to state that these conditions are of any importance, or if so, of how much importance. The seasonal variations of scarlet fever vary widely in different parts of the world. In such places as New York, Michigan State, and England, in which the seasons of the year do not differ very vastly, the period of maximum prevalence of scarlet fever falls in a totally different month in each, being in April in the first, in January in the second, and October in the third.

The influence of race is also uncertain. People so ethnologically distinct as the Chinese, the natives of South Africa, and the inhabitants of the principal European countries all suffer considerably from the disease; and it has been observed both in the aborigines of Australia and in the North American Indians. But it is certain that some races are more susceptible to the infection than others. The statistics of recent censuses in the United States of America tend to show that the disease is less prevalent and less fatal among the negroes and Red Indians than among the whites. It has already been pointed out that of the few cases of scarlet fever observed in India, almost all have occurred among Europeans, and a very small number indeed in natives of the country. The difficulty, however, of recognising the disease in persons with a dark-coloured skin must be taken into account, and it is certain that the natives of India are not entirely immune to the infection.

The effect of improved sanitary conditions and especially of increased care in isolating patients in hospitals has been already alluded to. These measures have certainly seemed to be of value in England in lessening the mortality of this disease and perhaps its prevalence also, and it is probable that they have had the same effect in other countries. But scarlet fever appears to have varying periods of increased or diminished prevalence independently of these conditions, and the recent reduction of mortality from the disease in England and elsewhere cannot wholly be ascribed to improved sanitation or to the increased use of isolation-hospitals, or to the other conditions named earlier in this chapter as possible explanations.

This great reduction in mortality from scarlet fever within recent years in a very large number of countries is one of the most remarkable phenomena in connection with the disease. It has been observed to a striking extent in the British Isles, in Belgium, Germany, Denmark, Norway, Sweden, Switzerland, Italy, and the States of Michigan and Massachusetts; and to a less extent in Austria and Hungary. In Russia, on the other hand, there has been no decrease of the malady, at least between 1887 and 1895. In the countries named the decrease, it must be pointed out, has extended over a very varying number of years. In many the material at my disposal only covers a comparatively short series of years, while in others it goes back for half a century or more. The two sets of figures are not strictly comparable, and a reduction observed in the course of five or ten years may be a merely temporary phenomenon of no great significance, as the disease has admittedly a tendency to wax and wane for periods of years. In England and Wales the fall was steady from 1860 to 1900 (the date of the most recent report); in Belgium from 1870–1897; in Norway from 1881–93; in Massachusetts from 1871 to 1890 (with a slight rise in 1891–4). Elsewhere the periods of observation were much shorter.

Scarlet fever is a disease with a high degree of infectivity, and is hence the cause of somewhat frequent epidemics in schools and similar institutions. Epidemics of a wider character, over whole communities and even over many countries, have also been recorded. In all countries, as just stated, there are periodic rises and falls in scarlet fever prevalence, but they show no tendency to regularity.

There is little doubt that the infection of this fever can be retained for long periods and carried considerable distances in fomites. It is doubtful if it can be carried to any extent by the air, nor is it believed to be transmissible by water. Milk on the other hand is perhaps capable of containing the infective material and giving the disease to susceptible persons. Power and Klein investigated an outbreak of scarlet fever at Hendon in 1885 and formed the opinion that the infection in that case was to be traced to a certain milk supply. The cows from which the milk was taken suffered from a vesicular affection of the udder, and a streptococcus apparently identical with that of scarlet fever was obtained, both from the cows, from the milk, and from the scarlet fever patients. The exact nature of the " Hendon disease " in the cows still remains somewhat doubtful, but the possibility of the infection of scarlet fever being carried by milk is quite strong enough to justify all reasonable precautions to prevent its occurrence.

SCURVY.

General Characters and Etiology. Scurvy is mainly characterized by great general debility, spongy swelling and bleeding of the gums, subcutaneous and deeper hæmorrhages in other parts of the body, and occasionally severe joint and other affections.

Many views have been held as to the causation of scurvy, but most are agreed in regarding it as a disease of mal-nutrition. Prolonged deprivation of fresh vegetable and animal food, particularly if combined with exposure to cold, fatigue, want, and other depressing conditions, has been the most usual precursor of an outbreak of scurvy. Hence it has been most commonly seen among the crews of imperfectly-equipped Arctic or Antarctic expeditions ; among shipwrecked sailors ; in times of famine and of war ; and among semi-civilised races whose dietary has been at fault in quantity or in quality. It is not infrequently seen in children whose regimen is faulty. An excess of salted or preserved articles of food, or the consumption of flour, meat, or other food in a state of commencing decomposition through being improperly kept, have been thought by many to be contributory, if not the main causes of the development of the scorbutic condition. It must be added, however, that the scorbutic condition, or one closely resembling it, has at times developed under circumstances where none of the causes just named were present, and hence some observers are of opinion that the real cause of scurvy yet remains to be discovered. Some have suggested that the disease is of microbial origin, but no specific micro-organism has been found with certainty in association with the set of symptoms grouped under the name of scurvy. Frerichs, Frein, and others claim to have found such organisms, but their exact relation to the disease is doubtful.

History. It is probable that scurvy has existed from time immemorial, but the references to it by early writers are scanty and vague. Hippocrates, Celsus, Avicena, and other Græco-Roman and Arabian physicians perhaps mention it. It seems probable that the Crusaders, just as certain European armies in recent wars, suffered from it. Vasco da Gama and other early explorers, particularly those who attempted to discover the North-West passage, had to contend with its ravages. For centuries no long voyage, whether to the Arctic regions or to any other part of the world, was undertaken without the certainty that a large proportion of the crew would fall victims to scurvy. It is difficult at the present day to realise the terrible tribute in lives that naval ships and merchantmen had to pay to the disease. It is only by studying the actual records of the time, such as the narratives of Drake's, Hawkins', or Anson's voyages, that any idea of its ravages can be gained. It was no unusual thing for one-half, two-thirds, four-fifths, or even a still higher proportion of the crew to die of scurvy in the course of a single voyage.

The classical works of Dr James Lind, physician to Haslar Hospital, first drew attention to this appalling and preventable loss of life on board ship, and from his time the disease has largely lost its terrors. Throughout the last century scurvy at sea made its worst ravages in the Arctic expeditions; and even so recently as 1875–6, and in so expensively-equipped an expedition as that of the "Alert" and the "Discovery," it was one of the worst ills with which the crews had to contend. Nansen's great expedition, on the other hand, has shown that the malady is not a necessary accompaniment of even a long stay in Arctic regions.

On land a large number of epidemics of scurvy or of an allied disease have been recorded as occurring in many countries in all parts of the world. A complete chronological record of 143 epidemics is given by Hirsch, and of these no less than 35 occurred in Russia alone.

Geographical Distribution in the Present Day. *Europe.* Scurvy, at one time a formidable disease in our prisons, is now rarely seen in the British Isles. Occasional cases are admitted to the Seamen's Hospital at Greenwich, but these are imported, and an indigenous case is probably unknown. In Ireland scurvy was severely prevalent at the time of the great potato famine of 1846–7. In France and Germany the condition is now very rarely met with ; and the reasons for its disappearance from these countries are the same as for its disappearance from the British Isles—namely, a vast improvement in the dietary and general hygiene on board ships, in the army, and in prisons. In Denmark cases of the disease are occasionally seen, and each year from one to six deaths are returned as attributable to this cause among the Danish urban population. In the northern portions of Sweden and Norway scurvy is said to still exist as an endemic disorder, attacking not only the Lapps but also the other inhabitants. It is also found to a considerable extent in Iceland.

But of all European countries Russia is the most deeply affected by this disease. In the past a larger number of epidemics have occurred here than in any other country, and at the present day Russia is still one of the most important homes of the malady. It is always present to a greater or less extent in all parts of European Russia, and every year a considerable number of deaths ascribed to it are returned from each of the "governments" into which the country is divided. Those returning the smallest proportion of cases and deaths are the Polish governments and the Baltic provinces, while that returning the highest is the government of Archangel. The city of Archangel and the northern portion generally of European Russia have always had an unenviable reputation for the prevalence of scurvy, and they still maintain it. The disease is endemic on the shores of the Kola Peninsula, north of the White Sea, including parts of Lapland and the Murman coast, and at times becomes epidemic there[1]. But while it is most prevalent in the northern portions of Russia, it may be questioned whether it is not almost equally so in several other parts of the country. In St Petersburg and its neighbourhood the scorbutic condition is certainly far from rare. The returns from the government of Voronezh, on the Volga, also indicate an exceptional frequency of the disease, and many other governments—on the Volga, in the Ukraine, in the far north-east—return high figures of scurvy mortality.

In the Crimea scurvy is not unknown at the present day, and its ravages at the time of the Crimean War, both among the Russian and the allied troops, have become historical. In the early months of the war the French were believed to have lost more men from this disease than from the guns of the enemy.

---

[1] Sivré, *Kazan University Med. Society*, Oct. 1901.

In some parts of Austria-Hungary scurvy is known, and has at times become epidemic. In Italy it is less rare than in some other countries, and is the cause of an annual death-rate which varied in the years 1887 to 1896 (inclusive) between 9 and 18 per million inhabitants living. Scurvy in Italy is still a disease of fortresses and prisons, and is especially met with in Venetia and along the valley of the Po. It is said to be endemic also in Roumania. In European Turkey it is now very rarely seen, only an occasional case being met with from time to time. In Greece, too, it is an exceptional malady.

Asia. The official returns of scurvy from the Caucasus proper and from mountainous Transcaucasia indicate that the disease is far from rare in all parts of both regions. It is met with also in most parts of Russian Central Asia and Siberia, and it is interesting to note that in many parts of the latter, notably in the Amur province, it is the well-fed upper classes and the Russian soldiers and peasant-immigrants who suffer most, while the native tribes, such as the Samoyeds, the Giliaks, the Golds, Ainus, and nomadic Tunguses suffer little[1]. The fact is a remarkable one, as these tribes are mainly flesh-eaters, though also eating bilberries and other wild berries, while those who suffer from scurvy are well fed and eat a large variety of vegetables of all kinds.

Among the Bashkirs, about the Ural River, scurvy is very rarely seen in ordinary years, but after the famine of 1891–92 they suffered greatly from it. In Asia Minor and Syria the disease is rarely met with, and there is little evidence to show that it is at all common in Arabia. On the Yemen coast, however, it is said to be endemic, and at Aden a scorbutic taint is believed to account for a large number of the cases of bowel complaint which come under treatment[2].

In India the malady has become comparatively rare among the European troops stationed there. At one time it was a scourge in Indian prisons and caused a vast amount of mortality, but improved dieting and hygiene have largely reduced its ravages.

---

[1] Zeland, *The Russian Army-Medical Journal* (*Voënno-meditzinskii Journal*), 1882, pp. 217–18. The explanation advanced by this author is that the Russian soldiers and immigrants live in houses with damp and rotting wooden floors, while the floors of the native huts are of damp-proof clay, and are constantly dried and "disinfected" by the smoke from their fires.

[2] *Report of Sanitary Commr. with the Govt. of India*, 1897.

It still, however, occurs in gaols in India, where it is especially found associated with a malarial taint, but it is always an imported disease and never developes among the prisoners in consequence of the prison régime[1]. In the native population it is said to prevail extensively at certain times, and in the north-western portion of the Indo-Gangetic valley—in Sindh and Marwar—it is said to be endemic. At times it has become so widely prevalent as to deserve to be called epidemic. In Rajputana it is rarely seen in years of plenty[2]. In Ceylon the disorder seems to be rare; an occasional death from it is, however, from time to time reported. In Siam it was formerly said to be entirely unknown, but Rasch reports having seen two characteristic cases—not in inhabitants of Bangkok itself but in persons from the island of Kosichang in the Gulf of Siam[3].

In Cochin China scurvy is, or was, not uncommon; and in China itself it was reported some decades ago (in 1870) to have been rather frequently seen among the poor in the northern portion of that country, and particularly in Peking[4]. Recent writers in the Medical Reports of the Chinese Maritime Customs make no mention of the presence or absence of scurvy in China. The evidence for its existence in Japan is rather conflicting. Sollaud[5], writing in 1882, spoke of the disease as common among the poorer classes in that country, but Iskerski, a Russian writer, stated in a paper published in 1899 that scurvy was unknown in Japan, either in the fleet, the army, or the general population[6].

Australasia. It must be doubted whether scurvy exists to any extent in Australia or the Pacific Islands generally. It is true that in Australia itself severe epidemics of the disease formerly attacked exploring parties who penetrated into the interior of the island-continent, and at one time it was spoken of as endemic among the shepherds on the Darling Downs, in the north of New South Wales. But at the present day the disease is probably a rarity in Australia. In recent Health Reports from Queensland and New South Wales I find no mention of it. In South Australia not a single death attributable to it occurred between the years 1889–1898 (both inclusive), and it is only in the Health Reports

[1] Col. K. McLeod, M.D., LL.D., *Journ. Trop. Med.* Sept. 1898.
[2] Adams, *op. cit.*   [3] *Janus*, 1896–7, p. 445.
[4] Morache, quoted by Hirsch.   [5] Quoted by Hirsch.
[6] Botkin's Hospital Gazette (*Bolnitchnaia Gazeta Botkina*), August, 1899.

of West Australia that a small number of deaths from scurvy are annually reported.

In Tasmania and New Zealand the disease is probably equally rare, if not unknown : and I find no mention of its occurrence in any of the Pacific Islands.

Africa. The negro races of the African continent appear to possess a high degree of susceptibility to scurvy. In many parts of the continent the disease is common. Filippo Rho, in describing the pathology of Massowa and its neighbourhood, states that scurvy rages everywhere among the natives, though few of the Italian troops stationed there were attacked[1]. In Central Africa also the scorbutic condition is very frequently met with among the native inhabitants[2]. In the neighbourhood of the great African lakes the disease is perhaps less common. Pruen, at least, who spent several years in that part of the continent, never saw a case of it; but adds at the same time that he is not sure that it does not exist there. Scurvy has at times been epidemic in Egypt, and was, and perhaps is still, commonly seen in Abyssinia among foreigners or among slaves, but not in the native population.

On the northern shores of Africa the only reference to the existence of scurvy is the record of some epidemics of the disease in Algeria in the past. In recent reports from Algiers I have come across no mention of the disease. On the West Coast of Africa it was at one time very frequently seen among the natives, but is probably less common at the present day.

In South Africa scurvy is a rather prominent disease. The workers in the gold-fields are said to suffer particularly, and shortly before the outbreak of the war attention was called to the need of altering and improving the dietary of the miners on this account. At the Frontier Hospital, Queenstown (Cape Colony), several cases are treated every year. At the Butterworth Hospital also scurvy accounts for a large percentage of the cases annually admitted for treatment. At Barkly West the disease was spoken of as "prevalent" in 1897, and it seems to have given rise to a small epidemic there. A few isolated cases were also reported from Engcobo in the Native Territories; the patients were natives who had returned from working at Capetown and Johannesburg,

---

[1] *Janus*, 1896–7, p. 598. [2] *Ibid.* 1896–7, p. 389.

and the disease did not spread to others. During the war the amount of scurvy among the South African miners is said to have increased considerably, owing to the supply of fresh vegetables and of lime-juice running short. The disease prevailed to some extent in the garrison of Kimberley during the siege of that town in the winter of 1899–1900.

Occasional cases of scurvy—sometimes, possibly always, imported—are recorded from Mauritius and from St Helena.

America. The extreme north of the North American continent and particularly the shores of Hudson Bay have always had an evil reputation for the prevalence of the malady. It must be remembered, however, that scurvy has always been associated in the past with the peculiar conditions found in connection with Arctic expeditions, and that by far the larger number of the latter have been those in search of the North-West passage, and therefore in this very part of the North American continent. The natives of these regions are said to enjoy an almost complete immunity from scurvy, and it is only under the conditions of absence of fresh vegetable food, together with fatigue and privation, such as explorers, and formerly the lumbermen of northern Canada were exposed to, that scurvy has been found to prevail in these countries.

In the United States scurvy is now a rare disorder, but cases are sometimes seen among sailors in the seaport towns, and in the mining districts of Pennsylvania the Hungarian, Bohemian, and Italian settlers are not infrequently attacked (Osler). Scurvy in infants, caused by improper dietary, is said to be increasing in the States. In the past it caused a considerable amount of sickness among the troops taking part in the American Civil War, and also in the Mexican War; and at the time of the famous "gold rush" in California not a few of the immigrants suffered from it. It is probably rare at the present day in the West Indies; cases of it are however occasionally mentioned in recent reports from some of the Leeward group.

The evidence for or against the existence of scurvy in South America is excessively scanty. Four cases of the disease were mentioned in the Health Reports for British Guiana for the year 1898; but with one other exception this is the only mention I have come across of this disease in the South American continent. The other exception is a reference by Hirsch to a report—as far back as the year 1844—of imported cases of scurvy in certain negroes brought from Africa to the Brazils.

Outbreaks at Sea. While outbreaks of scurvy at sea were at one time the rule in all long voyages they are now quite exceptional. There can be no question that this is due to the compulsory provision of lime-juice or other anti-scorbutics on board all ships, to the shorter duration of voyages, and to the greater ease with which fresh vegetable and animal food can be obtained and preserved—even throughout the longest voyage. In the absence of fresh food and of anti-scorbutics, however, an outbreak of scurvy occasionally takes place at sea even in the present day. Thus in 1891 a Dutch ship arrived at Mauritius from Akyab with a number of cases of scurvy on board ; many of the patients were in the last stages of the disease when the ship arrived, and many died[1].

Factors determining the Geographical Distribution. Scurvy is a disease which, for obvious reasons, has no permanent geographical distribution. Dependent as it probably is upon a faulty dietary, it is found only under the special and usually temporary conditions where the particular error in diet prevails, and disappears when the error is corrected. But these conditions appear to be more or less permanently existent among certain communities, as for example among the Russian peasantry and Laplanders, in certain gaols and barracks in India and elsewhere, and among the natives of many parts of Africa.

It will be observed that scurvy is not dependent upon temperature for its development. It has been found in the tropics, in temperate zones, and near the poles. Some writers have believed that cold and exposure to wet are strong predisposing factors in the production of scurvy, but it would seem that they are so only in the sense that they weaken the resisting powers of a person exposed to them, and make him more liable to develope scurvy when the other special conditions needed to produce the disease are present ; and the same statement appears to apply equally well to prolonged exposure to a warm, relaxing climate, such as is met with in the tropics.

The soil seems to have little, if any, influence on the prevalence of scurvy on land. Overcrowding, insanitation, and all that they

---

[1] Mauritius. *Report of the Chief Med. Officer on the Med. Depart. for the year* 1891.

imply, are probably favouring conditions. But there is a very general consensus of opinion that scurvy is essentially due to a defective dietary. It would be out of place here to enter into a discussion of the medical and physiological problems involved in the question of the true etiology of scurvy, and particularly as to the exact chemical constituent of diet the lack of which produces the scorbutic state. It will suffice to say that the evidence is practically conclusive that in most, if not all "outbreaks" of scurvy, an absence of fresh vegetable food or some substitute for it, as lime-juice or fresh meat, has been the main cause of the disease. Other conditions, such as those of climate, soil, or hygiene, of general inanition, of the consumption of decomposed food or of salted meat or fish—all of which have been urged at different times as the real cause of scurvy—are probably to be regarded as only contributing to the production of the disease[1].

It will be seen, therefore, that the scorbutic state is one which can be prevented from developing by proper attention to diet. The disease, if it may be truly called a "disease," is, in fact, a preventable one. Its complete disappearance from the earth's surface is only a question of the right regulation, at all times and in all places, of the dietary. In getting rid of it permanently there would be no question here, as in the case of "infectious" diseases, of getting rid of a specific virus—of destroying a prevalent and dangerous micro-organism. Some authors, as we have seen, have hazarded the opinion that scurvy is due to a micro-organism and that it does belong to the group of infectious diseases, but there is little to give colour to the suggestion. It is not likely, then, that it will ever permanently disappear. Errors of diet, and particularly lack of fresh vegetables and meat, and the consumption of badly kept and tainted food, are conditions which must from time to time recur here and there, and whenever they do recur, it is more than probable that they will be accompanied or followed by the development of the scorbutic state in the individual, and, if many are so affected at one time, by the occurrence of a so-called "outbreak" of the malady.

---

[1] Jackson and Harley, however, as the result of experiments on monkeys, believe that the main factor in the production of scurvy is not the absence of fresh vegetables or lime-juice from the dietary, but that the malady is mainly due to improperly preserved food, and particularly to tainted meat. (*Proc. Royal Soc.*, March, 1900.)

SLEEPING SICKNESS, or
NEGRO LETHARGY.

General Characters and Etiology. Sleeping sickness or negro lethargy is a remarkable and extremely fatal disease, which occurs as an endemic, and sometimes as an epidemic, in certain limited geographical areas. Its principal feature is an increasing tendency to sleepiness, until at last the patient becomes completely comatose, and after a longer or shorter period dies. The premonitory symptoms—such as a peculiar expression, a puffiness of the face, occasional attacks of headache, vertigo, fever, or diarrhœa—are usually well-marked, and during the later stages a number of severe nervous symptoms, such as spasms, paralysis, maniacal outbursts, etc., may develope. The disease may last for a few months or even for as long as two or three years, and it appears to be invariably fatal.

Innumerable theories have been advanced to account for the causation of sleeping sickness. Many of these have been abandoned, and now possess a historical interest only. Thus the suggestions that the disease is caused by the smoking of Indian hemp, by the excessive drinking of palm wine, or by criminal poisoning ; that it is the result of "depressing emotions, particularly those associated with the slave-trade"; that it is due to the presence of enlarged scrofulous glands in the neck pressing upon the cervical vessels and so lessening the blood supply to the brain ; or that it is the outcome of a cerebral sclerosis of scrofulous origin—have all been shown to be untenable. Recently Scheube has suggested that negro lethargy may be an intoxication analogous to ergotism, and caused by the parasites of maize, rice and other cereals. Still more recently Manson has pointed out the possibility that the disease may be in some way due to the

filaria perstans.   It is certain that the geographical distribution
of this filaria and of sleeping sickness very closely correspond,
and as the filaria has been found in several cases of this disease
the theory that they are causally related to each other finds much
to support it[1].   In 1901 a scientific committee, sent by the
Portuguese Ministry of Marine to study this disease, came to the
conclusion that it is due to a micro-organism to which they gave
the name of *diplo-streptococcus*.   In 1902 a Commission was sent
to Uganda for the same purpose under the auspices of the Royal
Society, the Foreign Office, and the London School of Tropical
Medicine, and it has lately been stated that the members of it
have been successful in discovering the cause of the disease.

History.   The earliest description of negro lethargy dates
from the year 1800, when Winterbottom, a surgeon in the British
service, called attention to its occurrence in natives on the shores
of the Bight of Benin.   Later, in 1840, Clarke published a more
detailed account of it as observed in Sierra Leone.   Since then a
very large number of observations on the disease and its causation
have been published.   Nothing, however, is known as to the
history of the disease itself before the nineteenth century.   During
that century and until quite recently it has shown a very constant
distribution, being strictly confined, as an indigenous disorder, to
the tropical portions of Western Africa, and when appearing
elsewhere—as in the West Indies, Brazil, and possibly British
Guiana—doing so only as an importation from Africa, and mainly,
if not solely, in negroes from the African coast.   With the
disappearance of the traffic in slaves between the eastern and
the western hemispheres sleeping sickness has almost, if not quite,
disappeared from the latter.   But in Africa it has recently become
much more prevalent than formerly and has spread widely to the
east and south, though still remaining a tropical disorder.

Recent Geographical Distribution.   Sleeping sickness,
so far as is known at present, is met with as an indigenous disease
principally along the West Coast of Africa.   Its northern boundary
appears to be the Senegal river, while to the south it is limited by
the southern border of the Portuguese colonies.   It is particularly
common in the French colony of Senegambia, and in some of
the ports here it has at times been so severe that the extraordinary

phenomenon has been seen of a negro garrison being replaced by white troops (French marine infantry) on account of the losses caused among the former by its ravages.   This has happened at Portudal and Joal, ports facing the island of Goree[1].

The disease is also common along the shores of the Gulf of Guinea.   In Liberia, at Grand Bassam on the Ivory Coast, on the Pepper Coast, along the Gold Coast, and at Old Calabar in the Bight of Biafra it is frequently seen[2].   On Fernando Po and other islands in the Bight itself cases are met with.   In Principe Island it was "raging" in 1901, and had been very active for the previous fourteen years[3].   Along the valley of the Congo it has long been known as an endemic disorder.   Centres of the disease are irregularly scattered through the Congo Free State[4], and Sims has recorded cases of it at Stanley Falls, in the very heart of equatorial Africa.   Further east, in Uganda, sleeping sickness has now established itself as a very serious endemic disease.   During the past year or two it has become extraordinarily prevalent there.   Thus writing early in 1901 on the diseases of Uganda, Dr A. R. Cook merely stated that "sleeping sickness is said to be common in Busoga to the east of Uganda," but that he had only had two cases in the hospital at Mengo[5].   A little later Dr J. H. Cook reported an extension of the disease in Uganda[6], and in December, 1901, the latter observer wrote that "in the districts of Kyagwe, in Busoga, and latterly round Mengo, it is literally slaying hundreds[7]."   In the Mengo dispensary he was seeing from four to six new cases every week.   It had spread to the islands on Lake Victoria Nyanza, and on the island of Burruna alone had caused the death of over 200 natives.

To the south sleeping sickness is found with considerable frequency in the Portuguese colony of Angola, in certain parts of which it is so prevalent as to be spoken of as a serious bar to negro colonisation.   It has been particularly active recently in the northern districts, where whole villages are said to have been depopulated by it.

---

[1] Corré, and Brault.   [2] Brault.

[3] Report of the Portuguese Scientific Committee.   Ref. in *Journ. Trop. Med.*, May 15, 1902.

[4] *Congrès d'Hygiène et de Climat. Méd. de la Belgique et du Congo*, 1897.

[5] *Journ. Trop. Med.*, June 1, 1901.   [6] *Ibid.* July 15, 1901.

[7] *Ibid.* Feb. 15, 1902.

While sleeping sickness is only found as an indigenous disorder in the tropical portions of Africa, secondary centres of the disease have from time to time developed in other regions of the world, though it is almost certain that these secondary centres have always owed their origin to importation from Africa. The malady was formerly far from rare among negroes transported from Africa to the West Indies, the Bahamas, and Brazil. Guérin about the middle of last century saw as many as 148 cases of this kind during twelve years' practice in Martinique ; and Nicolas believed that one per cent. of the deaths among negro immigrants on the voyage from Africa to the West Indies were due to sleeping sickness. But at the present day the disease is said to be unknown among the black population of the West Indies or of Brazil. In the Guianas, however, there is some evidence pointing to the possibility that it is now endemic. Ferguson reports that he has seen well-marked examples of the disease in British Guiana[1]. His patients were not all negroes, and from the fact that the *ankylostomum duodenale* was present in them, he came to the conclusion that negro lethargy was nothing else than ankylostomiasis, the symptoms being modified by individual or racial peculiarities. This opinion is, however, combated by Ozzard, who states that he never saw a case of the sickness in British Guiana, and believes that Ferguson must have mistaken advanced cases of ankylostomiasis for the sleeping sickness[2].

Factors concerned in the Geographical Distribution.
Negro lethargy is entirely a tropical disorder. As an endemic it is strictly limited to the continent of Africa, between the 16th degree of north latitude and the 18th degree of south latitude. It is not confined to the coast nor to the western half of the continent, but is seen in the very heart of the Congo Free State and in Uganda—countries which have no coast line.

In its area of endemic prevalence the disease is very irregularly spread both in regard to space and time. It occurs in the form mostly of scattered centres, consisting of single villages or groups of villages. These centres are not, however, permanent, and the disease would seem to migrate from one area to another. Having destroyed half the population, perhaps, of one village, it passes on to another village or group of villages that had been hitherto

[1] *Brit. Med. Journ.*, 1899, Vol. I. p. 315.
[2] *Ibid.* p. 964.

free from it. It is sometimes the cause of the most deadly epidemics, and whole villages have been known to be swept away by it. It is rightly dreaded by the negro inhabitants of those regions where it may occur, for no certain case of recovery from an attack of the sickness has apparently been recorded.

The relations of the disease to race are unusually well-marked. It is almost wholly confined to the African negro. In its endemic home in tropical Africa it is practically the native black alone that suffers from sleeping sickness. When the disease was formerly seen in the West Indies it was almost solely among negroes imported from Africa, and it never attacked those born on the plantations, or who had become domiciled on American soil. Rare cases have been seen in half-castes and in Moors, and Corré has mentioned a doubtful case in a white man. In British Guiana Ferguson believed he had seen the malady in persons other than negroes, but for the present considerable doubt must attach to his cases. It is noteworthy that, in those areas of Western Africa where negro lethargy is endemic, negroes coming from other districts are particularly liable to it, the local inhabitants appearing to become in a sense acclimatised[1].

[1] Corré, *Arch. de Méd. Navale*, 1877.

SMALL-POX.

General Characters and Etiology. The characters of an attack of small-pox do not require description here. The cause of the disease is unknown. It presents all the features of an infective fever of microbial origin, but no micro-organism has yet been found that can with certainty be regarded as specific to it. Klein and Copeman have, however, described a minute bacillus, occurring in small-pox pustules, which may perhaps prove to be the organism sought for.

History. Small-pox seems to have been known in India and China from time immemorial. In both countries the disease has been placed under the special protection or patronage of a goddess, to whom temples were dedicated and in whose honour sacrifices were performed. Ebers believes that he has found a reference to the existence of small-pox in ancient Egypt (about 3730 B.C.) in the papyrus that bears his name. Throughout classical and medieval times references to the occurrence of small-pox are comparatively numerous. The oldest known purely medical account of the disease seems to be that of the Arabian writer Rhazes, of the tenth century. In more modern times there are records of small-pox epidemics in almost all parts of the inhabited world. After the introduction of vaccination by Jenner at the end of the eighteenth· century the prevalence of the disease diminished to a remarkable extent in all countries where the practice was adopted. Small-pox, however, still causes local or general epidemics from time to time. The former seem to be largely due to the neglect of vaccination, the latter to some unknown factor in the natural history of the disease which leads it at intervals—like plague, cholera, or influenza—to spread widely over the earth's surface. The last great epidemic, or pandemic, of this kind occurred between the years 1868 and 1873, when the disease overran the greater part of Europe and North America and caused a large number of deaths.

The practice of "inoculating" for small-pox, that is to say of introducing the matter from a small-pox pustule into the tissues of a healthy person, in order to bring about a mild attack of the disease, and so render him immune from a severe attack, is said to have been first practised in China about the year 1000 B.C. In India the custom is perhaps older. It is still resorted to in both countries, as also in Tunisia, Persia and elsewhere. At what period the practice was introduced into Europe it is difficult to say, but it appears to have come there from the East. It was made known in England early in the 18th century by the writings of Lady Mary Wortley Montagu, the wife of the British Ambassador at Constantinople, where it was at that time largely performed. Unfortunately, inoculating the small-pox not only often led to fatal results, but it multiplied the number of persons suffering from the disease—and consequently the sources of infection to others—to an enormous extent. Hence it was generally abandoned on the introduction of vaccination.

Jenner's admirable essay on the protection from small-pox afforded to human beings by the inoculation of the cow-pox was published in 1796. Few great discoveries have found such immediate or such general acceptation, and in the course of a few years vaccination had been introduced into most civilised countries. It was not until some time later that it began to be enforced by law. In England gratuitous vaccination was provided in 1840, but it was still optional; in 1854 vaccination was made obligatory, but it was not until 1871 that public vaccination officers were appointed and the practice more or less rigidly enforced. In 1899, owing largely to the clamour of a section of the population opposed to the practice, the law was again relaxed, and parents who have registered a "conscientious objection" to vaccination are now allowed to leave their children unprotected against the attacks of this still deadly disease.

Recent Geographical Distribution. *Europe.* In England and Wales small-pox gave rise to an annual mean mortality of 45 per million living in the decade 1881–1890. It had been very much higher in the preceding ten years, when the British Isles did not escape from the pandemic extension of the disease which marked the earlier years of that decennium. The mortality figure from this cause was exceedingly low in the years 1889–91, but there then followed a considerable rise in the death-rate. The deaths in 1892 were equal to 15 per million, in 1893 to 49 per million, and in 1894 to 27 per million inhabitants. A marked fall then ensued, and in the years 1895–1900 the mortality ratios were respectively 7, 18, less than 1, 8, 5, and 3 per million. Severe epidemics within recent years in Middlesborough, Gloucester, London, and elsewhere, have aroused public attention to the immense power for evil which this disease still possesses when not kept in check by systematic vaccination.

In Scotland small-pox caused an average of 4 deaths per million inhabitants each year in the decade 1881–1890. In subsequent years the deaths from this disease have for the most part been few in number. Ten were registered in 1897 and only 1 in 1898.

In Ireland small-pox was epidemic. in 1894–5, but between 1895 and 1900 it almost disappeared.

In France the disease has become much rarer within the last ten years ; in towns of over 10,000 inhabitants the deaths from it steadily fell from a total of 1897 in 1891 to only 105 in 1898. The death-rate has also decreased considerably in recent times in Belgium. The mean for the years 1871–1880 was, as in most other countries, very high, as many as 5,080 deaths per annum being recorded from this disease ; but in 1881–1890 this figure fell to 1361 ; in 1895 only 298 deaths were registered, and in 1897 only 140. The last figure shows a ratio of about 22 per million inhabitants.

In Holland the prevalence of small-pox has varied greatly in recent years. A marked rise in the deaths from this cause occurred, just as in England and Wales, in the period 1892–94. In each of these years the mortality rates were 10, 40, and 130 per million respectively ; a rapid fall then set in and in 1897 and 1898 only 1 and 7 deaths from small-pox respectively were registered.

In the German Empire generally the malady is now almost extinct. This happy result has been brought about, almost if not quite entirely, by the enforcement of universal vaccination and re-vaccination. The deaths from the disease in 1894 were returned as 2 per million inhabitants ; the mean ratio for each of the preceding 7 years having been 3 per million. This insignificant degree of mortality is seen in all parts of the country. In Prussia the figures are the same as for the whole empire; in Saxony the deaths in 1894 were only 1 per million inhabitants; in Würtemburg the mean for 1887–1893 was as low as o·2 per million. In Bavaria the ratio fell from 2 per million in the septennium named, to 1 per million in 1894. The disease is not, however, quite extinct in Bavaria or Swabia, for 249 cases of it have been reported in the ten years 1889–98 in Bavaria, and 35 in the same period in Swabia.

In Denmark the disease has been almost unknown in recent years, only two deaths from it having been reported in the urban population since 1892 (1 in 1894 and 1 in 1896). In Norway it has not been mentioned in the list of fatal diseases during recent years, and has therefore presumably disappeared from the country. In Sweden small-pox caused an annual mortality of 25 per million in the years 1880–1886, of 1 per million in the next septennium, and of 4 per million in 1894.

Small-pox is more prevalent in Russia than in almost any other European country. In the years 1893–95 the mortality ratio from this cause was 530 per million in European Russia, the next highest ratio occurring in Hungary (350 per million in the same period). No part of the country is free from it, the degree of prevalence in different governments or provinces varying widely in different years. The number of registered cases of the disease has in some recent years (1893–4) been considerably lower in the Don Cossack Territory than elsewhere, but this has not always been the case, and in 1895 it was actually higher here than in almost any other division of the country. Small-pox is met with in the far northerly governments of Archangel and Olonetz, in the Baltic Provinces, in the Ukraine—in brief, in every division of European Russia, and it is almost everywhere a cause of considerable mortality. It was particularly rife in Polish Russia in 1893. Vaccination, it may be added, is not compulsory in Russia, though it is encouraged by various government and local authorities, by the parish priests, and by certain voluntary societies and agencies. Among the Bashkirs on the European-Asiatic borders of Russia, north of the Caspian, small-pox is said to be not common; these people have no prejudices against vaccination and it is to some extent practised amongst them.

In Austria small-pox was some years ago a very frequent and fatal disease. In the Austrian empire the recorded deaths from this cause in the seven years 1880–86 were 638 per million per annum; in the next septennium they fell to 366 per million.

In Hungary small-pox has of recent years been more active than in Austria. Thus, in the two successive septennia named above, the mean small-pox mortality rates in Hungary with Fiume were 653 and 439 respectively, and in Croatia with Slavonia as many as 588 and 812.

In Switzerland the disorder is apparently a rare one. In the 15 principal towns 1 death only was recorded in 1895, against 18 in 1894, 11 in 1893, 8 in 1892, and 3 in 1891.

A very marked decline has recently taken place in the recorded small-pox mortality in Italy. In the four years 1887–1890 the deaths from this cause in any single year varied between 233 and 610 per million, but between 1891 and 1896 the ratio never exceeded 97 in a single year, and in 1896 was only 65 per million inhabitants.

Small-pox is very widely spread in Spain and Portugal. Davidson, writing in 1892, stated that it caused an annual mortality in the principal towns of 1307 per million. The lower classes are said to be much opposed to vaccination. It sometimes appears in Gibraltar, where it caused serious epidemics in 1883–84, and again in 1889–90, owing, it is said, to neglect of re-vaccination in the civil population.

In the countries of the Balkan Peninsula small-pox is rife. In Servia it caused 4233 deaths in 1894–5, or 1830 per million living. In the other Balkan States the disease is also far more prevalent than in the west. In Turkey in Europe it was at one time a conspicuous cause of sickness and death. In Constantinople, however, it has become very much less common in the last few years than it was formerly, and is now indeed a rare disorder. A few years ago a house-to-house visitation was made, and every inmate compelled to undergo vaccination *nolens volens*, and the diminution in the disease is directly attributable to this. In Salonika and some other provinces of European Turkey small-pox is said to have become comparatively rare since vaccination and re-vaccination have been more widely practised. A severe epidemic of the disease occurred in the autumn of 1901 at Kavala on the Ægean shores[1]. In Greece the malady occurs, but with no great frequency.

Small-pox is occasionally introduced into Iceland, but in recent times little of it has been seen, thanks in part to the efficient vaccination of the people and in part to prompt isolation of cases. In the Faröe Islands no case of small-pox has been recorded since 1856, a result also ascribed to careful vaccination.

As to the prevalence of small-pox in some of the islands of the Mediterranean accurate information appears to be wanting. In Malta it is at times epidemic; it was widely so in 1898, owing to importations of infection from Tunis and Benghazi, 109 cases in all occurring.

Asia. Small-pox was more prevalent in the Caucasus and Transcaucasia in the years 1887–1895 than in any other part of the Russian empire, so far as may be judged from the returns of registered cases of the disease. It appears to be rife throughout this portion of the empire.

In Russian Central Asia, on the other hand, the recorded cases and deaths were remarkably low in the same period— averaging one-seventh, or even less, of the cases and deaths in the same unit of population recorded in the Caucasus, and about one-fourth of those in the Russian empire as a whole. How far this is due to imperfections of registration in this part of the country cannot be said.

In Siberia the registered cases of the disease in the same series of years bore almost the same ratio to the population as those in the Russian empire as a whole. Small-pox at times causes severe epidemics in some parts of Siberia. The indigenous races of that country are particularly liable to it, and even as far north as the shores of the Arctic Ocean serious epidemics sometimes occur among the natives.

Throughout the greater part of Arabia small-pox is the cause of a high mortality. In the Yemen district it is rife and almost uncontrolled by vaccination, and I am informed by eye-witnesses who have travelled in that country that the remains of villages may be seen, every inhabitant of which has been carried off by this scourge. Among the Mussulman pilgrims at Mecca a very large number are yearly attacked by the disease, which accounts for more deaths among them than any other affection.

[1] Reports of Sanitary Officers to the Ottoman Board of Health.

In Syria, in Mesopotamia, and in Persia small-pox is endemic. It was epidemic in Baghdad in November 1898, and in the Lebanon provinces in the same year. At Basra at the head of the Persian Gulf it is far from rare. In Persia it is said to be of milder form than in Europe; it is truly endemic in Teheran, where the old custom of variolisation is still practised. Epidemics from time to time occur in the provinces, and recently (November 1901) outbreaks of the kind occurred at Tabriz and at Enzeli, attacking both natives and Europeans.

The disease is no less endemic throughout the Indian peninsula. Its degree of prevalence varies widely, but in the past few years it has been unusually active. In 1894 and 1895 the recorded mortality was lower than in the preceding years, some 44,000 or 45,000 deaths occurring in each year; but in 1896 the recorded deaths suddenly rose to 141,443 and in 1897 to 167,318. In 1896 the increase had been general throughout all parts of the country, but was greatest in Ajmir-Merwara, and in the Punjab, which also showed the highest mortality ratio. In 1897 the increase was most marked in the North-West Provinces and Oudh. In Assam and Mysore in 1896 the disease caused mortalities exceeding 1000 per million, while in Ajmir-Merwara the mortality was no less than 7050 per million inhabitants. In 1897 Assam, the North-West Provinces and Oudh, and Mysore all returned mortality ratios of over 1000 per million. It is of interest to note that small-pox has appeared to have "little or no connection with famine; certainly it did not particularly affect the distressed districts[1]."

In the Rajputana States, where small-pox was once very frequent, it is said to have much diminished since the introduction of vaccination.

The disease is common in India in the north and in the south, in the hills and in the plains. In the Hindu-Khush it has from time immemorial raged as an epidemic at intervals of from five to twenty years. In 1891 it spread gradually up the Indus valley to Chilas, and in the autumn kept the Chilasis quiet during the Hunza-Nagar expedition. It reached Gilgit by Christmas, and thence was apparently distributed widely by a body of released prisoners. It caused in the end a very severe and fatal epidemic in Nagar

---

[1] *Report of the Sanitary Commissioner of the N.W. Provinces and Oudh* for 1896.

(where no vaccinations had been performed), and a mild one in Hunza (where several vaccinations had been performed in the previous year). It is of interest to note that, in the absence of vaccine lymph, the old practice of small-pox inoculation was revived on a large scale, and—it is claimed—with remarkable success, the epidemic subsiding in a fortnight[1].

In Ceylon small-pox is at times epidemic; it was so in 1891, when it caused 452 deaths; in other years it is quiescent, as in 1894, when only 3 deaths from it were recorded; in 1897 the number was 25.

In Farther India the affection is very common. In the southern Shan States in Burma it is one of the principal causes of death in childhood. Of Siam almost the same may be said; it was very prevalent here in 1898. The Siamese make no attempt to isolate the sick, and it is said that persons ill with the disease may be seen in the streets picking off the scabs, or bathing in the streams while still in the suppurative stage. It is endemic in the Malay Peninsula, and is frequently seen in the British Straits Settlements. In 1898, 84 cases with 46 deaths occurred in Singapore town alone, 2 deaths were recorded in Penang and 1 in Province Wellesley in the same year, and as many as 93 cases with 23 deaths in Malacca. In Annam and Tongking it is at times epidemic and fatal in an unvaccinated population.

In China small-pox is one of the commonest diseases. It is met with almost everywhere, particularly in the winter months. It is believed to be largely spread by the practice of variolisation, which is still performed by inserting a pledget of small-pox crusts into the nostrils. In recent years it has been very frequently epidemic in many places, and is constantly mentioned in the reports of the Customs Medical Officers. At Wuhu it is said to have caused the deaths of 2000 children under twelve in the winter of 1890–1891[2]. In Shanghai it was unusually prevalent in 1893; at Ningpo it was severely epidemic in 1893–4; it raged in and near Hoihow (Hainan Island) in 1894–5, in Hankow in 1896, in Canton in 1896–7, and in Chung-king in 1897–8. In Hongkong it was epidemic through a great part of 1898. In the Yünnan province it appears intermittently, and in Manchuria it is often very prevalent.

[1] Capt. H. B. Luard, I. M. S., *Trans. 1st Indian Medical Congress*, 1894.
[2] *Med. Reports of the Imp. Chin. Mar. Customs*, No. 42.

In Corea it usually rages in the winter months.

In Japan small-pox is constantly present in some part of the archipelago, with varying degrees of severity. In 1893 as many as 41,898 cases of the disease with 11,852 deaths were recorded, but in 1894 these figures fell to 12,418 and 3,342 respectively. The last figure was equivalent to a ratio of about 80 per million inhabitants.

The increased frequency of vaccination in many of the islands of the Malay Archipelago is said to have diminished the prevalence of small-pox considerably in recent years. This has been the case in Sumatra, Java, and Celebes. In Java the disease is now almost unknown. In Borneo it from time to time becomes epidemic. In Labuan, off the coast of British North Borneo, it is occasionally seen, but no cases have been recorded in recent hospital reports from that colony. Among the Dutch troops in Netherlands-India only a few cases of the disease, all of a mild type, were recorded in 1897.

Australasia. In British New Guinea small-pox appears to be unknown, though it is spoken of as endemic in German New Guinea and the Bismarck Archipelago[1]. In Australia it has been extremely rare in recent years; it has been several times imported into Victoria in the last decade, but has never spread or become epidemic. In South Australia a single death in 1898 was the first recorded in the colony for at least ten years. In Queensland no death from the disease was registered in any of the years 1893–1897 inclusive. In Western Australia 7 deaths from small-pox occurred in 1893 and 2 in 1894, none occurring in any other year of the decade 1889 to 1898. No mention is made of the disease in recent reports of the Board of Health of Tasmania[2], nor in recent reports from New Zealand, and it may therefore be assumed that both these colonies can still claim—as they could when Hirsch wrote twenty years ago—that they have always remained free from small-pox. In former years the disease was terribly prevalent among the Australian aborigines, and is said to have carried off from one-third to one-half of some of the tribes. Its absence from New Zealand is, therefore, all the more a matter for congratulation,

[1] *Colonial Report of the British Solomon Islands*, 1897.

[2] It has been conjectured that in the past, before the colonisation of Tasmania by the white man, the aborigines may have suffered severely from small-pox, but I do not know whether the conjecture is well founded.

as the Maori race, which has shown itself to be so susceptible to some other diseases, as for example tubercle, could not fail to suffer severely from it.

Some of the Polynesian islands have equally escaped small-pox. It has never been known, for example, to occur in Fiji or in the Samoan Islands. In some other groups however, as the Society Islands and New Caledonia, it has been frequently epidemic.

Africa. Small-pox is endemic in the countries on the northern shores of Africa. It is very common in Morocco, and Dr Guillemard informs me that a severe and very fatal epidemic occurred in Mogador in 1892. In Tunisia it is the commonest of all diseases, and variolisation is practised here, as in China, with disastrous results. The disease caused 1645 deaths in Tunis alone in 1888, and 870 in 1894—both epidemic years[1]. In Algeria also small-pox is extremely prevalent, the mortality from the disease some ten years ago averaging 1685 per million inhabitants per annum. Oran has suffered greatly from its ravages. It was severely epidemic in 1896, when it caused 343 deaths in the six principal Algerian towns, but in the following year this figure sank to 35, and in 1898 to 29.

Egypt has been comparatively free from small-pox since the introduction of vaccination in 1827, before which date the mortality from the disease was excessively high. Now the larger number of cases which occur are said to be due to importations from the surrounding countries, where vaccination is almost or quite unknown.

On the western side of the African continent small-pox is rife. In Senegal it is frequently epidemic. In Sierra Leone 165 cases with 11 deaths occurred in the colonial hospitals in 1897, and in Lagos 189 cases with 6 deaths were recorded in the same year; these were among the highest figures for any single disease in the former, and quite the highest in the latter. In the Gold Coast colony it is endemic in the natives; 261 cases of the disease with 48 deaths were recorded in 1897, and many others were known to have occurred. In the Cameroon district, on the other hand, it is said to be unknown[2].

---

[1] In the intervals between these two years the deaths from small-pox in Tunis were numbered by units.

[2] A. Plehn.

Throughout the greater part of Central Africa small-pox is extremely common. In Belgian Congo and in French Congo and elsewhere it is often very widely prevalent and fatal. The disease is largely spread in Central, Eastern, and Western Africa by the movements of traders. There are said to be few caravans in which a case or cases of small-pox do not occur; variolisation, which is occasionally practised, may also aid in spreading the infection. Pruen states that the natives often suffer from very mild attacks, and many recover from the worst confluent forms of the disease.

On the eastern side of Africa the disease is widely endemic. In Abyssinia it is very common, and it is reported to have ravaged the Abyssinian army under Ras Makonnen in 1898. At Massowa, on the Red Sea, it is rife, yet the well-vaccinated Italian army escaped without a single case during the campaign in 1895. In British East Africa it is believed to have been rare in the last twenty or thirty years, as pock-marks are only met with in persons no longer young[1]. On the Upper Zambesi also the disease only occurs at long intervals in epidemic form. In Unyoro it is said not to be endemic, though it is sometimes imported from the coast, but in Uganda, on the other hand, it is very deadly, slaying its hundreds[2]. In Zanzibar and along the Mozambique coast the disease is thought to be endemic.

Throughout South Africa small-pox is met with; usually in the form of scattered cases, but occasionally as severe epidemics. It is particularly common in the native territories and often spreads from them to other parts of the country. In 1898 it was epidemic among the natives in several places in Tembuland, the Transkei, and Pondoland, and it appeared also in the Transvaal, the Orange Free State, Natal, and Cape Colony. Three outbreaks occurred in Griqualand West in this year. In many large tracts of South Africa no vaccination whatever is performed. In Cape Colony vaccination is done by the Cape Police, who travel from place to place and are said to do a vast amount of good and useful work in this way; elsewhere vaccination is resorted to solely or mainly when small-pox threatens.

In Madagascar and Mauritius small-pox is probably not endemic, but at times has been imported from the African

---

[1] Kolb.   [2] A. R. Cook, *Journ. Trop. Med.* June 1, 1901.

continent and become seriously epidemic. In Madagascar it has often in the past caused terrible epidemics, the outbreaks being usually traceable to the importation of slaves from Africa. In Mauritius it was epidemic in 1891, and on two previous occasions in the 19th century.

North and Central America. The malady has been epidemic in Greenland on at least four occasions in history. In Alaska it carried off almost one-half of the population in 1838–39. As to its behaviour in these northern regions in recent years I have no information.

In Canada small-pox is met with, but has not been very active in the last decade. In the province of Quebec the last severe epidemic was in 1885–6, when almost the whole province, and especially the city of Montreal, was ravaged by it. Since that date occasional small outbreaks of the disease have occurred here and there[1]. In Ontario the mean annual number of recorded cases in the period 1882–86 was 128, in the next quinquennium it fell to 16, and in that ending in 1891 to only 7. Ontario is by its inland position largely sheltered from the importation by immigrants of small-pox and other diseases. In Manitoba imported cases occasionally occur; and also in British Columbia, where in 1892 the disease caused a severe epidemic in the city of Victoria.

In the United States small-pox is now, thanks to vaccination and other measures, a disease of much rarer occurrence than formerly. In the year 1897 not a single death from it was recorded in the cities and towns of the large majority of the States, and from 1 to 5 deaths only in each of the following:— Alabama, Georgia, Massachusetts, Missouri, New Mexico, Ohio, Pennsylvania, and Texas. In New York State the deaths numbered 28. In Massachusetts a limited outbreak occurred in 1893–4, but since then very few cases have been seen. In Michigan the small-pox mortality has been extremely low since

---

[1] An outbreak of this kind occurred in Quebec in 1895, and was entirely traceable to the movements of one person—a young girl, with small-pox on her, who fled across country, by train and boat, from Quebec to Carleton (county of Bonaventure). A large number of persons with whom she came in contact either caught the disease or carried it to others, with consequent local outbreaks in many places. Her track was indeed marked by small-pox cases to the number of 151, and deaths to the number of 32, occurring in 44 different houses. *Report of the Board of Health of the Prov. of Quebec,* 1895–6.

the great epidemic year, 1872. At that time the United States were suffering very severely from the pandemic of the disease to which reference has already been made. In former years the malady literally decimated many of the Indian tribes on the North-American continent. At the present time it seems principally to attack the negro population. It 1898 it was very prevalent among the negroes of Kentucky, and early in 1899 was widely epidemic in many of the southern States. In some places what were called "shot-gun quarantines" were established by the excited populace. In Georgia and Ohio it was very rife, Alabama is said to have been for long severely affected, while in other States the prevalence of the disease was traced to importation from Cuba, where during the war it was terribly active.

In Mexico the disease is common, particularly among the Indians; it was very rife throughout the country in 1898. In British Honduras it was epidemic on the last occasion in 1891. In other Central American countries it has at times become epidemic, especially attacking the Indian tribes.

In many of the West Indian Islands small-pox is far from rare. In Cuba it is endemic, and during the recent troubles prevailed in a malignant form and to a wide extent. In January, 1899, it was said that over 5000 cases of the disease were then under treatment at Holguin, and many hundreds elsewhere. In Jamaica it is at times epidemic; in Haiti it is spoken of as endemic; in Trinidad outbreaks occur at intervals. In the Leeward Islands on the other hand, and in many other British West Indian Islands, small-pox appears to have been quiescent or absent in recent years. In Puerto Rico it was seriously epidemic in the winter of 1898–9, but after the vigorous enforcement of vaccination —over 800,000 persons being vaccinated in three months—the disease completely disappeared.

South America. Along the northern coasts of South America small-pox appears to be comparatively rare, and of milder type than in some other countries. In the Guianas little mention of the disease has been made in recent years. It has been absent from British Guiana for many years, thanks to strict precautions for preventing its importation. In Venezuela it was severely epidemic in 1898; in Valencia alone, high up in the Cordilleras, 2000 cases occurred, with 600 deaths. In Brazil it has in times past been very fatal, particularly to the coloured population. It is still very rife among the negro inhabitants of the dirty, over-crowded portions of some Brazilian cities, such for example as Bahia. In Uruguay it caused 96 deaths in 1897, which was equivalent to about 120 per million inhabitants. In the Argentine Republic it was epidemic in 1888, 1890, and 1891, and the mean mortality from the disease in the Buenos Aires province in those years was 470 per million inhabitants. In 1897 the recorded deaths in the same province were 94 in number, of which 44 occurred in the northern region, 19 in the central, 30 in the southern, and 1 in the Patagonian region of the province. In Chile small-pox has in the past caused very violent epidemics among the Indian and coloured populations.

In Peru the disease is far from uncommon; in Callao a severe epidemic occurred in 1891, causing 238 deaths, and in 1896 another caused a still higher mortality in that city, the deaths numbering as many as 309.

Factors concerned in the Geographical Distribution. Small-pox has occurred in almost all portions of the inhabited globe. It has shown no regard for lines of latitude or isotherms, and has been just as severe and widespread in the tropics and in sub-arctic regions as in the temperate zones. It is true that in recent times the disease has been little seen in some remote northerly regions, as Iceland, the Faröe Islands, and apparently Greenland and the far north of North America; but this can in no way be attributed to climatic conditions, for in the past it has caused serious epidemics in each of the regions named. Their recent immunity from the disease would appear to be largely due to their remote position and rare intercourse with the rest of the world—conditions which greatly lessen the risk of chance importation of infection—while in Iceland and the Faröe Islands, if not elsewhere, the practice of vaccination and pre-cautions to prevent carriage of the infection to their shores have had a large share in keeping the islands free from the disease. The entire immunity from small-pox hitherto enjoyed by Tas-mania, New Zealand, Fiji, Samoa, and some other islands in the Pacific Ocean must be attributed in great part to the same causes.

There is, however, no doubt that small-pox is to some extent influenced by changes of temperature. This is shown by its seasonal variations in countries where it is endemic or epidemic. In almost all parts of the world the disease has shown greatest activity in the cooler portion of the year. This has been observed with remarkable uniformity not only in both the northern and southern hemispheres, where, of course, the seasons are reversed, but in tropical, subtropical, and temperate climates. In England, for example, the small-pox curve is high from January to May, and low in the summer. In India the disease is most prevalent in the cool season, between January and April, but its prevalence and its virulence are said to increase as the temperature rises, and in some parts of the country the maximum occurs in the hot month of May, or even in June. In Brazil the small-pox season is in October, November, and December, or the spring months in those latitudes, and in Chile it lasts during the winter and spring (July to November). The same seasonal relation of small-pox is observed in most European countries, in North America, in China, in Tunisia, in Egypt and elsewhere. The recent serious outbreak of the disease in London also fell in the winter months. On exceptional occasions, however, small-pox has been severely epidemic in the summer.

Small-pox possesses a high degree of infectivity. When introduced into a community that has not been protected from its attacks, either artificially by vaccination, or naturally by recent prevalence of the malady, it is one of the most deadly of diseases, and may give rise to the most fatal epidemics. Many historical examples of this kind are on record. In the sixteenth century, when the infection was carried to the New World, it attacked the indigenous races with extraordinary violence and swept away whole tribes in the West Indies, Mexico, Brazil, and Chile. In more recent times it has been scarcely less disastrous to many of the native tribes in Australia, Canada, and the United States, as well as in Siberia.

Like measles, scarlet-fever, and other diseases, small-pox dies out from small, isolated communities, where the supply of susceptible persons is not sufficient to keep the disease going, and it may remain absent from such communities for a long period until a fresh importation of infection occurs. This has been the case in Iceland and the Faröes, in Mauritius, and elsewhere. In continental countries, on the other hand, it seems to remain endemic, so long as there is a sufficiently numerous susceptible and unprotected population to keep up a succession of cases. In such countries it is not equally prevalent in all years; the disease is more active in some years, less in others. In most civilised countries this behaviour of small-pox is now largely obscured by the practice of vaccination; but it was clearly seen in the centuries before vaccination was introduced. The records of the disease in many countries in the 17th and 18th centuries seem to show that it had a certain tendency to periodicity, its activity waxing and waning every three or four years in some countries, in others at longer intervals. This periodicity in small-pox prevalence—as also in the case of scarlet-fever, diphtheria, and other diseases—has never been quite regular, the intervals varying not a little in length. It is a phenomenon worthy of attention, and one that has not yet been quite satisfactorily explained.

The virus of small-pox is mainly spread by the movements of human beings. It can, however, exist for some time outside the human body, whether attached to articles that have been near the sick—as bedding, body-linen, or even the furniture of the sickroom—or in the air. It seems capable even of being carried considerable distances through the air without losing its virulence. Attention was drawn to this mode of spread of small-pox—the possibility of which many had vaguely surmised before—by the investigations of Mr Power, of the Medical Department of the Local Government Board, in 1882 and subsequent years. Certain small-pox hospitals were taken as centres, and careful plans of the neighbourhood being prepared, concentric circles were drawn on the plans at distances representing one-quarter, one-half, three-quarters, and one mile from the hospital respectively. The number of small-pox cases occurring during fixed periods in each of these concentric areas was then determined, and it was found that they diminished regularly from within outwards. The conclusion drawn was that the hospital must have acted as a centre, from which the infection had been carried by the air and had been the cause of the large number of cases in its vicinity. Accepting this view it would seem that the virus of small-pox can be wafted through the air over a distance of some half-a-mile or more and still retain its vitality. The conclusion was further supported by an inquiry into the frequency of small-pox in certain parts of London, respectively before and after the opening of small-pox hospitals in their neighbourhood. It was found that the disease had become considerably more prevalent in these localities after the hospitals had been opened than it had been before. The recent epidemic of small-pox in London seems to have furnished still further confirmatory evidence on this point, and to have shown that the virus of the disease was carried from the small-pox ships on the Thames over nearly half a mile of water to the Essex shores, where it accounted for a large number of cases. This aerial diffusion of the infection is believed, however, to take place only or mainly when a number of acute cases are aggregated together, and perhaps only under certain—as yet undefined—atmospheric conditions. While the theory of the aerial convection of small-pox has been contested by some observers, it is now generally accepted, and has led to the practice of removing small-pox patients either to land-hospitals in sparsely inhabited districts, outside the limits of large towns, or, as in London, to floating hospitals. It has further shown that even this measure is not without its dangers, and that it will not adequately control the disease unless supplemented by general vaccination and re-vaccination of the surrounding population.

Small-pox is the most striking example of a disease which can to a very great extent be kept in check by prophylactic measures. The adoption of universal compulsory vaccination and re-vaccination—an ideal not yet even approximately attained in any country with the exception of Germany—would, there is good reason to believe, exterminate the disease altogether. The partial measure of more or less general, and more or less compulsory vaccination with occasional resort to re-vaccination, which is in vogue in most civilised countries, has already succeeded in diminishing the prevalence of the disorder to an extent that can with difficulty be realised by those who have not studied its history. It would be out of place here to enter fully into the whole question of vaccination—recently the subject of so much controversy—or to add unnecessarily to the already enormous bulk of literature upon this subject. It must suffice to state that at the present day no single group of factors is of such importance in determining the distribution of small-pox over the earth's surface as the presence or absence, the efficiency or inefficiency, of general vaccination and re-vaccination. For it is now clearly established that inefficient vaccination, and vaccination in infancy only, do not render the vaccinated person immune to the disease for life.

Innumerable illustrations of this statement have been already brought forward in this chapter. In few European countries, for example, are vaccination and re-vaccination so thoroughly carried out as in Germany, and in few countries is small-pox so rare. In Russia, in Austria, in Spain and Portugal, where both are largely neglected, small-pox is constantly present. In Arabia, in China, in Central and parts of Southern Africa, where vaccination is practically unknown, the disease is extremely rife. Nor is the difference less striking when lesser units are taken; and the incidence of small-pox on different communities in a given country is found to vary inversely with the extent to which vaccination is practised amongst them. Two recent examples will suffice to illustrate this fact. In Tunis serious epidemics of small-pox occurred in 1888 and in 1894. The population consisted of some 100,000 Arabs, entirely unvaccinated, 30,000 Europeans, among whom vaccination was practised but neglected, and 30,000 Jews, spoken of as "well-vaccinated" as a community. In the first of the two epidemics mentioned, out of 1645 deaths from small-pox, 1384 occurred in Arabs, 160 in Europeans, and 101 in Jews; and in the second, of 870 deaths, 712 occurred in Arabs, 128 in Europeans, and only 30 in Jews[1]. The second example is even more conclusive; while small-pox raged among the unvaccinated native population of Massowa, the well-vaccinated Italian troops stationed there during the recent war in Erythræa escaped without a single case.

While vaccination is thus one of the leading factors in the present day distribution of small-pox, it is not the only one. When once a community or a country has succeeded in ridding itself of the disease it can continue for long periods together more or less free from it by strict precautions at its ports and frontiers to prevent the importation of infection, and immediately to detect and isolate any infected person arriving from outside. This has been the case in recent years in some of the Canadian provinces, where small-pox has occurred mainly in the form of imported cases in Chinese or other immigrants; in Madagascar and Mauritius, where the infection is always imported from the

---

[1] *La Tunisie, Histoire et Description.* Paris, 1896.

African mainland; in Iceland, which has usually received the infection from Denmark; in British Guiana, which has been kept free from the disease for many years; and in the Faröes, where no small-pox has now been seen for nearly half a century. Whether the remarkable immunity from small-pox which Fiji and the Samoan Islands are said to have enjoyed up to the present is due to measures to prevent its importation or to some other cause I am unable to say; but there seems to be practically no doubt that the fact that New Zealand and Tasmania have never had an indigenous case of the disease is entirely the result of such measures. Cases of small-pox have, it is said, been brought to their shores, but they have been promptly dealt with, and the disease has not spread to others.

Small-pox has shown no definite relations to soil or to altitude above sea-level. Few diseases are indeed more indifferent to all external physical conditions than this. Race has possibly some slight influence on its distribution. A high level of susceptibility to the infection of small-pox seems to exist in most, if not all races, but all observers agree that it is particularly high in the African negro. This race, when negro slavery was practised, was frequently the means of carrying small-pox from the Old to the New World. They still suffer greatly from it and succumb to it more easily than white men. The fact, already mentioned more than once, that Fiji and the Samoan group have never been invaded by small-pox, raises the question whether the inhabitants of these islands possess any immunity from the disease. This question does not yet appear to have received an answer.

The relation of small-pox prevalence to general insanitary conditions has been the subject of much controversy. A low level of general hygiene and public health administration undoubtedly favours small-pox prevalence in many ways. Insanitary surroundings usually imply a diminished power of resistance in individuals to the attacks of any infective disorder. Overcrowding greatly favours the chances of the infection being transmitted from person to person. These conditions, moreover, are constantly found associated with a lack of the knowledge needed not only to recognise the early cases, but to grasp the necessity of at once reporting and dealing with them when discovered. They are often too, but by no means always, met with among classes of the population who have failed to understand the value of vaccination, or, through carelessness, inertia, or a fatal readiness to accept the misguided teaching of those opposed to the measure, have not had recourse to it.

Among general measures of public health administration, the provision of good water-supplies and of drainage-systems, the inspection of food-supplies and the like, have never been shown to have any direct influence over small-pox diffusion. On the other hand there are few diseases so directly amenable to certain public health measures of another character. It has already been shown that an administrative area can be kept free from small-pox for long periods together by measures directed to prevent the introduction of the infection from outside, and to isolate immediately any imported case. No small-pox-free community can safely ignore a measure of this kind. But it requires to be supplemented by others. Of these the most important is the provision of means for general vaccination and re-vaccination, so that even should the infection be imported—a contingency certain to happen sooner or later—it will find a sterile soil and must die out. Other measures are the construction in suitable places of hospitals for the immediate isolation of an imported case; arrangements for the prompt detection and removal of any person suffering from the disease; and provision of means for the disinfection and destruction without delay of the infective material in the patient and his surroundings. If these measures are of the greatest importance in maintaining a freedom from small-pox in communities already free from it, they are *à fortiori* of the first necessity in endeavouring to get rid of the disease in a community or a country where it is already established.

SYPHILIS AND OTHER VENEREAL DISEASES.

General Characters and Etiology. The group of venereal diseases—syphilis in its various manifestations, gonorrhœa and chancroids—may be most conveniently dealt with as a whole. Though each is due to a distinct and separate cause—either certainly or presumably a micro-organism—the means by which they are spread and the social and other conditions which tend to make them common or rare are practically the same for all, and it would serve but little purpose from the view-point of medical geography to deal with them separately.

It must at once be confessed that no materials exist for an exact comparative study of the frequency of these diseases in different countries. With the exception of syphilis they are almost never immediate causes of death, and consequently mortality statistics are lacking for comparison ; while even as regards syphilis it is often rather a remote than an immediate cause of death, so that the statistics of deaths registered as due to this cause are of only approximate value. Moreover in many countries where syphilis is extremely rife death registration is unknown. It becomes necessary, therefore, for the most part to fall back upon general statements, upon impressions derived from hospital returns, or upon the observations of travellers in remote countries.

In general terms these diseases may be said to have an almost world-wide distribution. But, as they are essentially of a preventable character, it is found that they are most common among those nations and among those classes of society whose moral views do not prevent promiscuous and uncontrolled sexual intercourse, and who are ignorant of the great infectivity of these diseases and of the need of preventing their spread either by venereal or non-venereal means. Of these two means of spread the former is of course the most usual and is perhaps the only one among educated communities, but the latter is very frequently seen both among uncivilised and uneducated races, such as the Chinese or the Russian peasantry, and among the lower classes of most civilised nations.

History. Descriptions of diseases of the generative organs associated with impure sexual intercourse are found in all times of the world's history. The Mosaic Law contained a large number of provisions that seem almost certainly to have been directed to the prevention of this class of diseases. In India and in China they seem to have been known at very remote periods. But while this group of maladies as a whole dates back to the earliest times, much controversy has arisen over the question whether syphilis as now seen is of equal antiquity. Its origin, like that of all other diseases, is quite unknown, and it is not until the Middle Ages that the descriptions left us of diseases of the generative organs can be positively affirmed to refer to this and not to some other form of venereal affections. There can be no doubt that at the end of the fifteenth century syphilis became enormously more prevalent in Europe than it had been before, and indeed it spread very much like an epidemic of some zymotic. Hirsch, in summing up this branch of the subject, came to the conclusion "that the venereal diseases had occurred in Europe, and, so far as we know, also in various parts of Asia, from the earliest times; that syphilis overran a large part of Europe towards the end of the fifteenth century in epidemic-like diffusion, when it attracted the general attention of the profession for the first time and was first recognised by its peculiar features; that after the extinction of that epidemic, which lasted about thirty years, the disease fell again to its former level; that it was imported from Europe to other parts of the globe as a consequence of commerce between countries; but that even at the present day there are some places, remote from the general stream of traffic, which it has not reached." The truth of these statements will be illustrated in the following pages, where mention will also be made of some epidemics of syphilis later than that of the fifteenth century.

Recent Geographical Distribution. *Europe.* Throughout Europe these diseases are found with very varying degrees of frequency and severity. In the British Isles they are common enough, but the deaths attributed to syphilis have slightly declined within the last decade.

In France venereal diseases are found principally in the larger towns, where, however, a system of control of prostitutes and houses for immoral purposes tends to restrict their spread. Germany is believed to be rather more free from venereal diseases than any other European country, but the statement must be accepted with caution. In Denmark, Norway, and Sweden they are very widely spread, and in Norway at least the deaths from syphilis have slightly increased in recent years. In Lapland this disease is rare. In Russia, on the other hand, syphilis and venereal diseases are extremely prevalent, and constitute in many parts of the country a veritable scourge. From returns recently published[1] it appears that in the years 1889–94 the proportion of population in the different "governments" of European Russia treated for syphilis varied between 700 and 10,300 per million inhabitants, while the proportion treated for other venereal disorders varied between 420 and 4270 per million. The figures are only approximate, and are probably much below the truth, and it is certain that in some districts a very large percentage of the peasantry are syphilitic. There is no portion of European Russia free from this disease. It is met with in the extreme north in the government of Archangel, and in the extreme south along the shores of the Black Sea. It is common in Poland, on the banks of the Volga and among the Cossacks of the Don Territory. It is somewhat less common in Finland. In recent years there has been a steady increase in the number of reported cases of syphilis in Russia.

In Southern Europe these diseases are widely prevalent. In Spain they are said to be exceedingly common, but somewhat less so in Portugal. In Italy they appear to be about as frequent as in the British Isles; the mortality from syphilis in the peninsula has varied from 64 to 78 per million living in the years 1887–96, while in England and Wales it has ranged over very much the same limits. In the Balkan countries venereal diseases are very common. Lombard states that in Roumania almost the entire population is infected with syphilis, but the statement is probably

---

[1] *Report of the Conference on Syphilis*, held in St Petersburg by Imperial command in 1897. St Petersburg, 1897. (In Russian.)

exaggerated. In Turkey in Europe these diseases are very prevalent. They are largely distributed by soldiers and others returning from the capital to their homes, and in this and other ways syphilis has been carried from Constantinople to all parts both of Turkey in Europe and Asia Minor. In Montenegro syphilis is not rare; it is said to have been imported by Montenegrin emigrants returning from Constantinople or from more western capitals, where they have contracted the disease.

In many parts of Greece all disorders of this class are very prevalent, and in the northern part of the country a form of syphilis of the most inveterate kind is met with, which is locally known as *spirokolon*.

In Iceland and the Faröes syphilis and other venereal diseases are exceedingly rare, owing, it is said, to the moral habits of the islanders rather than to the rigours of the climate.

Asia. Throughout the Caucasus and Transcaucasia venereal diseases are widely prevalent, but if we may judge from the returns of a single hospital (that at Tiflis) it would seem that the Russian inhabitants are more affected than the native races. In the country to the north of the Caspian Sea, in the valley of the Ural river, and among the Bashkir and other Tatar tribes on the borders of European and Asiatic Russia, syphilis appears to be comparatively rare. It was at one time more prevalent than now among these races, but at present it seems to be mainly introduced to the steppes by the younger members of the community returning from Russian towns visited by them. The Bashkirs have a high reputation for sexual morality, and prostitution is said to be unknown amongst them[1]. Among the Kirghiz hordes syphilis is almost unknown and for similar reasons[2].

From Central Asiatic Russia and Siberia the information as to the prevalence of these diseases is far from full or trustworthy, but it indicates that they are more or less common throughout. They are not, however, universal, and the returns from a considerable number of gold-mines and other mines in the Yakutsk government, in the Amur and Transbaikal provinces, and elsewhere, show a comparatively small proportion of syphilitics among

[1] Nikolski.

[2] Matzkevitch, *The Kirghiz Steppes of the Turgai Province in the Cholera Epidemic of* 1892. St Petersburg, 1893. (In Russian.)

the persons employed in them. From two mines in the Kirghiz steppes of Semipalatinsk and Semiretchinsk information is forthcoming that not a case of venereal disease of any kind had been met with there during the course of five years (1889–1893). Some of the aboriginal Siberian inhabitants, such as the Samoyeds, Yakuts, Tunguses, and Buriats, are on the other hand said to be seriously infected with syphilis. In Central Asia this disease is also extremely rife in parts. It is found everywhere in Bokhara, and in consequence of the entire absence of treatment the most horribly severe forms of the malady are met with—forms that are said to be unknown in more civilised countries[1].

Throughout Asia Minor and Syria syphilis is met with, and in many parts of Armenia it is exceedingly common. The great extent and severity of the disease in Asia Minor have attracted much attention, and a systematic attempt to bring it under control has recently been made by Prof. von Düring under the auspices of the Sultan. In Arabia it is spoken of as "frightfully common," and it is probably by no means rare in Persia. At Basra, at the head of the Persian Gulf, it is a frequent disorder.

Venereal diseases are met with with great frequency almost throughout the Indian peninsula, and certain parts of the country are notorious for the amount of syphilis and other diseases of the class among the inhabitants. In Kashmir, for example, syphilis is the greatest scourge of the country. The majority of the tribes living on the southern slopes of the Himalayas are more or less deeply infected with it. Carleton[2] states that, as the result of "careful and extensive inquiries," he believes that not five per cent. of the population in the Simla Division have escaped infection; and this extremely serious spread of the disease has all occurred within the last thirty years. Many of the villages in the Hindu-Khush are also largely infected by syphilis, although other forms of venereal disease are said to be comparatively rare both among the natives and among the troops stationed in that region. In Rajputana all forms of venereal disease are very commonly met with.

The frequency of this class of disorder among British troops in India has long been notorious, and the enormous increase of cases of the kind within recent years has justifiably given rise to

[1] Grekof.
[2] *Transactions of the First Indian Medical Congress*, 1894.

no little alarm, and to a demand for more stringent preventive measures than have hitherto been adopted. The increase in these diseases, and especially in primary syphilis, which has been observed in the British troops in India in the last thirty years is seen clearly from the following table. The figures indicate cases per thousand strength.

Year	Primary Syphilis	Total Venereal Disease	Year	Primary Syphilis	Total Venereal Disease
1872	61·2	179	1886	157·9	389·5[1]
73	52·4	166·7	87	142·1	361·2
74	68·3	192·7	88	142·1	370·6
75	67·1	205·1	89	225·1	481·5
76	59·8	189·9	90	220·7	533·5
77	65·2	208·5	91	159·2	400·7
78	95·3	271·3	92	161·1	409·9
79	79·2	234·8	93	213·6	466
80	87·9	249·7	94	248·1	511·4
81	92	260·5	95	239	522·3
82	87·6	265·2	96	226·4	511·6
83	87·2	270·3	97	201·7	485·7[2]
84	90·2	293·9	98	—	362·9
85	122·1	342·7[1]	99	—	313·4

From these figures it will be seen that while the invalidings for venereal diseases of all kinds had nearly trebled in number in this period, those for primary syphilis alone were nearly four times as numerous in the later years of the period as they were in the earlier. Since 1895 a marked fall has set in. This fall is thought to be "partly unreal" and due to treating more of the men as out-patients, and "partly real" and the result of certain judicious measures of prevention.

The native troops in India are much less severely affected by these diseases, and the annual number of cases among them since the year 1877 has varied between the extremes of 26 and 41 per thousand. The ratio has been remarkably steady, and the increase just referred to as observed among the British troops was not accompanied by a similar increase in the native army.

[1] Including troops on active service in Burma.

[2] This reduction was then considered more apparent than real, as among troops in cantonments the ratio was 507·8, and the lower figure was obtained by including troops on active service. The figures of the next two years, however, show that there has been a very real diminution.

In Ceylon syphilis appears to be much less common than on the Indian mainland. Many of the East Indian Islands are, on the other hand, very seriously affected by this and other forms of venereal disease. Thus in Borneo, according to Nieuwenhuis, syphilis prevails to a terrible extent and "causes ravages in individuals and in tribes in a way that Europe has no idea of." Gonorrhœa, too, is excessively frequent in the island, and often assumes alarmingly severe forms. In Java, on the contrary, syphilis is said to be rare. The Indo-Dutch army, both native and European, suffers considerably from venereal diseases, and it is noteworthy that here, as in British India, these maladies have become much more common in recent years among European troops than formerly, while in the natives they have remained stationary.

Throughout the countries of the south-eastern corner of the Asiatic continent venereal diseases are excessively rife. In Singapore and the Straits Settlements syphilis is one of the commonest of maladies, and the venereal group as a whole accounts for more admissions to hospital than any other group of diseases. They appear to be on the increase[1]. At Saigon, in French Cochin China, syphilis gives rise to more than one-fifth of the total admissions to hospital (Davidson). In Siam and among the Laos it is extremely prevalent, and in Tongking it is not only prevalent but is said to be increasingly so.

From all parts of China come reports of the extreme frequency of both syphilis and gonorrhœa. Soft chancres are, perhaps, not very frequent here, one observer at least regarding them as of less common occurrence than in his own country, the United States[2]. But of the terrible prevalence of the other forms of venereal disease in China there can be no question. No attempt is made to control them, and many writers speak of the disgusting carelessness with which they are allowed to spread by venereal and non-venereal channels, while no effort is made to conceal their existence from the eye of the passer-by or to treat them in the individual. In many places a high proportion of the total population is syphilitic; and in Pakhoi, according to one writer, who is perhaps not to be taken quite literally, the entire population is

---

[1] *Medical Report for the Straits Settlements* for 1898.

[2] *The Chinese, their Present and Future, Medical, Political, and Social.* By R. Coltman, Junior, M.D., 1891.

syphilised[1]. Corea seems to be no less deeply ravaged by syphilis than China, and soft sores are spoken of as common. In Japan the disease is also very widely prevalent, and it is said to be the principal scourge of the great middle class of the population[2].

Australasia. Syphilis is common enough in the Philippine Islands, both among natives and Europeans. In New Guinea venereal diseases have been introduced on several occasions, but usually in the less severe forms. Quite recently there was a fresh introduction of syphilis in Murua, and the infection has spread rapidly. Sir William MacGregor states that the effect of these diseases on the inhabitants of New Guinea from the Solomon and New Hebrides Islands is often disastrous, while among the autochthones, on the other hand, the disease does not seem so far to have assumed the same malignant type. From other sources it may be gathered that syphilis is a very frequent disease in some of the Solomon Islands, especially in those, such as Ugi, which have much intercourse with the outside world[3]; but in the New Hebrides it seems that venereal disease is not common, and one observer saw cases there in the white inhabitants only[4].

Australia appears to be comparatively free from the class of disorders now under consideration; as a cause of death, at least, syphilis may be regarded as an insignificant disease, so far as the white population is concerned. Among the aboriginal inhabitants, however, the story is a very different one, and the malady, once unknown in the island before the advent of the European, has played no small part in the extermination of the native population, and is still frequent among them. In New Zealand venereal diseases are as rare as in Australia.

Africa. Syphilis is widely distributed in Egypt and the Sudan. It is found in all parts of the Nile Valley, and British troops stationed in the country are more tainted by it than at home. Among the Baggaras, now a scattered and shattered race, but a few years ago the ruling people in the Sudan, venereal diseases were rife to a terrible degree. Father Ohrwalder, who passed ten

[1] *Chinese Customs Medical Reports.* No. 45.

[2] *Archives of Surgery*, 1898, p. 380.

[3] *The Solomon Islands and their Natives.* By H. B. Guppy, M.B., F.R.G.S. London, 1887.

[4] "Notes sur les Nouvelles Hébrides," par le Dr Bernal. *Arch. de Méd. Navale*, Aug. 1899.

years of captivity in the Mahdi's camp, has given a graphic picture of the ravages of these maladies in the Baggaras and their slaves. "The immorality of the slaves," he writes, "is quite beyond description, but it cannot be the fault of the unfortunate creatures themselves, for in their own savage homes it is not so. They learn all the vices of their masters, and indeed are forced to participate in them or submit to a flogging; consequently disease of the most loathsome kind is everywhere prevalent, and to be free from it is thought to be the mark of a poor creature....At first the Baggara were not affected to any large extent; but contact with the inhabitants of the Nile valley has communicated the pest, which is now eating into the constitution of this, the most powerful and warlike tribe in the Sudan." Things have changed greatly since those words were written, and "the most powerful and warlike tribe in the Sudan" is such no longer.

In Tunis and Algeria syphilis and other venereal diseases are far from rare. In Algeria soft chancres are remarkably common, and the French troops stationed there suffer more from this affection than French troops in any other part of the world[1]. Syphilis has also been exceedingly widely prevalent since the French occupation. Throughout Morocco syphilis of a malignant type is widely met with, and there can be little doubt that the population of the greater part of the West Coast of Africa is seriously infected by this and other venereal diseases. In Senegal very virulent types of these disorders are common. In Sierra Leone and Lagos many cases of the kind are annually treated in the hospitals. Among the natives of the Gold Coast Colony "one of the gravest features" it is said, in regard to the principal diseases from which they suffer, "is the prevalence of untreated syphilis." These people refuse to submit to the lengthened course of treatment required for its control, and consequently the disease is practically left untreated, with the result that a large proportion of the population are seriously crippled and the economic loss to the colony is considerable. Here, as elsewhere among ignorant races, the malady is spread largely by non-venereal means. In the Cameroon district, on the other hand, syphilis is said to be very rare, and was at one time quite unknown.

In the Belgian Congo this disease has apparently been imported by European and Arab intruders, the aboriginal inhabitants having

[1] Brault.

formerly been quite free from it[1]. Syphilis is spoken of as rare in German Central Africa, while gonorrhœa, on the other hand, is extraordinarily prevalent there.

On the eastern side of the African continent venereal diseases are widely met with. One of the principal centres of syphilis here is found in Nubia, whence it has been spread by soldiers to the country bordering the northern lakes[2]. In Abyssinia the disease is universally existent. The natives of Uganda and of Manyuema also suffer greatly. In the southern lake region syphilis appears to have been formerly little known, but it has increased of late years owing to more frequent communication with the coast[3]. On the upper and middle Zambesi, on the other hand, syphilis is said to be rare, and in British East Africa it is exceedingly rare in the interior, though not uncommonly seen on the coast. Gonorrhœa is, on the other hand, frequently seen in this possession[4]. In Zanzibar syphilis is or was excessively common; and one author, writing some years ago, even went so far as to assert that five-sixths of the population were probably syphilitic[5].

Venereal diseases of all kinds are excessively widely spread throughout South Africa. In Cape Colony syphilis is a common disorder; the natives suffer greatly from its ravages, and in some districts, as for example that of Molteno, it is said to be distinctly increasing. Similarly reports from Prieska, from the Durban Hospital, from Mafeking, and from Mount Frere for the year 1898 spoke of a serious degree of prevalence of syphilis among both the native and the white populations there, and in most of these places it was becoming more, rather than less prevalent. In the large native locations in the Vryburg district the disease was also described in the year preceding the outbreak of the war as excessively common. So also in the native territories of Tembuland, the Transkei, and Pondoland syphilis was then very widely met with, and no measures could be enforced to check its spread. In Namaqualand the natives suffered considerably from the same disorder, but gonorrhœa seems to have been rare. In Bechuanaland syphilis is, or was, no less frequent, and in Basutoland—

[1] Congrès Nat. d'Hygiène et de Climatologie Médicale de la Belgique et du Congo, 1897. *Comptes Rendus*.
[2] Davidson. [3] *Ibid*. [4] Kolb.
[5] Lostalot-Bachoué (1876), quoted by Hirsch.

where, as generally in South Africa, the disease is known by the name of *mocaula*—it was said to be "extremely prevalent." Among the native populations of South Africa this group of diseases appears to be spread to a very great extent by non-venereal methods.

In Mauritius syphilis is far from rare, and the Indian population in particular is said to suffer considerably from its ravages.

America. Little is known with accuracy in regard to the degree to which these diseases prevail in the far north of the American continent. In Greenland syphilis is said to be un-known, in spite of the visits of whaling vessels and the existence of prostitution. In Sitka, Alaska, and in the Aleutian Islands, on the other hand, the disease is by no means rare. Whether in the more temperate provinces of Canada it is common or rare I have no data for determining. This and other venereal diseases find occasional mention in the Health Reports from most of the Canadian provinces, and neither the colonists nor the native population are quite free from them. Syphilis was at one time, and is perhaps still, very common and malignant among the natives in British Columbia and Vancouver Island, and to a less extent in Ontario. The "Ottawa disease" described by Stratton in 1849 as affecting the Indians on the banks of the Grand River, Lake Erie, is believed to have been syphilis.

In regard to the degree of frequency or infrequency of these diseases in the United States of America I have also no accurate data at hand. All writers are, however, agreed that venereal diseases, and especially syphilis, have in the recent past ravaged to a terrible extent the native Indian population of the States. As among the aboriginal tribes of Africa, so among those of North America, syphilis was unknown before the coming of the white man, but once introduced amongst them it spread like an epidemic, and has no doubt been an important factor in the gradual disappearance of these peoples before the advance of so-called civilisation. In the Southern States the infection appears to have been introduced among the native tribes from Mexico, and the various tribes suffered in proportion as they held intercourse with the Mexicans. In California the opening of the gold-fields led to a serious diffusion of the disease. Mexico itself is, or was some decades ago, one of the most intense centres of syphilis in the Western hemisphere. In the

Central American States also the disease was widely spread; but in British Honduras at least syphilis is at the present day not at all common.

In the various islands of the West Indies, to judge from recent Health Reports, syphilis is common in some, less so in others, but is probably not entirely absent from any. In Jamaica it is very prevalent and appears to be increasingly so; the same statement holds good for St Vincent, and apparently for most of the Leeward Islands. In the Virgin Islands, however, syphilis is said to be rare.

The northern regions of the South American continent are not a little affected by syphilis. In British Guiana[1] primary venereal sores are said to be greatly on the increase; and all races are reported to suffer alike. In Dutch Guiana, however, syphilis is not frequent and is generally benign, "except among the negroes in the forests, amongst whom it is more common and severe." But it is in Brazil that the most serious prevalence of this disease is met with, and it seems doubtful whether any other country in the world, with the possible exception of Mexico, is so deeply infected by it. The low state of morals, the indulgence in vices of an indescribable character, and the general ignorance and carelessness of the people are mainly responsible for this extraordinary prevalence of syphilis. The disease is met with in some of its worst forms, and the mortality it causes is excessively high.

The remaining South American States are apparently little less ravaged than Brazil by this and other venereal diseases. In Chile, Bolivia, and Peru, in ·Paraguay and the Argentine Republic, syphilis is, or was within quite recent times, prevalent to a truly terrible extent; and, if reports from these countries are not exaggerated, it must be accepted that South America as a whole is incomparably more deeply ravaged by this disease than any other continent.

Factors concerned in the Geographical Distribution. It will be seen from this rapid review of the distribution of venereal disorders that they are found in almost all parts of the world. While this statement is true of the group of diseases as a whole, it seems to be no less true of the most important

[1] *Report of the Surgeon-General for* 1898–99.

member of the group. Almost the only regions on the earth's surface where syphilis has not been seen are Greenland, and certain remote parts of North and South America and South Africa, inhabited by tribes who have had little or no intercourse with the rest of the world.

Syphilis has shown itself to be quite regardless of lines of latitude or isotherms. Its absence or rarity in such northerly latitudes as those of Greenland and Iceland cannot be attributed to distance from the equator, for it has been exceedingly rife in Alaska, British Columbia, and Vancouver Island. It has been very generally thought that the type of the disease was more severe in cold countries than in warm, but even this must be doubted. Some of the worst forms of syphilis appear to be met with in such countries as Bokhara, China, Brazil, Morocco, Senegal, and Western Africa generally—all countries with warm or temperate climates. The severity of the disease, indeed, seems to be less a question of climate than of treatment. In all the countries just named the cases among the native populations are left to a large extent without any treatment at all, and hence they assume a most malignant and virulent type, such as is rarely seen in more civilised communities.

Altitude above sea-level has also apparently little or no influence upon the distribution of these diseases. Racial conditions may possibly have something to do with the frequency or infrequency of syphilis in a community, and the belief has been expressed that the inhabitants of certain countries,—such as those of Iceland, Newfoundland, Greenland, and Madagascar— were relatively insusceptible to the malady. But it is questionable whether its rarity in those countries is to be ascribed altogether to racial causes, or whether it is not rather due to other factors, among which—at least in some instances—a higher level of morals and an absence of irregular and uncontrolled intercourse must be granted some share. A study of the history of the disease also shows that certain races have for a time remained remarkably free from it, giving rise to the belief that they possessed a racial immunity, while subsequent events have shown that no such immunity has existed. This, for example, has been the case in regard to the natives of Bechuanaland, who thirty years ago were thought to be relatively insusceptible to syphilis, but who are now seriously ravaged by the disease.

It was pointed out in the beginning of this chapter that of the origin and antiquity of syphilis nothing is known with certainty save that it has probably existed from very remote times. At the present day the disorder is generally associated with a certain degree of civilisation. The history of recent centuries has shown that it is not found among the aboriginal inhabitants of newly explored countries, and that it has been invariably introduced by the soldier, the trader, and the settler, who, while being the direct means of importing the actual virus of the disease, have been too often the indirect means of introducing those particular vices which encourage its spread. In North America, in Brazil and the other South American States, in Siberia, Australia, and Africa, syphilis has been communicated to the native populations by the white man, and in many instances has committed terrible havoc amongst them. Of the many factors which taken together lead to the gradual extermination of the more primitive races when brought into contact with the more advanced, this disease is probably one of the most important.

Venereal diseases are, as already stated, in their essence controllable, and it has been proved over and over again that their degree of prevalence can be lessened greatly by police regulations and other measures. The outcry raised by a section of the public at home over the enforcement of the Contagious Diseases Act may have been based on a perfectly genuine and legitimate sentiment, but it should not be allowed to blind the eyes of even the most sensitive of purists to certain obvious facts. It is an undeniable scientific fact that these diseases have always been least prevalent where measures such as those prescribed by the Acts in question have been stringently enforced, and most so where such measures have been in abeyance.

If left uncontrolled by any kind of adequate precautions to prevent its spreading, both by venereal and non-venereal channels, syphilis may become so common a disease as to justify the use of the term epidemic in regard to it. War with its unrestrained license—when war was conducted on less humane and civilised principles than it is at the present day—has been in the past the most frequent and powerful cause of these so-called syphilis "epidemics," or "epidemo-endemics." A considerable number of well-known historical examples of this occurrence have been recorded. The *Sibbens* or *Sivvens*, a form of syphilis which

appeared in the south-western parts of Scotland in the middle
of the seventeenth century and later spread to the Highlands,
was apparently introduced to the country as the result of the
invasion by Cromwell's troops. The *Radesyge*, an endemic
manifestation of syphilis which occurred in Norway early in
the eighteenth century, was carried to Sweden in 1762 by the
Swedish troops returning from the Seven Years' War; and later,
in 1790, syphilis was again imported to Sweden in like manner
at the end of the war in Finland. The syphiloids of Jutland,
Lithuania and Courland—endemic or epidemic outbreaks which
occurred in the latter half of the eighteenth century—were
attributed to the landing of Russian troops in those countries.
Similarly syphilis became endemic in Servia in 1810 and subsequent years, after the occupation of the country by Russo-
Servian and Turkish troops; and the Russo-Turkish war of
1828–9 led to a wide spread of the disorder. But perhaps the
most striking example of this nature was the disastrous epidemic
of the disease through a great part of Europe in the fifteenth century, to which reference has already been made. "It was especially
Charles VIII's army of mercenaries," writes Hirsch in relation
to this event, "returning from Italy, relaxed by licentiousness,
broken up into lawless bands and overrunning France, Switzerland,
the Netherlands and Germany, which carried with them, as we
are expressly told by many medical writers and chroniclers of the
time, the germs of syphilis over the whole country, wherever an
adventurous life led them."

But syphilis is also at times spread on a large scale by the
movements of bodies of men other than troops, and in times of
peace. The well-known Dithmarschen or Holstein disease, which
was a very severe endemo-epidemic of syphilis lasting from the
latter half of the eighteenth century well on to the middle of
the nineteenth, was certainly spread and was perhaps originated
in this manner. It was at any rate associated at its origin with
the arrival of a large number of foreign navvies, especially from
East Friesland, who flocked into the Süder-Dithmarschen for the
work of embanking the Crown Prince Dyke. In exactly the same
manner in Russia at the present day, few things aid more in the
spread of syphilis than the movements of bodies of labourers who
annually leave their homes to find work in other parts of the
country, and on returning bring the infection of syphilis with them.

TETANUS AND TRISMUS
NEONATORUM.

General Characters and Etiology. The clinical features of an attack of tetanus are too familiar to need description here. The disease may be "idiopathic" or it may follow on injury. In all cases it must be supposed that some solution of continuity of the skin exists, and that the tetanus bacillus gains access to the tissues by this means. The specific causal relation of this bacillus to the disease has been finally proved. The bacillus exists principally in the soil and in manure heaps, and hence tetanus is particularly liable to occur in association with wounds that have been contaminated with one or other of these materials. Puerperal tetanus is occasionally seen after childbirth, and in newly-born children tetanus or trismus neonatorum is by no means rare in some countries. All forms of tetanus have become much rarer since the introduction of antisepsis and asepsis in the treatment of wounds and in midwifery.

The lower animals are liable to suffer from tetanus; the horse is the most susceptible, but oxen and sheep have also been attacked by it.

History. Tetanus was known to the earliest medical writers from Hippocrates downwards. There is nothing to show that the disease has varied in frequency or in its general geographical distribution at different periods of the world's history.

The infective nature of the disease was first shown in 1884 by Carle and Rattone, who reproduced the symptoms in animals by inoculation with the secretions of a wound in a tetanic patient. Nicolaier, a year later, produced the disease in mice by inoculating them with garden earth, and he observed the tetanus bacillus in their tissues. Kitasato, in 1889, first succeeded in obtaining the bacillus in pure culture.

Geographical Distribution. *Europe.* In the temperate countries of Europe tetanus is not a particularly common disorder. In the British Isles it is only occasionally seen. Trismus neonatorum was at one time extraordinarily frequent in St Kilda, in the western Hebrides, but is said to have been brought to an end since a trained nurse was sent to the island to teach the midwives to apply antiseptics to the navels of young infants. It is noteworthy that this form of tetanus is said to be quite unknown in the Faröes, while in Iceland, on the other hand, it is, or was some decades ago, most disastrously prevalent and fatal. Traumatic and idiopathic tetanus would, on the whole, appear to be somewhat more frequent in southern Europe than in the north. In Russia the public health returns seem to indicate that the disease is remarkably infrequent in all the European provinces. In Italy it is commoner, and causes an annual mortality of from 20 to 35 per million. The tetanus of infants is comparatively frequent in Italy, and in former times it was very common in some parts of Spain and in the Balearic Isles. Tetanus occurs in Turkey, but with no very striking frequency.

Asia. The returns from the Caucasus, Central Asia, and Siberia, if trustworthy, would show that tetanus was a very rare disorder in all these regions. But, as has been repeatedly pointed out before, the statistics from these parts of the Russian Empire are too imperfect to be taken as the basis of any final conclusion. In Arabia and Syria tetanus is said to be common. It is much more frequent also in India and in Ceylon than in most European countries, though its degree of prevalence varies considerably in different parts of the Indian peninsula. Cases of the disease are not infrequent in the Singapore prisons. In the East Indies and in Cochin China, on the other hand, it does not seem to be a particularly common disorder.

Australasia. Regarding Australia information seems to be lacking, but in some of the Pacific Islands, as for example New Caledonia, tetanus is said to be remarkably frequent. Trismus neonatorum is a common cause of death in many of the islands.

Africa. In Egypt tetanus is a common disease. In Algeria it is particularly frequent in the provinces of Oran and Constantine. Senegambia is spoken of as one of its principal seats, and certainly in the colonial hospitals both of the Gambia and of Lagos a considerable number of cases of the disease are yearly seen. In the

Cameroon territory, however, Plehn states that he never saw a case. At the Cape tetanus is said to be decidedly common, and in Madagascar and Mauritius it seems to be particularly frequent. In the latter island about two-thirds of the deaths ascribed to tetanus are due to trismus neonatorum.

America. Tetanus does not seem to be a very common disorder in Canada nor in the northern United States. But in some of the Southern States, and also in Mexico, Central America, and some of the West Indies it is peculiarly rife. Thus in British Honduras it is said to be exceedingly frequent, and here, as elsewhere, its frequency is ascribed to the great amount of soil pollution by stable manure[1]. In Guatemala also it is the cause of a considerable mortality.

But it is in some of the West Indies that this disease attains its highest degree of prevalence. In Cuba it is extraordinarily common, and the utmost care is needed in treating wounds to prevent tetanus supervening. The same appears to be true for San Domingo, Jamaica, and many of the other islands. In Barbados a remarkable epidemic of tetanus occurred in the General Hospital, between March and June, 1897. In some of the islands, however—St Lucia and Trinidad—the disease is rarer.

In British Guiana and generally in the tropical portions of Brazil, the Argentine Republic, Peru and Ecuador, tetanus is exceedingly prevalent; while trismus neonatorum accounts for an enormously high mortality.

Factors concerned in the Distribution. Tetanus, it will be seen, is very much more often met with in tropical and subtropical countries than in the more temperate zones. This may perhaps be explained by the conditions being more favourable in a hot than in a cold climate for the growth and multiplication of the tetanus bacillus in the soil, this bacillus being essentially a saprophytic organism. The absence or scantiness of the clothing worn by the natives of tropical countries, their habit of going about bare-footed, and their frequent neglect to cover up or treat skin abrasions, ulcers, or injuries, may also be held to account for their greater liability to the disorder.

Any skin discontinuity, however small, may suffice for the entrance of the bacillus to the tissues. Some have believed that

[1] Report of the Colonial Surgeon for 1898.

chill alone, without any skin abrasion, suffices to bring on an attack, but this view can scarcely be held now. The abrasion may be merely the prick of a thorn, or the bite of a mosquito. In newly-born infants the site of entry of the bacillus is probably usually at the navel, during the separation of the stump of the umbilical cord. Whatever may be the nature of the abrasion, the more it is exposed to the chance of dust, soil, or manure gaining access to it, the greater is the chance of an attack of tetanus. Hence, as already stated, ordinary tetanus, trismus neonatorum, and puerperal tetanus have diminished greatly wherever the principles of asepsis, whether in surgery or midwifery, have been understood and applied. Hence, also, in all probability the frequency of infantile trismus in some cold countries, such as Iceland, in reversal of the general rule in regard to the tetanus of adults, which, as just stated, is most common in hot latitudes. It is probable that wherever the tetanus bacillus exists, if infants are not kept scrupulously clean, or are allowed to roll about on the dirty floors of crowded and insanitary huts, trismus neonatorum will be a common disease. Its absence from the Faröes is remarkable, and it would be interesting to know whether the tetanus bacillus exists in the soil of those islands.

Of all races the negro is said to be most susceptible to tetanus. In the tropics the natives suffer much more than the European residents; and this is particularly true in regard to infantile trismus. But whether this is due to a real racial difference in susceptibility, or whether it is caused by the greater exposure of natives to the risk of soil-contamination of wounds, is open to question.

TUBERCLE.

General Characters and Etiology. The large group of diseases to which the term tubercular is applied has a distribution little short of world-wide. In this respect it is equalled or perhaps even surpassed by some other diseases, but in the enormous sum of human mortality to which it gives rise this class of disorder stands without a rival. In many countries it is quite common to find that one-twelfth or one-tenth of the deaths from all causes are due to it, and in some smaller areas and communities the proportion may rise to one-eighth, one-sixth, or even one-fifth.

While the manifestations of tubercular disease are very varied, depending principally upon the organ or tissue primarily affected, they are rightly treated as a single, etiologically distinct group, for all are associated with the presence and action of a specific bacillus, the *bacillus tuberculosis* of Koch. The principal forms of disease usually included in statistical returns under the heading of "tubercular" are pulmonary consumption or phthisis, tubercular meningitis, tabes mesenterica or abdominal tuberculosis, and general disseminated tuberculosis. In a few countries the figures for each form of the disease are available separately, in others they are returned together in one total, while in others again only the figures for pulmonary tuberculosis or phthisis are to be obtained. In the following discussion of the geographical distribution of these diseases all forms of tubercular affection are, whenever possible, dealt with together, as neither facts nor figures are available for a comparison of the distribution of each form separately. It is scarcely necessary to add that in every case the form of tubercular disease which contributes most largely to the total sum of tubercular mortality at all ages is the pulmonary form. Tubercular meningitis and tabes mesenterica, on the other hand, are the most fatal forms of the disease in infancy.

29—2

History. It is believed that pulmonary consumption is one of the oldest of known diseases. Mention of it is found in some of the earliest of medical writings. It would seem to have become commoner in most civilised countries in modern than in earlier times, but the evidence on this point is very uncertain. Within quite recent decades there is reason to believe that it has become rather less frequent than it was half a century ago in some European countries, while in others its prevalence has increased.

A diminution in phthisis has been observed in England and Scotland, and in Prussia, Denmark, and Italy. In Russia, Ireland, Würtemburg, and Norway the frequency of "consumption" appears on the other hand to be on the increase. Outside Europe there has been a slight decrease in the prevalence of the disease in Queensland, and in some at least of the United States, as for example, Massachusetts and Pennsylvania; while there has been an increase in the colony of Victoria, in the Caucasus, among the Bashkirs, in Greenland, and in some parts of Canada.

The history of other tubercular affections is practically unknown. The tubercle bacillus was first identified by Koch, who published his discovery in 1882.

Recent Geographical Distribution. *Europe.* While tubercular diseases are the cause of a very high degree of mortality in the British Isles, it is some satisfaction to note that the deaths due to them have of recent years become distinctly fewer than formerly. During the twenty years ending in 1890 the deaths under this heading in England and Wales diminished by as many as 25 per cent., and the fall has been steadily maintained since. But "consumption" and the other forms of the disease produced by the action of the tubercle bacillus are still excessively prevalent, and cause a total mortality which in 1900 was no less than 1899 per million living. London maintains an unenviable reputation for a high death-rate from these causes, but they are found not less active in such rural counties as Hampshire and Devonshire than in districts mainly devoted to manufactures, such as Lancashire, Northumberland, and South Wales. In Scotland the diminution in mortality from these diseases has been less marked than south of the Border, and the deaths they cause are still considerably over 2000 per million inhabitants each year; while in Ireland, far from decreasing, tubercular affections are yearly causing a greater mortality. Thus the ratio in the sister isle rose from an average of 2600 per million in the decade 1871–80, to one of 2700 in the next decade, and in the years 1897–1900 it was not less than 2900.

Tubercular diseases are also excessively rife in France. Statistics for the entire population are not available, but in the towns they gave rise to a mortality which in the years 1887-90 was as high as 3480 per million living, and in 1898, though somewhat lower, was still over 3400 per million. Of this total pulmonary phthisis was responsible for over three-fourths. In Belgium this group of affections causes a mortality exceeding that produced by any other disease or group of diseases, and the same would appear to be true of Holland[1]. In Germany tubercular diseases are equally, or possibly more prevalent. In Prussia they are perhaps more so than elsewhere, but in the last twenty years the mortality from this cause has diminished considerably, while in Würtemburg the figures, though showing a somewhat lower degree of prevalence than in Prussia, have indicated an increase rather than decrease of these diseases.

In Denmark tubercle seems to be less active as a cause of death than in France or Germany; the deaths in the Danish towns in the years 1892–97 have been less than 1800 per million each year, and show a tendency to decrease. The great prevalence of tubercle in Norway and Sweden is shown by the fact that in the former the deaths from tubercular diseases of the lungs alone account for 15 or 16 per cent. of the deaths from all causes, while in the latter these diseases are particularly fatal. In Norway the mortality from all forms of tubercular affections averages about 2300–2500 per annum, and appears to be increasing. Lapland, on the other hand, is comparatively free from tubercular disease.

Throughout European Russia tubercle is found with great frequency. It is, perhaps, on the whole more prevalent in the north than in the centre, and in the centre than in the south, but it is most prevalent of all in the Baltic provinces. Recently

---

[1] No useful statistics of the tubercular death-rate in Holland are available; the returns published in the *Annuaire Statistique des Pays-Bas* contain the following three headings, each of which probably includes part of the mortality from this group of diseases, viz :—"debility and phthisis," "laryngeal and pulmonary phthisis, haemoptysis and diabetes," and finally "chronic diseases of the respiratory organs."

published statistics indicate that it is increasing in most parts of the country.

In Austria-Hungary this group of diseases appears to be more fatal than in any other European country, pulmonary tubercle alone causing a mortality rate of 3682 per million per annum in the years 1887–93 in Austria proper, and one of 3008 in Hungary (with Fiume) in the same period. In Switzerland these diseases are less frequent, but the death-rate from phthisis in the Swiss towns is by no means inconsiderable. The pure mountain air of the Swiss Alps has long been known as peculiarly inimical to the causes of "consumption" and peculiarly favourable to its treatment. Tubercle is said to be exceedingly common and fatal in Spain and Portugal, while in Italy, though still very frequent, it appears to be decreasing, the mortality from this cause having recently fallen to less than 2000 per million. In the countries of the Balkan peninsula pulmonary phthisis is one of the most fatal of diseases. In Servia this variety of tubercular affection alone caused a mortality of some 2400 per million in the years 1894–5. In European Turkey the mortality from this cause among the poorer classes of the population is also very high; in Constantinople itself, in Prevesa, and in other places tubercular disorders are a greater cause of death than any other variety of disease. In Montenegro, in spite of the mountainous nature of the country, "consumption" is exceedingly common. So also in Greece tuberculosis causes a high annual mortality, particularly in the towns, but it becomes rather less prevalent and fatal as the mountainous part of the country is approached[1]. In the Greek prisons tubercle is extremely rife.

In the Faröe Islands it is a remarkable fact that pulmonary phthisis appears to be almost unknown. Other forms of tubercular disease have been met with there in recent years, but it is believed that these cases can be traced to the importation of the germ in the meat of diseased Scotch or English cattle. Not less remarkable is the comparative rarity of consumption in Iceland.

Asia. In Arabia tubercular diseases are common enough on the coasts, though less so on the rocky table-lands in the interior. At Medina "consumption" is said to be a prolific source of death among the female population, and this is attributed to the

[1] *Reports of the 1st Pan-Hellenic Medical Congress,* 1901.

custom of early marriage, and consequent early child-bearing, which prevails there as in most Moslem countries. In many parts of Syria and Asia Minor this class of disease is also very prevalent and fatal, particularly on the plains, while in the higher-lying districts it is less common. In Mitylene and in Rhodes tubercular affections cause more deaths than any other form of disease. Erzerum and the mountainous country in its neighbourhood are spoken of as remarkably free from them, though when they do occur there they run a rapid course. The same is to a great extent true of the mountainous parts of the Caucasus, the flat plains of the Caucasus proper (the provinces of Stavropol, of the Kuban and of the Terek) returning comparatively high mortality-rates from tubercle, while in many parts of the mountainous region of Transcaucasia the tubercular mortality is exceedingly low. At one time (in the sixties) it was even believed that the natives of these regions were quite free from these diseases; and that they were very rare seems to be proved by the fact that among the autopsies performed in a large hospital—that at Kutais—in the years 1863–5 tubercle was found to be the cause of death in but one instance. In quite recent years, however, these diseases have, it is said, become very much more prevalent, and in the last ten years have taken a serious hold of many of the Transcaucasian towns, such as Tiflis and Kutais. The villagers and hill-dwellers are still comparatively free from their ravages[1].

The official returns from Russian Central Asia (Transcaspia, Turkestan, Turgai, Uralsk, etc.) also indicate a remarkably low degree of prevalence of tubercular diseases in these regions, but the figures must be accepted with much reserve. Further north the immense stretches of steppe country which extend eastwards from the Ural river have acquired a reputation for a high degree of immunity from tubercular affections, and large numbers of tubercular patients from European Russia annually migrate thither to undergo the so-called "Koumiss treatment," which, together with the extreme purity of the steppe-air, and the open air life enjoined, has produced very favourable results in many instances. The Bashkir inhabitants of these regions were at one time believed to be quite free from consumption, but a Russian physician,

[1] Pantiukhof.

Nikolski, observed or collected records of 92 cases of tubercular disease among them between the years 1880 and 1895, and he asserts that its prevalence here is increasing, in spite of the fresh pure air and the koumiss diet to which they are accustomed.

In Siberia, if the official statistics may be accepted as even approximately accurate, tubercular affections are considerably more active than in the Caucasus and Central Asia, though much less so than in European Russia. The disappearance of the native Siberian tribes, the Buriats, the Yakuts, the Tunguses and others, which is rapidly being brought about by many social, political, and other causes, is, it is said, not a little hastened by the prevalence of tubercular disease amongst them.

In Persia, Afghanistan, and Baluchistan, this form of disease is said to be comparatively rare, but the information hitherto available in regard to these countries is scanty, and fuller knowledge may reveal that it is commoner here than is supposed. At Suleimaniyé and other places on the Turkish side of the Turco-Persian frontier it is certainly far from rare.

Throughout the Indian peninsula tubercular diseases are met with, with varying frequency, both among the natives and in Europeans. In general terms it would seem that Assam and Lower Bengal suffer the least from the ravages of the tubercle bacillus, while in Southern India, on the west coast and in Rajputana, the group of affections which it gives rise to are exceedingly common. In Indian gaols these diseases are the cause of a considerable annual mortality; and among the troops in India, both native and British, they are not rare. Of the native troops the Gurkhas are said to be particularly liable to consumption, and it is the principal cause of death amongst them. It is to be noted, however, that in Nepal, the home of the Gurkha race, phthisis is said to be "far from prevalent[1]," and the high mortality from this cause to which Gurkha troops appear to be liable elsewhere, must therefore be ascribed to other than racial reasons. On the southern slopes of the Himalayas consumption is met with, even at considerable heights, but it is less common here than on the plains. In Kashmir it is said to be rare, and to occur only among the shawl-weavers and others who lead an indoor life. Returns from Ceylon show that tubercular diseases, though

[1] *Report of the Sanitary Commis. with the Government of India,* 1897.

the cause of nearly 3000 deaths annually, are much less prevalent in the island than in any European country.

In many of the East Indian Islands phthisis appears to be rare, though certainly not unknown. In Borneo, however, it is said to be almost unknown among the natives, Malays or Dyaks, and Nieuwenhuis, who travelled across the island in 1894, only saw one case of incontestable tubercle in a native. In Java phthisis is said to occur to a considerable extent both in Europeans and natives. In the Malay Peninsula, however, phthisis and tubercular disease generally are comparatively uncommon. The number of patients treated for these maladies in the hospitals of the Straits Settlements is remarkably small in comparison with the numbers admitted for other disorders. Lupus also appears to be a particularly rare affection in the Straits Settlements. In French Indo-China tubercle is perhaps commoner, and in Siam tubercular diseases are said to cause a considerable mortality (Rasch).

Through the whole of China these diseases are frequently met with. The Chinese appear to be very susceptible to the action of the tubercle bacillus, and cases of consumption with them often run a very rapid course. Corea shows no freedom from the disease; the Japanese inhabitants of Chemulpo and other centres being particularly liable to its attacks. In both Coreans and Japanese, just as in the case of the Chinese, the course of the malady is often extremely acute. In Japan itself tuberculosis is widely met with; it is regarded as the principal scourge of the rich class of nobles, while syphilis is the commonest disease among the middle classes, and leprosy among the outcasts[1].

Australasia. In New Guinea phthisis is rare, and was probably unknown to the original Papuan inhabitants. Many parts of Australia, on the other hand, present a moderately high degree of tuberculosis prevalence. In Queensland phthisis is the most fatal of all diseases; the death-rate from this cause is however decreasing, and has fallen from 1115 per million in 1893 to 874 per million in 1897. The Pacific Islanders resident in the colony seem to be particularly liable to tubercular diseases, and nearly one-half of the deaths amongst them are due to some affection of this nature; moreover, while they form less than

---

[1] *Archives of Surgery*, 1898, p. 380.

2 per cent. of the total population, they contribute nearly 22 per cent. of the total deaths recorded from phthisis. In South Australia the degree of prevalence of phthisis appears to be about the same as in Queensland. In Victoria this form of tubercular affection is, however, much more active than in any other of the Australian colonies, and its frequency is said to be increasing[1]. Tasmania, on the other hand, seems to present conditions much less favourable to the ravages of the tubercle bacillus, and the phthisis mortality in 1896 varied between 450 per million in the midland districts and 800 per million in the north-eastern. In New Zealand, as in Australia generally, pulmonary phthisis is the cause of more deaths each year than any other disease. The mortality from this and all other forms of tubercular affection was remarkably constant during the 1889–98 decade, always slightly exceeding the thousand per million. It will be observed from these figures that both in Australia and New Zealand tubercular diseases, although in many parts of these colonies the most fatal of all maladies, are nevertheless much less potent as a cause of death than they are in Great Britain or in many other European countries. Some parts of Australia are regarded as presenting remarkably favourable conditions for the relief of consumptive patients, and in several places sanatoria have been opened for their reception and treatment.

In many islands of the Pacific, phthisis and other tubercular disorders are met with with considerable frequency. In the Fiji group, in the Society Islands, in the New Hebrides, and in the Hawaiian or Sandwich Islands phthisis causes a deplorable mortality among the natives, and in other groups it is very prevalent.

Africa. Egypt appears to possess a remarkable degree of freedom from the ravages of tubercular diseases. Phthisis is met with to some extent in Lower Egypt, about the mouth of the Nile and at Cairo, but in Middle and Upper Egypt it is very rare. The air of the desert, or such part of it as is not annually inundated by the Nile, being quite clear and free from a trace of humidity, offers conditions exactly the reverse of those needed for the existence of the tubercle bacillus as a cause of endemic disease; and in certain forms of pulmonary tubercle the climate here seems capable of

[1] McAdam, *Intercolonial Medical Journal*, Jan. 20, 1899.

arresting the process in the lungs, at least for a time. Upper Egypt has for some time taken its place among health-resorts for consumptives. Abyssinia is also reported to enjoy considerable immunity from tubercular affections.

In Tunisia there is reason to believe that tubercular diseases are also rare. Exact information on this point is difficult to obtain, but in spite of the occurrence of cases of phthisis among the natives, the French authorities believe the disease to be far from common. The Tunisian *" Corps d'occupation "* presents fewer cases of tubercular affection than any other corps of the French army; and the fact is also noteworthy that in the cattle killed at the Tunis abattoir cases of tubercle are quite exceptional[1].

In Algeria, on the contrary, tubercular diseases are far from rare. Gros believes that they have become much more frequent since the war which ended in the French occupation, though not unknown before[2]. The number of cases of consumption is said to be largely augmented by the arrival of phthisical subjects from France, and also of immigrants from Spain, many of whom are tubercular. Among the Arab population of Algeria the women suffer to some extent, but among the Jews phthisis is very rare. The nomad Arabs of Morocco are seldom the subjects of tubercle, but cases of phthisis are seen in the towns of that country.

Turning to the West Coast of Africa it is found that there is a considerable amount of "consumption" among the natives of the Gambia, where in the months of December and January the winds are often very cold, and the ill-clad natives easily succumb to the attacks of the tubercular micro-organism. European settlers on this coast do not suffer greatly from phthisis. In Sierra Leone tubercular affections are apparently not common, though not altogether absent. The same statement is true of Lagos and generally of the coasts of Guinea. In the Cameroons phthisis is said to be unknown among the natives[3].

The great susceptibility of the black races to consumption is notorious, and when negroes are removed from their African homes to the cooler climates of Europe or America this is the

---

[1] *La Tunisie. Histoire et Description.* Paris and Nancy. 1896.

[2] *Janus,* 1899.

[3] While pulmonary tubercle appears to be not a common affection in West Africa generally, very marked cases of destructive lupus are at times met with. (MacGregor.)

most prolific cause of death amongst them.   It seems certain that
they also suffer from it in their own country.   It has just been
shown that in some parts of the West Coast this disease is far
from rare, and in the interior of the African continent it is also
met with, though with what degree of frequency it is difficult to
say.   In the Congo Free State consumption is not uncommonly
seen.   In German Central Africa cases of the disease are met
with, though not with great frequency.   On the east side of the
continent it is found to be rare, or at least comparatively rare, on
the coast region.   More inland, in the neighbourhood of the great
lakes, it is perhaps commoner.   Pruen, though not specifically
mentioning consumption, states that lung affections are very
common among the natives of this part of Africa.   He points
out that the sudden changes of temperature in the rainy season,
with no corresponding change in the garments worn, notoriously
induce these conditions, and lung affections secondary to malarial
fever are a fruitful source of death.   It may be reasonably
supposed that part of this mortality from lung disease is due
to pulmonary phthisis.   In Uganda all forms of tubercular
affections seem to be common.

  In the cooler climate, or climates, of South Africa the natives
are found to suffer from the diseases now under consideration to
a still greater extent than in the tropical regions of the continent.
In the native territories of Tembuland, Pondoland and the
Transkei phthisis is spoken of as "alarmingly prevalent" in spite
of the open-air life led by the people[1].   Among natives living in
towns, such as Kimberley, East London, and Grahamstown, the
disease is said to be making rapid strides, no doubt on account
of the conditions under which the people live, but from wholly
rural districts reports of the great ravages caused by this form of
disease are also forthcoming.

  On the other hand the white population of South Africa
appear to show a lesser tendency to suffer from tubercular
disease than the natives.   Phthisis is said to be commoner along
the coast than inland, and in some towns it is a cause of high
mortality.   At Port Elizabeth, Grahamstown, King William's
Town, Cape Town, and Durban, a considerable number of
phthisical subjects come under treatment every year, and in

  [1] *Cape of Good Hope Reports upon the Government and State-aided Hospitals
and Asylums for* 1898.

some of these towns phthisis causes more deaths than any other disease. It is to be noted that a certain proportion of these cases dealt with in South African hospitals are imported from England or elsewhere; and in recent reports upon the health of the South African colonies may be found repeated protests against the sending of consumptive persons, often in a most advanced and hopeless state, to a country where many of them have no friends or means of support, and eventually become a charge on the community.

While tubercular diseases, at least of the lung, are common enough in the towns in southern Africa, there are large stretches of country where they are comparatively rare. Thus in Bechuanaland and Mashonaland, in many parts of Natal, in the Transvaal, and in Basutoland, phthisis is rarely met with. In Matabeleland it is also apparently not a common disorder, and only five cases were admitted to the Memorial Hospital at Buluwayo in the year 1897–8. In Mafeking phthisis was not very frequently seen before the war. The Boers in the Transvaal were rarely affected by tuberculosis[1].

The islands off the coast of Africa are not free from tubercular affections. In Mauritius phthisis is common, and very rapid in its course; it causes a mortality of about 1300 per million per annum. In Rodriguez it is less common, and causes a mortality of only 750 per million. In Madagascar it is far from rare in the interior province of Imerina, but is less prevalent on the coasts. Among the western African islands it is found that phthisis is not at all exceptional among the poor in Madeira, though rare among the upper classes of the natives. In the Canary Islands it is prevalent to some extent, but in the Cape Verde Islands it is rare. Both Madeira and the Canary Islands have been found suitable for the treatment of certain forms of consumption, though their reputation in this respect is not so high as in former years.

America. In the northern part of the continent of North America tubercular diseases are very prevalent. In far away Greenland we have it on the authority of Nansen, the Arctic traveller, that tuberculosis is the great scourge of the inhabitants. "Of late," he writes (in 1890), "the disease has manifested itself to an alarming extent, especially in the form of pulmonary

---

[1] Dr Alfred Hillier, *Lecture at the London Polyclinic*, Feb. 7, 1900.

tuberculosis, which is the great scourge of Greenlanders. It would be difficult to find a community with a proportionally larger number of inhabitants suffering from this plague. While we were in Godthaab, two died, and ten or twelve more were merely struggling for life; and this out of a total of something like one hundred. The disease was so common that it was almost easier to reckon up those who were free from it than those who had it[1]."

In like manner consumption is the principal cause of death among the Eskimos and Indian tribes throughout Alaska, Hudson Bay Territory, and Labrador.

The more temperate climates of the other Canadian provinces afford no immunity to the ravages of tubercle. In Quebec pulmonary phthisis is very widely spread, and, after "diarrhœal disorders," is the greatest factor in the general mortality, one-twelfth of which is caused by this disease. From almost all parts of Ontario come reports showing that more deaths are caused by phthisis and other tubercular diseases than by any other single disorder, and the same statement holds good for the province of New Brunswick. The prevalence of these affections is said to be increasing in Nova Scotia, and they are of frequent occurrence in Newfoundland to the east, and to the west in British Columbia. In Manitoba the mortality from tuberculosis exceeds that caused by all other infectious diseases.

A recent writer on the public health conditions of Canada remarks that this class of diseases is more rife both in Canada itself and in the United States than in Great Britain, and the facts just cited above, taken as they are from the official reports of each province, go far to bear out this statement so far as it concerns Canada, while in regard to the United States it can be fully confirmed from other sources. The excessive frequency of these diseases throughout the Republic formed the substance of a number of resolutions passed by the Thirteenth annual meeting of the Conference of State and Provincial Boards of Health of North America, held at Detroit in 1898; and one of these resolutions was to the effect that tuberculosis causes more mortality in the States than all other contagious diseases put together. It is some satisfaction, however, to note that, rife

---

[1] *The First Crossing of Greenland.* By Fridtjof Nansen, 1890.

as tuberculosis is, it was once much more so. In the urban population of Massachusetts, for example, the "phthisis" mortality has fallen from 4080 per million inhabitants in the year 1856 to 2190 per million in the year 1895. Similarly the "phthisis" mortality in Philadelphia has fallen in the last fifteen years from about 14 per cent. of the deaths from all causes to about 10 per cent. The distribution of consumption through the States is, of course, not uniform. Davidson gives the following as the areas of greatest and least prevalence. The areas of greatest phthisis mortality are, he says, the Pacific Ocean region, the North Atlantic coast, the Ohio river belt, the North-Eastern plateau, the central plains of Kentucky and Tennessee (Indiana being little affected), the middle Atlantic coast, and the interior plateau; while the regions where consumption is least fatal are the western plains, comprising West Texas, Kansas, Nebraska, and Western Dakota, the Cordillera region, the South Atlantic coast, and the South Mississippi river belt.

In Mexico phthisis prevails to some extent on both coasts, but on the high table-lands of the interior it is rare. In Guatemala it is known, but is apparently much less fatal than in European countries; in 1894 it caused a mortality of less than 250 per million inhabitants. In British Honduras on the other hand "phthisis of a peculiarly rapid and fatal character" is "fairly common." In Nicaragua, San Salvador, and Costa Rica phthisis is found along the coast to a greater or lesser extent, while it is rare in the high-lying lands of the interior.

The population of many, indeed of most, of the West Indian Islands are widely affected by tubercular disease. From Barbados, from St Lucia, Grenada, and Trinidad, all in the Windward group, reports are forthcoming showing that pulmonary phthisis is there, as in so many other parts of the world, the most deadly of scourges. In the Leeward group it is scarcely less prevalent, and in most of the larger islands of the West Indies it causes a very high annual mortality. Thus in Jamaica in 1897 the phthisis mortality was equal to 2760 per million living, the prevalence of the disease being closely associated with the unhealthy habits and overcrowding of the population. In Cuba consumption is equally rife, and in some years has caused nearly twenty per cent. of the deaths from all causes. Similarly, from the Bahama hospital reports it appears that more phthisical patients are admitted to hospital than any other kind, while in Bermuda no other disease causes so high a mortality.

In British Guiana while the pulmonary form of tubercle is, according to a statement of the Surgeon-General of the colony, not a very prevalent disease, affections included under the general heading of "tubercle" cause no little sickness and mortality each year, the figures under this heading being only second to those under "malaria" as regards cases, and far ahead of all other figures as regards deaths. In Georgetown itself phthisis has in recent years become far more prevalent and fatal than it was formerly. In Dutch and French Guiana there is reason to believe that this disease is common in low-lying districts, but less so in the more elevated interior.

Along the coasts of Brazil tubercular diseases are met with with great frequency. They are excessively rife in Rio de Janeiro, where in the years 1886–1897 the deaths from this cause varied between 9·5 and 19·4 per cent. of the deaths from all causes. The people of northern and central Brazil are said to be peculiarly disposed to phthisis, and this tendency is probably favoured by the warm, moist climate and the conditions of life which prevail there. Some localities in Brazil, however, appear to be free from it, as, for example, Campos do Jardao, Barbacena, etc. It is noteworthy, also, that tubercle is said to be found with great frequency in cattle in Brazil. The coasts of the Argentine Republic and of Uruguay appear to be as favourable to the prevalence of tubercular diseases as those of Brazil, while the high-lying regions towards the Andes range are much less affected. In the province of Buenos Aires no other disease is so fruitful a source of death, and the same appears to be the case in Monte Video, where the Spanish population are said to be especially affected. In the Falkland Islands, on the other hand, phthisis is extremely rare, and those who have contracted the disease elsewhere improve by a residence in the islands.

Turning to the west coast of South America it is found that along the shores of Chile phthisis is far from rare, and that in Valparaiso it is exceedingly common, while in the Bolivian sierras, on the contrary, it is said to be entirely unknown, both among the Indian and the white population. A similar distribution of the disease appears to hold for the countries of Ecuador and Peru, where on the low-lying coasts consumption is very commonly seen, while the mountainous interior is relatively or absolutely free from it.

Factors which determine the Geographical Distribution. It will be observed that tubercular diseases have, as was stated at the beginning of this chapter, an almost world-wide distribution. Their degree of prevalence varies, it is true, within somewhat wide limits, but there appears to be no geographical unit of any considerable size in which they are quite unknown. They are consequently found under an extreme variety of conditions as regards temperature. Phthisis is as great a scourge among the inhabitants of semi-arctic Greenland as among those of the temperate shores of Great Britain, or of the tropical coasts of Brazil and Ecuador. Nearness to or distance from the equator, indeed, seems to be without influence upon this form of tubercular disease; for though it is mild or comparatively rare in some warm countries, such as Egypt, Tunisia, Guatemala, the Malay Peninsula, the Straits Settlements, the East Indies, New Guinea, and some parts of Brazil, it is exceedingly common in many other countries in the same latitudes as those just named. In like manner, although the Falkland Islands, Iceland, the Faröes, and Lapland, are examples of cold countries comparatively free from tubercular diseases, there are other regions equally cold, as, for example, Greenland, the Hudson Bay territories, and Labrador, where they are rife.

It was at one time believed, and there is indeed still a large body of evidence to show, that cold is the principal predisposing or determining cause of the onset of "consumption." But whatever the influence of cold may be, it is less a question of an absolutely than of a relatively low temperature. Thus it is found that on the coasts of Africa the natives, accustomed for most of the year to a warm climate, fall victims to tubercular disease when the cooler weather of winter sets in; and the same class of people, when removed from the tropics to a cooler country, easily succumb to the malady. Humidity probably plays a still more important *rôle* than temperature in the distribution of tubercle. A moist atmosphere seems the most favourable to the activity and spread of the bacillus. The well-recognised fact that consumption is rare or unknown in the Swiss Alps, in the South American sierras, and in other high-lying districts is no doubt partly to be explained by the relative dryness of the atmosphere in these regions, and this is borne out by the equal rarity of the disease in the plains of Upper Egypt, where the air is almost entirely devoid of humidity.

But more important than temperature or humidity, whether absolute or relative, is the purity of the air. It appears almost certain that the freedom from consumption of communities dwelling at great heights is to a very great extent due to the absence of organic impurities in the air constantly breathed by them. In addition to the instances just named of high altitudes where tubercle is little known, many other equally striking examples are forthcoming from all parts of the world. In Greece, in Arabia, in Asia Minor, in the Caucasus, in Mexico, in the Central American Republics, in the Guianas, in the Argentine Republic and Uruguay, in Ecuador, and in Peru, tubercular diseases are found to be rife in the low-lying coast districts or plains, and rare or unknown in the adjoining mountains or table-lands. But even this very general rule is not without exceptions. In Madagascar, for example, phthisis is spoken of as "far from rare" in the Imerina province on the central plateau, and as "by no means prevalent" on the coasts[1]. In the mountainous countries of Norway and Montenegro, and on the slopes of the Himalayas, even at considerable heights, pulmonary phthisis occurs with great frequency; while, on the other hand, the Russian steppes on the borders of Asia and Europe, the desert plains of Egypt, and the western plains of the United States of America, are all examples of low-level areas where this disease is comparatively rare. In spite of these exceptions, however, the general rule holds good that a high altitude is less favourable, at least to pulmonary tubercle, than a low one. No doubt many factors are at work in producing this result. Among these must be mentioned the lowered barometrical pressure, the consequent rarity of the air and more active respiratory movements needed, and the healthier life led by some of these communities, all of which would tend to increase their resistance to the attacks of the tubercle bacillus. But, over and above this, the greater purity of the mountain air over that of the plains below has an importance that can scarcely be exaggerated. At the other end of the scale phthisis is found prevailing to its greatest extent among overcrowded communities, dwellers in

[1] Davidson.

insanitary cities or villages, the air of which is constantly contaminated by organic exhalations. The infective nature of tubercular disease has long been fully recognised, and the condition of overcrowding directly favours the passage of the infecting organism from one individual to another, while at the same time it is associated with a lowered condition of vitality and a greater amount of organic impurity in the air.

Among general sanitary defects, overcrowding and deficient ventilation are apparently the most powerful aids to the diffusion of tubercular disease. A recognition of this fact, and the introduction of improvements in the construction, ventilation, and management of barracks, gaols, and schools, has led to an immense reduction in phthisis prevalence in institutions of this kind, where at one time it was excessively high. Some other sanitary and public health measures, as, for example, improved water-supply and drainage systems, cannot be regarded as having any direct influence upon tubercular diseases. The supervision of food supplies, on the other hand, was, until recently, believed to be of the first importance in their control. That some of the lower animals, particularly horned cattle, can and do largely suffer from a disease closely allied to tubercular disease in man, and associated with a bacillus closely resembling, if not identical with, the human tubercle bacillus, has long been recognised. So like does the one disease appear to be to the other, that from a public health point of view it seemed necessary to treat them as identical; and every precaution to prevent the consumption of meat or milk from tubercular cattle was, and, until the question is finally settled, still is, amply justified. Doubt, however, had from time to time been thrown upon their identity, and the question was brought to a head by Prof. Koch in a communication made to the British Medical Association in 1901. In this paper the author stated his conviction that tuberculosis in man and tuberculosis in cattle are two distinct and separate diseases, and that the disease in the latter cannot be communicated to the former. The announcement gave rise to considerable controversy at the time, and led in England to the appointment of a Government Commission to inquire into this very important question. Until the results of this inquiry shall be made known the matter may be regarded as *sub judice*.

The distribution of tubercular diseases is influenced to some extent by racial differences. All races suffer, but the negro appears to be particularly liable to them. The ease with which negroes fall a prey to consumption when removed to a cooler clime and to other social conditions than they have been accustomed to is notorious. The Chinese and Japanese are also highly susceptible to tubercle   The Jewish race, on the contrary, is said to be comparatively immune to it. The contact of the civilised with the uncivilised races has nearly always led to the introduction among the latter of certain diseases, of which tubercle has often been one of the most deadly. Its spread among native races under these conditions is no doubt largely due to changes in their habits, and to the substitution of town-life in overcrowded ill-ventilated houses for the more healthy, open-air life formerly led by them. Drink and other demoralising vices of so-called civilisation no doubt aid in the process. This disastrous effect of tubercular diseases among native races after the advent of the white man has been seen in modern times among the Kanakas of New Caledonia, the natives of Hawaii and other Pacific Islands, the Maoris of New Zealand, the aborigines of Australia, and the indigenous tribes of Siberia.

The nature of the soil has probably considerable influence on phthisis prevalence. It was shown by Bowditch in America in 1862, and by Buchanan in England in 1867, that the disease is greatly favoured by a damp, impervious soil. Buchanan, indeed, gave to the condition of soil the first position among all factors concerned in the distribution of phthisis. His conclusions were that (at least in the three English counties where his observations were made) phthisis is less common on pervious than impervious soil; that it is less common on high-lying pervious soils than on low-lying pervious soils; that it is less common on sloping impervious soils than on level impervious soils; that wetness of soil conduces to phthisis in people living on it; and that no other circumstance can be detected that coincides on a large scale with the greater or less prevalence of phthisis than the one condition of soil.

TYPHOID or ENTERIC FEVER
(and TYPHO-MALARIA).

History. The early history of typhoid fever is practically unknown. In the seventeenth and eighteenth centuries there is frequent mention of a class of fever which may fairly be regarded as enteric; but it was not until the beginning of the nineteenth century that a true conception of the specific character of the disease was formed. It was for long confused with various other fevers, of which typhus was the most important. The distinction between typhus and typhoid fevers, though suspected previously by various observers, was finally and fully established by Sir William Jenner in 1849–51. The typhoid bacillus, which is believed to be specific to the disease, was first described in the years 1880–81, by Eberth, Koch, and Klebs; and was first cultivated successfully on artificial media by Gaffky in 1884.

Recent Geographical Distribution. *Europe.* In England and Wales the mortality from enteric fever has shown a most gratifying diminution within the past thirty years. In the years 1871 to 1875 the deaths annually recorded from this cause were equal to a rate of 374 per million living, but this figure has steadily fallen, until in 1892 a minimum rate of only 137 per million was reached. There has been a slight rise again in the years since 1892, the highest mortality being in 1893, when it was equal to 229 per million. There are certain more or less well-defined areas in England and Wales where the disease shows a special tendency to prevail year after year. One such area consists of eighteen registration districts occupying the eastern portion of the counties of Northumberland and Durham; the deaths from enteric have been constantly higher in this area than in the country as a whole for a long period of years. The county of Lancashire forms another area in which the disease is persistently prevalent. In Nottinghamshire also enteric fever seems to be endemic in certain contiguous districts (Nottingham, Basford, and Mansfield), which together contain about three-fourths of the population of the county. Other areas of unusual prevalence are found in the West and East Ridings of Yorkshire and elsewhere. It will be observed that all these are in the main essentially manufacturing areas. While the general distribution of enteric fever in England and Wales shows on the whole a considerable degree of constancy from year to year, temporary variations in its distribution, owing to the occurrence of severe local epidemics, are not infrequent. Among the most notable examples of such epidemics may be mentioned those in the Tees Valley in 1890 and 1891, at King's Lynn in 1891 and 1892, and the extremely severe outbreak at Worthing in 1893. Epidemics of this kind are almost invariably traced to a definite pollution of the drinking-water supply with fæcal matter, and, in most instances, with the excreta from persons suffering from typhoid fever.

In Scotland the mortality from enteric fever, which had been 231 per million per annum in the decade 1881–1890, fell to 180 in 1895, and in each of the two following years was 160 per million. The degree of prevalence of the disease, as judged by the mortality, is about the same in Scotland as in England. In both countries typhoid fever is far more prevalent in towns than in country districts. In Ireland the disease caused an annual mortality of 161 per million in the years 1881–90 and of 192 in 1890–99; but here, as elsewhere, it seems probable that some of the deaths returned under the headings of "simple continued" and "ill-defined" fevers should in truth be ascribed to enteric fever.

In the larger French towns of over 10,000 inhabitants typhoid fever has caused a diminishing mortality in recent years. The ratio of deaths from this cause in 1886–90 was 530 per million per annum; this figure fell to 370 in 1891, and has further diminished to 260 and 270 in the years 1897 and 1898 respectively.

The returns of typhoid fever in Holland are still grouped with those of typhus. The deaths from the two diseases together fell from 140 per million inhabitants in 1889 to a minimum of 90 per million in 1897. What proportion of this mortality was due to enteric alone cannot be stated, but it is clear that the disease is very much less common in Holland than in the British Islands[1]. Recently attention has been called to the remarkable fact that this relative freedom from typhoid fever is maintained in Holland in spite of the method in vogue there of removing excreta—or, perhaps, it should be said because of the care with which the method is carried out. The excreta are removed dry, and after being mixed with peat-ashes and street-sweepings are converted into manure which is spread over the fields. "When we consider," says Dr Poore, "that dairy farming is the staple industry of these provinces [Gröningen and Friesland], that milk, which is most sensitive to typhoid fever infection, is everywhere abundant, and that human excreta containing a certain proportion of typhoid fever excreta are used, and have been used for ages, to manure this fertile district which almost lies in the water, the comparative freedom from enteric fever seems very full of instruction[2]." In two large towns, Gröningen and Leeuwarden, with populations respectively of 63,863 and 31,598, not a single death from typhoid occurred in the year 1897.

A very marked reduction in enteric prevalence has come about in recent years in Belgium. In the decade 1871–80 the mean deaths from this cause each year numbered 4161; in the next decade they fell to 2807; in 1895 the deaths from typhoid were only 1843, and in 1897 only 1598. The last figure was equivalent to a mortality of 246 per million inhabitants. The disease, though decreasing, is therefore very much more frequent in Belgium than in Holland.

In Denmark the deaths from typhoid fever have shown somewhat less tendency to decline in recent years than in the other countries already named, at any rate in the urban population. In 1892 the deaths from this cause among the dwellers in Danish towns were 109; in 1895 they rose to 122; and in the following two years they fell to 94 and 87 respectively. If, as is happening in other countries, the urban population in Denmark is annually increasing at the expense of the rural, the diminution of mortality

---

[1] Prof. Saltet, of Amsterdam, has shown that the mortality from typhoid fever in Holland was reduced by as much as 72 per cent. in towns in the years 1875–95, and by 60 per cent. in rural populations in the same period. (Epidemiological Society, May 19th, 1899.)

[2] *The Milroy Lectures*, 1899. By G. V. Poore, M.D., F.R.C.P.

in the last three years named may be more striking than it seems. The deaths in 1895 were equivalent to a ratio of 155 per million living. In Copenhagen itself there has been a very marked and steady decrease in the deaths from enteric during the last sixty or seventy years[1].

The mortality from enteric fever in the German Empire is relatively low, and the same diminution in the prevalence of the disease which has been observed in other countries has also been seen here. The deaths in 1894 were equivalent to 131 per million inhabitants; in the preceding seven years (1887–1893) the annual mortality was 165 per million. The prevalence of typhoid fever has varied considerably in various portions of the empire, but all have shown a very marked decline in its prevalence within recent times. Thus in Prussia the deaths from this cause fell from 445 per million per annum in the years 1880–1886 to 215 per million per annum in the next septennium, and to 151 in 1894. Saxony has recently had a remarkably low death-rate from typhoid; in 1894 it was only 60 per million, as compared with rates of 260 and 132 per million respectively in the two septennial periods just referred to. In Würtemburg the fall has been somewhat less striking, the mortality rates for the two septennia and for the year 1894 being 199, 107, and 104 respectively. In Bavaria the corresponding rates in the same time-periods were 206, 119, and 83 per million.

In Norway the enteric fever death-rate is low. It steadily decreased until the year 1892, when the mortality from this cause also reached its minimum in England. Since that date there has been a slight rise in the number of deaths. The enteric fever mortality in Norway in 1891 was approximately 105 and in 1894 about 70 per million. The disease is perhaps even less prevalent in Norway than in Holland, and considerably less so than in England.

In Sweden typhoid fever appears, on the other hand, to be very much more prevalent. In the seven-year period 1887–1893 the mortality in that country was equal to 278 per million; it fell in the next septennium to 218, and in 1894 was 186 per million living.

In European Russia there is no doubt that enteric fever is the cause of a very much higher mortality than in most

1 Schierbeck, *Trans. Epidemiol. Soc.* 1900–1.

other European countries. In the three years 1893–1895 the deaths from it were equal to an annual mortality of 880 per million living. It is to be noted, however, that in the Russian returns "undetermined" forms of fever are included under the heading of "abdominal typhus" (*briushnoi typh*), and that the true enteric mortality is therefore perhaps lower than this figure would indicate. Unlike the other countries already named, Russia has shown comparatively little diminution in typhoid fever mortality in recent years. The returns from various parts of the country for this disease, as for all others, are approximate only, but they indicate a very wide-spread prevalence of typhoid fever throughout all parts of European Russia. The frequency of this and other acute abdominal diseases in Russia is closely associated with the imperfect character of the public water-supplies, of the drainage-systems, and generally of the public health administration in the smaller towns and villages. Typhoid fever is much less prevalent in the northerly government of Archangel than in Russia generally ; the other parts of the country in which it has been least prevalent in recent years (1887 to 1895) have been the Baltic Provinces, the governments on the Vistula—that is to say, Russian Poland —and the Don Cossack territory. In these the number of registered cases of the disease (not deaths, as have been hitherto dealt with for other countries) varied between 700 and 1100 per million, whereas in the rest of European Russia the registered cases varied between 1900 and 2800 per million. Much caution is needed in accepting these figures as representing the total prevalence of the disease, but they have a certain value for purposes of comparison *inter se*[1].

The Austro-Hungarian empire has a very high annual mortality from enteric fever. In Austria the deaths from this cause were 724 per million per annum in the septennium 1880 to 1886, and 522 in that ending in 1893. This reduction in mortality has apparently been maintained. In Hungary the typhoid fever deaths in the two successive septennial periods just mentioned

---

[1] The figures are taken from the *Annual Reports of the Medical Department of the Russian Ministry of the Interior*. There is an obvious discrepancy between a general "enteric" death-rate of 880 per million and the recorded case-rates quoted in the text, and it can only be explained by supposing that a large number of cases which recover escape record.

were at the rate of 698 and 504 per million inhabitants respectively; in 1897 they had sunk to 480 per million. Considerable caution is probably necessary in accepting these figures as more than an approximate indication of the true mortality from this malady, but it is extremely unlikely that they are above the truth, and they fully justify the assertion that in the Austro-Hungarian empire typhoid fever is excessively prevalent. The same appears to be the case in Croatia and Slavonia (annexes of the Crown of Hungary, but under separate administration and laws), where the mortality from this disorder has in many years been not less than 600 per million per annum.

The deaths from enteric fever in Switzerland have recently diminished very considerably, and are now fewer in proportion to the population than in the British Isles. This is, at least, the case in the fifteen Swiss towns with a population exceeding ten thousand inhabitants. In these towns the mortality from this cause fell from 243 per million in 1889 (which was a considerably higher figure than that for the preceding year) to a minimum of 75 per million in 1896. In the following year it rose again slightly, to 100 per million, but this was still a low figure, and Switzerland is comparable with Holland in its relative freedom from the disease.

In Italy the deaths from enteric fever have considerably diminished in recent years, but are still high; the mortality ratio fell steadily from 925 per million in 1887 to 439 in 1894. In Spain and Portugal the disease is said to be particularly frequent.

For the countries of the Balkan peninsula accurate figures of the prevalence of, or mortality from, enteric fever are not available. In Servia all forms of "typhus," including the abdominal form or enteric fever, are dealt with under a single heading. In Turkey typhoid fever is a very common disorder, and is constantly present in Constantinople and its neighbourhood. In 1898 it caused a severe epidemic in Macedonia, where it was said to have been imported from Thessaly. During the Turkish occupation of Thessaly after the war enteric was excessively prevalent. From many parts of the province of Salonika cases are from time to time reported.

In the Faröe Islands typhoid fever is said to have been unknown for a considerable number of years. On the other hand the disease is met with as far north as Iceland, and has repeatedly caused epidemics in that island.

Enteric fever is the cause of a considerable mortality in Malta.

In Crete the disease appears to be endemic, and the English troops, both officers and men, suffered considerably from it during the troubles in that island a few years ago. In Cyprus enteric fever at times becomes epidemic and causes much mortality, as, for example, in 1889.

Asia. In the Caucasus and Transcaucasia it would appear from the official returns that typhoid fever is a very much less frequent disease than in any other part of the empire, with the exception of Central Asiatic Russia. The registered cases during the years 1887–1895 varied between 220 and 270 per million per annum, in contrast to the much higher ratios for European Russia quoted above. It is noteworthy that the disease appears to be, if anything, less prevalent in the three flat, marshy provinces of the Caucasus proper than in the mountainous regions of Transcaucasia. The figures are however too untrustworthy to bear much weight, and it is to be noted that a recent writer states that typhoid has, within the last ten years, taken considerable hold in the Caucasus, particularly in the towns; though nowhere causing a severe epidemic[1].

In Russian Central Asia the reported cases of typhoid fever in the same period were lower still, and were equivalent to a ratio varying between 50 and 120 per million per annum. To what extent these very low ratios in Transcaspia, Turkestan, and the other provinces of Russian Central Asia are to be ascribed to defects in diagnosis and registration, or to a real infrequency of the malady in those regions, can only be a matter of surmise. The difference is, however, so very great between the ratios for these provinces and for European Russia, and also for Siberia, that it cannot all be due to defective registration and diagnosis (which are known to obtain in the other parts of Russia, though perhaps to a less extent), and it may be asserted with considerable confidence that enteric fever is a comparatively infrequent disease in the regions lying beyond the Caspian Sea. Among the Bashkirs, on the other hand, who live mainly in the regions lying north of that sea, in the basins

---

[1] Pantiukhof, *op. cit.*

of the Volga and the Ural rivers, it is not uncommon, and it occasionally becomes epidemic throughout an entire village.

Siberia occupies an intermediate position between Central Asia and European Russia, the case-ratio varying in the years 1887–1895 between 990 and 1330 per million of the population. The disease appears to be especially prevalent in the governments of Irkutsk and the Amur. It is found in the northerly province of Yakutsk, and rare cases are returned from the most northerly district of that province, that of Verkhoyansk, the greatest portion of which lies within the Arctic circle. At Blagoveshtchensk and at Vladivostok it has prevailed to a considerable extent.

Typhoid fever is certainly known in Arabia, and in the early months of 1898 it was epidemic at Hodeida on the Red Sea coast. It is said to be common along the coasts, but absent from the Arabian plateau. The latter statement is certainly far too sweeping, for of recent years it has caused an excessive degree of mortality among the highlanders of Assyr. In 1900 a severe epidemic of the disease was reported as prevailing among them. At Mecca, too, it is a frequent cause of death among the pilgrims.

At Aden 3·5 per thousand of the British troops stationed there were annually attacked by typhoid in the ten years ending 1895; in the year 1897 the proportion fell to 1·81 per thousand.

In Asia Minor the disease is common; it is very prevalent in Smyrna in the winter of most years; at Trebizond it is often seen; at Erzerum it becomes epidemic every spring[1]; and in Chios, Samos, Mitylene, Rhodes, and other islands off the coast it is of common occurrence.

The disease is frequently seen at Aleppo, Damascus, Jaffa, Tripoli, and other places in Syria. It undoubtedly exists in Persia, usually as a sporadic disorder, but occasionally in the form of intense epidemics, as, for example, at Bidjar, a village six days north of Kermanshah, in the autumn of 1900. It is common enough in Teheran itself, where both natives and Europeans suffer.

At no distant period the existence of enteric fever in India

---

[1] It was seriously epidemic in Erzerum after the earthquakes there in 1901, and this was attributed to an enormous mass of mud which spread over the streets and led to serious pollution of the water-supply.

was denied, but later knowledge has shown the extraordinarily inaccurate nature of this view, and it is now recognised as one of the most fatal of all diseases among European residents in that country. The European races suffer to a far greater extent than the natives, and for a considerable time it was even doubted whether the natives of India were not immune to the disease; but it has now been proved beyond question that they can and do suffer from it, though to an incomparably less extent than Europeans[1]. In 1897 only 51 cases with 15 deaths were recorded in the whole native army; in the previous year the figures were 19 cases with 5 deaths. It would seem that the disease is becoming somewhat more common in the native troops, or else improved diagnosis is causing an apparent increase in the cases and deaths. The admission-rate rose from an annual average of 0·2 per thousand strength in the decade 1886–95 to 0·4 in 1897, and the death-rate from ·09 to ·12. In Indian gaols, on the other hand, the admission-rate remained the same in the two periods (0·3 per thousand strength). In the Central gaol at Nagpur 25 cases of enteric in natives were treated between the years 1894 and 1899. It is noteworthy that the Gurkhas seem considerably more prone to typhoid fever than other native races, and hence the disease is commoner where Gurkha regiments are stationed than elsewhere[2]. Its prevalence in the native troops in Assam is directly ascribed to the number of men of this race stationed there.

The blood 'of natives over the age of childhood has been repeatedly found to react to Widal's test, which would indicate that the persons whose blood was examined had suffered from a previous attack of the disease. Evidence of a similar character, which has, however, been questioned, has been adduced in regard to the negroes in Africa, and the apparent immunity of

[1] The Sanitary Officer of South Canara stated, in 1897, that enteric fever is endemic in the town of Bangalore. This was characterised by the Sanitary Commissioner with the Government of India as an "unusual acknowledgement of the constant presence, as a matter of observed fact, of enteric fever in an Indian city."

[2] Dr Andrew Duncan (Major I.M.S. retired), on the other hand, stated in a discussion before the Brit. Med. Assoc., Ipswich, 1900, that, contrary to popular belief, enteric fever is not a common disease among Gurkhas. He had only seen one case among them.

other native races—in Algeria and elsewhere—has been quoted, and the view tentatively put forward that these races may owe their apparent immunity to their suffering from unrecognised attacks of the disease in childhood[1]. There is no question that the virus of typhoid fever is very widely spread throughout India; it attacks Europeans in all parts of the peninsula, and the typhoid bacillus has been found by Hankin and others in the drinking-water supplies of such towns as Lucknow, Fatehgarh, Meerut, Muttra, Agra, Sabathu, Mean Mir, Amritsar, Cherat, Mhow, and Deesa, and in milk from Fatehgarh and Dagshai. The native population must be constantly and specially exposed to its attacks from their habit of frequently bathing in and drinking the water of foul tanks, streams, and rivers, and it is probable that large numbers acquire an immunity to the disease by constant exposure to its cause. It has also to be remembered that an enormous proportion of natives are not attended by skilled observers, and that it is therefore possible that large numbers of cases of and deaths from enteric fever occur in native children which are regarded as due to some other cause. Support is lent to this view by the readiness with which English soldiers newly arrived in India contract the disease. Fuller information is, however, required before the degree of liability of the natives of India to enteric fever can be determined with accuracy.

Among European troops in India enteric fever is the most prolific source of sickness and death. The degree of its prevalence in any given station varies very greatly in different years, but no portion of the peninsula seems to be free from the disease. The following table shows the admission-rates under this heading in European troops stationed at the places named, firstly in the decennium 1886-95, and secondly in the year 1897, when the disease was excessively prevalent throughout India.

---

[1] Lt.-Col. A. Adams, I.M.S., who has seen undoubted cases of enteric fever in natives in Rajputana, where the disease even at times becomes epidemic, also expresses the belief that most of the children of the country suffer at some time from enteric fever. *The Western Rajputana States*, London, 1899.

ENTERIC FEVER IN EUROPEAN TROOPS IN INDIA.

Station	Annual mean, 1886—1895		1897	
	Admission-rate per thousand strength	Death-rate	Admission-rate	Death-rate
Agra	24	7·66	166·4	45·04
Umballa	21·7	5·41	64·1	15·32
Mhow	27·5	6·78	49	16·95
Lucknow	37·3	7·07	48·5	11·93
Meerut	26·8	7·83	47·4	18·58
Allahabad	24·2	4·84	44·8	14·34
Peshawar	18·7	7·88	44·2	26·93
Bangalore	16	3·47	40·2	7·2
Secunderabad	23·7	6·04	36·2	8·77
Rawalpindi	24·5	6·85	32·2	7·79
Rangoon	5·4	2·55	22·6	6·6
Quetta	15	3·15	22·5	4·77
Poona	15·4	5·06	18·1	2·52
Bareilly	39·8	9·41	14·7	5·5
Wellington	7·5	2·46	11·7	2·94
Karachi	13·8	3·07	8·2	3·64
Colaba	4·9	1·83	3·5	0·89
Fort Dufferin	6·8	2·47	2·0	1·99
Moultan	16·8	4·18	1·9	0·96
Fort William	4·3	2·13	—	—
Aden	3·5	1·81	—	—
	19·6	5·28	32·4	9·01

In Ceylon enteric fever is certainly met with. Cases are annually recorded in most parts of the island, and in the Report upon hospitals in the Northern Province for 1897 attention was called to "the constant and increasing occurrence of typhoid cases in town and district." In Colombo the disease is not infrequently seen, but it has been on the decline since the introduction of a pure water-supply[1].

In Farther India and the Malay Peninsula enteric fever is also seen to some extent, but far more among Europeans than among natives. In Siam it is known, though evidently far from common. In the Malay Peninsula it is, perhaps, more frequent. In the town of Singapore 186 cases of the disease were recorded in 1896, 55 in 1897, and 82 in 1898 ; and deaths annually occur from it

[1] *Medical Report for Ceylon for* 1898.

among the prisoners in the Singapore gaol. But the disease is not mentioned in the list of the principal causes of death in the Straits Settlements generally.

In French Cochin China enteric occupies the fifth position in the list of principal causes of mortality among European residents.

In the Dutch East Indies enteric fever is essentially a disease of Europeans and not of natives. In the year 1897 among 17,000 European (Dutch) troops stationed in this part of the world 29 were attacked with the disease, while in nearly 25,000 Asiatic troops only 8 cases were recorded.

That enteric fever exists in China there can now be little doubt. Europeans frequently suffer from it, and probably the native Chinese are also attacked. Some doubt as to the possibility of the natives contracting it seems to have existed until recently, because it is practically never possible to obtain a post-mortem examination of a Chinese patient, and observers have hesitated to diagnose the disease upon the clinical appearances only. On at least one occasion, however, its nature has been proved by an autopsy on a Chinaman in Canton. At Shanghai enteric fever occurs, according to some authorities rarely, according to others with great frequency. At Hankow, on the Yangtze-Kiang, the disease every year attacks the foreign residents, the steamer population and visitors, and at times the sailors on the naval boats stationed there contract it. A "severe epidemic" of enteric prevailed there in 1894. The natives of this place are believed also to suffer from the disease. At Swatow 4 cases of typhoid were treated in the Seamen's Hospital in 1896, presumably in Europeans. Three cases were seen at Canton in the same year; the disease was said in 1894 to have become less frequent there than formerly, but many cases were believed to have occurred in the south-eastern corner of the city in that year. Several cases were recorded at Foochow in the same year. At Hongkong 33 cases of enteric were treated in the Government Civil Hospital in 1898. On the other hand some observers elsewhere in China furnish negative evidence. Thus MacCartney doubts if true typhoid is ever seen at Chungking, and the same doubt is expressed by others at Ningpo and many other places. In the northerly province of Manchuria the disease certainly exists; and the Russian troops stationed there have suffered from it considerably.

In Corea enteric would seem to be a rare disease, though not unknown. In the island of Formosa no genuine case has, it is said, been seen; but here, also, improved methods of diagnosis may, as in China, show that the disease does exist.

In Japan 34,069 cases of typhoid fever with 8183 deaths were registered in 1893, and 36,667 cases with 8054 deaths in 1894. The mortality from this cause therefore oscillated about 200 per million in the years in question.

Australasia. In British New Guinea enteric fever is said to be quite unknown.

In Australia, on the other hand, not only is the disease very widely prevalent, but it has, at least in many parts, become much more so in recent years than it was formerly. In Queensland it is the most frequent and fatal of all the so-called "miasmatic" diseases. It caused a mortality here of 255 per million in 1893, which fell in each of the following two years to a minimum of 159 in 1895; but then a marked rise set in, the ratios in 1896 and 1897 being 279 and 382 per million respectively. In New South Wales the disease is also very prevalent. In Victoria the cases of enteric annually reported usually exceed those of any other zymotic disease. In South Australia the recorded deaths from typhoid fever fell from 128 in 1889 to 62 in 1893, but since that date the figures have risen, and in 1897 the deaths numbered 106, and in 1898, 145. The last figure was equivalent to a mortality of about 400 per million of the population. In Western Australia a most marked and excessive rise in enteric mortality was observed in the years 1895–1897. In the years 1889–1891 the recorded deaths from this cause had been 13, 3, and 19 respectively; in the next three years they rose to 55, 28, and 73; but in the following three years the very large numbers of 325, 400, and 407 respectively were recorded, followed by a fall to 296 in the year 1898. This excessive rise in typhoid fever mortality is, of course, out of all proportion to the increase in the population at the same time, and seems to indicate an enormously increased prevalence of the disease. The figure for 1897 was equivalent to a mortality of not less than 2420 per million inhabitants.

In Tasmania typhoid fever is a conspicuous disease, causing yearly between 40 and 50 deaths (between 200 and 300 per million inhabitants). It is essentially a malady of towns, and frequently becomes epidemic in Hobart, Launceston, and other places. It is said to be particularly rife among the mining population.

In the Pacific Islands enteric fever is now met with, though before the year 1875 it is believed to have been unknown there. In that year a case was observed in one of the islands in a gentleman newly arrived from Australia, and in this way the infection is thought to have been introduced to Polynesia[1]. It is now no great rarity in some of the islands. It prevailed extensively among the United States troops stationed in Honolulu in 1898. In Fiji two cases of the disease in Europeans were treated in the Colonial Hospital at Suva in 1898 and six in 1897; but in neither year were cases recorded in the natives. The disease exists among the natives of New Caledonia, observers differing, however, as to the extent of its prevalence. In the Sandwich Islands the natives are said to suffer considerably, and it is stated to be endemic in the Society and the Marquesas Islands.

In New Zealand the recorded deaths from enteric fever have not varied greatly in recent years. In the decennium 1889–1898 they showed no definite tendency to increase or decrease; in the first year of the period 118 deaths were recorded, and in the last 120; in the year of greatest prevalence in the interval 145 deaths from this cause occurred, and in that of the least 94. The mortality in 1898 was equivalent to about 160 per million, indicating a degree of prevalence of typhoid fever in New Zealand about equal to that of the disease in England in recent years.

Africa. In Egypt enteric fever is particularly prevalent. It is believed to be rife throughout the Nile valley and especially in Lower Egypt. There is, it is said on good authority, no station abroad in which British troops serve where the sickness and mortality rates for typhoid fever are as high as in Egypt. Among our soldiers returning from Khartum to Cairo, after the defeat of the Mahdi, the disease was very active, although on the advance up the Nile the health of the army had been exceptionally good. It was suggested that the relaxation of the strict discipline enacted during the advance may have favoured the development of the disease on the return.

---

[1] Sir William MacGregor, K.C.M.G. etc., *Address delivered to the London School of Trop. Med.* Oct. 3, 1900.

In Tunisia the disorder is now much less frequent and severe than it was formerly. At the time of the French occupation in 1882 it was particularly prevalent. In Algeria it is of common occurrence among Europeans, but until recently it was believed to be rare among natives. Of the total number of typhoid patients treated in the French hospitals at Mascara and Mostaganem the natives formed only from 6 to 8 per cent. But here, as was shown to be the case in regard to the natives of India, the belief has been expressed that the immunity of the natives may be only apparent and not real. The Arabs in Algeria apply for medical treatment with great unwillingness, and there is reason to believe that many of them may have suffered from typhoid fever in childhood or youth, and thus acquired immunity to the disease.

Some uncertainty existed until recently in regard to the frequency or even the existence of typhoid fever among the native negro races of the central zone of the African continent. It is now, however, generally admitted that the disease certainly exists here, though with what degree of frequency is still un-determined. Pruen states, for Central Africa generally, that typhoid fever, as well as dysentery and diarrhœa, prevail most at the beginning of the rainy season. He speaks of the disease as occurring on the shores of the Victoria Nyanza. Undoubted cases of enteric have also been seen in Uganda by others[1]. In the Cameroon district a small epidemic of typhoid fever was observed among the natives in 1894, and in two cases the presence of intestinal ulceration and gastro-intestinal catarrh was demonstrated by post-mortem examinations; in 1895 several sporadic cases occurred in the natives, and two Europeans were attacked. It is probably, however, a comparatively rare disease on the West Coast of Africa generally. It is not mentioned in some recent Annual Reports on the health of the Gold Coast Colony, of Lagos, of the Gambia, and of Sierra Leone, and several writers upon the principal diseases of the Belgian Congo make no reference to typhoid fever, though acknowledging the existence of typho-malarial fever in that country. In Senegal it is said to be met with occasionally both among Europeans and natives. In Eastern Africa it is probably also rare. At Zanzibar it does not appear to have been recognised, and in British and German East Africa it is not included among the diseases met with.

[1] A. R. Cook, *Journ. Trop. Med.* March 1, 1902.

In Madagascar typhoid fever is stated to be a very common disease in the central plateau and rather less so on the coast; the Hovas are said to suffer from it severely.

In Mauritius it is also frequent, especially in the Rivière du Rempart district in the north of the island. It is very prevalent on the great sugar estates, owing to the insanitary habits of the natives and their entire indifference to the pollution of soil and of drinking-water. It has been endemic here since 1838.

It certainly exists in St Helena, one case being recorded in the Colonial Hospital Report for the island in 1898, and seven in 1897.

The wide prevalence of typhoid fever throughout South Africa is in marked contrast with its comparative rarity in the central portions of that continent. In both colonial and other territories in the south it is everywhere, almost without exception, the predominating disease. In Cape Colony it is widely endemic, and not infrequently becomes epidemic. Its frequency is clearly shown by the hospital returns and by the reports of the health officers in the various districts of the colony. In the Somerset Hospital at Cape Town, the Frontier Hospital at Queenstown, the Albany Hospital at Grahamstown, the Provincial Hospital at Port Elizabeth, and in many others, typhoid fever occupies the first or nearly the first position among the diseases dealt with. The medical officer of the Grey Hospital at King William's Town wrote in 1898 that enteric fever was not only the most prevalent disease, but that it was increasing and that "from a public health point of view such a state of matters is scandalous." In Kimberley the disease was very rife in the hot weather of 1897, but considerably less so in 1898. In Gordonia it was epidemic in 1897–98, in the Strydenburg sub-district it caused a severe and alarming epidemic in 1898, and the mortality among the natives was very high. From Fraserburg, Cradock, Queenstown, Richmond, Victoria West, Vryburg, and from most other places whence reports are received, there is the same evidence of the frequency of enteric fever. In Namaqualand it is constantly present. In the native territories of Tembuland, the Transkei and Pondoland it is a less conspicuous disease, but is at times epidemic in several places. In Griqualand West it was epidemic in 1898. In Mashonaland it appears to occur only sporadically.

In Natal also typhoid fever is widely prevalent. In the Durban

Hospital it is one of the most frequent diseases dealt with, and at Pietermaritzburg it is constantly present.

In Basutoland the malady is frequently epidemic in the native villages, and in Rhodesia it is known, a few cases of it coming under treatment in the Memorial Hospital in Buluwayo both in 1897 and 1898.

North and Central America. Enteric has been observed in epidemic form as far north as Greenland, Labrador, and Sitka.

In Quebec typhoid fever is of much less frequent occurrence than it was some years ago. Outbreaks occur here and there in the province from time to time, but they have become rarer as greater attention has been paid to the protection of water-supplies. In the province of Ontario it prevails more or less in the autumn months, but here also it is diminishing. In 1897 the disease was prevalent in a few districts, particularly in the mining centres in the west of the province. The town of Kingston, where typhoid in former years was common, is now almost free from the disease, thanks to improved sewering and water-supply. In New Brunswick the disorder is frequently seen ; 49 cases occurred in the city and county of St John in 1897, and 19 cases at Fredericton in the same year, and it is mentioned in the reports from most other places in the province. In Nova Scotia sporadic outbreaks occur at times in different parts of the province. In Manitoba the disease is met with and appears to be far from infrequent, though somewhat less so in recent years than formerly. The cases recorded in the province in the four years 1894 to 1897 were respectively 568, 187, 147, and 274. In British Columbia enteric is frequent; in 1896 epidemics of the disease were reported from Kamloops, Rossland, Nelson and other places, in the last-named in the depth of winter when snow lay deep on the ground. In Vancouver Island it is also met with. .

Typhoid fever appears to be prevalent in greater or less degree throughout the United States. In 1897 there was not a single State in which deaths from the disease were not included among the mortality statistics of the principal towns and cities, with the doubtful exception of Idaho. In the urban population of the other States the mortality varied between 100 and 500 per million inhabitants. In some States this has considerably diminished in recent times. Thus in Massachusetts the mean annual death-rate from typhoid fever fell from 820 per million inhabitants in the quinquennium 1871–75 to 450 in the next quinquennium. In the next five years it rose again to 500, but fell to 410 and 320 respectively in the two succeeding quinquennia. The mortality rates in the chief cities of this State showed an exactly parallel course in the same period, the figures for the cities being almost identical with those for the State as a whole, but certain places, as for example North Adams, have a constant bad reputation for the prevalence of typhoid. It is interesting to note that in Massachusetts the decline in mortality from this disease has coincided with the introduction of public water-supplies, and an instructive diagram was recently published by the State Board of Health showing an exact parallelism between the line of typhoid fever mortality and the line indicating the percentage of population *not* furnished with public water-supplies. As the latter has diminished, so with unfailing regularity has the former.

In the State of Michigan enteric fever is believed to prevail in successive waves, the principal of these being some twelve years apart and one or two minor waves occurring in the intervals. The mortality from the disease fell in this State from 526 per million in 1881 to 214 in the following year, and from 1882 to 1895 it has fluctuated irregularly between 200 and 400 per million per annum.

Mexico is said to enjoy a remarkable immunity from typhoid fever, and in British Honduras the disease is said to be unknown. In Guatemala, on the other hand, it is probably not rare, for in 1894 the recorded deaths—229 in number—were equivalent to a mortality of 160 per million.

In many of the West Indian Islands enteric fever is frequently seen. In Jamaica it is particularly prevalent. In the year ending March 31, 1898, the recorded deaths in Kingston were equal to 719 per million of the population, and the disease is spoken of as a serious menace to the health of the island, "the conditions for its dissemination and consequent permanent continuance and increase" existing everywhere. In Trinidad a few cases were recorded in many districts, *e.g.* those of Port of Spain, Chaquana and Guaracara, and in many hospitals, *e.g.* that of San Fernando, in 1898. In Cuba enteric was terribly prevalent during the war of 1897. In Barbados cases were reported from many places in these two years, and in 1898 53 cases of the disease were treated at various hospitals and institutions throughout the island. In Golden Ridge it is spoken of as endemic. In the Leeward Islands generally, though the disease is apparently not a very conspicuous one, some cases are always mentioned in the annual reports of the Colonial Hospitals. In the Bahamas an occasional case is mentioned in the like reports for these islands. In the Bermudas typhoid fever seems to be much commoner than in the more southerly islands; 60 cases of the disease were registered in 1898, and many others were thought to have escaped diagnosis and record, particularly in the coloured population, who are said to suffer frequently from it in a mild form.

South America. In British Guiana the recorded cases of typhoid fever are low in comparison with those of many other acute diseases; it is said, however, to prevail considerably among the coloured races (mostly negroes). In French Guiana it is observed at all seasons and among all races. In Brazil it is common; it is very prevalent in Rio de Janeiro, and in many towns and villages of the interior a severe type of the disorder is said to be met with.

In the Argentine Republic typhoid fever is exceedingly prevalent. In the Statistical Annual for the province of Buenos Aires for 1897 it is stated that the disease there has a grave and permanent character. In the ten years 1888 to 1897 the mortality from this cause was equal to 650 per million inhabitants per annum, or about four times as great as the corresponding mortality in England within recent years. In 1897 the recorded deaths from this cause in the province were 534 in number (about 533 per million), of which 270 occurred in the northern region, 182 in the central, 77 in the southern, and 5 in the Patagonian region of the province.

In the Uruguay Republic the disease is also widely spread. In 1897 it was the cause of 302 registered deaths—a figure far higher than the returns of any other of the zymotic diseases, and equivalent to a mortality of little short of 400 per million inhabitants. The disease is known in Peru and is said to be common in Callao; the statistics, however, for that city return enteric and malaria under one and the same heading.

Factors which determine the Distribution. Few diseases are more widely spread over the earth's surface than enteric fever. It is almost ubiquitous. It is met with in the tropics, in the temperate zones, and in sub-arctic regions. No latitude is altogether free from it, for though it is rare or even unknown in a few tropical and subtropical countries, such as British Honduras, Mexico, New Guinea, Central and Western Africa, it is endemic in many other countries in the same latitudes and with similar climates. That it can occur and even remain endemic in the presence of extremes both of heat and cold is shown on the one hand by its persistence in India, in Ceylon, and many of the West Indian Islands, and on the other by its appearance as an epidemic in Greenland and Labrador, and its constant presence in Northern Siberia. It is, however, mainly in the temperate zones that typhoid fever is most widely and permanently prevalent.

In its seasonal relations typhoid fever, in many parts of the world, prevails most in the autumn and least in the spring and early summer. In tropical and subtropical countries, however, it is said to be most common in the hot season. In Central Africa it is most prevalent at the beginning of the rainy season, "when the showers first moisten the refuse lying about and thus cause its decomposition, and then wash the decomposing materials into the nearest stream." Elevation above sea-level appears to have little determining influence in the distribution of the disease, and it flourishes just as well in the Transcaucasian mountains as in the plains to the north of them, in the Arabian Highlands as in the flat coast-line of the Hedjaz, and on the central plateau of Madagascar as on the shores of that island. Its relation to different kinds of soil is uncertain, but it is held by many that fluctuations in the height of the sub-soil water may be a powerful factor in typhoid prevalence. This view, first advanced by Pettenkoffer and Buhl, is briefly as follows :—typhoid fever cases increase in number as the sub-soil water falls, and decrease as it rises, and the wider the fluctuations in the sub-soil water the greater will be the frequency of typhoid. The rule has been followed in a considerable number of epidemics in various countries, but there have been, on the other hand, many epidemics in which it did not hold good.

It is now almost universally admitted that enteric fever is a water-borne disease ; and though it may possibly be spread in other ways, it is certain that its most frequent method of diffusion is by means of sewage-contaminated water. As in the case of cholera, so in that of typhoid fever, the specific virus appears to be contained in the *dejecta* of persons suffering from the disease, and when drinking-water is directly or indirectly contaminated with such *dejecta* it becomes the most potent, though probably not the sole means of diffusing infection. Innumerable epidemics have been traced with certainty to a water-supply thus specifically contaminated. This does not exclude the recognition of other modes of diffusion of the virus. Direct infection from person to person appears to be possible; infected dust may almost certainly spread the disease; and there is not a little evidence to show that the typhoid fever poison may be conveyed by—some even believe it may be generated in—the air of sewers and drains. The escape of sewer-gas into dwellings has certainly in many instances been associated with the occurrence of typhoid fever, and the possibility of the specific virus thus gaining access to the dwelling can scarcely be denied. Other hygienic defects are very commonly found in association with a high degree of typhoid prevalence, and the disease may be justly regarded as one to a great extent dependent upon insanitation. In other words it is largely, though perhaps for the present not wholly, a preventable disease, and has been found to diminish in a striking manner as the general level of sanitation has risen in a city or country. Of all great public measures for improving the health of the community the most powerful in combating typhoid prevalence have been the provision of pure water-supplies and the construction of good systems of drainage. In England the late Sir George Buchanan showed clearly how the deaths from typhoid fever had steadily declined with the increase and improvement of sewerage systems. In innumerable places at home and abroad care to prevent contamination of water-supplies has led to equally satisfactory results. Some striking examples have been already quoted in the course of this chapter. Milk and other articles of food may also, it is believed, become contaminated by the typhoid fever virus and be the means of its diffusion, and no doubt a better supervision over the purity of food-supplies has had its share, among other public health measures, in diminishing the prevalence of the malady.

The very noteworthy diminution in typhoid fever mortality in many European countries in recent years may justly be attributed to a general improvement in the level of sanitation and public health administration. Diminution of this kind has been observed in the last few decades in the British Isles, in France, Holland, Belgium, Denmark, Germany, Norway, Austria, Hungary, Switzerland, and Italy.

Race is probably a factor of considerable importance in the distribution of typhoid fever. All races are perhaps capable of suffering from it, but all are not equally liable to its attacks. It has already been seen how much more susceptible the European in India is to the disease than the native. Among the natives themselves also there are different degrees of susceptibility, and the Gurkha is believed to be particularly liable to it. A similar excessive susceptibility of the European over the native is seen in the Malay Peninsula, in the Dutch East Indies, in China, Algeria, and elsewhere. No doubt in many of these instances the racial element is not the whole explanation of the difference; the Europeans attacked by the disease are often new-comers, exposed for the first time to a particular form of typhoid infection to which the native has long been "acclimatised," if he has not been actually "immunised" by a recognised or unrecognised attack of the disease. In the United States the white races are said to suffer most from the disease, the coloured races slightly less, and the Indians least of all; the difference between the three races is not, however, very striking. In French Guiana all races, Europeans, negroes, and Indians, are liable to it.

Of the many diseases which war brings in its train none is perhaps so frequent and fatal as typhoid fever. There is scarcely a war of recent times in which thousands of lives have not been sacrificed to it. The South African war has been no exception, and, indeed, in the great outbreak which attacked our men in and around Bloemfontein in 1900, this war has furnished one of the most striking examples of modern times of the deadly effects of this disease upon troops. The frequency of enteric fever during military operations is to be attributed, perhaps wholly, but certainly mainly, to the ease with which streams and other water-supplies become contaminated with sewage-matter, and to the absence or inadequacy of the measures taken to prevent the men drinking such polluted water. A movement has been recently set on foot to induce military authorities to pay more attention to this question, and to evolve some scheme of providing troops with pure water under all conditions of military life. Any experiment of this kind will be watched with great interest, and should it succeed in abolishing, or even in greatly diminishing the typhoid scourge in future wars, the recent terrible experiences of our men in South Africa will not have been in vain.

TYPHO-MALARIA.

Typho-malaria and malarial typhoid are names that have been employed by many writers to indicate a form of disease presenting the symptoms of both malaria and typhoid fever. That the two diseases can exist in the same subject at the same time there can be no doubt. Lyon[1], in New York, has demonstrated the malarial organism in the blood and the Eberth bacillus in the spleen and intestine of five patients presenting symptoms of both diseases. Washbourn in South Africa, during the war[2], A. R. Cook in Uganda[3], Alvaro in Naples, and others have made precisely similar observations. In the same way small-pox and malaria, plague and malaria, plague and leprosy, typhus fever and scurvy, scarlet fever and diphtheria, measles and chicken-pox, have been observed on several occasions to attack a person at the same time, and the list of diseases which may occur in pairs in the same individual could no doubt be lengthened considerably[4].

The name typho-malaria has, however, come to be used by many writers to describe a disease which they believe is not due to a combination of the two infections, but distinct from either—a third disease, in fact. If this view should prove correct, and if from the many "unclassified" fevers which certainly exist, one should come to be clearly differentiated, both clinically and etiologically, and take its place in the list of named fevers, it may be hoped that some other name will be found for it than this. At present no other name has been suggested, and the term typho-malaria or malarial typhoid will therefore be used here to indicate a group of cases, some of which may ultimately prove to be of the nature of a distinct and separate disease. The confusion existing between Mediterranean fever and typho-malaria has been pointed out in an earlier chapter[5]. Whether the many obscure fevers met with in the near east—such as "Cyprus fever," "Bukowina

---

[1] *American Journal of Med. Sciences*, Jan. 1899.

[2] *Brit. Med. Journal*, April 20, 1901.

[3] *Journ. Trop. Med.* March 1, 1902.

[4] See a paper on this subject by F. Foord Caiger, M.D., B.S. *Transact. of the Epidemiological Society.* 1893–4.

[5] See p. 282.

fever," the "bilious typhoid" described by Griesinger in Egypt, the "bilious typhus" seen every summer in-Alexandria, "Smyrna fever," "Levant fever," and many others—are or are not of the nature of so-called typho-malaria there is no scientific evidence to show.

Recent Geographical Distribution. The evidence of the existence of this disease in Europe is not very conclusive. In Greece, where both malarial and typhoid fevers are extremely prevalent, cases presenting difficulty of diagnosis, and apparently on the border line of the two diseases, are often seen. Dr Cardamatis of Athens, who has discussed at length the fevers of that country, does not, however, include typho-malaria amongst them[1]. He concludes that the large majority met with there belong to one or other of three distinct groups—paludic fevers, fevers due to auto-intoxication (from gastro-intestinal disturbance), and typhoid fevers. In Constantinople and the neighbourhood a class of obscure fevers is very commonly seen in which some of the features of both malaria and typhoid are present, but it would be difficult to assert that these are identical with the cases described by American writers as typho-malaria.

It is principally in America and in Africa that the disease known under this name has been seen. In America it was first described by Woodward as observed by him in the Federal army on the Potomac in the year 1861. According to that author over 57,000 cases of this disease, with over 5000 deaths, occurred in the American army; but it seems probable that these figures included a large number of cases of, and deaths from, other disorders. At the present day the term typho-malaria is frequently used by American medical writers, and cases of the disease known by this name appear to be not infrequent in many parts of the United States. In New York, Pennsylvania, parts of New England, and Ohio the term is most commonly met with. In Michigan it is accepted by the State Board of Health, and in some parts of the State the malady thus indicated appears to be one of the fifteen diseases most prevalent in certain years[2].

In British North America this form of disease is also not unknown. Thus at Greenwood city, in British Columbia, a mild

[1] *La Grèce Médicale*, 1900.
[2] Prof. G. Dock, M.D., *New York Medical Journal*, Feb. 25, 1899.

outbreak of "typho-malarial fever" occurred in 1897, and, in the opinion of the Medical Officer of Health, was caused by soil disturbance. The course of a stream called the Boundary Creek had been changed within the limits of the city, and the attacks of this disease were distributed among those who used the water drawn from the channel just below the point where its course had been changed. In the preceding year a "fever of typho-malarial type which ran a typhoid course" was met with at Richmond, also in British Columbia; in this outbreak 26 Japanese, 12 Indians and 18 white persons were attacked.

A disease of like character appears to be endemic in some of the West Indian Islands. In the hospital reports of the Leeward Islands occasional mention is found of cases of typho-malarial fever; and in Cuba during the recent disturbances many cases of the kind were observed among the American troops employed in the war.

In many parts of the African continent fever described as of typho-malarial character has been met with. Vincent saw 17 cases in Algeria. On the West Coast it has been seen in the Gambia, in Senegal, and in the Belgian Congo. On the Red Sea coast the Italian troops suffered from it in Erythrea in 1897. In South Africa this type of fever appears to be common. It is occasionally epidemic in some parts of Cape Colony, where it is locally known as "River Fever" or "South African Fever." Epidemics of it were reported from Gordonia in 1897–8, and from the Xalanga district in 1898, and here as elsewhere the suspicion arose that the fever in question might possibly be true Mediterranean fever. In the Galeka-Gaika campaign of 1877–8 and in the Zulu war of 1879 it is also to be noted that the British troops suffered heavily from a mixed infection of enteric fever and malaria (Sambon). In the recent war Washbourn saw a case in which "a well-marked attack of remittent malaria with parasites in the blood recurred during the incubation period of enteric fever[1]."

Finally from some parts of the continent of Asia reports of the existence of this form of fever have been published. The occurrence in Jerusalem of a typho-malarial disease bearing some resemblances to Mediterranean fever has been already mentioned[2]. In China typho-malarial fever is apparently far from rare, and

[1] *Brit. Med. Journal*, April 20, 1901.   [2] See p. 286.

mention of its occurrence in the reports of the medical officers in the Chinese Customs Service is comparatively frequent. In quite recent years it has been reported from Hankow, from Canton, from Chungking, and from Mengtsze in Yünnan. In India it seems probable that this form of disease is met with, but here again doubt has arisen as to whether it is not of the nature of Mediterranean or undulant fever.

TYPHUS FEVER.

History. Uncertain references to this disease are found in the records of some of the oldest writers and historians, and there is good reason to believe that in the earliest ages of the world's history typhus fever accompanied famines, sieges, and wars, just as it has in modern times. The first definite mention of the disease has been assigned to the eleventh century. Throughout the Middle Ages and down to the last century typhus fever was exceedingly prevalent in most European countries, and was the cause of frequent epidemics. In the nineteenth century there was a very general lessening in the activity of the disease. The Napoleonic wars had led to a most serious and deadly extension of typhus fever over a great part of the continent. But when these came to an end in 1815, and an era of peace began, there was a marked subsidence in typhus prevalence. Since that date the disease has remained endemic in certain countries only, but has elsewhere either disappeared altogether or become very rare. In 1846-7, however, there was again a wide diffusion of typhus fever over Europe, and serious epidemics of it are constantly recorded from different parts of the world where famine and want and other conditions exist to bring them about.

Recent Geographical Distribution. *Europe*. In England of recent years typhus fever has been gradually becoming extinct. It still lingers, however, in certain areas in the north-west and northern registration divisions of the country, where the over-crowding and other conditions which favour its development are most markedly present. This was very clearly seen in the records for the decade 1881–90, when considerably over half the deaths from typhus recorded for the whole country occurred in districts containing not more than one-eighteenth of the entire population.

These districts were situated in Cheshire, Lancashire, Durham, and Northumberland; the mortality here from this disease in the decennium in question was as high as 139 per million, while in the rest of England it was only 7 per million. In the last decennium a still further decrease in the mortality from typhus has been observed, and in 1899 and 1900 it was less than 1 per million.

In Scotland occasional localised outbreaks of the disease occur from time to time. In the decade 1881–90 the average annual mortality from this cause was 33 per million, or some 126 deaths per annum; in 1895 only 34 deaths were recorded; in 1896 the deaths from typhus numbered 38, and in 1897 they fell again to 29.

In Ireland the disease is very much more common, though in recent years there has been a tendency to a steady decline in its prevalence. Thus in the three decades 1871–80, 1881–90, and 1891–1900, the mean mortality rates from this malady were 141, 110, and 47 per million respectively. Ireland was formerly regarded as one of the most important European centres of typhus fever infection, and was the principal source of epidemics in England and Scotland in the past.

The degree of prevalence of typhus fever in France has varied widely in recent years. Thus in 1893 as many as 200 deaths from the disease were recorded in the group of towns with over 10,000 inhabitants. The deaths from this cause then fell to 63 in 1894, to 11 in 1895, to 7 in 1896, and to 4 in 1897; and in 1898 the towns were apparently free from the disease, no deaths from typhus fever being recorded in them in that year. The malady appears to be endemic to some extent in Brittany, and the epidemic of 1893 just mentioned, which prevailed principally in Paris, and between that city and Havre, was thought to have originated in the Breton centres of the disease[1].

In Germany typhus is said to be commoner in the east than in the west, in consequence, it is thought, of the greater proximity to Russia and Austria, where the disease is endemic. In Silesia and parts of Prussia it is also perhaps endemic. In Saxony, Bavaria, and Würtemburg, on the other hand, it is said to be almost unknown. In Holland it appears to be

---

[1] *Contribution à l'Étude de l'Épidémie de Typhus Exanthématique du Nord de la France en* 1892–3, par le Dr Netter, Paris, 1893.

exceedingly rare, and in Denmark it must be equally so, for only four deaths from the disease have been recorded as occurring in the Danish urban population in the six years 1892–1897, and all four occurred in the same year (1893).

In Norway and Sweden a few cases of the disease occur every year, but it is apparently not a common malady in either country. In Russia, on the other hand, there can be no doubt that it is exceedingly frequently met with. There is no portion of the Russian Empire, whether in Europe or Asia, from which cases of and deaths from this disease are not yearly reported. Its distribution varies from time to time, but so far as may be judged from the published statistics of cases and deaths it is only in the Central Asiatic provinces of Russia, in the Baltic provinces, and in the Caucasus that the disease is rare. The following table shows the recent returns of typhus fever in the various territorial divisions of the Russian Empire, the figures representing the numbers of registered cases per million inhabitants.

Year	EUROPEAN RUSSIA					ASIATIC RUSSIA			Total
	Governments without Zemstvos [1]	Governments with Zemstvos [1]	Polish Governments	Baltic Governments	The Don Cossacks	The Caucasus	Siberia	Central Asiatic Provinces	
1887–92 (mean)	410	1020	360	160	390	40	680	50	740
1893	660	1860	600	120	280	60	730	70	1230
1894	550	1290	230	100	170	100	570	10	860
1895	440	830	200	50	540	30	280	10	570

It will be observed from this table that typhus fever is most common in European Russia, and particularly in those parts of the country in which government by *Zemstvos* has been introduced. These, it is to be added, constitute by far the greater part of European Russia proper. It will further be noted that the disease is somewhat less common in Siberia, still less so in Polish Russia and in the country of the Don Cossacks, and markedly less so in the Baltic provinces; while, as stated already, in the Caucasus and Central Asia it is, so far as these figures may be accepted as an

---

[1] That is to say in which the system of local government by bodies known as *Zemstvos* has, or has not, been introduced.

index, comparatively rare. This fever was exceedingly prevalent in Russia during and after the great famine of 1891, particularly in Saratof, Kazan, Tambof, and other districts where the suffering from the famine was greatest. Many cases were seen in which a severe form of typhus was combined with scurvy; and it is noteworthy that the great Russian surgeon Pirogof states in one of his essays that the same combination of diseases was very common among both the besiegers and besieged during the siege of Sevastopol in the Crimean war.

The frequency or infrequency of typhus fever in Austria-Hungary is a matter of some uncertainty. In the year 1896 some 1009 deaths from this cause were recorded in the country, but the large majority of them occurred in Galicia. In the other provinces no deaths from this disease were registered, with the exception of a few in Carinthia, Silesia, Bukowina, and Dalmatia. Epidemics have not been infrequent in the past in the Austrian Empire, particularly in Galicia.

In Italy typhus has of recent years been exceedingly rare. In 1887 and 1888 the recorded deaths from the disease in that country were equal to a mortality-rate of 65 and 71 per million respectively; but after 1888 the mortality-rate fell, and from 1890 to 1896 it never exceeded three per million, and in some years was less than one per million. Formerly typhus fever was excessively common in Italy, which was ranked next to Ireland and the Slavonic countries among the leading endemic centres of the disease in Europe. In Spain and Portugal there is little certain information as to the frequency of this disease, but it is said to be rare in the Peninsula. In Turkey in Europe it is rare at the present day. After the Russo-Turkish war of 1877 and the Turco-Greek war of 1896 there was a considerable amount of the disease in some parts of the country, but it has since almost disappeared. In Constantinople it may be doubted whether it is ever seen now as an indigenous disorder.

The malady has been epidemic on more than one occasion in the past in Iceland, but it does not seem to be endemic there, and in the Faröes it has never been met with.

Asia. The prevalence of this disease in the Russian portions of Asia, *i.e.* the Caucasus, Central Asia, and Siberia, has been already discussed.

In Asia Minor, Armenia, and Syria, typhus fever is said to

occur, not as an endemic disease, but in the form of occasional epidemics. In Mesopotamia, on the other hand, it is said to be endemic. There can be little doubt that it is also an endemic disease in Persia. In Teheran it occurs in the form of occasional epidemics; and, elsewhere, recurring outbreaks of the disease appear to be not uncommon. In 1880 and 1881 it raged with great severity in some localities that had been devastated by a Kurdish invasion.

As regards the prevalence of typhus fever in India and Baluchistan it appears that there is a somewhat extensive area in which the disease is truly endemic. This area includes the Trans-Indus districts from Baluchistan to Yusufzai, extending through the Hazara and Rawal-pindi districts, and including the Himalayan hill-tracts[1]. The Yusufzai valley, near Peshawur, is a well-known home of the disease, in which epidemics are far from uncommon. In Kumaon and Garhwal, on the southern slopes of the Himalayas, it seems that it is also more or less endemic. But a certain amount of confusion has arisen between typhus and plague in these provinces. It appears that both these diseases occur here at times in epidemic form. The terms *sunjar* and *mahamari*, in general use among the Himalayan hill tribes, have been usually held to mean typhus fever and plague respectively. But an epidemic of *sunjar* was investigated in 1899 by Dr Leonard Rogers, and proved to be one of relapsing fever[2]; and it appeared from inquiries made on the spot that mild outbreaks of any epidemic disease in these hill-tracts are termed *sunjar*, and severe ones *mahamari*. In some parts of India, as for example in Rajputana, typhus fever is said to be unknown. It is interesting to note in this connection that the fever which has prevailed in the Central Provinces in consequence of the recent famine, has not been of the nature of typhus fever. Typhus is also unknown in Ceylon.

Concerning the East Indian islands and the Malay Peninsula the information is more scanty. In Borneo there appears to be no certain record of the occurrence of this disease, while in Java it is

---

[1] *Transactions of the Indian Medical Congress*, 1894. In 1892–3 the disease became epidemic among the Ghilzai labourers (from Jellalabad) employed in building the Mushkaf-Bolan Railway.

[2] *Annual Report of the Sanitary Commissioner of the North-Western Provinces and Oudh*.

definitely said to be unknown, at least among the hill tribes, and as it is among hill tribes that typhus fever generally finds favourable conditions for its prevalence, it may be assumed that the disease is absent from the whole island. On the other hand typhus is said to have been observed in Celebes. I find no mention of the disease in recent reports from British North Borneo, nor in any recent reports or writings that I have consulted upon the pathology of the Straits Settlements, Siam, Cochin China, or Tongking.

In China typhus fever would seem to be rare in the south, while in the north it is commonly met with. At Peking and throughout the north generally the disease is endemic, and tends to become epidemic each recurring spring. It is known to the natives as *Yen-ping*. The fever known as " Peking fever " is said to be no other than typhus. In China and elsewhere the disease is closely associated with conditions of overcrowding and misery, and it is consequently mostly among the natives that it is found to prevail. The European residents usually escape its attacks, unless they, too, are living under conditions resembling those under which the natives live. This has sometimes happened in the case of foreign missionaries, who then appear to have suffered no less than the natives themselves. In some years the disease has been so severe among the Chinese that the hospital attendants have fled panic-stricken to escape the infection[1]. Reports of the occurrence of typhus are also forthcoming from Chingkiang, Hankow, and Taiping-fu. At Shanghai, on the other hand, the disease seems to be rare, and at Amoy it is said to be unknown in spite of the presence of a large number of typical fever dens. In general terms it seems to become comparatively rare to the south of Shanghai, and in Canton its existence has been denied altogether, though an outbreak of fever called *chut-pan*, or spotted fever, which occurred in Canton in 1879, may have been one of typhus.

The existence of typhus fever in Japan has also been denied, but it appears that the disease was epidemic in Tokio in the spring of 1881. A large number of convicts in the prisons were attacked and the disease spread more or less over the whole city, more than one hundred cases being reported to the police[2]. Since that date

---

[1] Matignon. *Chinese Imperial Maritime Customs Medical Reports*, 1895–6.
[2] *Sei-i-kwai.* October 11, 1898.

a few cases of typhus have been recorded every year in some part of Japan, and particularly in the Hiogo province.

Australasia. In the Philippine Islands and New Guinea I find no mention of typhus fever, and in Australia the disease appears to be rare, if indeed it is known there. In New Zealand, and throughout the islands of the Pacific it seems to be unknown.

Africa. Whether typhus fever be endemic in Egypt, or whether it is always imported from Nubia, where according to Hirsch it is endemic, there can be no doubt that the disease is met with in Egypt, and is no great rarity there. It has been known to break out with considerable severity in Egyptian prisons, as, for example, in that of Tourah in 1886 and again in 1890. In the latter year typhus was unusually prevalent in many parts of Egypt; it was particularly severe at Tantah, Zagazig, and some other places.

It has often been said that typhus is endemic in Tripoli. This does not appear to be true of the Cyrenaic plateau, where the disease is not permanently endemic, but has on certain occasions broken out and caused epidemics of extreme intensity. An outbreak of this kind occurred in 1892-3. Dr Biednavski, who investigated it on behalf of the Turkish sanitary administration, informs me that it succeeded several years of serious famine, and that the inhabitants, mostly nomad Bedawin, had collected in the three principal towns, Benghazi, Merdj, and Derna, under conditions of overcrowding, filth, and mal-hygiene which need not be dwelt upon. Typhus broke out, and in the course of some three or four months caused the death of about 15,000 persons in a population which perhaps did not exceed some 40,000 souls. Since that date, so far as is known, there has been no epidemic recurrence of the disease.

In Algeria it is seen, though rarely. Hirsch, however, says that it is—or was—endemic and occasionally epidemic in Kabylia. In Morocco it is said sometimes to occur in the train of famines caused by the ravages of locusts.

On the West Coast of Africa the disease appears to be almost unknown. It is not mentioned in most recent reports or publications on the prevailing diseases of the Gambia, Lagos, the Gold Coast, Sierra Leone, or the Cameroons. It appears to be in like manner absent from the list of diseases met with in the Congo Free State. Equally remarkable is the absence or extreme rarity of typhus fever in South Africa. Doubt has been expressed whether the disease does exist at all in this part of the world, and, if it does, it is certainly among the rarest of all disorders. In recent official reports from the Cape Colony, from Natal, and from the other British possessions in that region there appears to be no mention whatever of typhus fever, at least up to the time of the outbreak of the late war. In 1898 three isolated cases of the disease were reported from the sub-district of Petrusville, in Cape Colony, and this would seem to be almost, if not quite, the only instance in which typhus fever has appeared in this part of the world in recent times.

The disease appears to be equally rare or unknown on the east coast of Africa, and in the islands of Madagascar, the Seychelles, and Mauritius.

North and Central America. At the present day typhus fever must be regarded as a rarity throughout the greater part of British North America and the United States. In both it has in times past prevailed to a greater or less extent, usually in the form of localised outbreaks, which, in the majority of instances, were traceable to importation by immigrants from Ireland. "In many cases," writes Hirsch, "the disease was confined exclusively to immigrants; in others it spread among the population of the seaports, particularly in the filthy and crowded quarters, but without ever attaining the dimensions of a great epidemic." In neither Canada nor the United States has typhus ever become truly endemic, and in recent reports from the Canadian provinces, and in such reports from the States as I have had access to, the name of this disease is conspicuous by its absence. Osler states that it is occasionally seen in New York and Philadelphia, and mentions an epidemic of it in 1877 in the House of Refuge in· Montreal.

In Mexico, on the other hand, it is truly endemic, and has apparently been so for more than three centuries. Prof. Monjuras of San Luis Potosi states that the disease is exceedingly common in that town; no season of the year is free from it, and he even adds that there is no day in the year on which cases are not recorded. As many as 851 deaths from this cause have been reported in the city in five years. The disease here, as elsewhere, is closely associated with poverty and over-crowding. Typhus appears to be common enough throughout the whole Mexican Republic, but especially so in the high plateaux of the interior. In British Honduras and Guatemala, however, it appears to be as little seen as it is further north in the United States; but in Nicaragua at least one epidemic of it has been recorded in the past (in 1851). The West Indies and Bermuda have enjoyed an immunity from its ravages for some decades.

South America. Typhus fever has not, in the last few years, been seen in British Guiana, nor is it apparently known in the other Guianas, Colombia, or Venezuela. In Brazil, however, it certainly occurs at times, as, for example, at Rio de Janeiro. In Peru it is still more common, and in the sierras especially has at no remote period caused great ravages among the Indian inhabitants. It is believed that typhus was originally introduced to Peru from Spain, and probably at a very early period. The disease is apparently endemic, and at times epidemic in Chile and Bolivia[1]. Further south, in the Argentine Republic, Paraguay, Uruguay, and the Falkland Islands it is said to be unknown.

Factors determining the Geographical Distribution. Typhus fever is an excellent example of that group of diseases which have become rarer as civilisation has advanced. It is essentially associated with ill-nutrition, overcrowding, misery, and insanitation. But while all these conditions are of common occurrence all the world over, the same is not true of typhus fever, and it seems that another factor, in all probability a specific germ, is necessary to its development in addition to the conditions already named. The micro-organism, if there be one associated with the disease, has not, however, yet been discovered with certainty.

The recent distribution of the disease has been a rather remarkable one. In Europe, where it was at one time almost universal, it has remained endemic to any notable extent only in the British Isles, Brittany, Russia, and parts of Austria and Germany. In Asia it appears to be endemic in Mesopotamia, Persia, the Himalayan belt of India, and Siberia. In Africa it is probably endemic only in Nubia, and perhaps in Egypt; in the New World only in Mexico, Peru, Chile and Bolivia. But while the truly endemic range of the disorder is thus restricted at the present day, in the past, and indeed in quite recent times, it has prevailed as an epidemic in many other parts of the world.

[1] Brault.

The countries in which, so far as is known, it has never been seen either in endemic or epidemic form are Australia, New Zealand, the Pacific Islands, the greater part of Central and Southern Africa, a considerable portion of the United States, the Argentine Republic, Paraguay, Uruguay, and the Falkland Islands.

Typhus fever shows unusually definite relations to atmospheric temperature, and this accounts in part, though by no means wholly, for its geographical distribution. In all countries where it is endemic it is most active in the cooler months of the year. It is, indeed, essentially a disease of temperate and cold climates. In the tropical and sub-tropical countries named above in the list of its endemic areas, the disease is found to occur principally at considerable heights above sea-level, where the temperature conditions are practically those of the temperate or cool zones. In like manner the larger number of epidemic outbreaks in all countries have occurred in the cooler portion of the year. No doubt this is mainly due to the increased chances of infection in the winter, when people are apt to be more crowded together in insanitary dwellings than in the summer.

The relation of typhus fever to altitude above sea-level, of which mention has just been made, is apparently in the main a question of temperature. In the temperate zones the disease is found prevailing principally in low-lying areas, or indifferently in flat and in mountainous regions. In the flat plains of European Russia, for example, it is met with more or less throughout, and in the Caucasus there does not appear to be any striking difference in its prevalence between the low-lying Caucasus proper and the highly-mountainous Transcaucasia. But as the tropics are approached, if typhus fever occurs at all as an endemic disease, it is mainly at considerable heights, as, for example, in the Himalayan belt of India, in the high plateaux of Mexico, and in the sierras of Chile and Peru.

But of all the external conditions favouring the prevalence of this zymotic none are so important as those of insufficient food, overcrowding, and insanitation. The names " famine fever " and " hunger typhus " often given to the disease are a recognition of its close dependence upon insufficiency of food. The historical examples of typhus fever epidemics following on great famines are numerous. In Russia during recent years of bad harvest the disease has been widely prevalent, and in 1892 it was one of the worst and most fatal results of the great famine of that year. Quite recently, in November, 1902, a serious outbreak occurred in Podolia as the result of dearth. In Algeria and Tunisia in 1868, in Tripoli in 1893, in Ireland in 1846–7, typhus fever epidemics of extreme severity were associated with failure of the harvest and consequent famine and misery. But it is important to note that by no means all famines have been accompanied by such outbreaks. The recent famine years in the Central Provinces of India, for example, have been marked by unusual prevalence of "fevers" and perhaps other diseases, yet among the sicknesses vaguely classed as "fevers" typhus does not seem to have been one. Nor, on the other hand, does famine appear to be essential to the prevalence of this disease. Virchow drew attention to this fact in connection with the typhus outbreak in Upper Silesia in 1847–8, and a quarter of a century earlier Graves had clearly shown in Ireland that typhus could prevail in years of plenty. The occurrence of serious famine without the appearance of typhus is no doubt to be explained by the statement made above, that the disease is most probably due to some micro-parasite. It was formerly very generally believed that it could be generated *de novo* under conditions of want and misery, but this must now be considered extremely doubtful; and in all probability in the absence of the specific micro-organism no degree of dearth or famine can give rise to the disease.

If starvation is one of the most favourable conditions for the prevalence of this malady, overcrowding and misery are scarcely less so. The frequent occurrence of typhus fever epidemics in the past in besieged cities, and in connection with military operations generally, has apparently been largely favoured by these latter conditions, although in many instances no doubt the former—starvation or at least an insufficiency of proper food—has also been present. The wide extension of typhus fever in Europe during the Napoleonic wars has already been noted, and the marked reduction in its prevalence over the whole continent after Napoleon's fall shows clearly how important a factor war may be in determining the geographical distribution of this disease. In the Franco-Prussian and Russo-Turkish wars there were serious outbreaks of typhus. Both sides suffered from it severely in the Crimean war. Among the Russian troops its prevalence and severity have been attributed by Russian writers as much to insufficiency of the food supplies as to overcrowding[1]. The English troops also suffered greatly from the disease in the winter of 1854–5 ; it was less active in the following summer, and in the winter of 1855–6 the troops were much less affected by it, as they had now been provided with better quarters and were less overcrowded. In this second winter of the war, however, the French troops who were encamped in close tents on wet ground suffered from typhus. In the Franco-German war the disease broke out among the population of Metz during the siege, yet the troops encamped in the outworks, who were better fed and less crowded, almost escaped it. The Germans do not appear to have had any typhus in their army during the war, and this has been ascribed to the strict military hygiene enforced by them in the field. In the Russo-Turkish war of 1877–8 there was a good deal of the disease among the Russian troops both in the Caucasus and in European Turkey. Innumerable other instances of typhus in war time might be quoted, and they all show the close correlation of the disease with want and overcrowding.

These conditions are unfortunately found in many countries at other times than those of war. Few of the infective fevers have shown so direct and positive a relation to sanitary deficiencies as this, and particularly to such deficiencies as overcrowding and bad ventilation. Dirt, squalor, and misery generally, accumulations of filth and refuse, and imperfect removal of sewage are sanitary defects so commonly found together with the others just named that they may also have something to do with typhus fever prevalence. But it is probable that overcrowding and ill-ventilation are the most important. They have been universally found in the great epidemics and endemics of typhus. In former days, when the gaols and workhouses in all European countries were shockingly overcrowded, and ventilation almost unknown, the disease was rife in these institutions; and it was more or less rife just in proportion as the buildings were more or less overcrowded and more or less badly ventilated. When Howard made his memorable visitation of British and Continental gaols in the eighteenth century, "the prisons were the central seminaries and forcing-houses from which the typhus-contagion of those days was ever overflowing

---

[1] *Golod i Vuizuivaemuia im boliézni.* (Famine and the Diseases caused by it.) By A. A. Lipski, St Petersburg, 1892.

into fleets and barracks and hospitals, and was a constant terror to courts of justice and to the common population[1]."

In more modern times the disease has lingered among peasant populations in Ireland, Brittany, Russia and elsewhere, among whom conditions of overcrowding, ill-ventilation, and squalor are only too commonly found. In its recent epidemic outbreaks also (other than those in connection with wars, already discussed) overcrowding and ill-ventilation have been the constant accompaniments of the disease. In Algiers, for example, in the great epidemic of 1867–8, the people had flocked to the towns for food, and were crowded together in temporary buildings run up to receive them, when the disease appeared amongst them. Exactly the same conditions preceded the outbreak in Tunis in 1868; and the terrible epidemic in the Cyrenaic plateau in 1893 was mainly fostered by the densely overcrowded state of the few towns there, the result of a preceding famine.

The reduction in typhus fever prevalence in many European countries in recent times may, with a probability amounting almost to certainty, be attributed to improvements in sanitation. Of these improvements, by far the most important have been the reduction of overcrowding; the opening up of the over-built, over-populated, filthy slums and rookeries of large cities; the provision of better house accommodation for the lower classes; the introduction of proper systems of ventilation, and the provision of more space per head in gaols, reformatories, workhouses, barracks, and ships. To these may perhaps be added a general rise in wages, and an improvement in the social and domestic conditions under which most classes of the population live. Where typhus fever now lingers in the capitals and large cities of civilised countries, as for example in Berlin and in Dublin, it is solely among the poorest classes of the population, who live under those conditions of overcrowding and squalor, and perhaps of want, already described.

Other public health measures, as the provision of water-supplies, the construction of drains and sewers, and the inspection of food-supplies, can scarcely affect the prevalence of this disease directly. For the infection of typhus is not spread by water; it is almost certainly not spread by sewer-gas; and it does not, so far as is

[1] *English Sanitary Institutions.* By Sir John Simon, London, 1890, p. 140.

known, attack any of the lower animals. The infection is indeed probably spread solely by the air or by infected fomites. Typhus fever is one of the most directly infectious of all diseases. The atmosphere around a typhus fever patient appears to contain the infecting agent, whatever this may be, in large quantities and in a high degree of virulence. Those in immediate attendance upon persons suffering from the disease often become infected. To what distance the virus can be spread by the air is quite unknown; but there seem to be no observations showing that it can be carried over long distances like that of small-pox. Most probably the infecting agent is highly virulent in the air immediately around the patient, and rapidly becomes innocuous after travelling only a short distance through the atmosphere. It can, however, attach itself to linen and clothes, the walls of rooms, and furniture : and in this way the infecting material of typhus is believed, on some occasions, to have remained active for some time after leaving the patient's body, and even to have been carried long distances over the earth's surface.

Typhus fever has shown no very definite relations to race, and it is probable that racial differences have had little or no share in determining its geographical distribution. The European races, the Himalayan hill tribes, the Indians of the plains, the Chinese, the Kabyles and Arabs of Algiers and Tripoli, the African negroes, the Indians of Chile and Peru, and the mulattos of North America have all suffered from it; and though there does not seem to be any mention of typhus occurring among the native tribes of Australia and New Zealand, the Pacific Islands, the United States, or the southern portion of South America, this is more likely to be due to the fact that the infection has never reached them than to any racial immunity.

WHOOPING-COUGH.

History. The first certain mention of whooping-cough relates to an epidemic in Paris in the year 1578. In the next two centuries it was rarely referred to by medical or other writers. From the 18th century onwards there are numerous records of epidemics in many parts of the world. Hirsch, however, believes that this disease is really of comparatively modern origin, and that it has spread over the globe within recent times.

Recent Geographical Distribution. *Europe.* In England and Wales whooping-cough is very frequently seen. As a cause of death it has diminished to some extent in recent years. The mean annual death-rate per million from the disease was 527 in the years 1861–70, 512 in the next decade, and 450 in the years 1881–90. It has since been still lower, but has fluctuated considerably; the lowest ratio since registration began was observed in 1895, when it was only 316 per million. Whooping-cough has a distribution not unlike measles; it is most fatal in large centres of population and in mining districts. In the decennium 1881–90 the lowest ratios were met with in Westmoreland (193 per million), Dorset (230), and Hereford (231); and the highest in Lancashire (520), Essex (537), and London (690 per million).

The disease has recently caused a higher mortality-rate in Scotland than south of the Tweed. The ratio of deaths from whooping-cough in Scotland in 1881–90 was 605 per million per annum; in each of the years 1895 and 1896 it was 470, and in 1897 it rose again to 690 per million.

In Ireland the mean annual mortality from this disease in the decade 1881–90 was only 290, and in 1890–99 only 280, per million.

In French towns of over 10,000 inhabitants the deaths from whooping-cough varied between 90 and 170 per million in the years 1886-1898.

In Belgium the deaths from this cause remained at about the same figure from 1870 to 1895; the annual mean of recorded deaths fluctuating but slightly above and below 3800 (something over 600 per million inhabitants); in 1897 they fell to 2809, or approximately 430 per million. In Holland the disease is a less conspicuous cause of mortality, the ratios of deaths per million inhabitants varying between 220 and 360 in the years 1889–98 inclusive. The figures for Holland show no tendency to diminution in recent years.

In the German Empire there was a slight increase in the recorded deaths from this cause in the years 1892–4, but the records for a number of years show considerable fluctations in the mortality-rates. This is true of all the principal divisions of the empire.

In Denmark whooping-cough has caused a mortality among urban residents varying between 184 (in 1894) and 490 (in 1895), respectively the lowest and highest rates between the years 1892 and 1897. In 1897 the ratio was about 190 per million. Equally wide fluctuations have been observed in Norway, with a tendency in the four years 1891–1894 to an alternation of years with high and years with low mortality. The mortality-rate in 1893 was about 85, and in 1894 about 220 per million inhabitants. In Sweden the deaths from this cause fell considerably between 1891 and 1895, the mortality figure for the latter year being approximately 115 per million; but in 1896 it rose to 222.

Whooping-cough is the cause of a higher mortality in Russia than in most other European countries excepting Hungary. The deaths from this disease in European Russia in the years 1893–95 were equivalent to an annual ratio of 660 per million. Very wide fluctuations are noted from year to year in the prevalence of the malady in various parts of the country. It is commonly met with throughout, from north to south and from east to west. The Polish governments have of recent years returned a somewhat lower proportion of cases than have others, but returns of cases are notoriously unreliable in any country.

In Austria the deaths from the disease in 1896 were equal to a mortality-rate of about 410 per million; in the years 1892–94 they are said to have been at the still higher annual rate of 650 per million. In Hungary whooping-cough appears to be a most prolific cause of mortality; in 1897 the death-rate from it was as high as 1060 (approximately) per million. In the years 1892–94 the mean annual mortality from the disease was 660 per million. In Switzerland the mortality varies widely; in the fifteen large Swiss towns the lowest rate in recent years occurred in 1893 (150 per million); it rose in the next year to nearly double, but subsequently fell to 157 in 1897.

The returns of whooping-cough deaths in Italy have shown a steady fall in recent years. The mortality-rate rose from 257 per million in 1868 to 435 in 1890, but has since declined to a minimum of 213 in 1896. In Spain and Portugal whooping-cough is said to be only moderately prevalent; and Davidson states that the death-rates from this cause, both in town and country, were (the years are not named) only about half those which obtained in England. It is probable, however, that the disease fluctuates widely here, as elsewhere.

For the Balkan countries full information is lacking. In Servia the disease must have been either extremely malignant or extraordinarily prevalent in 1894 (the only year for which I have had access to statistics); for in that year the deaths from it were 4227 in number, or approximately 1800 per million inhabitants; and it accounted for 6·63 per cent. of the deaths from all causes. In Constantinople the disease is always present, and reports of it are by no means rare from provincial Turkish towns and villages.

Whooping-cough "is not endemic in Iceland, and, what is still more remarkable, it has never been observed to spread even when introduced into the country"—an observation which becomes still more striking when the very different behaviour of measles and other infectious disorders after importation to the island is recalled. In the Faröes it has been epidemic on three occasions only.

It is certainly known in some of the islands of the Mediterranean. Eight cases of the disease were registered in Malta in 1898.

Asia. So far as may be gathered from the admittedly imperfect returns of cases of disease from the Caucasus, whooping-cough appears to be much less common there than in European Russia. In the years 1887–95 the returns were constantly from one-third to one-fifth of those in the same unit of population in the European governments. The recorded deaths from this cause in the Caucasian provinces in 1895 were at the rate of about 71 per million. In Russian Central Asia the returns of cases are still lower, averaging only from one-third to one-fourth of those received from the Caucasus. The recorded deaths in the provinces beyond the Caspian in 1895 were only 19 in all, or about 3 per million inhabitants. To what extent this figure really implied that the disease was very rare here in that year there are no data to show. In 1894 the corresponding figure had been about 5·7 per million.

In Siberia the registered cases of whooping-cough have varied widely in recent years; in some years they have been much lower in proportion to the population than in European Russia, while in others they have been higher than in the majority of the governments on this side of the Urals. In 1895 the recorded cases in Siberia were only one-third as many as in 1894. The disease is certainly met with in the remote northerly government of Yakutsk; it appears to be quite common in the province of the Amur, and it is included in the returns from the government of Tchita, on the Mongolian borders, as well as from all other Siberian "governments."

The disease is occasionally epidemic in Mesopotamia and Kurdistan; of its prevalence in Syria, Asia Minor, and Persia little seems to be known. In India it becomes epidemic from time to time, but rarely with great severity. It is seen not only in the plains but also in the hills; in 1893 it was epidemic, at the same time as measles, in the Hindu-Khush, especially among the country people in the hilly parts of Gor and Gilgit.

In Ceylon it is occasionally epidemic; it was so in many villages in the summer months of 1897.

Of its prevalence in Farther India little seems to be known. It is not mentioned by recent writers on the pathology of Siam. Davidson makes no reference to it among the diseases of the Malay Peninsula, Annam, or Tongking. I find no mention of it in recent reports from the Straits Settlements. In Burma, however, it has been seen recently; a severe epidemic of the disease occurred in Rangoon in 1901, and necessitated the closing of the schools. This was the first known epidemic of the kind in Rangoon during the last quarter of a century.

In China, on the other hand, whooping-cough is apparently not a rare disease. In Shanghai it is said to be endemic, attacking both natives and Europeans[1], it was epidemic among the Chinese inhabitants of the town in May—July, 1891[2]. It was very prevalent also at Wenchow in 1892[3]. In the Tientsin district it "went the round of the children," both foreign and Chinese, in the summer of 1897[4].

The disease is probably rare in the Malay Archipelago. It is mentioned among the epidemic diseases seen in Borneo. It has not been seen among the hill tribes in Java, and it is not named among the prevalent diseases of Sumatra, Celebes, or the Philippine Islands. The absence of mention of the disease is, of course, not proof that it does not occur in these islands.

Australasia. In British New Guinea whooping-cough appears to be unknown among the aborigines. In Australia the disease is commonly prevalent, but varies widely in its activity from year to year. In Queensland the deaths from this cause in 1893 were equal to 299 per million inhabitants, in the following year they were scarcely more than one-fourth as many, and have fluctuated widely since, another epidemic occurring in 1895. In New South Wales it is occasionally epidemic. In Victoria it has apparently not been very prevalent in some recent years; in 1896 it was the cause of only 4 deaths in Melbourne, and, in 1895, of 7 deaths. In South Australia it seems to have been more active than usual in 1890, 1893, and 1898; in the last-named year causing 112 deaths, as contrasted with no deaths whatever in the preceding year. In West Australia the deaths in the years 1893–98 were respectively 0, 4, 21, 9, 3, and 8.

In Tasmania the disorder was epidemic along the north-west coast in 1895–96, but it disappeared completely in 1897. In New Zealand it was epidemic in 1891, when it caused as many as 242 deaths; but in the three years 1896–98 the recorded fatal cases were only 24, 2, and 6, respectively.

In the Polynesian Islands whooping-cough occurs, rarely in some, in others in epidemics. It is not unknown in Fiji, where occasional cases of the disease are met with in the natives; it has also been seen among the Kanakas in New Caledonia, and is occasionally epidemic in the Sandwich Islands.

Africa. Whooping-cough was some years ago of very considerable frequency in Egypt, attacking especially the children of

---

[1] *Chinese Customs Med. Rep.* No. 41. [2] *Ibid.* No. 42.
[3] *Ibid.* No. 44. [4] *Ibid.* No. 54.

the natives, and less often those of foreigners living in the country. More recently the mortality from the disease is said to have diminished very considerably, at least in the towns of Lower Egypt, and this is attributed to increase of knowledge on the part of the people, to the greater care they have learnt to take of their children, and to their greater readiness to seek earlier treatment. In the Western Sudan and Algeria the disorder is said to be by no means rare.

On the West Coast whooping-cough not infrequently prevails. In Senegal it is said often to become epidemic. In the Cameroon district it was epidemic in the autumn of 1895.

For the East Coast of Africa information is incomplete. On the Middle Zambesi the disease becomes epidemic from time to time, and may possibly exist throughout Central Africa; I find no positive mention of it, however, by some recent writers on the pathology of British East Africa and of Abyssinia, although in Uganda it is spoken of as common.

Whooping-cough is not unknown in Mauritius, and a few cases of it are occasionally mentioned in the annual medical reports of the colony. In the Seychelles Islands it is said to be of rather frequent occurrence and to cause a high mortality.

In South Africa the disease is common enough and finds frequent mention in the colonial reports. It was epidemic in many parts of the Cape Colony in 1896, 1897, and 1898; at Aberdeen it was very severely prevalent in 1896, and less so in 1897. It was widely spread all over Namaqualand in 1898, and early in the same year it was reported as causing serious epidemics at Elliot, at Elliotdale, and at Willowdale, all in the native territories. At Elliotdale it led to a large mortality among native children. In Griqualand West many cases occurred about the same time. In British Bechuanaland the disease becomes from time to time epidemic, but it is said to be milder than in England. Many cases have been seen in both children and adults among the Kafir inhabitants of Mashonaland.

North and Central America. Of the existence or prevalence of whooping-cough in the far north of the American continent little can be said with certainty. Lombard states that it appeared in Labrador for the first time in 1875–76, and was probably imported from the south, where it was then epidemic. It has apparently not been seen in Greenland since 1849, before which date it had been several times epidemic. In Canada it is far from rare. In the provinces of Quebec and Ontario it occurs in occasional epidemics. In New Brunswick it is described as coming in "epidemic waves." In Manitoba a considerable number of cases of the disease were registered in 1896, and it finds frequent mention in the annual health reports from Nova Scotia on the east and British Columbia on the west.

Throughout the United States of America whooping-cough is more or less frequently met with. In the returns of deaths in the principal cities and towns of the Republic for the year 1897 this affection appears as a cause of death in every State, with the exceptions of Arizona, Arkansas, Idaho, Nevada, New Mexico, and North Carolina. It was nowhere, however, a cause of high mortality. In Massachusetts the mean mortality from whooping-cough fell from 220 per million per annum in the years 1856–75 to 140 per million in the years 1876–95.

Information is largely lacking in regard to the behaviour of the disease in Mexico and the Central American Republics. In Costa Rica it is said to have been epidemic in 1866, but to be of less frequent occurrence than in the West Indies. In British Honduras and Guatemala the disease is occasionally epidemic. It was so in the former, in the Toledo district, in October 1898, apparently as the result of an invasion of the infection from Guatemala. It prevailed in Belize in November, and in 1899 spread to Honduras.

In some of the West Indian Islands whooping-cough has of recent years been very prevalent. In Jamaica it was epidemic in May and June, 1897, throughout the eastern district of the island; at Port Antonio it prevailed during 1896 and 1897; in West St Andrew large numbers of children were attacked between August, 1897, and February, 1898. In Grenada the disease was epidemic at the end of 1898. In St Lucia, in Trinidad, and in Barbados it prevailed at the beginning of that year.

South America. I have failed to find any recent information as to the occurrence of whooping-cough in many of the countries of the South American continent. In Uruguay it was the cause of 53 deaths in 1897, or 4·34 per thousand of the deaths from all causes. This figure is equal to a ratio of about 66 per million inhabitants. In the Buenos Aires province of the Argentine Republic 126 deaths from whooping-cough were recorded in the same year; of these 54 occurred in the northern regions, 52 in the central, 20 in the southern, and none in Patagonia. The disease is certainly met with as far south as the Falkland Islands, where it "prevailed to a sad extent for so small a community" on East Island at the end of 1890[1]. Influenza was epidemic at the time, and the mortality among children was very high.

Factors concerned in the Geographical Distribution. It will be seen that whooping-cough has an extremely wide distribution. No latitude has entirely escaped it. But while the disease has shown that it can exist and even cause epidemics in tropical, sub-tropical, temperate, and sub-arctic regions, it is not equally common nor equally severe in all. It is mainly a disease of temperate and cool climates, and is both rarer and less malignant and fatal in warm or hot countries. This relation of the disease to latitude is to some extent shown by the mortality rates quoted above for the different European countries. The more northerly countries—Norway and Sweden, Denmark, Russia, Germany, the Low Countries, and the British Isles—have all returned in recent years higher mortality rates from this disease than the more southerly countries of France, Spain and Portugal. The only exceptions to this rule have been Austria-Hungary, Italy, and perhaps Servia. Similarly in Africa, although statistics are wanting, whooping-cough seems to be decidedly more common and severe in the cooler south than in the tropical zone. In Asia, in like manner, the disease is both less frequent and less severe as a rule in India, Farther India, and the Malay islands than in Siberia and northern China. Similar evidence is forthcoming from America, where it is more prevalent and of a worse type in Canada and the Northern States than in the tropical and sub-tropical portions of the continent.

But while this rule generally holds good it has not been without marked exceptions. Some have been already mentioned, as for example the recent occurrence of a severe epidemic of the disease in Burma (Rangoon); and to this may be added the mention of "pernicious epidemics of whooping-cough" in the past in many islands of the West Indies. Moreover the remarkable rareness and mildness of the disease in Iceland and the Faröes show that distance from the equator does not of necessity imply increased prevalence or severity of whooping-cough.

[1] *Falkland Islands Consular Reports*, 1890.

In its seasonal variations the disease has shown much the same relation to temperature as in its geographical distribution. That is to say, it has as a rule been most common and most severe in the cool seasons of the year. The difference between the seasons, however,—at least as regards prevalence of the disease— has not been very marked, and a large number of summer epidemics have been recorded. As regards the character of the malady, on the other hand, it seems to have been usually distinctly more severe and fatal in winter epidemics than in those occurring in the summer, at least in England. This rule, however, has not been observed in all countries.

Neither altitude above sea-level nor the character of the soil has had any influence upon the distribution of whooping-cough. Racial susceptibility also appears to be a factor of little importance, for all races are affected, though some, perhaps, have suffered from it more than others. Thus in the United States, for example, the negro inhabitants fall victims to it much more readily than those of other races.

Whooping-cough is a highly infectious disorder. It is mainly, but not wholly, a disease of childhood, and frequently gives rise to epidemics in schools or other collections of susceptible individuals. Epidemics of a wider character, affecting whole towns, whole countries, or even several contiguous countries, have occurred, and, in every country where it is endemic, its degree of prevalence varies greatly from year to year. Like all other infective diseases, however, the "cycles" of greater and lesser prevalence are very irregular and appear to conform to no law.

The infection can apparently cling to inanimate objects, such as clothes, fomites generally, and the furniture of rooms, but the principal means by which it is spread appear to be the movements of infected individuals. Whooping-cough is not known to attack any of the lower animals, nor is it thought to be spread by the agency of water or milk. It does not seem to be in any way dependent upon sanitary or insanitary conditions, and when epidemic it attacks all classes alike.

YELLOW FEVER.

General Characters and Etiology. Yellow Fever may be defined as an endemo-epidemic disease, infectious in character, the most prominent symptoms of which are extreme depression, jaundice (whence its name), suppression of urine, and hæmorrhages from mucous surfaces. The frequency with which the hæmorrhage occurs into the stomach, causing the vomiting of partly digested blood, is indicated by the name of "*vomito negro*" or "black vomit," commonly applied to it in its western endemic home.

The disease is believed by some observers to be constantly associated with the presence in the tissues, and later in the blood of the patient, of a micro-organism to which its discoverer, Prof. Sanarelli of Bologna, gave the name of *bacillus icteroides*. The evidence as to the relation of this bacillus to the disease is contradictory. A Commission appointed by the Marine Hospital Service of the United States in 1898 believed that it was the specific cause of yellow fever. Sternberg, on the other hand, and other more recent observers have made experiments and observations which seem to show that this bacillus has no causative relation to the disease.

A series of important investigations have recently been made in America, by Surgeons Walter Reed and James Carroll, and other members of the United States Army Commission, upon the part played by mosquitoes in the spread of yellow fever. Space will not permit here of a full account of their carefully planned experiments, for the details of which the reader is referred to the papers enumerated in the footnote below[1]. It will suffice to say here that several non-immune individuals were exposed for long periods together to the most intimate contact with clothes ˙or

---

[1] "The Etiology of Yellow Fever. A Preliminary Note." *Philadelphia Medical Journal*, Oct. 27, 1900.

fomites worn and soiled by yellow fever patients, without developing the disease; while, if bitten by mosquitoes that had twelve or more days before bitten yellow fever patients, these same and other non-immune individuals developed the disease, almost without exception. The particular mosquito believed to be the carrier of the yellow fever virus is *Culex fasciatus* or *Stegomyia fasciata*. The following were the conclusions arrived at by the authors of the above experiments:—

1. The mosquito, *C. fasciatus*, serves as the intermediate host for the parasite of yellow fever.

2. Yellow fever is transmitted to the non-immune individual by means of the bite of the mosquito that has previously fed on the blood of those sick with this disease.

3. An interval of about twelve days or more after contamination appears to be necessary before the mosquito is capable of conveying the infection.

4. The bite of the mosquito at an earlier period after contamination does not appear to confer any immunity against a subsequent attack.

5. Yellow fever can also be experimentally produced by the subcutaneous injection of blood taken from the general circulation during the first or second days of this disease.

6. An attack of yellow fever, produced by the bite of the mosquito, confers immunity against the subsequent injection of the blood of an individual suffering from the non-experimental form of this disease.

7. The period of incubation in 13 cases of experimental yellow fever has varied from 41 hours to 5 days and 17 hours.

8. Yellow fever is not conveyed by fomites, and hence disinfection of articles of clothing, bedding, or merchandise supposedly contaminated by contact with those sick with this disease, is unnecessary.

9. A house may be said to be infected with yellow fever only when there are present within its walls contaminated mosquitoes capable of conveying the parasite of this disease.

"The Etiology of Yellow Fever. A Supplemental Note." *American Medicine*, Feb. 22, 1902.

"The Etiology of Yellow Fever. An Additional Note." *Journal of the American Medical Association*, Feb. 16, 1901.

"Experimental Yellow Fever." *Trans. Association of American Physicians*, vol. xvi. 1901.

"Recent Researches on Yellow Fever." By Wm. Reid, M.D. *The Journal of Hygiene*, vol. ii. No. 2, April 1, 1902.

10. The spread of yellow fever can be most effectually controlled by measures directed to the destruction of mosquitoes and the protection of the sick against the bites of these insects.

11. While the mode of propagation of yellow fever has now been definitely determined, the specific cause of this disease remains to be discovered.

History. The first certain records of yellow fever relate to the early part of the seventeenth century. The natives and settlers in the West Indies suffered from severe pestilences at an earlier date, but it is doubtful if these were of the nature of yellow fever. From 1635 to the present day epidemics of it have been very frequent in and near the West Indies. The disease is said to have reached South America only on two occasions (1740 and 1842) before the year 1849. It was then introduced to Brazil, and from that date to the present day has been endemic there, and has occasionally spread thence to other parts of the continent.

The antiquity of the African centres of yellow fever is uncertain. There is mention of an epidemic of the disease in St Louis, Senegal, due to importation from Sierra Leone, in 1778. From 1810 to the present day it has been repeatedly epidemic in a portion of the West-Coast region which will be more accurately defined later. Its extensions to Europe will also be discussed in a subsequent part of the chapter.

Geographical Distribution. The area of endemic prevalence of yellow fever is on the whole a remarkably constant one. It is practically confined to tropical or sub-tropical regions, and to but a limited portion of these. The shores of the tropical and sub-tropical sections of the Atlantic Ocean can alone be regarded as permanent seats of the disease. It may, indeed, be questioned whether the truly endemic area of infection is not still more limited, for, as will be shown later, it cannot be regarded as certain that outbreaks of the disease on the eastern or African coasts of the ocean are not the result of carriage of the infection from the western or American coasts. So far as a somewhat imperfect knowledge of the history of the disease enables us to judge, it seems certain that originally yellow fever was essentially an American disorder, and that though the infection has to some extent become domiciled on the African shores, it is more than probable it was originally brought there from South America or the West Indies. It is to be noted that some writers' have held exactly the contrary view, and believed that yellow fever was originally imported to the West Indies by means of infected negroes from Africa. The facts of history are, however, against this view, and there is much evidence to show that this fever was known in those parts of America where it still prevails even before the discovery of the continent by Europeans, and long before the transport of negroes from Africa to the West Indies began to be practised.

In recent times yellow fever has rarely spread beyond an area which may be roughly defined as follows. In America its northerly limit of epidemic occurrence has been 46° 50′ (Quebec in 1805), and its southerly limit about the 35th degree of south latitude (Buenos Aires and Monte Video). In Africa the limits are much more restricted, the disease never spreading inland and being confined to certain points of the coast between the Equator and the Tropic of Cancer. It will be shown later that the disease has on several occasions passed beyond these limits, but always as a temporary epidemic outbreak, and it has never become truly endemic outside this area. Within the area it is found to be constantly present, and is perhaps only truly endemic, within still more restricted limits.

Endemic Centres and Epidemic Extensions in the New World. The Greater Antilles group of the West Indian Islands may be regarded as essentially the home of yellow fever, and it is almost certain that the prevalence of the disease on the shores of the Gulf of Mexico and of Brazil is to be traced originally to a spread of infection from the West Indies. In certain of the Antilles yellow fever is constantly present. Cuba is, or was, one of the most severely affected islands of the group. Before 1901 the fever used to become epidemic here every summer, but no period of the year was free from it, and in many years it was a veritable scourge. Here, as elsewhere, the behaviour of the disease was irregular and almost capricious in character; in some years it was mild and the cause of but few deaths, in others it was extraordinarily active. In 1893 it caused 26·8 per cent. of the total deaths in Cuba during the six summer months. More recently it has been widely prevalent again; the disturbances in connection with the rebellion and the Spanish-American war gave it an immense impetus, and in 1897 as many as 2583 deaths from this disease occurred in the military hospitals of Havana and Regla alone, while in the whole island no less than 6034 deaths from it were recorded—a mortality which may be held to imply the occurrence of about 30,000 cases. The recent history of the disease in Cuba is discussed at greater length below.

Haiti must be regarded as a centre of yellow fever. In Guadeloupe epidemics have frequently occurred, and it is possible, though perhaps not quite proved, that these are due to the revival of an endemic infection and not to repeated importations. In Jamaica it is believed by some that the epidemics which break out at intervals are always the result of an importation of infection. Morse, however, in a recent report on the health of this island[1] takes the opposite view, and believes that the severe outbreak in 1897 was not due to an infection brought from elsewhere but to the renewed activity of an infection remaining dormant in the island itself. The latter explanation seems to apply also to the marked recrudescence of yellow fever in Martinique in the years 1895–98.

In most of the other islands of the Lesser Antilles yellow fever is apparently not truly endemic, and when outbreaks of the disease occur here they can usually be traced to an introduction of the virus from elsewhere. The same statement holds good for the Bermudas and Bahamas, and for Trinidad and Barbados.

It was formerly believed that the shores of the Gulf of Mexico were deeply and permanently affected by this disease, and it is certain that for long periods together several places, such as Vera Cruz and New Orleans, were ravaged by it. But the disappearance of the fever from these coasts for intervals of some length, and the apparent feasibility of keeping it away from ports where strict measures against its introduction were enforced—as at New Orleans, Mobile, and Charleston—seem to show that it is not truly endemic on these shores, but that its occurrence there is due to introduction from without. *A fortiori* is this the case in regard to the occurrence of the disease in the interior of Mexico and even on its western shores, and its extension northwards up the Mississippi valley and along the Atlantic coasts of the States, which will be discussed later.

The epidemic extensions of yellow fever in the immediate neighbourhood of its endemic home in the Greater Antilles, some

---

[1] *Report of Island Medical Department*, for year ending March 31, 1898.

of which have just been referred to, are irregular both in time and in extent. As an example of a moderately extensive epidemic of the disease the last severe outbreak in these regions may be very briefly described. As already stated, Cuba, the real home of yellow fever, was under special conditions in the year 1897—conditions of war and general disturbance such as favour the spread of any infectious disorder. The result was not only a severe recrudescence of yellow fever in the island itself but a wide extension of it elsewhere, which, if not wholly, may at least in part be ascribed to this exceptional cause. In Jamaica it began to be epidemic in October and November of that year, and caused no little mortality in the island. As early as August the disease was already present at Ocean Springs on the Louisiana coast, and in September at Mazatlan in Mexico, and in that and the following year it prevailed in many parts of that country and of the States bordering the Gulf of Mexico. Vera Cruz, Tampico, San Fernando, New Orleans, Mobile, Edwards, Biloxi, Galveston, Houston and Montgomery in Alabama were visited by it. The Mississippi valley was infected as far north as the town of Cairo in Illinois (about 37° north). Isolated outbreaks or cases were also reported from places on the Atlantic coast of the southern States, and single cases occurred in the Colombia district of Washington and in New York. About the same period yellow fever was epidemic at Panama in Central America and spread thence to Colon, but in this case the infection seems to have been brought from the south and not from Cuba. This outbreak will be referred to later.

The epidemic extension of yellow fever just described was a fairly characteristic example, but others in the past have been even more extensive. Hirsch mentions the following States as having at some time or other been the scene of an epidemic :—Texas, Louisiana, Arkansas, Mississippi, Tennessee, Alabama, Florida, Georgia, South and North Carolina, Virginia, Maryland and Delaware, Pennsylvania, New Jersey, New York, Connecticut, Rhode I., Massachusetts and New Hampshire. Canada has been visited by yellow fever only on two occasions, in 1805 (Quebec) and in 1861 (Halifax, Nova Scotia).

It is noteworthy that many of the Central American States have suffered remarkably little from yellow fever, in spite of their nearness to the West Indies. In Nicaragua, San Salvador, and British Honduras but one epidemic of the disease has been observed. None of these seem to have taken any share in the epidemic of 1897–8. The comparative rarity of yellow fever in Central America is the more noteworthy in that the adjoining countries of the northern part of South America have not at all infrequently suffered from outbreaks of the disease. The Atlantic coast of Colombia has been frequently visited, and so also have Venezuela and the Guianas. The fever is possibly even endemic in British Guiana.

The coasts of Brazil are now seriously infected with this disease. It is believed that the infection was brought thither so recently as the year 1849. The North-American brig *Brazil* arrived at Bahia in that year from New Orleans, where yellow fever was prevalent. She lost two of her crew on the voyage, and after her arrival the disease appeared among those who had communicated with the ship, and later upon other vessels in the harbour. From Bahia it spread to Rio de Janeiro, and from 1851 to 1861 it was more or less severely active each year on the Brazilian coasts. In 1862 its activity diminished remarkably, only twelve deaths from the disease being recorded that year in the town of Rio de Janeiro; in the following year 15 deaths occurred there, and then for four years not a single death was recorded. But in 1868 it reappeared, and from that day to this Rio de Janeiro has been annually ravaged by an outbreak of greater or less severity. The following are the returns of yellow fever deaths in Brazil since the year 1868; the larger number of these deaths occurred in the city of Rio de Janeiro itself.

Year	Deaths	Year	Deaths	Year	Deaths
1868	18	1878	1177	1888	754
1869	274	1879	974	1889	2155
1870	1118	1880	1623	1890	719
1871	9	1881	257	1891	4456
1872	295	1882	502	1892	4313
1873	3659	1883	1606	1893	742
1874	841	1884	640	1894	4715
1875	1308	1885	445	1895	818
1876	3476	1886	1446	1896	2909
1877	283	1887	135	1897	159

Peru seems to have escaped in the recent period of yellow fever activity in South America[1], but it has been visited by the

[1] Three deaths from the disease occurred at Callao, in Peru, in 1889, but none have been recorded since, at least up to the year 1896.

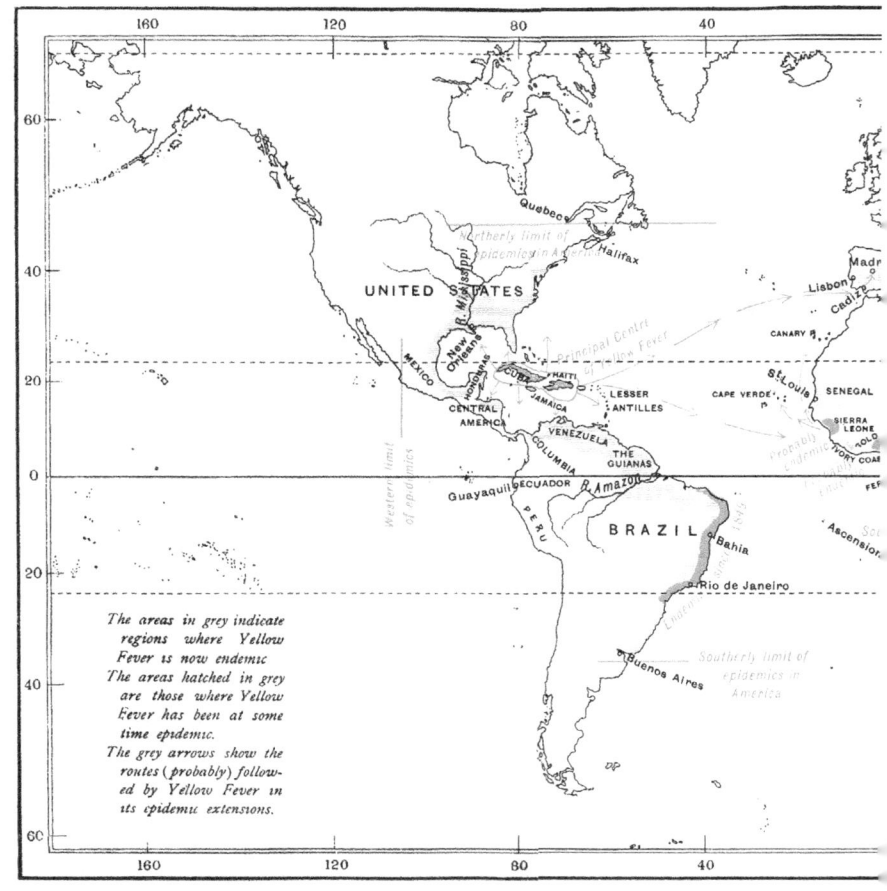

The areas in grey indicate regions where Yellow Fever is now endemic
The areas hatched in grey are those where Yellow Fever has been at some time epidemic.
The grey arrows show the routes (probably) followed by Yellow Fever in its epidemic extensions.

DISTRIBUTION OF YELLOW FEVER

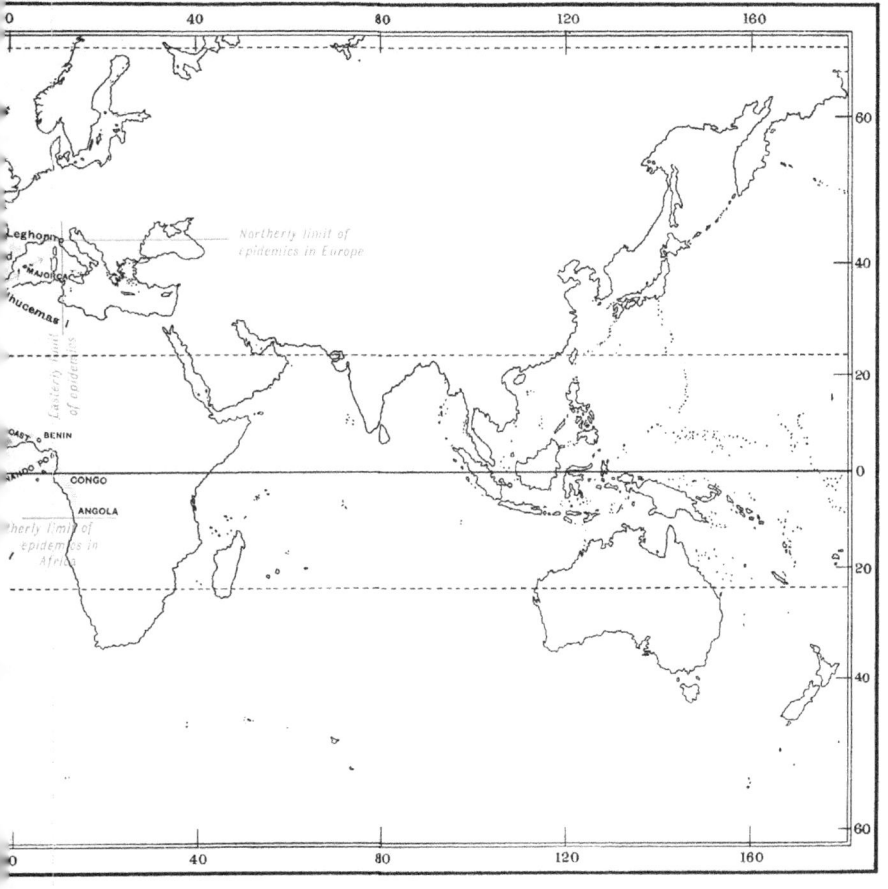

Northerly limit of
epidemics in Europe

Leghorn

MAJORCA

hucemas

Eastern limit
of epidemics

COAST   BENIN

RINCE DO

CONGO

ANGOLA

therly limit of
epidemics in
Africa

disease on several occasions, and Hirsch states that it was constantly present from 1854–1869. Ecuador has been less frequently visited, but it appears that the disease was epidemic at Guayaquil as recently as the year 1897, and the outbreak of yellow fever in Panama alluded to above is said to have been due to infection brought from Guayaquil. These extensions of the disease to the western coasts of South America must, however, be regarded as exceptional, and as an endemic disorder yellow fever is found in South America only on the eastern shores.

Further south Asuncion, Monte Video, and Buenos Aires have all been visited by yellow fever at different times, but only exceptionally, and Buenos Aires must be regarded as forming the southern limit of its epidemic spread.

Endemic Centres and Epidemic Extensions in Africa. While the regions just described form the western and more important area of yellow fever prevalence, a second area of great epidemiological interest is found on the eastern side of the Atlantic, on the shores of Western Africa. This area is of much smaller size than the American one; it is confined to the Gold Coast, Sierra Leone, and portions of the Gambia and Senegal coasts. Of the original source of the disease here but little is known with certainty, but it seems probable that the infection was brought to Africa from the coasts of America or the West Indies. Whether each succeeding outbreak has been due to a fresh importation of infection must be doubted, and at the present day yellow fever appears to be truly endemic on the Gold Coast and at Sierra Leone, whence it has at times spread to Senegal and some of the West African Islands. The outbreaks have been most numerous in Sierra Leone, but Senegambia has also been frequently visited. Hirsch mentions only three outbreaks on the Gold Coast (in 1852, 1857, and 1862); but quite recently the disease has again appeared, and it seems possible that it is really endemic there. The Colonial Report for this colony for the year 1895 states that "the endemic fever assumed a pseudo-epidemic form of malignant type" which closely resembled yellow fever. The outbreak, which was limited to Axim, Chama, Elmina, Cape Coast, Salt Pond, and Accra, caused a very high mortality, and almost led to a panic among the European inhabitants. Dr Easmon, in his sanitary and medical report on the colony for the year 1897, describes this fever as "a pseudo-epidemic fever of a malignant type, closely approaching in its clinical manifestations the *vomito negro* or yellow fever of the West Indies," and he states that it was again epidemic at the Gold Coast Stations in the second quarter of 1897. At Salt Pond, of six Europeans who died, three presented the classical symptoms of yellow fever.

The dividing line between yellow fever and the severer "bilious" forms of malaria is clinically a very thin one, but the author of the report just quoted is inclined to regard this Gold Coast fever as, if not identical with, at least indistinguishable from yellow fever. In 1899 Grand Bassam, on the Ivory Coast, was visited by a severe epidemic of this disease, and in 1900 the same affection broke out in many places in Senegal, which had been "almost" free from the disease, it is said, during the preceding nineteen years. The towns of St Louis, Dakar, Goree and Rufisque suffered very severely, and there was almost a panic in the colony.

From these (endemic?) centres of yellow fever in Africa the disease has spread in the past to the Benin coast (in 1862); to the Congo Coast (in 1816, 1860, 1862, and 1865); to Ascension Island (in 1823 and 1837); to Fernando Po (in 1839 and 1862); to the Cape Verde Islands on three occasions, and to the Canary Islands on five occasions. The Cameroon district is said to have escaped it. The most southerly limit of extension of the disease in Africa has been St Paul de Loanda, in Angola. It seems probable that it has never spread very far inland, but has been confined mainly, if not entirely, to the coast-line.

Yellow Fever outside its Endemic Limits. It remains to refer to the occasional spread of yellow fever beyond its usual limits. In the form of single or multiple cases on board a vessel, the carriage of yellow fever to places outside the areas above described, to the coasts of North America and even to Europe, has been by no means rare. But the tendency in such cases is for the disease to disappear when the ship reaches cooler climes, and the development of an epidemic outside the tropical or sub-tropical regions above defined has been excessively rare. The recorded instances of such an occurrence date back to somewhat distant periods, and no European epidemic of yellow fever has been seen since the small outbreak in Madrid in 1878. In the eighteenth century several epidemics of the disease occurred at Cadiz, and one each at Malaga and Lisbon. In the nineteenth century a more extensive epidemic began in 1800 in Cadiz, and in four years spread over a great part of Granada and Andalusia, and northwards along the banks of the Guadalquivir to Cordova, and from Andalusia to the seaboard of Murcia, Valencia, and Catalonia. A second and more limited outbreak in 1810–12 was confined to Cadiz, Cartagena, and Gibraltar, and to several coast towns of Granada, Murcia, and Valencia. The third and last great epidemic in Spain, that of 1819–21, affected much the same areas, and since that date small outbreaks of yellow fever have occurred in Spain on four occasions only. The last was in 1878 when troops returning from Cuba brought the disease to Madrid. Lisbon and Oporto have also occasionally been visited by the fever.

The European outbreaks of yellow fever have been almost confined to the Iberian peninsula, the only exceptions being three small epidemics in Majorca, in 1804, 1821, and 1870; and an outbreak at Leghorn in 1804, which was due to an importation from Cadiz.

The shores of England and of France have, on rare occasions, been visited by ships having cases of yellow fever on board, but no epidemic of the disease has resulted, although on some of these occasions (as at Brest in 1856, at St Nazaire in 1861, and at Swansea in 1864) a certain number of persons who have had communication with the infected ships have caught the disease; and on one occasion, at St Nazaire, the crews of ships lying near another with cases on board became infected.

Factors concerned in the Geographical Distribution. The geographical distribution of yellow fever is remarkable both for its peculiarly limited character and for the great constancy with which the disease remains within its prescribed limits. The explanation of these facts is not wholly easy, but on the other hand the manner in which this fever is influenced by surrounding conditions is better understood and more easily demonstrated than is the case with many other infectious diseases. Thus in regard to temperature yellow fever shows very definite relations. It does not occur as an endemic disease in places where the mean winter temperature is much below 65° F. In its endemic areas it is always worst in the hot season and dies away as the cold weather advances; and, as already stated, it almost invariably disappears from an infected ship as soon as she arrives in cooler latitudes. It is to be noted, however, that, though epidemics of yellow fever seem to require a certain amount of warmth for their development, they can, when once developed, continue for a considerable time with a low thermometer. When the temperature sinks to freezing point, however, they invariably disappear.

Sea-coasts, and particularly low-lying and insanitary port-towns, are specially liable to the attacks of yellow fever within the zone. The courses of river valleys, such as the Mississippi and the Amazon, have also on some occasions been followed in the spread of the disease. It is comparatively rare to find the fever at any great height above sea-level. On some exceptional occasions yellow fever has been epidemic at a height of 2000 to 3000 feet, as on the east coast of Mexico, and even at a height of 4000 feet, as at Newcastle in Jamaica. But it has never appeared at a greater height than this, and in Mexico it has not occurred in places more than 3300 feet above sea-level. Further observations on the distribution of the mosquito which is believed to carry the infection will, no doubt, satisfactorily explain this relation of the disease to altitude.

Similar observations are needed to explain the relation of yellow fever to race. This factor has hitherto seemed to be one of no little importance in the distribution of the disease. The white races, the American Indian, and the Hindu coolie have all suffered from it, but the negro race appears to possess a remarkable immunity. Nor does this immunity seem to be an acquired characteristic from long exposure to the infection through one or more generations, for in many instances negroes newly introduced into an infected city have escaped the disease. There would seem to be something in negro blood which either tends to protect the individual from the disease, or (what is perhaps more probable) renders him less liable to the bites of infected mosquitoes. It is noteworthy that half-breeds, quadroons, and others of mixed black and white descent are no longer immune, but show a degree of susceptibility to the disease which varies inversely with the amount of negro blood in their veins.

While the negro race is relatively immune to yellow fever, almost all other races, with the exception perhaps of the Chinese, seem liable to contract the malady if exposed to it. Persons, however, who live in the yellow fever zone and have been exposed to one or more epidemics of the disease, seem to lose their susceptibility to some extent, and it is the new-comer and the stranger that are more particularly liable to attack. No doubt this is in great part the explanation of the violent outbreaks of yellow fever which have attended military operations or the movements of immigrants within the yellow fever zone. An example of this has been seen in the excessively high mortality from the disease in Cuba during the recent years of rebellion and war. Large bodies of unacclimatised troops were successively landed on the island, both from Spain and from America, and a considerable proportion of them of necessity fell victims to this most fatal disorder.

The character of the soil has, it seems, little if any influence on the distribution of yellow fever. The disease has prevailed on the most varied geological formations. A certain amount of moisture and an excess of decomposing organic matter in the soil seem to favour its existence. Dirt and filth and insanitary conditions generally are perhaps needed for the genesis of the disease, or at any rate for its development as an epidemic. As Hirsch pointed out, epidemics have always started "in seaports, in the immediate neighbourhood of the harbour and the wharves, and generally speaking *in the filthy quarters of the town*, the centres of poverty, misery, and vice, with their narrow and foul-smelling streets, their tenements densely crowded from cellar to garret, their taverns, dancing-saloons, and lodging-houses. It is after the epidemic has come to a head in these purlieus that it begins to spread, always in the first instance into the immediate neighbourhood; but not infrequently it remains confined to them, and the other parts of the town some distance off and better situated hygienically may be little troubled by the sickness or not at all."

The work of the American surgeons referred to above has established a strong *à priori* argument in favour of the view that this disease is spread mainly, if not solely, by mosquitoes. Still stronger support to it is found in the remarkable results that have followed a practical application of the mosquito theory in the attempt to deal with yellow fever in its worst endemic home, and it may be interesting to quote these results before briefly discussing the theory in some of its other bearings. In February, 1901, a systematic effort was inaugurated in Havana to get rid of yellow fever, by quarantining every patient, by

C. 34

protecting his room with wire screens so as to prevent mosquitoes becoming infected by sucking his blood, and by destroying as far as possible all mosquitoes within a given radius of each case of the disease. The accompanying charts, borrowed from the late Surgeon Reed's account of the experiment, indicate very clearly the admirable result of these efforts. They are best explained in that author's own words :—

"As an illustration of what has been accomplished by these newer sanitary regulations, I may state that counting from the date when they were put into force, viz. February 15th, 1901—Havana was freed from yellow fever within ninety days ; so that from May 7th to July 1st—a period of fifty-four· days—no cases occurred. Notwithstanding the fact that on the latter date and during the months of July, August, and September, the disease was repeatedly introduced into Havana from an inland town, no difficulty was encountered in promptly stamping it out by the same measures of sanitation intelligently applied both in the city of Havana as well as in the town of Santiago de las Vegas, whence the disease was being brought into Havana.

"As a further illustration of the remarkable sanitary victory accomplished over a disease whose progress we had heretofore been powerless to arrest, I will close this paper by inviting the reader's attention, first to the accompanying Chart I, which shows the average monthly mortality from yellow fever in Havana for the twenty years 1880–1899 inclusive, and also the mortality by month for the years 1900 and 1901. I will then ask him to examine Chart II, which shows the progress of yellow fever in Havana during the epidemic year ending March 1st, 1901, when the sanitary authorities were putting forth every effort known at that time to sanitary science in order to control the march of the disease ; and when he has satisfied himself that no effect whatever was produced upon the epidemic of that year, I will invite his attention to Chart III, which shows the occurrences of this disease in Havana for the epidemic year March 1st, 1901, to March 1st, 1902, during which year yellow fever was fought on the theory that the specific agent of this disease is transmitted solely by means of the bites of infected mosquitoes. By carefully comparing the figures both as to deaths and cases in these two Charts, and recalling that between the years 1853 and 1900 there have been recorded in the city of Havana 35,952 deaths from yellow fever, he will then be able to more clearly appreciate the value of the work accomplished by the American Army Commission."

The important bearing of these observations on the geography of yellow fever is obvious. It remains to discover the actual

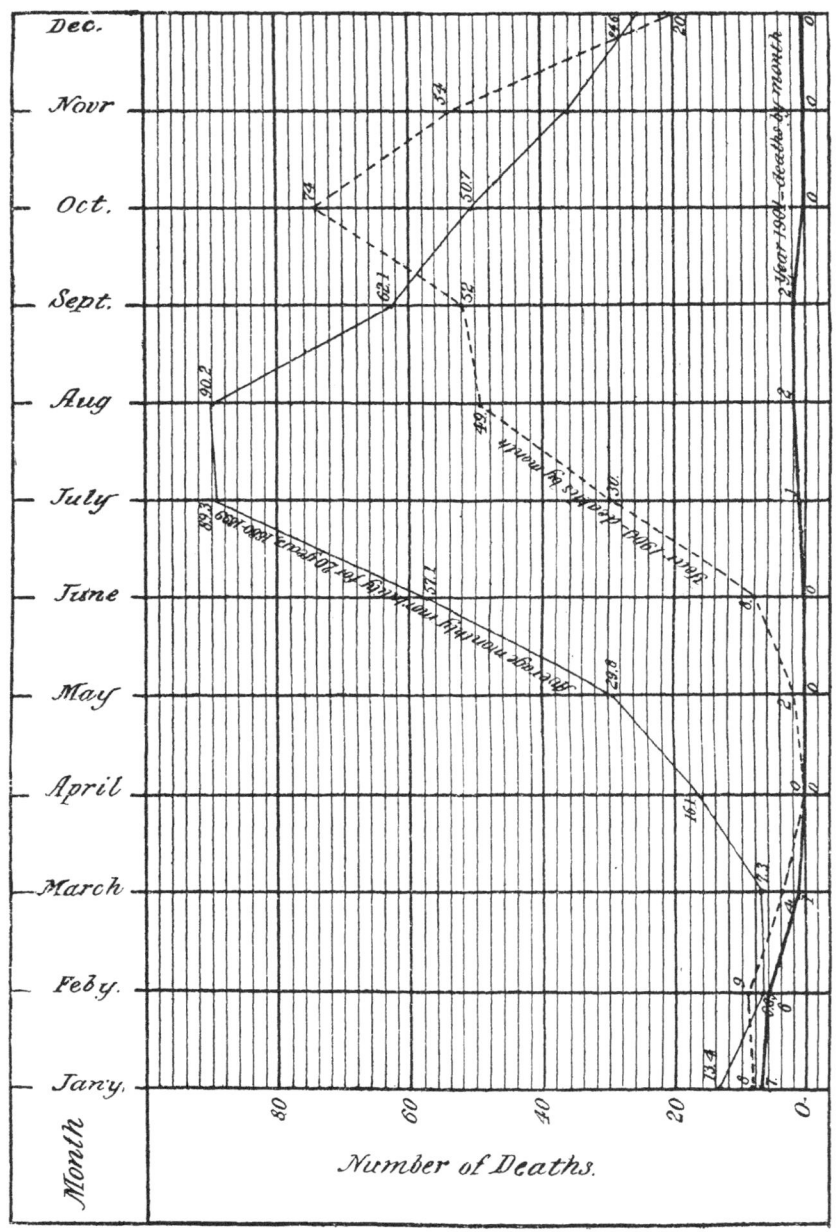

I. Chart showing the monthly mortality from Yellow Fever in Havana for the 20 years 1880—1899, and for the year 1900—1901.

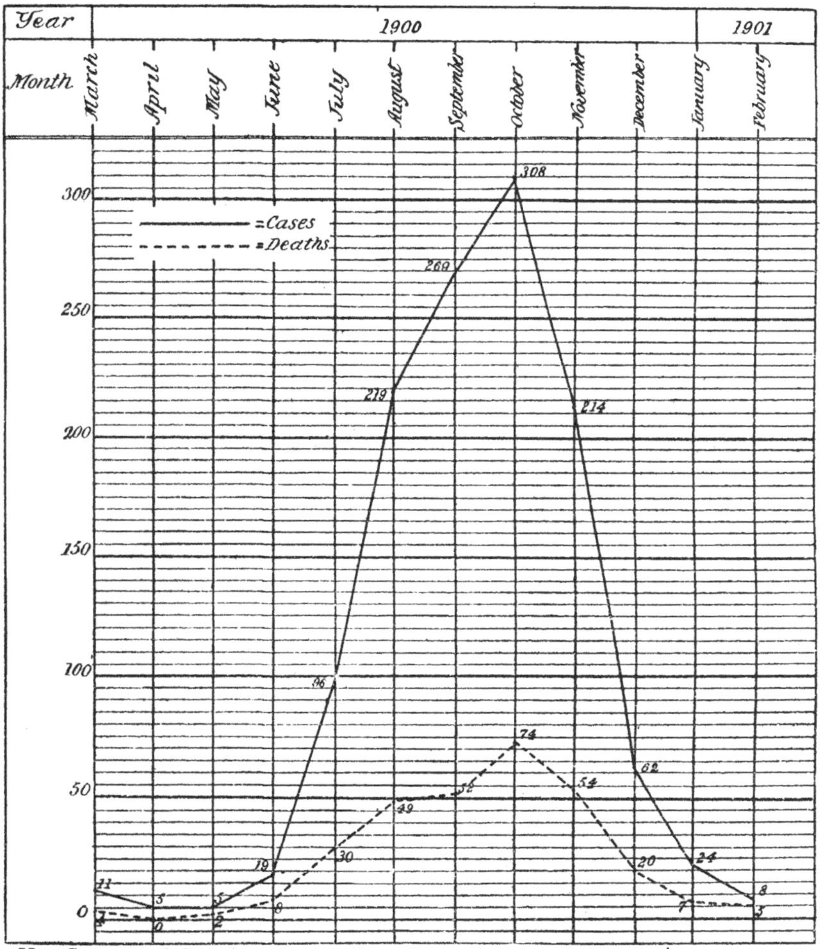

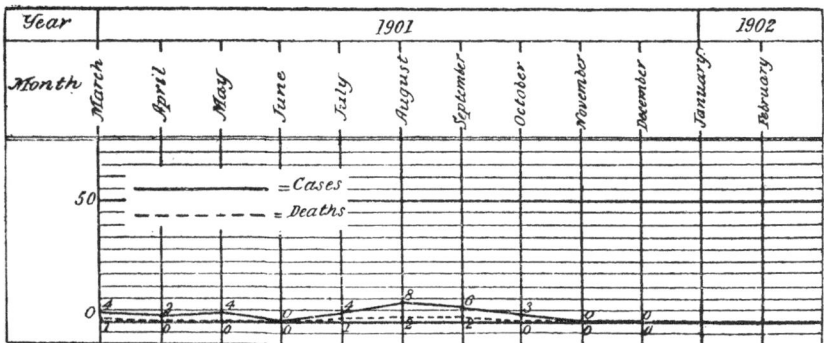

II. Cases and Deaths from Yellow Fever in Havana for Epidemic Year,
March 1, 1900—1901.

III. Cases and Deaths for Epidemic Year, March 1, 1901 to March 1, 1902.

infective material of the disease itself, whether a bacillus or other organism, to determine what particular species of mosquito are capable of acting as its intermediate host, and to ascertain the exact distribution of the latter over the earth's surface. In this, as in so many other parasitic diseases, three elements appear to be necessary for its production—the infecting organism, the intermediate host, and a susceptible human population. It must be supposed that all three elements exist at their fullest in the endemic centres of the disease in the West Indies and Brazil. In the secondary centres on the West Coast of Africa they must also exist, though probably to a lesser extent. One factor, at least—the susceptibility of the population—is probably less favourable here than in America. The negro race, as pointed out above, is remarkably immune to yellow fever, and the tropical and subtropical zones of Africa are peopled by negroes. Possibly in this fact, possibly in the absence of the yellow-fever-bearing mosquito from the interior of Africa, or possibly again in both these facts, is to be found the reason why the African continent has seemed to act as a barrier to the easterly extension of the disease. Whether the inhabitants of Morocco, Algiers, and the rest of the northern shores of Africa—with sea-coasts well within the limits of latitude of the American yellow fever zone—are or are not immune to the disease; or whether the specific mosquito is absent from those regions—has not yet been determined. From the point of view of latitude and climate there seems no reason why the infection should not have invaded these countries and spread by them to the east and even invaded Asia, and it is a remarkable fact that it never has done so. Possibly the infection has never been introduced to their shores; although during the epidemics in Spain in former years they must have been frequently exposed to the chance of invasion, and on one occasion at least (in 1804) the disease did break out as close to the shores of Morocco as the small island containing the Spanish fortress of Alhucemas.

Finally, it is well known that the virus of yellow fever can be carried over long distances by sea, and ship-epidemics of the disease are by no means rare. It has hitherto been believed that the infection could be thus carried not only in the bodies of persons suffering from the disease but also in infected articles. Many instances have been recorded in which linen and other personal effects have apparently retained the poison for a con-

siderable time and been the means of spreading the disease.
The American experiments referred to above seemed to prove
that such infected articles cannot of themselves give the disease
to susceptible persons, even after very prolonged and close contact with them.   But further observations may yet perhaps prove
that if, in addition to articles soiled with the blood or excreta of
yellow fever patients, the right kind of mosquito is also present,
it may be capable of taking up the infecting material from the
articles and later introducing it into susceptible human beings.

BOOK II.

DISEASES OF THE SKIN.

SKIN DISEASES IN GENERAL.

It must at once be confessed that materials do not exist for an individual study—to any useful end—of the distribution of the majority of skin diseases. Or it should perhaps be said that the materials exist, but not in an available form. Diseases of this class are rarely fatal, and are consequently scarcely mentioned in the statistical Health Reports of most countries; and there are in fact no general and accurate statistics by which their frequency in different parts of the earth can be judged of. Even when mentioned in the returns from any country they are more often than not spoken of as a whole, with nothing to indicate what particular skin diseases are meant.

Affections of the skin are, of course, exceedingly numerous and varied, and differ widely in their causation. A considerable number of them, however, are directly transmissible from person to person, and are largely favoured by dirt, by a low level of personal hygiene, and by ignorance or apathy leading to neglect of treatment. The result is that these particular diseases, and consequently the group of skin diseases as a whole, are found to be peculiarly common where conditions of this kind prevail. In European countries they are most prevalent among the poorer and less educated classes, and where overcrowding and sanitary defects are greatest. In tropical and sub-tropical countries they are often found to run rife amongst native races, who think little of personal hygiene or of taking steps to cure the disease or prevent its spread to others.

Many of these disorders seem to flourish best in warm countries, and hence the whole group is probably more prevalent in and near the tropics than in more temperate zones. But the difference does not appear to be very great, and "skin diseases" are not only extremely common among the peasant classes of many European countries, but are strikingly so as far north as Greenland, Iceland, and the Faröes. Some have an almost world-wide distribution. Eczema, for example, is found in all latitudes, if not in all countries. Psoriasis is almost universal, though it is thought to be most common in tropical and sub-tropical countries and among the coloured races. Itch is, perhaps, quite universal. Acne, seborrhœa, urticaria, the various forms of herpes including herpes zoster, and many other skin affections have a very wide diffusion, if not quite so universal as those just named.

The class of ringworm diseases—*i.e.* those due to some variety of trichophyton or tinea—are common in many countries, but especially so in and near the tropics. Ordinary ringworm (*Trichophyton tonsurans*), favus, and tinea versicolor have a very wide distribution in all latitudes. "Dhobi itch," a troublesome skin eruption in the axilla and about the genitals, is another example of this class. It is very frequent in many hot countries and few European residents in the tropics escape an attack of it at some time or other. Tinea imbricata and pinta are of the same character, and as they have a very distinct and interesting geographical distribution they will be separately discussed later.

Perhaps the commonest of all the skin diseases to which unacclimatised persons are liable in the tropics is "prickly heat," or lichen tropicus. This is a simple papular eruption as a rule, of no serious import except for the irritation and often intolerable itching it may give rise to. Most Europeans suffer from it more or less after arriving in a hot climate, but while in some it causes practically no trouble at all, in others it may be of a most violent character, causing really intense suffering and some constitutional disturbance. It is not confined to the tropics and may occur in any hot climate. It has been seen as far north as Sicily and Minorca in the Old World, and in the Mississippi Valley in the New.

Boils and carbuncles are forms of skin affection that occur more or less in all latitudes, but are most frequent in hot countries. They attack all races to some extent, but boils affect particularly the European resident in the tropics. Like prickly heat and dhobi itch, one or two boils or a crop of boils are almost certain to develope sooner or later during a prolonged stay in hot countries. When very numerous they may cause a long and painful illness and leave a good deal of debility behind. In some years they appear in so many individuals as to justify the use of the term epidemic. Such epidemics of furunculosis have been seen not only in India and other tropical countries, but in the United States, in England, Scotland, Ireland, France, the Cape, Hungary, and Italy.

Pemphigus contagiosus is the name given by Manson to a form of skin disease in the tropics which appears to correspond to the impetigo contagiosa of more temperate climates. It has been seen mostly in Southern China, where it is very common in the hot weather, and in some years almost epidemic. It is also a frequent disorder in the Straits Settlements and in Madras, and perhaps in some other tropical countries.

The following skin affections, some of which have been named already, have a distinct and peculiar geographical distribution and are deserving of brief sections to themselves:—craw-craw, oriental sore, pinta, tropical phagedæna, tinea imbricata, verruga peruviana and yaws. Mycetoma is sometimes classified among skin diseases, but as it affects the entire tissues of a limb it has been dealt with elsewhere.

CRAW-CRAW.

General Characters and Etiology. Craw-craw, or kra-kra, is the name given to a peculiar affection met with among natives of the West Coast of Africa and the French Congo. The exact nature of the malady is for the present uncertain. Probably the name is frequently applied by the natives indiscriminately to a variety of skin diseases. But the affection usually described as craw-craw appears to be distinct and characteristic enough. Attention was first drawn to it by O'Neil in the year 1875. The disease is characterised by the appearance of papules which pass on to pustular formation and ulceration. The parts most often affected are the hands, feet, and lower extremities; but the eruption may cover the whole body. A filaria which presents some resemblances to the filaria perstans has been found in the pustules, but its exact relation to the disease has not been determined. Filaria perstans being exceedingly common in those countries where craw-craw prevails, it is possible that their presence in these cases is a mere coincidence, and that they do not stand in any causal relation to it. Manson has suggested that the filaria found in cases of craw-craw, which shows some slight differences from the ordinary form of filaria perstans, may be an advanced form of that parasite; and he suggests that the skin may be the normal channel by which it escapes from the human body, after undergoing there a certain measure of development.

Recently Bennett, after experience of the disease in Southern Nigeria, has expressed the opinion that craw-craw is merely a pustular eczema; that it is not a specific disease; that the filaria is found in cases of so-called craw-craw in just the same proportion that it is found in the general population of the districts; and that the name craw-craw, which is given by the natives to almost all skin diseases excepting yaws, should be abandoned as indicating any distinct and separable malady[1].

Geographical Distribution. The disease usually known under this name is apparently confined to the African continent. On the West Coast it is very common, particularly in the district of Old Calabar and in the Cameroons[2]. It is also found widely in the French Congo and along the valley of the Ubangi river. Dr J. Emily, who accompanied the Marchand mission to Fashoda in the Sudan in 1898, states that he saw cases in the districts of Libreville, Loango, and Brazzaville, and at nearly all the stations along the Ubangi. The disease attacked many of the members of Major Marchand's expedition, and caused much suffering and subsequent anæmia. Two officers of the expedition who contracted it at Loango and Brazzaville respectively were still suffering from it a year and a half later[3].

[1] *Journ. of Trop. Med.* Feb. 15, 1901.
[2] Plehn, *Janus*, 1896–7, p. 383.
[3] *Arch. de Méd. Navale*, Jan. 1899.

ORIENTAL SORE.

General Characters and Etiology. A great variety of names have been applied to a skin affection which, though found in many widely distant countries, probably owns an identical cause in all. The "Aleppo button," the "Delhi sore," the "Penjdeh ulcer," the "clou de Biskra," the "bouton d'orient," are a few of the names under which the affection is known. It usually begins with a certain amount of itching, followed by the appearance of a small papule, which enlarges and becomes indurated. This period of induration, which may last from one to four months, is followed by a period of ulceration, during which an indolent, slowly-spreading ulcer forms. This ulcer, which is covered by crusts and with some discharge, itches, but is not as a rule painful. After another interval, of usually some months' duration, healing sets in. A depressed, pinkish or brownish cicatrix is left behind, which may cause no little deformity from contraction. Several such sores may be present at the same time, and may seriously affect the health of the patient, more especially if, as occasionally happens, they should be complicated with lymphangitis, adenitis, phlebitis, erysipelas, or sloughing phagedæna. Second attacks are rare.

The late Dr Tilbury Fox proposed the name Oriental sore for this disease, and, until a better can be found, this title seems more convenient than any other. If, however, a disease recently described by Juliano[1] as being common in Bahia in Brazil should prove to be identical with the one under discussion, it may be hoped some other name will be found for it.

History. The earliest known accounts of this affection came from Aleppo, where, at the end of the 18th century,

---

[1] Ref. in Manson's *Tropical Diseases*.

Russell first observed it, and described it under the name of Aleppo boil. Later it was seen in other parts of the Levant, but it was not until the latter half of the nineteenth century that it came to be recognised as having a much wider distribution than was at first supposed.

Recent Geographical Distribution. *Europe.* This sore is practically unknown at the present day in any part of the continent of Europe. Half a century ago Libert observed some cases presumed to be of this nature among Tatars in the Crimea, but there seems to be no recent evidence of its existence there or in any other part of European Russia.

Asia. Syria and Asia Minor have long been recognised as countries in which oriental sore is common. In Syria it is said to occur chiefly in the northern districts of the Orontes, between Killis and Aleppo; while it is rare in the towns and villages to the west of Aleppo, and unknown among the Bedawin and in the Kurdish highlands. In Asia Minor it is met with at Brusa. In the Caucasus it has been seen in the government of Elizavetpol. It is probably common in the Russian Central Asiatic provinces of Transcaspia and Turkestan. What is known in Tashkent as "Sart disease" by the Russians, or as "Afghan sore" by the natives, is most likely of this character. It is described as "a very disagreeable ulcer, which breaks out on the face or hands, spreading constantly, and eating deeper and deeper[1]."

In Bokhara it is particularly frequent at Patta-Hissar, Shirabad and elsewhere. All the inhabitants of these regions suffer from it, without distinction of sex, social condition, or nationality, and it is said that the Russian soldiers on the frontier guard rarely escape it[2]. In Mesopotamia oriental sore is particularly common; at Baghdad almost every inhabitant seems to have suffered from it at some time, and visitors are often attacked. It is not confined to Baghdad, and Hirsch states that it is found throughout the whole plain between the Tigris and Euphrates, from Diarbekr through Baghdad to Basra. It also occurs in Crete and Cyprus.

In Persia it is commonest in the central regions of Teheran, Kashan and Ispahan, less frequent in Hamadan, and unknown in the northern mountainous districts and on the shores of the Caspian. The name of Penjdeh ulcer indicates that it is common

[1] *Turkestan.* By Eugene Schuyler, London, 1876, vol. I., p. 148.
[2] Grekof.

at Penjdeh on the borders of Russian and Afghan territory. It probably occurs also in Afghanistan and Baluchistan. In India it is found over a large area which extends along the Indus from the Punjab (Lahore, Multan) southwards through Sindh, and eastwards through Rajputana and the North-West Provinces as far as Delhi, Meerut, Lucknow, and Gwalior. Bengal proper, Madras, Central India, and most of the Bombay presidency appear to be free from the malady (Hirsch). In recent years its frequency in Delhi is said to have much decreased. The so-called "Nepal button" is probably not of the character of the sore now under consideration.

There is no positive evidence that this affection exists in Asia to the east of the Indian peninsula, or in any portion of Australasia.

Africa. Cases of oriental sore have been seen in Egypt, at Suez and Cairo. In Algeria it has been observed at many places, notably at Tlemcen, near the Morocco frontier, at Ouaregla in the Oran province, at Algiers itself and Laghouat in the Algiers province, and at Biskra, Tugurt, and Liban in the Constantine province. In the Tunisian Sahara it is said to be very common. In Morocco it is seen principally on the banks of the Muluya river.

Factors which govern the Geographical Distribution. This disease is essentially one of warm countries. So far as is known at present it is confined to the sub-tropical regions and northern coasts of Africa, to the south-western portion of Asia, and to tropical and sub-tropical India. Moderate warmth rather than excessive heat must therefore be regarded as favouring its occurrence, and this conclusion is supported to some extent by the relations of the disease to seasonal changes. In India it most usually appears, not during the hot season, but at the beginning of the cold weather, after the rains have ceased. In Aleppo and Algeria the affection is commonest in the autumn. Neither the nature of the soil nor the height above sea-level seems to have any influence over the distribution of the malady, which is equally indifferent to the sex, race, and social condition of the people among whom it prevails.

Oriental sore is probably a contagious disorder. The discharges can cause fresh sores by inoculation both in the same individual and in others, and also in the lower animals. It is probable therefore that the affection is due to a micro-organism contained in the discharges, but the specific micro-organism, if it exists, has not yet been detected. It has been suggested that the infection may be carried from person to person by means of flies or other insects, but there seems to be no positive proof that this takes place. From the date of the earliest records of the disease down to the present day there has been a very general belief that oriental sore may be spread by means of drinking-water. In Algeria, in Aleppo, in India and elsewhere the behaviour of the disease has on some occasions seemed to show that the infection might perhaps be spread in this way. But on the whole the balance of evidence seems to be against this view. In Algiers and elsewhere an excessive consumption of dates has sometimes been held responsible for the appearance of the sore, but here again there seems to be no certain proof that they stand in the relation of cause and effect.

Oriental sore is essentially an endemic disease. There is little to show that its geographical distribution undergoes any considerable variation from year to year, or perhaps even from century to century. But its degree of frequency varies to some extent from time to time in those areas where it is endemic, and it has occasionally become so general as almost to deserve the title of epidemic.

It has already been pointed out that the sore can be artificially produced in the lower animals by inoculation. To this it must be added that certain animals seem to be capable of suffering from it under natural conditions. Thus in Syria and India dogs are said to have been subjects of the disease on several occasions; and in Syria oriental sore has been seen in the horse.

TROPICAL PHAGEDÆNA.

General Characters and Etiology. Phagedænic tropical ulcers are found in almost all tropical countries. While simple clean wounds heal with almost exceptional rapidity in the healthy natives of these countries, it is at the same time true that any small, jagged, contused wound, ulcer, or other solution of skin continuity in these countries may take on a malignant, phagedænic character and cause wide and destructive ulceration, seriously endangering the patient's life. Any small abrasion, even a mere insect bite, may suffice to start the process. A chigo ulcer or the site of a yaws papule may become phagedænic. Persons in a low state of health, depressed by attacks of malaria or other disease, are particularly liable to become subjects of these ulcers, which are commonest on the feet, but are also met with in other parts of the body. The symptoms of tropical phagedæna closely resemble those of hospital gangrene, at one time so common in all hospitals, and it is still uncertain whether the two processes are not identical. Tropical phagedæna is almost certainly due to some micro-parasitic infection, but the specific organism has not yet been certainly detected. Le Dantec, in 1884, described a bacillus found in these ulcers in Arab convicts in French Guiana, and a similar bacillus is said to have been observed by Clarac at Martinique, and by Peht in the Comoro Islands. Its specific relation to the disease has not yet, however, been finally proved.

Geographical Distribution. *Europe.* Europe being outside the tropical and sub-tropical zones, true tropical phagedæna is unknown on this continent. If hospital gangrene is pathologically identical with it, it also is now never seen in well-managed European hospitals.

Asia. On the Arabian shores of the Red Sea phagedænic ulcers are very common from Yanbo to Aden. In the province of Yemen the so-called Yemen ulcer appears to belong to this group. In India such ulcers occur with great frequency, particularly in Lower Bengal, Orissa, Arakan, the Malabar Coast, and Sindh. Possibly the so-called Jalkat sore is of this nature. This is a peculiar affection, met with at Jalkat and Chilas on the upper Indus, near the Afghan-Kashmir frontier, and also at Punial and Skardoo. It is said to be caused originally by the bite of the so-called Chilas fly[1].

In Farther India and the Malay Islands these ulcers are probably common. In Siam they are very frequently seen[2]. In British North Borneo large numbers of them are yearly treated, and in the Dutch possessions in the East Indies no complaint seems to be more common than ulcers among the natives. In the Straits Settlements they are very rife; in each of the years 1897 and 1898, for example, over 600 cases of this nature were treated in the hospitals, and in the Singapore prisons such ulcers very frequently come under treatment.

In Annam and French Cochin-China also the disease must be extremely prevalent. The French expeditionary corps suffered very severely from it. In Southern China ulcers of one kind or another are very common in the natives, but it is not clear whether these are of the tropical phagedænic variety.

Australasia. In many of the Polynesian islands the natives suffer greatly from phagedænic ulcers. In New Caledonia, for example, they are, or were at one time, a serious source of disablement not only to the natives but also to the crews of the French men-of-war stationed there. In Fiji at the present day between one-fourth and one-third of the cases treated in the Colonial hospital are usually cases of ulcer, and in the Solomon Islands large ulcerous sores, particularly on the feet, occur with great frequency[3].

Africa. This form of ulcer is also widely spread in tropical and sub-tropical Africa. Whether it exists endemically in Algeria is uncertain. Many cases of it were treated in the hospitals after the return of the Algerian (Arab and Kabyle) porters from the French expedition to Madagascar; but the men seem to have

---

[1] *Trans. 1st Indian Medical Congress,* 1894.
[2] Rasch, *Janus,* 1896–7. [3] Guppy, *op. cit.*

contracted the sores in that island, and evidence is apparently lacking of the presence of phagedæna in Algeria as an indigenous disease. In Abyssinia, the Sudan, and Egypt, however, it is far from rare. At Massowa severe forms of phagedænic ulceration are seen[1]. These ulcers are also frequent in Somaliland and Zanzibar, and the well-known "Mozambique ulcer" is probably of this character. On the West Coast phagedænic ulcer is extremely common on the Gold Coast, in Senegambia, and elsewhere. In Madagascar it is also constantly met with, and, as already stated, was a serious source of trouble among the porters employed in the French expedition in 1895.

America. Tropical phagedæna is very commonly seen in Mexico, in Guatemala, Nicaragua, Honduras, and other Central American States. In the West Indian Islands the natives suffer greatly from it. In Trinidad, for example, and generally throughout the Leeward Islands, few diseases more commonly come under treatment. At St Lucia in 1898 ulcers were so widespread as to be spoken of as an "epidemic," and they caused much maiming and disablement among the people. In Jamaica a certain number of cases are admitted to hospital each year.

In South America these ulcers probably occur throughout the tropical zone, but the fullest accounts of them have come from the Guianas. In French Guiana they seem to be particularly frequent and severe.

Factors governing the Geographical Distribution. As the name usually given to these ulcers implies, they are mostly found within the tropics. In Asia and in America they appear to be wholly or almost wholly tropical; in Africa they are found mainly in the tropical zones, but also in the sub-tropical regions of Egypt and—doubtfully—of Algeria. A high temperature may therefore be regarded as necessary to their endemic prevalence. They also prefer low-lying damp places, such as swamps, the alluvial beds of rivers, and sea-coasts, though they may develope at considerable altitudes, as in the Abyssinian highlands and the Himalayas.

The state of health of the individual is of importance as a predisposing factor in the development of the ulcers, and, as has been stated above, persons already weakened by other illnesses are very liable to become subjects of them. Whether there is any

[1] Rho.

true racial susceptibility or immunity to them is uncertain. All native races living in tropical countries, and Europeans migrating there, may become the subjects of phagedænic ulcers. Some, however, do seem to suffer more than others. Brault, for example, has shown that in the French expedition to Madagascar, while the Kabyle and Arab porters already alluded to suffered very seriously from sloughing ulcers, the negroes from the Sudan and the Somalis remained almost entirely free from them. He states that all went barefooted and passed through the expedition equally badly equipped and under exactly the same circumstances, yet the one race suffered severely while the others appeared to be more or less immune. This, however, can scarcely be accepted as an instance of true racial difference in susceptibility to the affection, for the Somalis and the Sudanese negroes are certainly not immune to these ulcers in their own countries. It is therefore possible that the porters of these races employed by the French in Madagascar had already suffered from them, or were more or less immune from constant exposure to the conditions which bring them about, while the Kabyles and Arabs from Algeria were readily attacked by a form of disease which does not appear to be endemic in their own country.

PINTA.

General Characters and Etiology. Pinta or *Mal de las tintas* is a remarkable epiphytic disease, which until recently was believed to be confined to Central America and the contiguous areas of Mexico and South America, but is now known to have a much wider distribution. It is essentially a dermatomycosis, commencing in the form of isolated spots of discoloured skin. There is more or less severe itching in the spots, which become rough and scurfy from desquamation of the epidermis. The spots enlarge and multiply, and when the disease is fully developed the whole surface of the body (including, rarely, the palms of the hands) may be mottled with patches of red, blue, black, or white discoloration, giving a markedly grotesque appearance to the patient. It causes little constitutional disturbance and is never fatal.

Pinta is caused by a microscopic parasite, with mycelium and pigmented spores. The mycelium measures from 18 to $20\,\mu$ in length by $2\,\mu$ in breadth; and the spores are from 8 to $12\,\mu$ in diameter. The difference in colour in the patches may depend on differences in the pigmentation of the spores or upon the mode of distribution of the parasite in the layers of the skin. The disease is almost certainly contagious. Montoza of Medellin (Colombia) claims to have found the pinta fungus in the waters of the Colombian gold-mines, and also in the bodies of mosquitoes which he believes to be carriers of the disease. It is noteworthy also that in Mexico the natives attribute the spread of pinta to the agency of mosquitoes.

History. Very little is known as to the history of pinta. It is said to have been prevalent in the southern parts of North America before the Spanish conquest of Mexico, and to have been imported to Mexico only as late as about the year 1775.

The disease has only recently been recognised in other parts of the world than Central and South America.

Recent Geographical Distribution. *Europe.* No case of pinta has, up to the present, been observed in any European country or in any of the islands adjacent to Europe.

Asia. In Asia it is still perhaps open to question whether the disease exists. Recently, it is true, cases have been described as occurring in Perak in the Malay States which bear great resemblance to the accepted descriptions of pinta. In the *Journal of Tropical Medicine* for February 15th, 1901, Dr P. G. Edgar has described in detail a case of this kind. Doubt, however, has been thrown upon the nature of the case on account of the absence of coloured, red or blue, patches and the incomplete description of the fungus found[1]. Leucoderma, or white patching of the skin, is by no means rare in natives of India, and it has been suggested that this case, where the patient was from southern India, was not one of true pinta, but an example of leucoderma. For the present the question is undecided, but if this and similar cases seen in that part of the world are indeed instances of pinta, then the disease must have a much wider distribution than was formerly believed, for the author just quoted writes :—" Less typical cases of pinta than the one described here are not uncommon among the Malays, Chinese, and Tamils in Perak and the Straits Settlements. Among the Malays the disease is known under the name of *sopah*; it is fairly common in the Kuala Kangsar district in Perak, occurring among several members of the same family."

There is no mention of pinta or a disease resembling it in any other country of Asia, nor in any part of Australia, New Zealand, or the Pacific Islands.

Africa. In Africa pinta or a similar disease is said to have been met with in the north (Manson), in Egypt (Sandwith), and on the West Coast. Some uncertainty still exists as to the identity of the African malady with that observed in America, but Dr Osborne Browne, who had seen pinta in British Honduras and was subsequently transferred to the Gold Coast, is satisfied that the disease met with in the latter country is true pinta. He states that it is very common on the Gold Coast, where the natives erroneously attribute it to yaws, although the latter disease may have occurred years before the pinta. He adds also that the

---

[1] *Journ. Trop. Med.* April 15, 1901.

natives aver that pinta extends into Hausaland about Kano, where it is called *tungere*. "The fungus," he states, "is in all cases identical[1]."

North and Central America. But if of very limited distribution in the Old World, pinta is much more widely met with in the New.

In Mexico it is endemic along the west coast, particularly in the provinces of Guerrero, Valladolid, and Michoacan ; it is more rare in the interior, and has not been seen further to the east than the western districts of the Tabasco province. The east coast of Mexico appears to be entirely free from it. In some villages in Mexico as many as 9 per cent. of the population are affected by it, and in 1826 there was in the Mexican capital a whole regiment composed exclusively of "pintados." The Mexican name for the disease is *saltsayanolitzth*[2].

In Central America it is found, though rarely, in Panama. In British Honduras it is exceedingly common, and as many as 60 per cent. of the adult Caribs in that colony are said to be affected by it ; while in some places the proportion is even higher, and nearly every adult Carib is said to suffer[3]. The Carib name for it is *wall-wall*, meaning a surface not uniform in colour.

South America. There is no mention of pinta in any of the West Indian Islands, but in several parts of South America the disease is seen. In Venezuela it is endemic in the provinces of Barquisimeto and Merida. In most parts of Colombia it is very prevalent, and particularly so in the province of Santander. It has been seen at one or two places in Peru and Chile, and finally it has been reported from Brazil, in the district between the Juciparana and the Sant' Antonio rivers.

Factors which govern the Geographical Distribution. In addition to the presence of the specific parasitic fungus, certain other conditions seem to be necessary for the prevalence of pinta. These are a warm climate, want of cleanliness and general hygiene, and an absence of precautions to prevent its spread from person to person.

That warmth is one essential for its existence is clear from the fact that pinta is only met with in tropical and sub-tropical latitudes; that in Mexico it is no longer found as an endemic above an

---

[1] *Journ. Trop. Med.* June 15, 1901.

[2] *Mexico as I saw it.* By Mrs Alec. Tweedie, 1901.

[3] Osborne Browne, *Journ. Trop. Med.* January, 1900.

elevation of 1500 feet; while in Colombia it is seen only in places with a mean temperature of from 68° to 85° F. A damp soil is said also to favour its development, and it is particularly frequent along the banks of damp or swampy rivers.

Want of personal hygiene and particularly want of cleanliness are also favouring factors. The poor and the dirty are the classes of population most liable to it. Race has also probably some influence. In British Honduras, for example, it is observed that Europeans and white Spaniards escape almost altogether, and that Creoles are only occasionally attacked, while, as already stated, an immense proportion of the Carib population are subjects of the disease. To what extent this immunity or relative immunity of certain races is truly a racial characteristic, or to what extent it is due to a higher level of cleanliness, care, and personal hygiene is not, however, clear. In the Malay Peninsula not only Malays, but also Chinese and Tamils are attacked (if, indeed, the disease observed there is in truth pinta); and (with the same proviso) the true negroes of western Africa can also suffer from it.

There is considerable evidence that pinta is contagious, and that it is spread from place to place and from person to person by the movements of infected individuals.

TINEA IMBRICATA (TOKELAU RINGWORM).

General Characters and Etiology. Tinea imbricata is a skin affection characterised by a concentric arrangement of closely set rings of scaling epidermis. It may gradually spread over the whole surface of the body, with the exception, usually, of the palms of the hands and the soles of the feet. It is an objectionable, but apparently in no sense serious, skin disease, of wide distribution in the Far East.

It is caused by a trichophyton.

History. Nothing is known as to the origin and early history of this skin affection. If, as there seems reason to believe, the disease called *"gune"* is the same as the Tokelau ringworm, then its existence has been known since the year 1844. Under that name the medical officers of the United States Exploring Expedition under Commodore Wilkes described an endemic skin disease as seen by them in the Gilbert Islands, and the description tallies closely with that of tinea imbricata[1]. In 1870 Turner gave an account of the affection[2], and stated that it had been imported from the Gilbert Islands to the Tokelau group, and later to the Samoa Islands. In the latter it is thought the disease did not exist before 1860. In 1870 it was introduced to Fiji, and two years later had become pretty general there[3]. In 1874 Tilbury Fox described the fungus associated with the disease, as observed by him in some scrapings sent from Samoa[4]. More recently many observers have given accounts of it. All are agreed that its dis-

---

[1] *Narrative of the U. S. Exploring Expedition*, 1844.
[2] *Glasgow Med. Journal*, Aug. 1870.
[3] Daniels. [4] *Lancet*, Aug. 29, 1874.

tribution is a constantly expanding one. Within the latter half of the nineteenth century it has been carried to large numbers of the Pacific Islands where it was unknown before, and in many has taken a firm hold, which it seems little likely to relinquish for the present.

Recent Geographical Distribution. This form of skin disease is met with most commonly in the Pacific Islands. The inhabitants of the Tokelau and Solomon Islands are particularly affected by it. These two groups are some fifteen hundred miles apart, and, as might be anticipated, the disease is not confined to them, but is also seen in many of the intervening islands.

In the Solomon Islands it is thought that not less than two-fifths of the entire population are subjects of this skin affection[1]. Its degree of frequency varies in the different islands of this group. In Treasury Island, for example, as many as four-fifths of the people are affected by it. Half of the thirty wives who share the joys and sorrows of the chief of this island are said to be almost covered with the eruption. In one of the Florida Islands "quite half" of the population have the disease.

In the Tokelau or Union group the disease is particularly common in Bowditch Island.

Northwards it has been seen in the Sandwich Islands, and westwards and northwards in the Ladrones[2], in Mindanao, the most southerly of the Philippine Islands, and in the Pelew and Caroline groups.

The disease is found in New Britain, New Guinea, and some of the adjoining islands. In British New Guinea it is very prevalent both at the western and eastern extremities, and " is gradually closing in on the central district, which it will before long completely invade[3]." In the neighbouring groups of Ké and Aru, in Teste Island, and Woodlark Island, the disease is common. In New Ireland and the Duke of York Island it is no less so.

---

[1] See an interesting account of the disorder in *The Solomon Islands and their Natives*, by H. B. Guppy, M.B., F.G.S., late Surgeon R.N. London, 1887.

[2] Dampier, *Voyage round the World*.

[3] *British New Guinea, Annual Report*, 1897–8. By Sir William Mac-Gregor, K.C.M.G. Later evidence seems to show that this prophecy is being fulfilled. Cases have been reported at Port Moresby on the south coast of the island (*Journ. Trop. Med.* Oct. 1900).

Tokelau ringworm is found in many of the South Pacific Islands. In the New Hebrides it is widely diffused. Bernal states that in some of the islands of this group, and particularly where the population is dense, as on the east coast of Mallicolo, at Atchin, Vao, and Whala, almost every inhabitant is affected by it. In the Fiji group it has become very general since its intro-duction in the year 1870, and the Gilbert, Ellice, Tonga, and Samoa Islands have all been invaded by the disease. It has also been met with in the Loyalty Islands and in New Caledonia.

The ringworm has also spread to Formosa. While com-paratively rare among the Chinese inhabitants, it is fairly common among the Pepohoans of the Kapsulan plains, and among the Hakkas, who are brought most into contact with the aboriginal inhabitants of the island. The aborigines of the mountains suffer less from the disease than the more civilised aborigines of the plains. In the Kilai plain Rennie found quite six per cent. of the semi-civilised native inhabitants affected by it. While no rarity, then, in Formosa, it will be seen that tinea imbricata attacks a far smaller proportion of the population here than, for example, in the Solomon Islands.

A few cases have been recorded at Foochow and other places on the southern shores of China by Manson, but these occurred only in persons who had come from Malacca or elsewhere in Malaysia, where it appears to be a very frequent disorder[1]. Nieu-wenhuis found that many of the natives of Borneo suffer from the affection[2]. Tamson also refers to its frequency among the Dyaks of that island, who have given it the names of *kurab* or *babouron*[3]. It has also been seen in Sumatra[4], and in Gisser, Ceram, Ceram Laut, and Goram. In Burma the disease has been known to exist in Rangoon, Maubin, Tomeghu and Akyab. The only place in India where it has been observed is Chittagong, near the Ganges delta. It has been recorded from the island of Nossi-Bé, off Madagascar.

Factors governing the Geographical Distribution. Tinea imbricata is practically confined to the tropics. A damp, warm climate, with a minimum temperature of 70° Fahr., and

[1] Wallace, *The Malay Archipelago.*
[2] *Janus*, 1897–8, p. 205.
[3] *Archives de Méd. Navale*, Sept. 1899.
[4] *Illustrated Supplement to the Journal of Trop. Med.* Feb. 1899.

an average of from 80 to 90° Fahr. is the most favourable for the existence of the disease. In addition to heat and moisture, dirt and a general condition of bad hygiene seem to be, if not essential, at least conducive to its spread. It is usually seen in native populations who pay little attention to personal hygiene, and, as it causes but slight inconvenience in the way of irritation, and is not viewed with disgust by the majority of subjects, it is very often left untreated. As it is also extremely contagious, it rapidly spreads in a community of this kind, when once introduced among them.

In regard to race, the disease has hitherto been mainly confined to the Papuans, Malays, and the mixed races which people Oceania, but especially to the first named. These people have been the main carriers of the dermic fungus which causes the disease, from island to island, through a great part of Polynesia and the Far East. But how far this is to be ascribed to a true racial predisposition is uncertain. In Formosa the Chinese have suffered much less than the aboriginal inhabitants. The origin of the latter race in this island is somewhat obscure, and Rennie has suggested that the very fact that Tokelau ringworm is common among them and rare among the Chinese may help to solve this ethnological problem, by indicating that the disease was brought to Formosa by a representative of one or more of the present tribes whose original home was one of the Pacific Islands[1].

Tinea imbricata has shown no special relation to age or sex. It is very easily got rid of by appropriate treatment, but the inhabitants of the affected countries rarely submit to the necessary measures.

[1] *Chinese Customs Med. Reports*, No. 45.

VERRUGA PERUVIANA or
THE PERUVIAN WART.

General Characters and Etiology. Verruga or the Peruvian Wart is a disease resembling, but probably distinct from yaws. The whole course of an attack of verruga usually lasts for several months. The first few weeks, which may be called the period of invasion, are characterised by a condition of fever, acute rheumatic pains in the muscles, bones, and joints, and a general sense of ill-health, together with headache, giddiness, and a feeling of contraction in the gullet. The most characteristic feature of the disease is an eruption, consisting of raised, elastic, reddish-blue tumours, sessile or pedunculated, varying in size from that of a filbert to that of a pigeon's egg. These wart-like tumours appear to be of the nature of highly vascular granulomata. Warts may appear not only on the skin but also on the mucous linings of the mouth, the nose, pharynx, alimentary tract, bladder, uterus, and vagina. They bleed with considerable ease, and hence anæmia is often brought about by hæmorrhages from the external warts, or by loss of blood from the internal passages. It is possible that these hæmorrhages are favoured by the reduced atmospheric pressure found at the high altitudes where verruga prevails, as they are said to cease when patients descend to lower levels. The disease may end in recovery, by the drying and shrivelling of the tumours until they fall off, leaving a small ulcer which slowly heals; or (in from 6 to 10 per cent. of the cases) in death, either from profuse hæmorrhages or from some complication.

Its cause is unknown.

History. The earliest mention of verruga relates to the 16th century. Zarate, the Chancellor of Lima whose *History of*

Peru was published in 1543, speaks of the existence there of a disease which certainly resembles the modern descriptions of verruga. "The men here," he wrote, "suffer from a wart or small tumour like a boil, very malignant and dangerous, which appears on the face or other parts of the body, and is more destructive than the small-pox and almost as disastrous as the plague itself." When Pizarro, with his small army of 700 men, was conquering Peru, it is said that more than one-fourth of his men died from hæmorrhages following gangrenous ulcers of the skin. For the next three hundred years history seems to be silent as to the existence of this disease. But in 1842 Archibald Smith described it anew in the *Edinburgh Medical and Surgical Journal*, and since then there have been numerous references to it in medical literature.

Recent Geographical Distribution. Verruga, so far as is known, is now confined to certain Peruvian valleys on the western slopes of the Andes. Hirsch quotes authorities to the effect that its headquarters are at Santa Ulaya in the Huarichi district, Matucana and a few other villages in Cocachacra, the valleys at the foot of the Cerro de Pasco, inhabited for the most part by miners, and several valleys in Chiquiang and the mountainous districts to the south of Lima. Imported cases have occasionally been seen outside these limits, as on the littoral of Peru. No cases have been seen on the eastern slopes of the Andes, and the disease is apparently unknown in Chile, Bolivia, or Colombia.

The disease is endemic in these regions, but on at least one occasion it seems to have become epidemic. This was in 1874, when large numbers of labourers employed in making the railway between Lima and Oroya were attacked. This outbreak caused a considerable mortality. The disease is sometimes called Oroya fever, but this would appear to be a misnomer, as it is unknown as an endemic disorder in Oroya itself[1].

Whether this disease exists in any other part of the world it is difficult to say. Beaumannoir has described a case of illness in the island of Réunion which he believed to have been verruga; and de Havilland Hall mentions a similar disease which he was informed exists in Zaruma, Ecuador. Mr Hutchinson treated a

[1] *La Maladie de Carrion ou la Verruga Peruvienne*, by E. Odriozola. Paris, 1898.

case in London, but the patient had worked in a Peruvian silver-mine five years before, and the attack was presumably a recurrence of the disease contracted then.

Factors governing the Geographical Distribution. Unless verruga is, as some observers think, only a severe form of yaws, it has perhaps the most limited distribution of any known disease. It is found, with the uncertain exceptions just named, solely in certain *quebradas* or narrow deep valleys or ravines on the western slope of the Andes, between the limits of the 9th and 16th degrees of south latitude. The verruga valleys vary in height from 2500 to 8000 feet. The temperature in the daytime is often very high (95° to 103° F.) but the daily variations are great, and the nights are often cold. The inhabitants live under conditions of much poverty and misery.

The valleys in which verruga prevails are said to be formed of bare rock (granite or diorite) and are usually traversed by a stream. In some valleys it is to the water of these streams that the disease is attributed. In the valley of Huarochiri, for example, there is a stream which for this reason is called the *Agua de verrugas*. Whether this theory of the origin of the disease has any real basis in truth it is difficult to say. Some have thought that verruga is spread mostly by contagion. Its general characters lend great probability to the view that it is caused by some micro-organism, but none has ever been discovered that could be called specific to the disease. On the other hand it is known to be inoculable. In 1875, at the time of the outbreak on the Lima-Oroya railway already mentioned, doubt seems to have arisen as to the true nature of that disease, and a medical student named Carrion had the courage to inoculate himself from one of the cases. He developed verruga and thus showed the identity of the two diseases, but unfortunately he succumbed to the attack. His memory is kept alive by the name of *Maladie de Carrion* which is often applied to this disease, particularly by French authors.

While verruga is thus clearly inoculable, there is no certain evidence that it is usually spread from person to person by means of the discharges from the warts. The opinion of the inhabitants of the verruga valleys seems to be generally against the view that it is spread in this way. Its degree of "contagiousness," in the popular sense of that word, is in any case probably not very high, and cases of verruga are treated in the general wards of the Lima hospitals without any spread of the disease to others.

In its relation to race verruga has shown a much greater tendency to attack white persons than either the Indians or negroes living in or visiting the verruga valleys. Bourse has even stated that no white foreigner who has lived any length of time in one of these valleys escapes the disease. He adds that during the outbreak on the Lima-Oroya Railway all the foreign engineers (mostly English) employed in superintending the work were attacked and several died; while of forty sailors who had deserted from a British ship and taken work on the railway, thirty died of the disease in the course of seven or eight months. It appears to be much more severe and fatal in foreigners coming to the verruga valleys than in the natives, who seem to acquire a sort of immunity to its attacks.

Verruga is said to attack some of the lower animals, particularly horses and mules, while even dogs, pigs, cows, lambs, turkeys and chickens are all stated to suffer from the disease.

YAWS or FRAMBŒSIA.

General Characters and Etiology. The terms Yaws and Frambœsia are the most usual and therefore the most convenient titles for a malady which is met with in many tropical and subtropical countries, and to which a different name has been given in almost every country where it has been observed. It seems, at least, that the diseases known as *pian* in the French colonies, in the Antilles, and in Guiana; as *aboukoué* on the Gaboon river; as *buba* in Brazil and Venezuela; as *keisse* or *changon* in Madagascar; as *pateh* in Malaysia; as *bouton d'Amboine* in the Moluccas; as *poöroë* in Borneo; as *paranghi* in Ceylon; as *tonga*, *coko*, or *patita* in Polynesia; and by other native names on the African coasts and elsewhere, are in reality identical with the disease known in many English colonies as *yaws*.

The essential character of this disorder is the appearance of a papular eruption, turning to tubercles, which are prominent, covered with a crust, and present when the crust is removed a reddish "raspberry-like" surface,—an appearance which has given the disease the scientific name of frambœsia. The general health is often seriously affected, and occasionally deep lesions of the joints and the bones are present in addition to the characteristic skin eruption.

Much discussion has arisen over the exact nature of yaws. One school of observers holds that the disease is in truth a manifestation of the virus of syphilis. Others believe that several pathological conditions are included under the general name, the rash in some individuals being syphilitic, in others tubercular in origin, and in others again possibly due to some other not clearly known cause. A third view is that yaws is a specific disease, *sui generis*, without relation to syphilis, tubercle, or any other malady. The micro-organism, if any exists, associated with the disease has not been demonstrated.

History. The early history of yaws is obscure, and has been rendered the more so on account of its frequent confusion with other disorders—particularly syphilis. The earliest reference to it is by the traveller Oviedo, who records its existence in Hispaniola (San Domingo), in the seventeenth century. In the same century medical accounts of the disease were forthcoming from Brazil, from the East Indies, and from the West Indies ; and later it was recognised in Africa and some of the Pacific Islands. In most of the countries where it is now endemic it appears to have existed for a very long period, and in some "from time immemorial."

Much discussion has taken place as to the original habitat of the disease ; some observers contending that yaws was at one time an affection confined entirely to the negroes of West Africa, and that it has been carried by them to other parts of the world. But the fact that yaws was described in the West Indies and Brazil long before the transport of negro slaves to those shores from the coasts of Africa began to be practised, and that it is known to have existed for a very long time in the Pacific Islands —the inhabitants of which have had no dealings with West African negroes—suffice entirely to disprove this view.

Geographical Distribution. *Europe.* Yaws is at the present day apparently quite unknown in Europe. Whether it has ever occurred on the Continent in the past is uncertain ; but several writers are prepared to regard the historical epidemics of *sibbens* or *sivvens*, a disease which was introduced by Cromwell's army into Scotland in the seventeenth century, the *radesyge* of Norway and Sweden which prevailed especially after the Seven Years' War, and some others, as of the nature of frambœsia. These historical outbreaks of a disease which certainly in some ways resembled that now under consideration have been dealt with in the chapter on syphilis and need not therefore be more fully discussed here. A disease analogous to yaws is said to be endemic in Greece, and in the published accounts of it, its resemblance to the "radesyge" of Norway is dwelt upon[1]; and the so-called "button-scurvy" which prevailed last century in Ireland had not a little in common with yaws.

[1] *Conference of Greek Physicians*, 1882.

Asia. Central and Western Asia—including Persia, Arabia, Syria, Mesopotamia, and Afghanistan—appear to be free from this disease. There is at any rate no positive record of its existence in any of these countries, nor in the whole of Siberia, Mongolia, China, and Japan. In India it is probably exceedingly rare ; only occasional cases have been seen, though in Pondicherry it was at one time rather common among the Hindu population. In very striking contrast to its rarity on the mainland of the Indian peninsula is its great frequency in Ceylon. It seems to prevail in all parts of the island and to be one of the most common of all diseases there. It is seen in every village, attacking not only the poor but also the well-to-do classes. There is not a health report from any portion of the island in which *paranghi*, the name given to it by the Siñhalese, does not occupy a most prominent position. It is perhaps now diminishing either in prevalence or in severity, as the number of deaths attributed to it in recent years in Ceylon has steadily fallen [1].

In Farther India yaws is occasionally seen. It was epidemic in the Surma valley in Assam in 1894, and a disease called by the natives *kroc-na*, seen in Upper Burma, seems to be no other than yaws [2].

In the East Indian Islands yaws is frequently seen, but there is reason to believe that it is less prevalent here than in Ceylon. Nieuwenhuis, the first European to cross Borneo (1894), says that it is "common enough" in that island. Mention of it is found in the annual reports of the Dutch troops in the Netherlands Indies. It is endemic in Java, attacking the natives, Chinese, and creoles, in Batavia and Pasuruan, but not affecting Europeans. It is common in Celebes and the Moluccas, where it is known as *bouton d'Amboine*; less so, perhaps, in the Rhio group, and apparently rare in Sumatra and Bangka.

Australasia. Throughout Australia, Tasmania, and New Zealand yaws is, it seems, entirely unknown. But in many of the adjoining islands in the Pacific Ocean it is a common disorder. I find no mention of it in the Philippines, but in

---

[1] The deaths from "*paranghi*" in Ceylon in each of the seven years 1891–1897 were respectively 314, 266, 275, 282, 199, 193, and 138 ; the last figure in the series is considerably less than half the first.

[2] *Trans. Indian Med. Congress,* 1894.

British New Guinea[1] it is said to be "universally endemic," though of a milder character than in other parts of the Pacific, and not causing much trouble. The natives of New Guinea have not invented a name for it, for which we may at least be grateful. In the Fiji Islands yaws, here known as *coko*, is endemic, and apparently a cause of much sickness among children; the Indian population of this group seem to suffer especially. The Friendly or Tonga islanders are also affected by it, and it is met with in the Loyalty Islands, Samoa, New Caledonia, and the New Hebrides.

Africa. I am aware of no proof of the existence of yaws or frambœsia in Egypt; and the only references to its occurrence in any of the countries along the northern and eastern coasts of the African continent are brief statements that it is met with in Algiers, in the Nile basin, and at Mozambique. But on the west coast, and to some extent in the interior, yaws, or a disease closely allied to it, is met with, and with considerable frequency. Thus in Wadai, in the region of the Upper Ubangi, and just to the north of the Congo Free State, *pian* or yaws is said to be the chief cause of infantile mortality. On the Gaboon river it is equally common, and is known by the natives under the name of *aboukoué*[2]. In Bornu and at Timbuctoo it is believed to exist.

In the Gold Coast yaws is endemic "in all its forms," and it is noteworthy that the medical officers in this colony believe that a late affection of the bones and joints frequently met with among the natives there is a result of untreated yaws[3]. In Lagos the disease is not unknown; 7 cases of it were treated in the Colonial Hospital there in 1897 and 14 in 1898. It is said to be almost peculiar to the Ijebus and Aworis, tribes inhabiting the lagoon littoral near Lagos. It is probably seen in several other regions of the West Coast, but information upon its exact distribution is incomplete and occasionally contradictory. Hirsch states that it is endemic "from Senegambia to Angola"; Plehn, on the other hand, says it is unknown in the Cameroon district. It is possibly endemic in the Congo Free State, and it is exceedingly common in Uganda.

There is little evidence of the existence of yaws in South

---

[1] *Annual Report on British New Guinea* for 1897–8.

[2] Bestion, *Arch. de Méd. Nav.* 1881.

[3] *Sanitary and Medical Report of the Gold Coast Colony* for 1897.

Africa, but I find mention of a single case of the disease admitted to the Durban Hospital in 1898[1], and in the same year four patients—a mother and 3 children—were admitted to the Kimberley Hospital, suffering from a rash which closely resembled that of yaws[2]. Finally in Madagascar and the Comoro Islands yaws is met with rather frequently, and though not known with certainty to exist in Mauritius an affection of the feet is seen there, called locally crab-yaws, which may perhaps be of this nature[3].

America. In Canada and the Northern States of America yaws is entirely unknown. In the Southern States it is said to be common, but of its exact distribution or intensity in this region I have not come across any detailed evidence. In regard to the West Indies, however, all uncertainty disappears, and there can be no question that these islands form one of the most important centres in the world of this disease. The majority of the islands seem to be affected by it to a greater or less extent. In Jamaica it is excessively common, and causes "terrible havoc" among the children of the island—a havoc largely contributed to by the ignorance of the native parents and their reluctance to allow the disease to be treated. In Cuba it is well known. In the French Antilles, Guadeloupe and Martinique, *le pian* was so widely prevalent in the 18th century that "à chaque habitation était annexée une case à pian pour loger les malades" (Le Dantec). It is scarcely less common in the same islands at the present day. In Trinidad, St Lucia, and Grenada yaws has become such a serious endemic disorder that special hospitals for its treatment have been instituted, and yearly deal with a large number of patients. Antigua and Dominica are both deeply affected by it, no other disease accounting for so many cases treated in hospital. In the Virgin Islands, on the other hand, yaws is a rare disorder.

In Central America yaws is probably not a common disease, the only mention of it being that of its rather frequent occurrence in Punta Arenas, in Costa Rica.

In South America, yaws has a fairly wide distribution. A few cases of the disease are yearly mentioned in the Health Reports

---

[1] *Durban Hospital Reports*, 1898.
[2] *Journ. Trop. Med.* Feb. 1900.
[3] Davidson.

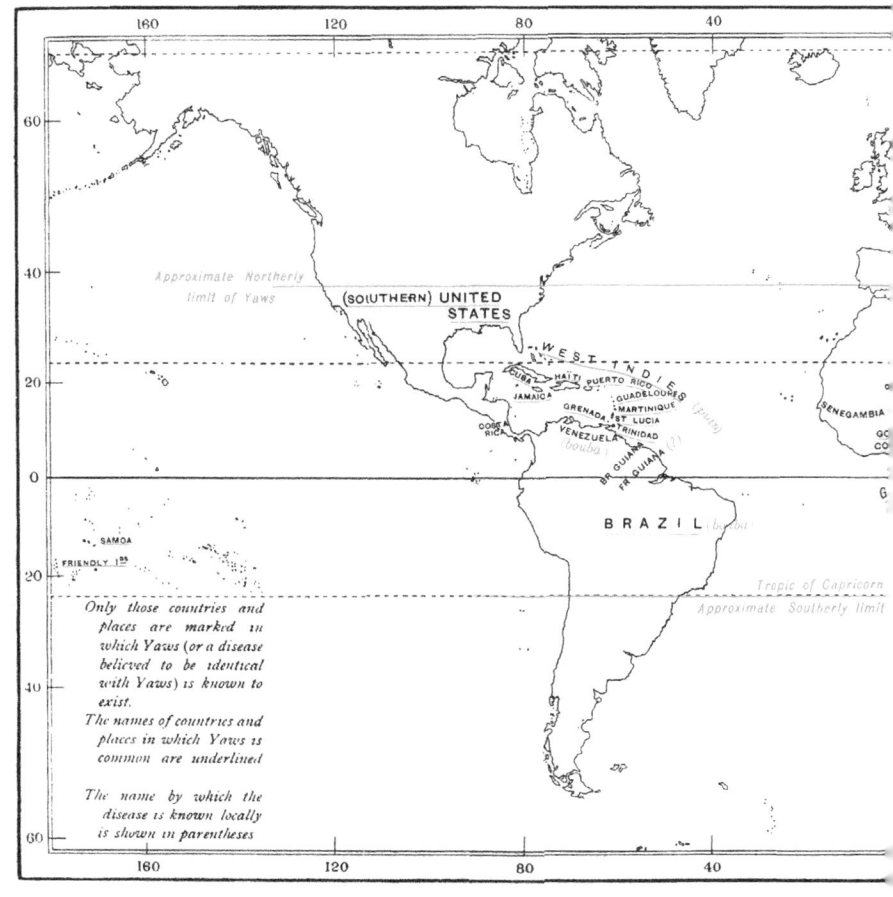

DISTRIBUTION OF YAWS OR FRAMBŒSIA

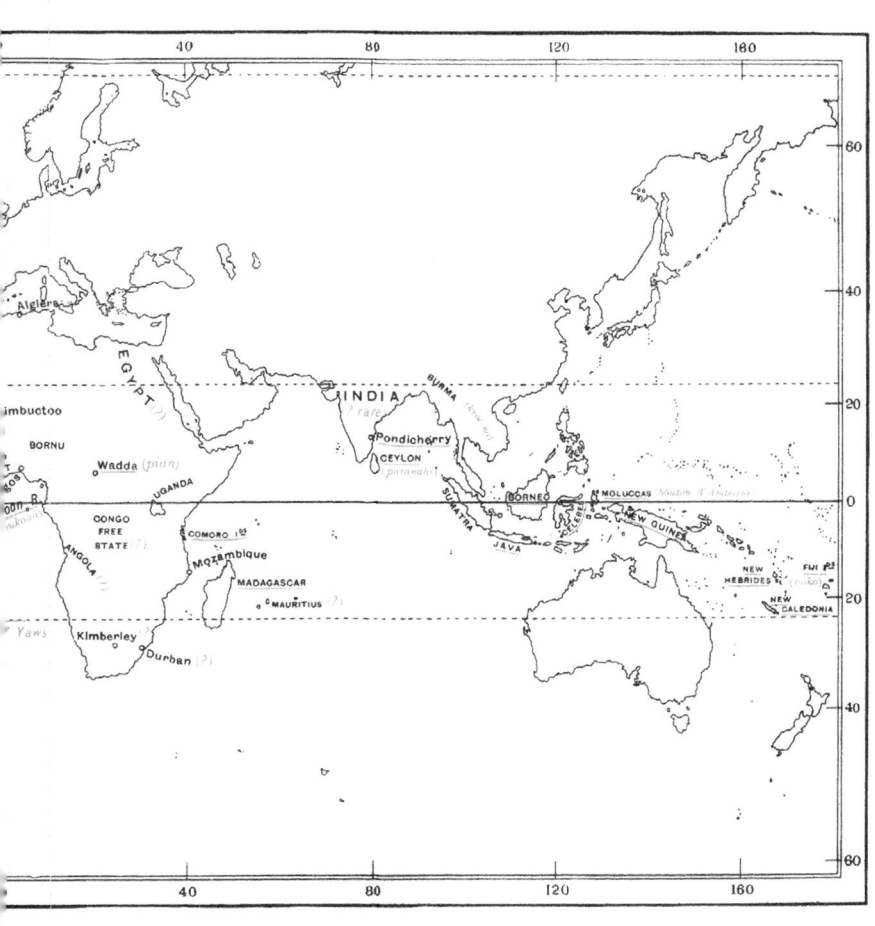

of British Guiana. In French Guiana Le Dantec saw several cases called *pian* by the negroes, but some were cases of leprosy, others yielded to syphilitic treatment, and others were a mixture of ecthyma and ulceration. In Venezuela yaws also probably exists, and is known by the name of *buba*. The same name is applied to the disease as met with in Brazil. It is said to occur in all the provinces of the Brazilian Republic, but the Indians living in the virgin forests far from the coasts are free from it[1]. In the other South American territories yaws appears to be entirely unknown.

Factors concerned in the Geographical Distribution. Comparatively few disorders are so strictly confined within limits of latitude as the malady under consideration. It is at the present day essentially a tropical disease. Even in the past—unless, indeed, we are to regard as yaws those epidemics of an apparently similar but more probably syphilitic disorder which occurred in Scotland, in Norway, and elsewhere, and have been already referred to—yaws has rarely been seen outside the tropical zone. The only other exceptions seem to have been the appearances of the disease in Assam (for Asia); in Algiers, possibly the Nile valley, and Durban (for Africa); and perhaps in some of the southern United States (for the Western Hemisphere). But, even if all these be genuine instances of yaws and not of some other disease, it will be observed that they have all occurred in districts closely bordering on the tropics.

Within the tropics yaws is found principally in certain great island groups,—notably in Ceylon, the Malay Archipelago, New Guinea, and the Pacific Islands in the Old World, and in the West Indies in the New World. It is also found on certain coasts of the corresponding continents—notably on the West Coast of Africa, on the coasts of Guiana and Brazil, and to some extent on that of the Indian peninsula.

In many of the islands just named yaws is a serious cause of disablement to the general population, and is among the most prominent of all diseases.

The occurrence of yaws almost solely within the tropics must be taken to indicate that it needs a high degree of temperature for its prevalence. All races have at times suffered from its attacks,

---

[1] Bréda of Padua, quoted by Le Dantec.

but the negro and other coloured races are particularly susceptible to it. Indeed, while the white man almost escapes in countries where yaws exists, it is found that the half-breeds, mulattoes and others, are attacked by it to a certain extent, and the coloured population most of all. Of coloured races the negro appears to be more susceptible than any. It would seem, then, that there is a direct relation between the prevalence of yaws and the depth of colour of the population. As the depth of colour of the skin is only a question of the quantity and nature of the pigment in the cells of the dermis, this fact raises the question whether yaws, in its causation, is not in some way dependent upon this element of the skin. The question appears to open up lines upon which further research might be carried out in those countries where the disease exists.

Both sexes and all ages suffer from yaws. Dirt and insanitation perhaps favour its development, but they do not seem to be essential to it. The disease is highly "contagious," and it is probable that direct infection from person to person is the principal way in which it is spread. It may also be spread, however, by means of infected articles, and the virus can apparently cling to clothes or sleeping mats that have been used by an infected person, and thus be transmitted to others. Sexual intercourse has been known to cause its spread; and it is believed to have been diffused at times, though fuller evidence on these points is wanted, by means of the common house-fly, and by vaccination. In its endemic homes yaws has frequently also been spread by the natives voluntarily inoculating their children with the crusts from another patient—a practice which is wisely discouraged. Heredity seems to play little part in its prevalence.

BOOK III.

ANIMAL PARASITES AND THE DISEASES ASSOCIATED WITH THEM.

CESTODES.

1. TÆNIA SOLIUM.

Geographical Distribution. *Europe.* This worm is found especially in Russia, East Prussia, some parts of Thuringia, of Belgium and of Switzerland; in Roumania; and among the non-Mussulman population of Turkey; and it is probably not entirely unknown in each or any of the remaining European countries. In Iceland, however, it is very rare, if it occurs at all.

Asia. Intestinal worms of all kinds are very common in the Caucasus and Russian Central Asia, but the statistics deal with the group as a whole, and not with each particular worm. Whether, therefore, *Tænia solium* is common here or not I am unable to say. The same remarks apply to the occurrence of such worms in Siberia and among the Bashkirs. In some Bashkir villages every inhabitant is said to harbour an intestinal worm of some kind. In India *Tænia solium* is never seen among the purely vegetable-eating Hindus, and is apparently not very common among the other races. The parasite is perhaps commoner in China than in any other Asiatic country. It abounds in and around Peking, in Wuchow, and in other towns[1].

Australasia. *Tænia solium* is said to be unknown in New Guinea. It occurs in the Fiji Islands.

Africa. It appears to be common in British East Africa. In Massowa it is frequent among the natives. On the other hand it is rare in Egypt. On the West Coast tænia is common enough, though in the Cameroons it is said to be less often met with than the Guinea-worm.

---

[1] Coltman, on the other hand, never saw a single case of the worm during six years' residence in Shantung.

America. This tænia is said to be unknown in Greenland. It is seen in Newfoundland. It is rare in the United States, and also in Guatemala and some of the West India Islands, but apparently less so in Mexico. It is uncertain whether this is the particular form of tape-worm met with in Brazil.

Factors concerned in the Distribution. The distribution of this worm depends almost wholly on the fact that part of its life-history (the cysticercus stage) is passed in the pig. Hence it is found in man only among those races and in those countries where pork is a common article of food. Hence also its extreme rarity or entire absence among orthodox Mussulman communities, who never touch the flesh of the pig, and among such purely vegetable feeders as the Hindus, Egyptians and others.

Since the alarm created a few years ago by the increase of trichiniasis led to a lessened consumption of pork in Europe this worm is said to have become considerably rarer.

2. TÆNIA MEDIOCANELLATA.

Geographical Distribution. *Europe.* This worm is comparatively common in most European countries, and in many it is far more frequent than *T. solium.* Its prevalence is believed to have increased in recent years.

Asia. The parasite is seen in Syria and Arabia, and occurs with some frequency in India, but solely, or almost solely, among European residents or flesh-eating Mohammedans. It is said to be common among the Buriats in southern Siberia, and also in China and Japan. In China and Cochin-China it is more prevalent than *T. solium.*

Africa. In Abyssinia this tape-worm abounds; and in Egypt, Nubia, Algeria, and the western Sudan it is exceedingly common. It is found also with great frequency on the West Coast, and seems to be by no means rare in South Africa.

America. In the United States *Tænia mediocanellata* is the form of tape-worm most usually met with. In South America the only mention of it relates to its occurrence in the Argentine Republic, where it is extremely prevalent.

Factors concerned in the Distribution. The cysticercus stage of this parasite is passed in horned cattle ; hence the worm is found in man only where imperfectly cooked beef is eaten.

Like *T. solium* it is never seen among strict vegetable eaters, as the Hindus. It is said to have become more common in France, Denmark, and Switzerland in recent years, owing to increased importation of beef from Algeria, and the consumption of it in a raw state as a strengthening article of food for invalids.

3. TÆNIA NANA.

The Parasite is the smallest of human tape-worms, varying from 10 to 15 mm. (about half an inch). It is identical with *T. murina* which affects rats, and is of little clinical importance unless present in large numbers, when it occasions gastro-intestinal disorder.

Geographical Distribution. It has been found in Egypt (where it was first detected by Bilharz in 1852), in Italy, Buenos Aires, and Bangkok, Siam.

4. TÆNIA MADAGASCARIENSIS.

The Parasite may be as long as 24 cm. (about nine or ten inches), and about $\frac{1}{8}$ of an inch broad. The intermediate host is unknown, but Blanchard suggests it may possibly be the cockroach.

Geographical Distribution. This worm has been seen in Mayotta in the Comoro Islands, in Mauritius, in Bangkok, in Georgetown, British Guiana, and in Nossi Bé off the Madagascar coast.

5. TÆNIA FLAVO-PUNCTATA.

The Parasite, which was first described by Weinland, is about 8 to 12 inches in length, and is characterised by the joints in the fore part being marked by a large yellow spot—the distended receptaculum seminis.

Geographical Distribution. This tænia would appear to be very uncommon, and two or three cases, from the United States and from Italy, have alone been recorded.

Nothing is known of the intermediate host.

6. TÆNIA CANINA.

The Parasite, identical with *T. cucumerina* and *T. elliptica*, is recognisable by its delicate form and small size (12 to 35 mm.), and also by the fact of its having two sets of reproductive organs in each mature segment. It occurs principally in the dog and cat. Its geographical distribution is imperfectly known, but it has been stated occasionally to affect man in the West Indies, and not infrequently in Scandinavia.

7.　TÆNIA ECHINOCOCCUS : HYDATIDS.

The Parasite. The adult tænia occurs in the dog, wolf, and jackal. The embryo form of the parasite gives rise in man to the well-known hydatid or echinococcus cysts, which may occur in any part of the body, but are most common in the liver, and (in order of frequency) in the intestinal canal, the lungs, the kidneys, and the central nervous system.

Geographical Distribution. *Europe.* Hydatids probably occur in all European countries with more or less frequency, but statistics are incomplete as to their relative degree of prevalence in different countries. They are said to be remarkably common in Iceland. Different observers have estimated that from one-fortieth to one-seventh of the whole population of Iceland are the subjects of hydatids. It is remarkable, on the other hand, that in the Faröe Islands the disease is quite unknown[1], in spite of the large numbers of dogs and sheep and cattle, and the similar general conditions here to those obtaining in Iceland. In Turkey hydatids are rather exceptionally frequent, and there can be little doubt that this is due to the large number òf dogs in that country and their close relations with the people.

Asia. In the Caucasus the disease is frequently seen ; the post-mortem records of the Tiflis Hospital show that 2 per cent. of all bodies examined have hydatid cysts in the liver.

Hydatids are of by no means rare occurrence in many parts of India. In China the disease must be rare, for Dr Cantlie only saw it once in 40,000 cases.

Australasia. Hydatids are peculiarly common in many parts of Australia, where, no doubt, they are to be associated with the large numbers of dogs employed in sheep-farming. Victoria was formerly spoken of as one of the principal centres of hydatid disease. In South Australia it was the cause of from 7 to 15 deaths in each of the ten years 1889–98 (say from 20 to 45

[1] *The Faröe Islands*, by J. Russell-Jeaffreson, F.R.G.S., London, 1898.

per million inhabitants). In Queensland it caused three deaths in 1898; I have not seen the records of previous years. In Tasmania it is also rather common, but in New Zealand it was at one time considerably less so; the average death-rate in the latter between 1882–88 was only 6·6 per million. In recent years it has, apparently, become much more common here, as in 1898 it was the cause of 20 deaths, or about 27 per million living. There seems to be no mention of the occurrence of hydatids in the Pacific Islands.

Africa. In Egypt and in Algeria hydatids are not infrequent. For the rest of the continent information on this point seems to be lacking.

America. In Canada and the United States hydatids are extremely rare and the great majority of cases occur in foreigners. Osler states that up to 1891 he had found records of only 85 cases in all in both countries. He adds that many instances have occurred in the Icelandic settlements of Manitoba, particularly since the year 1874, the date of the Icelandic immigration.

There seems to be no mention of hydatids in Central or South America, or the West Indies.

Factors concerned in the Distribution. As stated above, the adult worm occurs in the dog and some other animals. The ova escape in the intestinal excreta, and either pass directly to the human host, or enter the bodies of sheep and cattle, both of which are liable to the disease. Hydatids are therefore found most commonly among persons who have much to do with the rearing of these animals, and particularly among such classes as the Icelanders, and the shepherds and cowherds of Australia, who spend their lives in close contact both with their flocks and herds and with the dogs that help them in their work. In Iceland the dogs live in the same dwellings, eat off the same platter, and even share the same bed as their owners, and this would seem to explain the frequency of hydatids in that island. But it is less easy to account for their absence from the Faröe Islands, where the conditions are probably not very different, unless, indeed, the parasite has never been imported there.

8. BOTHRIOCEPHALUS LATUS.

Geographical Distribution. *Europe.* This worm is almost confined to the continent of Europe. It is most common

on the coasts of Sweden ; on the Russian and Finnish shores of
the Gulf of Finland ; in the neighbourhood of some of the Swiss
lakes, and in the adjoining departments of France. It is less
frequent, but still common, in Poland, East Prussia, Denmark,
Holland, Belgium, and about the Italian lakes. Occasional
examples of it are seen in parts of Germany, in Brittany, and in
Ireland.

Asia. This worm was formerly said to occur not infrequently
in children sent home from Ceylon to England ; and it has been
found in the crews of ships of war on the Dutch East India station.
It is stated also to have been met with in Japan.

Africa. The parasite is said to occur in British East Africa[1],
and (doubtfully) in South Africa.

America. At least one case has been recorded in Canada.
The worm exists also in Mexico.

Factors concerned in the Distribution. The ova of
Bothriocephalus latus develope in the peritoneum and muscles of
the pike and other fish. Hence they only pass into the human
body through the consumption of infected and imperfectly cooked
fish. The parasite is consequently found in greatest frequency
on the shores of seas, lakes, or other inland waters. The brief
summary given above of its distribution shows clearly this relation
of the worm to sea or inland coasts ; and a closer study of its
frequency in Sweden and Switzerland has elicited the fact that the
number of cases varies inversely with the distance from the water.

9. BOTHRIOCEPHALUS MANSONI.

The Parasite. The larval form only has been seen. It
measures from 30 to 35 cm. in length (12 to 14 inches). It has
been found in the sub-peritoneal fascia, and on one occasion
in the pleural cavity.

Geographical Distribution. Manson first discovered the
worm in a Chinese cadaver at Amoy. Scheube extracted a
similar parasite from the urethra of a Japanese ; and it has also
been seen at Bathurst in Australia. Daniels has observed a
similar worm in British Guiana.

[1] Kolb.

TREMATODES.

1. DISTOMUM PULMONALE vel D. RINGERI (Endemic Hæmoptysis).

The Parasite. This is a fluke-shaped worm, about one-quarter or three-eighths of an inch in length. Its life-history is unknown. One or more worms may be found in the lungs at a time. They burrow into the organ and give rise to serious symptoms, of which chronic cough, rusty sputum, and more or less extensive hæmoptysis are the most important. The fluke has also been found in the liver, peritoneum, testes, and brain.

Geographical Distribution. It is mainly if not wholly confined to Japan, Corea, and Formosa. In northern Formosa it is probably very common, but it is said to be unknown in the south of the island. Rennie states that he found it in 0·9 per cent. of patients seen at the Tamsui Hospital, but that it is probably more frequent than this figure would represent, and Manson records that in some parts of the island and in some Japanese villages as many as from 10 to 30 per cent. of the population are affected.

This, or an allied fluke, has been seen in the United States in the dog and the cat, but does not yet seem to have been observed there in man.

2. DISTOMUM CRASSUM vel D. BUSKI.

The Parasite is the largest fluke seen in man. It may be 2 or $2\frac{1}{2}$ inches in length. Its life-history is unknown. It inhabits the upper part of the small intestine in man, and may cause diarrhœa and other intestinal trouble.

Geographical Distribution. This worm was discovered in 1843 by Busk in a Lascar who had died in the Greenwich Hospital. It has been seen several times in China (Canton, Ningpo, etc.) and also in Borneo, Sumatra, the Straits Settlements, Assam, India, and on one occasion in an East Indian immigrant in British Guiana.

3. DISTOMUM SINENSE.

The Parasite. This fluke, first discovered in 1874, is about three-quarters of an inch in length. It inhabits the bile-ducts and gall-bladder, and sets up a serious disease which may prove fatal. The cachexia is said to resemble that of sheep-rot.

Geographical Distribution. This distomum has been observed in India, Mauritius, Japan, Corea, Formosa, China, and Tongking. It is said to be very common in the last named country, and in some parts of Japan 20 per cent. of the population have been found to be subjects of it.

4. DISTOMUM HETEROPHYES.

The Parasite is the smallest of the human flukes, measuring only $1\frac{1}{2}$ mm. in length; it is of a bright red colour. It occurs in the small intestine and is probably harmless.

Geographical Distribution. It was observed by Bilharz in Cairo in 1851, and has apparently only been found in Egyptian subjects.

5. BILHARZIA HÆMATOBIA (Endemic Hæmaturia).

The Parasite. *Bilharzia hæmatobia* or *Distomum hæmatobium* is a trematode worm, differing from most other of its allies in being bisexual, that is to say, the two sexes occur in different individuals. The male is about 15 mm. in length, and 1 mm. in breadth; the female is about one-third longer than the male. The adult parasites exist in the veins of various parts of the body, but in post-mortem examination are seen especially in those of the portal system.

The life-history of the parasite is not fully known, but it is believed, from analogy with other nearly allied worms, that it passes part of its existence in some extra-human host. It is conjectured that this is some form of living organism found in fresh water, possibly, Guillemard suggests, *Melania tuberculata*. The method of infection still remains unknown, but it seems probable that it is not by ingestion of the animal, but by the armed embryos directly penetrating the body during bathing. The pathological effects associated with the presence of this worm are brought about by the extrusion of the ova in the veins of the various tissues. Of these by far the most important are the mucous and sub-mucous tissues of the bladder and ureters, but eggs have also been found in the walls of the intestine, in the kidneys, generative organs, liver, heart, and even the lungs. The animal or its ova may give rise to the formation of urinary calculi in any part of the urinary tract.

The leading symptoms of *bilharzia disease* are most often referable to the urinary system. The urine is thick and contains blood, and microscopic examination reveals the presence of the ova. The presence of the parasite in other parts may lead to the appearance of other symptoms referable to those parts. But the bladder symptoms are almost always the most prominent, and hence " bilharzia disease " and " endemic hæmaturia " have come to be regarded as almost convertible terms. The parasites may be present in the tissues without causing any symptoms, but sooner or later, and particularly if the hæmaturia is persistent and prolonged, the health suffers more or less severely. The disease may last for months or years ; complete recovery is by no means rare, but death as a direct result of the infection is not unknown. For a fuller account of the pathology and clinical characters of this affection the reader is referred to the writings of Sonsino and others.

History. The parasite was first identified in Cairo in the year 1851 by Bilharz, whose name it bears. Of the history of the worm itself almost nothing is known. The occurrence of hæmaturia, and the frequency of urinary calculi in Egypt had long been recognised, and it is therefore probable that the parasite has long been domiciled in the blood of the Egyptian population. In 1864 Harley first observed the disease in persons coming from the Cape. It has since been recognised in many other parts of Africa, and in scattered portions of some other continents.

Recent Geographical Distribution. *Europe.* It must be regarded as doubtful whether bilharzia disease exists as an endemic in any part of the European continent.

Asia. Observations on the presence of the parasite in Asiatic countries are not numerous.

The bilharzia has been observed in Mecca, and on the Arabian coasts of the Red Sea. Hatch, of Bombay, who saw several cases there, found that Mussulmans returning from pilgrimage in the Hedjaz were particularly liable to be affected by the disease, and there is some reason to believe that it is common in Arabia. At least one case has been seen in Cyprus[1].

In Mesopotamia, in the valleys of the Tigris and Euphrates, the parasite has been observed by Sturrock.

In India bilharzia has been met with, especially on the west coast.

It is doubtful whether the parasite exists in China. Brault, of Algiers, examined the blood of a large number of French troops returning from Tongking, and sent to hospital at Dey, but never succeeded in finding either the parasite or its eggs.

Africa. In contrast with its comparative rarity in Asia this trematode has a wide distribution through the greater part of the African continent. Egypt has long been known as a country where endemic hæmaturia was common, and it was at Cairo, as already stated, that the parasitic cause of the disease was first detected. It is still extremely common in the country. Sonsino, as the result of 91 necropsies performed in Lower Egypt between 1875 and 1883, found bilharzia infection in 42 (or 46 per cent.). This was even a larger proportion than that recorded by Bilharz and Griesinger in the early 'fifties.' The latter made their observations in the Kasr-el-ain Hospital at Cairo, and found bilharzia in 117 out of 363 necropsies (or 32 per cent.).

In Tripoli there is no positive evidence of its presence, but in southern Tunisia its endemicity has been proved by the French observers Villeneuve and Brault. The parasite seems to be most common here, and its effects most severe, in Gafsa, "where it is quite as bad as in Egypt." It is also seen at Gabes. In Algeria it is described as not uncommon.

On the west coast of Africa the parasite is found on the Gold Coast in Nigeria and in the German Cameroon territory. At Chisamka, Angola, a spot 300 miles from the west coast and 400 south of the Congo, bilharzia has also been detected[2].

---

[1] *Brit. Med. Jour.* 1902, Vol. II. p. 956.
[2] Massey, *Journ. Trop. Med.* Feb. 15, 1901.

On the east coast the parasite has been met with throughout —from Egypt to the Cape. It is, however, apparently absent from Massowa on the Red Sea shores (Rho). In Abyssinia, on the other hand, it is probably common, as the Italian troops in the last Abyssinian campaign suffered considerably from it. In Zanzibar, on the Mozambique coast, and on the lower reaches of the Zambesi it is very often seen. Further south the parasite is found on the shores of Delagoa Bay, and St Lucia Bay. At Port Elizabeth and Uitenhage in Cape Colony it seems to be particularly frequent[1], and it is also very common in Natal. In Pietermaritzburg, according to Batho, "it seems as if the majority of the male youth were affected."

Generally through Central and South Africa bilharzia disease seems to be prevalent. In the north endemic hæmaturia, conjecturally due to this parasite, is said to be very common in the countries of the Tibbos, Lake Tchad, Darfur and Kordofan. Cases have been seen in Uganda, and in British Central Africa bilharzia is one of the most frequent causes of anæmia. In the Congo Free State it is common[2]. In the south it is met with in the Transvaal and in Kaffraria and elsewhere. Occasional cases have been seen in Mashonaland, apparently imported from elsewhere. The parasite has been identified in at least two cases at Bloemfontein, in the Orange River Colony[3], and it is thought that a large proportion of the cases of hæmaturia which occurred among the troops during the Boer War were of bilharzic origin.

In Madagascar, Mauritius, Réunion, and Nossi Bé there is more or less definite proof of the presence of this parasite. In Mauritius, Chevreau, Chazal, and Davidson have demonstrated its existence. In the case of the other islands named the statements as to its presence are less positive. Brault, after the return of the French troops from Madagascar, examined a considerable number for bilharzia, but neither in Europeans nor in natives did he find it present in a single case.

America. Until quite recently it was believed that bilharzia did not occur on the American continent. But early in 1900 Dr E. Walker reported a case of the disease as observed in Sparta in the State of Illinois. Great care is said to have been

---

[1] Sambon, *Journ. Trop. Med.* Nov. 1899.

[2] Brault, *loc. cit.*

[3] *Journ. Trop. Med.* Oct. 1, 1901.

taken to avoid error in the diagnosis, and this case must therefore be regarded as the first recorded example of the bilharzia disease in the New World. Recently Manson has reported a case in an English resident in the West Indies who had never been in any country where bilharzia was known to exist, so that, if not *B. hæmatobia* itself, it is probable that some closely allied species exists in the archipelago.

Factors governing the Geographical Distribution. This parasite is largely but not wholly confined to tropical and subtropical countries. To the north of the equator it seems, until quite recently, to have been observed nowhere to the north of the African shores of the Mediterranean (latitude 35° N.). The only recent exception has been the recognition of the parasite in the State of Illinois in North America. Southward, the limit of extension of the worm has similarly been the southern extremity of the African continent (about 35° S.). It is remarkable that the bilharzia disease is so frequent in Africa as to be spoken of by one authority as "the most prominent, prevalent, and important disease" in that continent; while in the other continents it is either unknown, or at most occurs as a more or less exceptional disorder in very limited areas. Further observations may show that this distome exists much more widely than is supposed, and that its distribution is much more continuous and less "patchy" than the observations hitherto published would seem to indicate.

It is more than probable that the great determining factor in the distribution of bilharzia will prove to be the distribution of the living organism, whatever this may be, which forms its intermediate host. Up to the present this organism has not been identified; but whether it prove to be some mollusc, crustacean, or arthropod, or something wholly different from any of these, it is clear that if it be an essential factor in the life-cycle of the bilharzia, neither this parasite nor the disease it produces will be found in those countries where the particular organism is absent. The true relation of such physical conditions as latitude, temperature, elevation, soil, etc., to the distribution of bilharzia will then become apparent; and it will probably be found that those factors have only an indirect influence on the parasite, though they may have an all-powerful influence in determining the presence or absence, the frequency or infrequency, of the intermediate host.

6. OTHER TREMATODES.

Distomum lanceolatum, found once or twice in patients in Germany and France; *Distomum conjunctum,* recorded by Professor McConnell in Calcutta; and the liver-fluke, *Fasciola hepatica,* which has occasionally occurred in man, are too rare to demand extended reference here. *Amphistomum hominis* also is a parasite of almost no clinical importance. It has been found solely in India (Lewis and McConnell); in Assam (Giles); and in an East Indian immigrant in British Guiana (Law).

NEMATODES.

1. ASCARIS LUMBRICOIDES.

The Parasite. The male of the round-worm, one of the commonest of the parasites of man, measures from 15 to 25 cm. in length by 4 mm. in breadth; the female 16 to 45 cm. by 6 mm. Both are cylindrical. The ova may be spherical, ovoid, or sometimes barrel-shaped, and it would seem that they are capable of hatching direct in man, no intermediate host being necessary. The life-history is, however, still uncertain.

Geographical Distribution. The round-worm has practically a world-wide distribution. It is, however, particularly common in tropical and subtropical countries. Recent writers have dwelt on its frequency in China, in India, in Java, in Mauritius, on the Gold Coast, in the West Indies, and in the Pacific Islands; and there are abundant references in the past to its prevalence in nearly all other warm countries. It is common enough, also, in the temperate zones, and even in Newfoundland and Greenland it occurs indigenously. It is only in Iceland that it would seem to be rare.

2. OXYURIS VERMICULARIS.

The Parasite. The male of the thread-worm is very much smaller than the female, measuring only from 3 to 5 mm. in length, while the female is from 9 to 12 mm. The eggs are oval and flatter on one side than on the other. No intermediate host is necessary, and auto-infection keeps up the disorder.

Geographical Distribution. The thread-worm has, apparently, quite as wide a distribution as *Ascaris lumbricoides*. Like it, it is most common in tropical and subtropical countries. It differs from it, however, in being extremely frequent in Iceland.

STRONGYLES.

3. STRONGYLUS GIGAS.

The Parasite. This very large nematode has been well described as resembling an overgrown *Ascaris lumbricoides.* The male measures from 14 to 35 cm. in length, and the female may be from 25 cm. to a metre in length, the breadth varying from 4 to 12 mm. The larval stage is possibly passed in some fish. The parasite has only been very rarely met with in man, but is not uncommon, according to Leuckart, in the dog, wolf, fox, and seal.

Geographical Distribution. The parasite has been recorded as occurring, in the lower animals, in most countries of Europe and the New World.

4. ANKYLOSTOMUM DUODENALE
(Ankylostomiasis and Pani-Ghao).

The Parasite. The male varies in length from 6 to 12 mm., the female from 7 to 18 mm.: the diameter of the former is about half a millimetre, and of the female about 1 millimetre. Both are cylindrical, tapering towards the anterior, or buccal end, where there is a curious mouth-capsule, provided with four claw-shaped hooks or teeth, two on the dorsal and two on the ventral aspect. The worm is greyish-white when empty, and reddish or reddish-brown when full of blood.

It is found usually in the jejunum, where it attaches itself by its toothed mouth-capsule to the mucous lining of the bowel, and sucks the blood of the host, it is believed, in considerable quantities. The eggs, escaping from the female ankylostomum, pass out in the fæces of the host. Some of the quite early stages of development of the embryo are passed in the body of the host, but for the most part development occurs after the fæces have escaped, and the circumstances favouring it seem to be a plentiful air supply, a warm but not too hot temperature, and deposition of the fæces containing the eggs on the soil. Sooner or later the ovum, or larva, is swallowed by a human being, in whose intestinal canal it again becomes the sexually mature parasite.

This worm and the disease associated with it assume a grave importance in the pathology of many countries, in consequence of the vast numbers of the labouring classes affected by them, and the resulting loss of labour.

Ankylostomiasis is the name applied to the group of symptoms associated with the presence of considerable numbers of this worm in the intestine. It is characterised by a severe and progressive anæmia, accompanied by dyspepsia and other symptoms of alimentary disturbance, and, when prolonged, with symptoms of serious affection of the circulatory system. Serous effusions may occur in different parts, and fatty degeneration of the heart may follow. It is not rarely a direct or indirect cause of death.

The mere presence of a few ankylostomes in the intestine can scarcely be said to constitute a condition of ankylostomiasis, and the parasite may, indeed, exist in considerable numbers without giving rise to any symptoms of its presence. In such cases it is only on the post-mortem table that it is found in the bowel-contents. The number of parasites required to produce definite symptoms in life varies very widely in different individuals, and perhaps in different races. Rogers has proposed to limit the term ankylostomiasis to that condition in which "some 500 of the worms have been present for upwards of six months, and have by themselves produced distinct anæmia and other well-known symptoms." As, however, it must be very difficult, if not impossible, in most cases to determine that 500 worms have been present for six months, or indeed that any definite number have been present for any definite period, it would seem better to employ the term in a more general manner, as indicating a condition in which the symptoms of ankylostomiasis are associated with the proved presence of a considerable number of worms, and are believed to be caused by them.

In Assam, Bengal and elsewhere persons have been known to harbour as many as a hundred or more of the worms without showing any marked symptoms of illness. Much appears to depend upon the general condition of health of the subject of these parasites. Any pre-existing disease of a debilitating kind, and particularly if associated with anæmia, appears to favour the development of the condition of ankylostomiasis. A person in good general health may be able to tolerate the presence of a large number of the worms without his health suffering, while in another person of indifferent general health, a smaller number of worms may give rise to the characteristic symptoms of the disease. The one is able to make good the loss of blood caused by the presence of the worms, while the other is not.

Pani-Ghao, or Ground-itch, is the name given to a skin affection which has been met with in Assam, Trinidad, and many other tropical countries. Its causation was for long unknown but evidence has recently been published tending to show that, at least in many cases, it is due to the penetration of ankylostomum larvæ into the skin of the patient. Looss and Sandwith, of Cairo, have both shown experimentally that such penetration is possible. Bentley, of Assam, has repeated these experiments, and has further discovered the larvæ in the skin of patients suffering from pani-ghao[1]. The disease affects only the lower extremities, and hence is often called the " sore-feet of coolies." It takes the form of an erythema, followed by vesicle and pustule formation, and may go on to ulceration and even gangrene. Dalgetty, of South Sylhet[2], has described an acarus found in the skin in some of these cases, and it may therefore be assumed that more diseases than one are included under the colloquial term of ground-itch or pani-ghao. This affection is by no means an essential accompaniment of ankylostomiasis. So far as is known at present its geographical distribution is much more limited than that of the latter, and it may therefore be supposed that the parasite most usually gains access to the human body by means of the alimentary canal, and only in some instances by penetrating the skin.

History. The parasite was first described by Dubini, of Milan, in 1843. It was later observed by Pruner and Bilharz in Egypt. Griesinger in 1854 first showed the true relation of the parasite to the disease then known as " cachexie aqueuse," or " Egyptian chlorosis," and now generally spoken of as "ankylostomiasis." In 1878 Grassi and Parona first showed that the presence of the worm could be detected in the living body by searching for the eggs in the dejecta. The St Gothard Tunnel epidemic of this disease in 1880 again drew general attention to the parasite and its effects, and within the last decade its wide diffusion and importance in certain countries have become more fully recognised.

[1] C. A. Bentley, M.D., *Brit. Med. Journ.* Jan. 25, 1902.
[2] *Journ. Trop. Med.* March 1, 1901.

Recent Geographical Distribution. The known distribution of *Ankylostomum duodenale* is a wide one, but it is probable that its actual distribution is still wider. It is only within quite recent years that its presence has been proved in many countries where it was not known before to exist. More detailed observations directed to this point, and particularly the systematic examination of the intestinal contents of living persons and of cadavera, will in the future probably show that this parasite is more widely spread throughout the world than is known to be the case at present.

Europe. The parasite was until recently unknown in the British Isles. But, in the latter part of 1902, Dr Haldane, in the course of an enquiry into the ventilation of Cornish mines, carried out on behalf of the Home Office, found that there was a serious outbreak of ankylostomiasis at the Dolcoath mine, Camborne. This appears to be the first mention of the disease in the British Isles[1].

In France it has been seen among miners at St Etienne in the Department of the Loire.

It has been observed also in Belgium and Germany. In 1899 ankylostomiasis was reported to be prevailing severely in the neighbourhood of Liège among the working population, and particularly among miners[2]. The source of the disease here was a matter of controversy. The most northerly places in Europe where the worm had been observed before were Cologne, and Dortmund in Westphalia; and some Belgian writers have believed that their country was infected by the parasite by means of workmen coming from the Cologne brickfields. German writers, on the other hand, assert that both Belgium and Cologne were invaded by means of infected labourers coming from the St Gothard tunnel. A third view is that advocated by Sonsino, who is of opinion that even before the famous St Gothard epidemic the ankylostomum was already "diffused among certain classes of workmen in the northern countries of Europe, such as France, Belgium, Germany, Holland, and Austria, as well as in Italy and in more southern countries[3]." I have not been able to confirm this mention of the existence of the parasite in Holland.

[1] *Brit. Med. Journ.* Dec. 13, 1902.
[2] *Revue Scientifique*, 1899, No. 22.
[3] *Janus*, March—April, 1900.

In miners working in the coal-mines of Hungary the affection is perhaps common. One Hungarian observer is quoted as having treated 470 patients of this class for ankylostomiasis, in a period the length of which is not stated[1].

The disease caused by this parasite has been observed on an extended scale in Italy. Hirsch has given references to its occurrence in many places in the northern part of the peninsula— as in the district of Treviso (Venetia), at Florence, Turin, and Cesena, and particularly in the provinces of Milan and Pavia in Lombardy. It was to this last centre of the disease that was ascribed the historical outbreak of ankylostomiasis among the labourers employed in making the St Gothard tunnel. On this occasion men working both on the Italian and on the Swiss sides of the mountain range were affected.

Asia. Information is entirely lacking as to the presence or absence of this worm in all the countries of Nearer Asia, as also in Siberia.

The parasite, on the other hand, is exceedingly common throughout the Indian peninsula and Ceylon. In recent years systematic inquiries have been made by many medical officers in India as to the presence of the worm in the patients under their charge, both in life and after death, and it has been found in a very large proportion of subjects so examined. Dobson, Rogers, and others have shown that in Assam, Bengal, and some other parts of India from 60 to 80 per cent. of the population harbour the parasite, in numbers varying from a very few up to one hundred or more, with apparently no symptoms of a serious nature. As the Sanitary Commissioner with the Government of India pointed out in a recent report, the apparent frequency of the ankylostomum in a given district depends largely on the keenness with which the medical officer in charge of the district prosecutes a search for its presence. Some observers have even succeeded in finding them in every patient examined for them. Thus Grün and Jacoby discovered them in the intestinal contents of every one of 78 Sinhalese whose excreta were searched; and in the case of six Madrasis the same results were obtained[2]. Close, on the other hand, as the result of 81 autopsies made at Budaon in the North-West Provinces, only

---

[1] *Janus*, 1899, p. 151.
[2] *Berliner Klin. Wochenschrift*, No. 36, 1896.

found the worm in 68 per cent. of males examined (28 cases), and 40 per cent. of females (16 cases); or 54 per cent. of both sexes[1]. In some of the hospitals in Ceylon it is believed that the majority of cases of dysentery, diarrhœa, and debility that come under treatment are due to the action of this parasite.

In convicts in the Andaman Islands the ankylostomum was found by one observer in the excreta of 75 out of 100 subjects examined for its presence[2]. In Burma the worm is widely prevalent. In Assam and the Malay Peninsula it has been detected, and, as it certainly exists in Java and Borneo, it is probable that it has a wide distribution in the East Indian Islands. It has also been found in Cochin China, in French Indo-China, and in Japan. In some parts of China it is probably very common.

Pani-ghao has been seen at Cachar, and Sylhet in Assam. The coolies employed in the tea plantations suffer from it to a very considerable extent.

Australasia. The parasite also exists in Australia, and ankylostomiasis is occasionally sufficiently severe to be a cause of death. Nine deaths from this cause occurred in Queensland in 1896. It has also been observed in some of the Pacific Islands; thus as long ago as 1876 it was found in some of the Fijian mountaineers and in recruits from the Solomon Islands (MacGregor).

Africa. Egypt has long been recognised as a country in which ankylostomum is a common parasite and ankylostomiasis a common disorder. It was as the result of work in a Cairo hospital that Griesinger first proved the relation of the worm to the anæmic and cachectic symptoms of the disease.

The disease is, on the other hand, said to be absent from the pathology of Massowa, on the western shores of the Red Sea. It is uncertain if it exists in Algeria, and there is apparently no mention of its occurrence in Tripoli or Morocco. The absence of any mention of this parasite, however, cannot be regarded as proof that it does not exist in those countries. Sonsino found that it was common in Gabes, in Southern Tunisia.

In Lagos, on the West Coast of Africa, ankylostomiasis is said to be common, and to be a cause of much of the mortality attributed to "dropsy" and "anæmia[3]." In Senegambia the

---

[1] *Indian Medical Gazette,* May, 1899.

[2] *Report of San. Commiss. with Gov. of India,* 1897.

[3] H. Strachan, L.R.C.P., M.R.C.S., *Journ. Trop. Med.* March, 1899.

disease is not rarely met with, and it is probable that later observations will show that it is widely diffused along the West Coast of Africa. The parasite is widely spread in the Belgian Congo, where it is a common cause of anæmia in both the white and black-skinned inhabitants. Guillemard also found it in Madeira.

The ankylostomum is known to occur in Mauritius, where it was first demonstrated by Lesur. It has also been observed at Mayotta, in the Comoro Islands, and on the Zanzibar coast.

The disease is not unknown in South Africa. The parasite was found to be present in the intestinal contents of some 45 European miners, working in the Kimberley and de Beers mines, during a period of ten months (1897–98). The previous record of the disease in this region is uncertain; but it is said to have been prevalent in Kimberley as early as 1890, beyond which date the statistics do not go. The disease here was much commoner among the men working underground than in those employed on the surface. Matthias, who described this outbreak, while regarding it as the first known instance of a serious epidemic of the affection in South Africa, was of opinion that it is probably endemic in many parts of that region[1].

North and Central America. It is apparently only in the southern portions of the United States that the ankylostomum has been observed in the continent of North America. Cases of the disease it produces have been reported from Louisiana, Alabama, and Georgia.

It has only recently been demonstrated that this nematode occurs in Central America. It had for some time been suspected by the medical officers in British Honduras that the parasite existed there, but it was only in August 1898 that at a post-mortem examination it was discovered in the intestinal contents. Later it was detected in living subjects. It is probably common in this colony. As the result of sixteen autopsies, it was found in abundance in eleven, in small numbers in two, and was entirely absent only in three. In the district known as Orange Walk it was almost as common as the ordinary round or thread-worm. Panighao or ground-itch is also very frequent in this colony.

The inhabitants of many of the West Indian Islands are affected by the worm. In Grenada it must be a very common

[1] *South African Medical Journal.* Quoted in the *Journ. Trop. Med.* October, 1898.

parasite, and a considerable number of cases of ankylostomiasis are yearly treated at the Colony Hospital. In St Lucia the disease caused by it is a very prominent one, and the Colonial Surgeon in a recent report (for 1898) expresses his belief that "every labourer in the island has either ankylostoma, ascaris lumbricoides, or oxyuris vermicularis." It is noteworthy, however, that a large proportion of the cases of ankylostomiasis treated in St Lucia occur in coolies imported from India. Thus in 1897, out of 252 cases of the disease no less than 208 were in Indian coolies[1]. In Puerto Rico anæmia is said to be the great scourge of the island, and it is believed to be due to the ankylostomum. In Antigua it is extremely common. In Trinidad both ankylostomiasis and pani-ghao are exceedingly prevalent and the cause of much sickness; and in Barbados and St Kitts the parasite is met with.

In Bermuda no cases of ankylostomiasis have apparently been seen, though owing to immigration from the adjoining West Indian Islands it is constantly threatened by an importation of the disease[2].

South America. Throughout British Guiana both the parasite and the disorder it gives rise to are widely spread. The affection takes a prominent position in the list of diseases observed in the colony, and is a not inconspicuous cause of death. The residents on the sugar estates are said to be principally affected by it. In French and Dutch Guiana the worm is also not uncommon. Hirsch has quoted a number of authorities for the existence of "malignant anæmia," apparently due to this parasite, in the past throughout Brazil, "where," he wrote, "it is prevalent over the whole of the country excepting the most southern or sub-tropical provinces—equally on the coast and in the valleys and elevated regions of the interior." It has also been seen in the valleys of the upper basin of the Marañon in northern Peru as well as among the natives of Sarayacu on the pampas of the Sacramento[3], and it is not unknown in Colombia and Venezuela.

Factors concerned in the Geographical Distribution. The ankylostomum is found mainly in hot or warm

[1] *Annual Report on the Hospitals and Charitable Institutions of St Lucia,* 1898.

[2] See the *Report of the Medical Officer of Health for Bermuda* for 1898.

[3] Hirsch, Vol. II. p. 314.

countries. It does not appear to have been observed as yet, even as an imported parasite, to the north of the latitude of Dortmund in Europe, or the Southern States in America; while as a cause of permanent, endemic disease it is almost confined to tropical or subtropical countries. The only country which forms an exception to this statement is Italy, where it is common. Within the tropical and subtropical zones the parasite has now been observed in a very large number of countries. Later observation may show that it is absent from none, as it requires looking for, and, as already stated, has recently been detected in countries where it was not previously suspected to exist.

The parasite has been met with at very varying heights above sea-level. It is said to be most common in marshy soil.

There appears to be little room for doubt that the soil plays an important part in the life-history of the animal and in its transmission to man. It was pointed out above that the ova of the worm leave the body in the intestinal excreta. Their further development takes place in muddy water, mud, or damp earth, while, for the final stage, and in order that they may again become sexually mature worms, they must enter the tissues of a human being. This probably takes place by means of foul drinking-water; or perhaps more commonly by being transferred in the act of eating from earth-soiled hands to the mouth; or, thirdly, by a direct penetration of the skin. Hence it is most common to find the affection in persons who work in the earth. In India and the colonies it is the labourers on tea, coffee, or sugar plantations who suffer most. In Belgium, Germany, Hungary, South Africa and elsewhere it has been the miners who have mostly harboured the worm. The outbreak in connection with the St Gothard tunnel was a marked example of the occurrence of ankylostomiasis in persons digging in the soil. Evidence of a similar character is forthcoming from nearly all countries where the parasite exists. In many instances, moreover, the entrance of the worm into the human stomach has been traced to the dangerous and disgusting habit of earth-eating or geophagy.

The spread of the worm and the disease caused by it over the earth's surface is brought about mainly, if not solely, by the movements of infected human beings. The St Gothard tunnel outbreak seems to have been due to the employment of Italian workmen from parts of Italy where the parasite is known to be common, and it has already been pointed out that the spread of the disease in Europe may plausibly be attributed to the movements of infected labourers coming from the St Gothard tunnel, or from the Cologne brickfields. It seems to have been frequently carried to the West Indies in the bodies of Indian coolies.

The distribution of the parasite is apparently little influenced by varying susceptibility in different races. So far as can be gathered it is probable that all races can become hosts of the worm in those countries where it occurs, and that any difference in the frequency of its occurrence between European and native, or between white and coloured races, is accounted for by difference in exposure to the risk of infection.

TRICHOTRACHELIDES.

5. TRICHOCEPHALUS DISPAR.

The Parasite. *Trichocephalus dispar*, or *T. hominis*, the whip-worm, is common in the human intestine, especially in the cæcum. The male and female are nearly the same size—from 35 to 50 mm.—but the latter are slightly the longer. Both have the long, filiform neck, expanding into the thickish body, but while in the male the latter is spirally coiled, in the female it is nearly straight. The oval, brownish eggs are easily recognisable by the clear blunt spines at each extremity. There is no inter-mediate host, the animal reaching the stomach of its human host while still in the egg.

Geographical Distribution. The whip-worm is of almost universal distribution, and is in some places very common. Mérat records that in ten years of post-mortem examinations in Paris he was always able to demonstrate the existence of this species to his pupils whenever he so desired, even in the case of subjects who had met with a violent death when in perfect health. It is known in the Malay islands equally with America; in Africa not less than in Europe. Accurate statistics, as in the case of many of the entozoa, are wanting as to the comparative frequency and distri-bution of this parasite, but it is perhaps more common in warm than cold climates.

6. TRICHINA SPIRALIS.

The Parasite. The adult worm lives in the small intestine. The embryos pass into the tissues and bore their way to the voluntary muscles, where they become encapsuled. They may be present in large numbers and give rise to serious and even fatal symptoms.

Geographical Distribution. *Europe.* Trichiniasis, the disease resulting from the presence of this worm in the tissues, has always been much more common in Germany than in any other part of the world, and particularly in North Germany. It has frequently been "epidemic" there. In South Germany it is less common but still by no means rare. In Great Britain it is not often seen now, but two small epidemics of it have been recorded in the past, one at Workington in Cumberland in 1871, and one on board a training ship in the Thames in 1879. It has been seen in France, as also in Denmark, Sweden, and Russia. In Russia it was prevalent to a considerable extent in St Petersburg, Moscow, Riga, and Lodz between the years 1873 and 1879. Small epidemics of trichiniasis have also occurred in Austria, Switzerland, and Spain; and isolated cases in Italy, Portugal, and Roumania. The worm is unknown in the Netherlands, and has apparently never been seen in Turkey.

Asia. From the very scanty records of trichiniasis in Asia, no more definite statement is possible than that it has been seen in Syria, in India, and in China.

Africa. The only mention of the parasite in Africa relates to its discovery at an autopsy made in Algiers in 1867.

America. A considerable number of epidemics of trichiniasis have been observed in the past in many parts of the United States, but mostly in New York and in other north-eastern States. Osler has stated that "the dissecting-room and post-mortem statistics show that from one-half to two per cent. of all bodies contain trichinæ," so that the parasite is apparently a far from rare one in America. There seems to be no mention of this entozoon or the disease caused by it in Central America, the West Indies, or South America.

Factors concerned in the Distribution. The trichina enters the human body solely from the consumption of raw or insufficiently cooked flesh of the pig. The encapsuled worm occurs in the muscles of the latter animal, as in those of man. When trichinous pork is eaten by man, the eggs escape in the alimentary canal, the embryos are set free, and if a sufficient number of them bore their way into the tissues they set up the disease known as trichiniasis.

Hence the disease is only found among pork-eating people, and principally among persons who eat the flesh—whether in the form of ham, of pork, or of sausages—in a raw state or imperfectly cooked. The frequent consumption of raw ham and of *wurst* in North Germany accounts for the frequency of the disease there. Thorough cooking is necessary to destroy the worm, and merely salting or smoking the flesh that contains it is not sufficient.

Trichiniasis has become a good deal rarer in the last twenty years, owing to increased precautions being taken to prevent the consumption of trichinous flesh. These were largely the result of the "scare" caused about the years 1875–1880 by the growing frequency of the disease in many parts of Europe and America. Systematic inspection of meat is now carried out in Germany, England, and elsewhere, and this has led to a marked diminution in its prevalence. In Holland, where very strict precautionary measures are carried out, the parasite is apparently non-existent.

FILARIÆ.

7. FILARIA MEDINENSIS (Guinea-worm).

The Parasite. The *Filaria* or *Dracunculus medinensis*, or Guinea-worm, varies in length from about one to three feet, or even longer. Only the female is known. She is found, in man, buried in the connective tissue of some part of the body—usually, but not always, the lower extremity. Each adult female worm contains an enormous number of embryos. These are ejected from her body at the sore where the creature "points," and probably live for a considerable time in water. Fedchenko has shown that the embryos enter the bodies of a certain fresh-water *Cyclops*, where they undergo part of their development.

The channel by which the larva enters the human system has been a subject of much discussion, and for the present it must be regarded as not finally determined whether it is swallowed in drinking-water, or penetrates the skin of the trunk or limbs, or (as seems probable) whether it may not follow both these paths. The presence of a living Guinea-worm in the tissues may not give rise to much trouble or discomfort for a considerable time. But sooner or later the worm pierces the skin, apparently in order to eject the young embryos. In most countries the natives who suffer from the filaria seize this opportunity to extract one end of the worm, and then proceed to wind it little by little, and day by day, round a stick or piece of wood until it comes away entire. Care is taken to prevent its rupture, for, should this occur, the myriads of embryos escape into the tissues and set up violent local and general reaction, often with disastrous results. Premature death of the worm, before she has "presented" at some surface point, may lead to abscess formation, or she may remain imbedded for years as an apparently harmless, often cretified, body.

History. The earliest positive reference to the existence of the Guinea-worm is apparently that of Plutarch, who quotes a description of it as occurring on the shores of the Red Sea by a writer who lived in the second century before Christ. A later account of it is that of Leonides, in the second century of the Christian era, who mentions it as met with in Ethiopia and India. Both these writers speak of the disease as caused by a living worm, but Galen, Soranus, and many of the Arabian physicians threw doubt upon this explanation of its causation, and suggested that the object extracted from the bodies of those affected was no worm at all, but a diseased vein, or nerve, or other tissue of the patient. But the right view as to its nature gradually prevailed, and, as a knowledge of zoology and the use of the microscope advanced, the exact characters of the worm and its position in the group of human parasites were finally established.

Geographical Distribution. *Europe.* The worm is unknown in Europe.

Asia. It has been seen in Syria, but there appears to be no record of it in Asia Minor. In Arabia it occurs with some frequency, particularly in Arabia Petræa, and on the coasts of the Hedjaz and the Yemen. In spite of the name *Filaria medinensis*, it is probably rare at Medina as an indigenous parasite, though common enough among the Mussulman pilgrims of all nationalities who yearly visit the Hedjaz.

It was formerly seen in the Kirghiz steppes and even to the north of the Caspian Sea, in so high a latitude as $47°$ N.[1] But recent writers on the pathology of the inhabitants of the Kirghiz steppes and of the Bashkirs make no mention of it   In Bokhara it was at one time very prevalent, and a not uncommon name for it was the Bokhara worm.   At the present day, however, it is much rarer than formerly.   Grekof, a Russian observer who has studied the prevailing diseases of the state of Bokhara, never saw a single person affected by the worm.   In many parts of Russian Turkestan it is frequently seen.   At Tashkent it is said to be rare, but in Jisakh and Karshi it is very common, or was a few decades ago, at which period it was apparently not very frequent in Samarkand[2].   It is endemic on the southern shores of Persia.

The Guinea-worm is very widely spread throughout the Indian peninsula, but it is almost entirely confined to the native inhabitants, and rarely attacks the European residents.   From the returns of the Indian native army and of Indian gaols some idea may be gathered of the frequency with which the parasite is met with.   In 1896 there were 534 cases of Guinea-worm affection in the native army, this being equivalent to 4·2 per thousand strength. In 1897 the corresponding figures were 632 cases, or 4·9 per thousand strength.   The largest proportion of the cases were reported from the south-eastern districts, from Rajputana, Central India, and Gujerat.   The parasite was also common in the Deccan, and it was not entirely absent from any of the geographical divisions of the peninsula.   The gaol returns indicated that the highest degree of prevalence was in the Deccan and Southern India.   In the years in question there were no cases in the European army.   The Rajputana States of Mewar and Marwar have always been known as regions where the worm is particularly common, and this appears to be the case at the present day[3]. But it is frequently seen in many other parts of India, and would appear to be rare only in the North-West Provinces, in Lower Bengal, and in the coast belts of the Madras Presidency (Northern Circars, Carnatic, and Cochin).

[1] Kämpfer, quoted by Hirsch; apparently the observation was made in the early years of the 18th century.

[2] *Turkestan*, by Eugene Schuyler, London, 1876.   Vol. I. p. 147.

[3] *The Western Rajputana States* (Adams).   Also Thornhill; *First Indian Medical Congress*, 1894.   The Mewar Bhil contingent of the Indian army have always suffered considerably from the worm.

In Ceylon the dracunculus is frequently seen, especially among the labourers employed on the coffee estates[1].

In Farther India, the Malay Peninsula and Archipelago, China and Japan, the worm is apparently quite unknown as an indigenous parasite. It is said to have been seen in Java in former years but only among troops imported from Africa, and as soon as the transfer of such troops ceased the worm disappeared from the island.

The parasite is apparently unknown in British New Guinea (MacGregor), and probably also in other parts of the island, and there is no mention of its occurrence in Australia, New Zealand, or any of the Pacific Islands.

Africa. As the name most usually given to this parasite implies, one of its principal centres of diffusion is on the shores of the Gulf of Guinea. It is found, indeed, throughout the greater part of the west coast of the African continent, but appears to be most common along the northern shores of the Gulf. In Senegambia, particularly in the basin of the Senegal river, it is not rarely seen. But it is still more generally diffused on the Ivory and Gold Coasts, in Dahomey, and in the basins of the Niger and Gaboon. In the Cameroon territory, on the shores of the Bight of Biafra, it occurs somewhat less frequently; and further south, towards the mouth of the Congo, is rare and perhaps unknown as a true indigenous parasite. In the Congo State the Congolese natives are said not to be affected by it, though it is well known among labourers imported to the Congo from Accra, Elmina, and Lagos.

In British East Africa the Guinea-worm is not infrequently met with. It abounds in the Nile Valley below the Albert Nyanza, but does not exist in Unyoro or Uganda, except in people who have lately come from the infected district. In Abyssinia it is said to occur only on the coast. In Nubia, Kordofan, and Darfur it is endemic; but it is not so in either Tunis, Egypt, or the Greater Sudan, and if it occurs in any of these it is only as an imported parasite. There is no mention of the worm among the parasites of South Africa.

America. Cases of Guinea-worm were at one time far more numerous in North and South America and the West Indies than they are at the present day. As the negro slave was the principal

[1] Thornhill.

carrier of the parasite from Africa to the west, so with the cessation of the slave-trade the worm has become exceedingly rare in America. Examples are, however, seen from time to time in the United States. Osler mentions two instances, apparently of indigenous origin, occurring respectively in Fortress Monroe, Virginia, and in Philadelphia; but he adds that a majority of the cases mentioned in American journals have been imported. Occasional cases, apparently also due to importation, are seen in Trinidad and other West Indian Islands.

The only mention of the disease in recent times in South America relates to the existence of a small endemic centre in the village of Feira de Santa Anna in the Bahia province of Brazil.

Factors concerned in the Geographical Distribution. From the above summary of the recent distribution of Guinea-worm it will be seen that the parasite is almost confined to tropical or subtropical countries. The only exception to this rule has been the mention, by a writer at the beginning of the 18th century, of the existence of the worm to the north of the Caspian Sea, in latitude 47° N. It is very doubtful if it still exists there at the present day, and its most northerly limit now seems to be between 40° and 45° N. in Russian Turkestan.

While mainly a parasite of hot or warm countries, it is not found in all such climates. It is most abundant in the great tropical belt of Africa, which is believed by many to be the original habitat of the worm. It is also fairly common in parts of Western and Central Asia, and in India, but further east than India it is unknown, at least as an indigenous parasite. In the New World, though it has been seen as far north as Philadelphia (40° N.), it is only in the tropical zone that it is anything but extremely rare, and in this zone it is found in a few places only, and more often than not as an imported parasite.

Warmth may be regarded therefore as necessary for the prevalence of Guinea-worm. A certain degree of moisture is equally needed. The parasite, as already pointed out, is believed to pass one portion of its existence in certain lowly fresh-water organisms, and its distribution over the world's surface is no doubt largely, though probably not wholly, determined by the distribution of these particular crustaceans. That is to say, that, although this particular fresh-water cyclops may have a much wider distribution than the Guinea-worm, the latter, on the other hand, cannot multiply and flourish as an indigenous parasite in the absence of the cyclops. As water, then, is a primary necessity for the cyclops, it is equally a primary necessity for the multiplication of the worm. It is also, there is good reason to believe, the principal means by which the parasite gains access to the tissues of man.

The relation between the frequency of the parasite in any district and the amount of rainfall is probably not, as some have thought, a direct one. It has varied greatly in different countries and at different times. In both Africa and India it has some-times been unusually frequent after heavy rains, and at other times in hot, dry weather. The former condition would no doubt favour the multiplication of the cyclops, and indirectly that of the worm itself. The latter condition would lead to a lessening of the amount of water in pools, water-courses, or other sources of water-supply ; and as the water became more concentrated the chances of any given quantity of it containing embryos of the worm would increase, and in like manner the chances of human infection would also become greater.

Dracunculus shows no definite relations to the geological character of the soil. It is probably little influenced by race. The white man always suffers, it is true, to a far less extent than the native in those countries where the worm is common, but this is apparently solely due to the fact that he is far less exposed to chance infection, in that his limbs are protected by clothes and boots, and that he is more careful in attending to the source of his drinking-water.

Guinea-worm has been spread over the earth's surface mainly by the movements of infected human beings, and these have most frequently been persons of the negro race. It is very probable, as pointed out above, that the worm is originally of African origin, and that if it is now endemic elsewhere, it is so only as the result of importation from Africa. Within quite recent times carriage of the worm from this continent has been frequently observed. Its appearance in Java as the result of the transfer of troops from Africa, and its disappearance from the island as soon as this military movement ceased, was mentioned above. In like manner the appearance of the worm in Egypt, Tunisia, the West Indies, Guiana, and Brazil, has been repeatedly traced to importation by infected negroes from those parts of Africa where it is common. At the time when negro slaves were being carried to America in large numbers such instances of importation of the Guinea-worm were far more frequent than now.

8. FILARIA BANCROFTI (*Filaria Nocturna, and Filaria Sanguinis Hominis*).

The Parasite. The *Filaria bancrofti* is believed to be the adult form of the *Filaria nocturna*. The latter name is now most commonly employed to designate a parasite which was formerly known as the *Filaria sanguinis hominis*—the parasite associated with a very large number of pathological conditions, all of which are characterised by more or less severe affection of the lymphatic system. Of these various diseases, due more or less directly to the *Filaria nocturna*, Manson has enumerated the following :—ulcers ; lymphangitis ; varicose groin glands ; varicose axillary glands ; lymph scrotum ; cutaneous and deep lymphatic varix ; orchitis ; chyluria ; elephantiasis of the leg, scrotum, vulva, arm, mamma and elsewhere ; chylous dropsy of the tunica vaginalis ; chylous ascites ; chylous diarrhœa, and probably other forms of disease depending on obstruction or varicosity of the lymphatics, or on death of the parent filariæ.

Of all these varied affections elephantiasis is by far the most widely spread, and will consequently be most frequently referred to here. It would be interesting to make a comparative study of the frequency and geographical distribution of each of the diseases caused by this filaria ; but, even if space permitted, it is doubtful if the necessary material exists for such a study ; and it must suffice here to point out briefly what is known as to the distribution either of the filaria itself, or of the group of diseases as a whole with which it is associated.

The embryonic form, or *Filaria nocturna*, is a microscopic worm, about $\frac{1}{80}$ of an inch in length and $\frac{1}{3000}$ or $\frac{1}{3500}$ of an inch in breadth. Its diameter is thus about that of a red blood corpuscle. For the structural characters of this worm the reader is referred to any recent text-book upon tropical diseases or helminthology. The parasite is found in the blood, as a rule, only during the night hours, say from 5 or 6 o'clock in the evening, to 8 or 9 o'clock the next morning. During the day it is absent from the general circulation, and Manson has recently made an observation which seems to show that during these hours the parasite retires to the larger arteries near the heart, and to the lungs.

Like most of the human blood-worms this filaria requires an intermediate host to complete its life cycle. The intermediate host in this instance is a certain species (or perhaps several species) of mosquito. The mosquito bites a filarial patient, and in the act of sucking his blood the filariæ pass into the stomach of the insect. Thence they migrate to the thoracic muscles. A mosquito's life is a short one; after depositing her eggs (it is only the female mosquito that bites and that can therefore carry this or other filariæ) on the surface of stagnant water, she dies. The filaria, now about $\frac{1}{16}$ inch in length, escapes from her dead body and sooner or later may be taken into the body of a human being again, by means of drinking-water. It is believed to pass by boring from the human stomach into the lymphatics, where it becomes sexually mature, and where new generations of embryo filariæ are produced. While this has hitherto been most generally believed to be the method by which the parasite gains access to the human frame, it has to be added that a second method is also possible. This is by the passage of the filaria from the mosquito to the human tissues while the mosquito is "biting," and support is lent to this view by the recent demonstration by Low and James of the presence of filariæ in the proboscis of the insect. Grassi and Noë and the Liverpool Expedition to Nigeria have also confirmed this observation. The mature parasite (called the *Filaria bancrofti* after Dr Bancroft, of Brisbane, who first observed them) is a long worm, from three to four inches in length, and about the thickness of a hair. It has been found in cyst-like dilatations of the distal lymphatics; in lymphatic varices; in the larger lymphatic trunks between the glands; in the glands themselves, and in the thoracic duct.

History. The early history of the many affections that are now known to be caused by the *Filaria sanguinis hominis* is obscure. Of the many filarial disorders elephantiasis is the only one for the history of which materials exist to any considerable extent, and even their value is largely diminished by the confusion which existed in the earlier records between leprosy (*elephantiasis Græcorum*) and the disease now under consideration (*elephan-*

tiasis Arabum). There is every reason to believe that the filarial elephantiasis is of considerable antiquity, and there seems no good reason to doubt that the other filarial disorders enumerated above are of the same age. So far as the scanty historical records go, they are said to show that the distribution of the disease in medieval times was very much what it is at the present day. The *Filaria nocturna*, which is now regarded as almost certainly the cause of elephantiasis and the other affections named, was first discovered by Demarquay in 1863 in a case of chylous dropsy of the tunica vaginalis. In 1866 Wücherer found the same parasite in the urine in several cases of chyluria. In the same year Lewis found it in India, and a little later proved that the normal habitat of the embryo parasite (which he called the *Filaria sanguinis hominis*) was the blood of man. Recently Manson and others have devoted much work to this and other filariæ, and it is now generally accepted that the *Filaria nocturna*, the *Filaria bancrofti*, and elephantiasis and the other affections named, bear to each other the definite relations which have been described above.

Recent Geographical Distribution. *Europe.* True elephantiasis and the other forms of filarial disease are not known with certainty to be indigenous in any parts of Europe. Examples of elephantiasis are occasionally seen in several European countries in persons who have contracted the disease elsewhere, and from time to time cases are recorded, even in subjects who have never left their own country, in which the clinical symptoms are indistinguishable from those of elephantiasis. But in these instances there is a lack of evidence of the filarial origin of the disease, and for the present there is no reason to believe that *Filaria nocturna* exists in any part of Europe. The countries in which cases of the kind just mentioned have been seen are England[1], Scotland, Ireland, the south of France, Greece, Turkey, Portugal, and Spain.

Asia. Filarial disease exists in Syria and in Arabia, where examples of elephantiasis are seen, though probably with no great

[1] Sir Dyce Duckworth, at a meeting of the Clinical Society on Jan. 13, 1899, showed a case of leg elephantiasis in a girl aged 20, who had never left Yorkshire. The blood in this case was practically normal, and no filariæ were ever found. Dr Toogood at the same meeting mentioned two similar cases occurring in anæmic girls.

frequency. I am aware of no positive evidence of the presence of the filaria in either the Caucasus, Central Asiatic Russia, or Siberia, but should not be surprised to learn that it exists in Transcaspia, Turkestan, and the other Russian possessions in Central Asia.

In India elephantiasis is a comparatively frequent disorder. It is said to be most prevalent along the shores of Lower Bengal and of Orissa, in Pondicherry, at a few places on the Coromandel Coast (particularly Tanjore at the extreme south), and most of all on the Malabar Coast. In Travancore and Cochin it seems to be extremely common, and different observers state that from 10 to 50 per cent. of the population are affected by it in the latter town. It is rarer in the Deccan and Upper India.

In many parts of Ceylon elephantiasis is a common disorder. Here, as in India, many of the village inhabitants live near tanks, drink tank water, and eat fish taken from tanks, and in this way no doubt the embryo filariæ gain access to their bodies. Elephantiasis, or *Barawa*, as it is called by the Sinhalese, has been shown to be most prevalent in villages near the swampy ground of old, abandoned tanks. The scrotal, labial, and pectoral forms of the affection are said to be rare here, and so also is chyluria[1].

These diseases are also met with in many of the East Indian islands. Elephantiasis is said to prevail most in the Lampong country of Sumatra, in Banka, the Nicobars and the Philippines, while in Java, Amboina and other islands it is less often seen.

Filarial disease is far from rare in the Malay Peninsula, and every year a certain number of cases are treated in the hospitals of Penang and Singapore. Among the Annamese it is common.

In China the *Filaria sanguinis* has a wide distribution, and in some parts affects a considerable proportion of the inhabitants[2]. Its degree of frequency appears to vary rather widely. It is common in the Shan-tung province; is "extremely prevalent" in Huang Yen, a division of the prefecture of T'ai-chou; and is rather frequently seen at Ningpo. In the Fo-kien province the disease is perhaps more general than in any other. At Amoy it accounts for 2 per cent. of the patients treated in hospital.

[1] Thornhill, *Journ. Trop. Med.* Dec. 1898.
[2] See *Chinese Customs Medical Reports*, Nos. 45 and 55. Also Coltman, *op. cit.*

In Japan elephantiasis is seen, but apparently is rather less common than on the southern and south-eastern coasts of China.

The observation, originally made by Myers in 1881[1], that the natives of Formosa, in spite of the proximity of the island to China, are to a great extent free from this parasite, has been recently confirmed. In the course of a seven years' residence in Formosa (1886 to 1892) Rennie only saw six cases of elephantiasis in the Tamsui native hospital. He adds that the disease appears to be exceedingly rare in persons born in, and never resident out of the island, and that this is probably to be explained by the curious fact that the filaria-nurturing mosquito is not native to Formosa.

Australasia. In British New Guinea elephantiasis is endemic, but is not apparently a very common disorder (MacGregor).

Filariasis is certainly no rarity in Fiji. Many cases of elephantiasis and of "lymphangitis" of filarial origin are yearly treated in the colonial hospital at Suva, the patients being mostly native Fijians, but occasionally Europeans are afflicted by the disease. In Samoa it is said that about every second individual is the subject of elephantiasis; in Huahine Island of the Society group the proportion rises to seven-tenths of the adult male inhabitants, and it is also of frequent occurrence in Tahiti and Raiatea.

In the Solomon Islands elephantiasis is not uncommon. It here usually assumes the form of lymph-scrotum, but cases in which the limbs are affected are sometimes seen. It is noteworthy that the natives of these islands attribute the disease to the water of certain streams[2]. In New Caledonia, Wallis Island, and the Gambier group, the disease is very prevalent; and it exists, but to a less extent, in the Marquesas and Hawaii groups. In the Tonga or Friendly Islands nearly one-third of the population are said to harbour the filaria.

It has been sometimes stated that elephantiasis does not exist in Australia or New Zealand as an endemic disease, but recent mention has been made of the occurrence of a case of filariasis in Queensland[3], and one death from it was recorded in the Report of the Registrar-General for this colony for 1896.

[1] *Chinese Customs Medical Reports*, No. 21 and No. 23.

[2] Guppy, *op. cit.*

[3] Bancroft, *Journ. Trop. Med.* Jan. 1900, p. 149.

Africa. Filarial disease is not infrequently seen in Egypt, and is most common in Lower Egypt, particularly at Rosetta, Damietta, and some other places. It is found also in swampy valleys in Abyssinia, and in parts of Tunis and Algeria. It is however apparently not met with at Massowa on the Red Sea. It is also said to be absent from the upper valleys of the Nile; but on the other hand it occurs as an endemic in some regions to the south of the Sahara, such as Bornu, Segu Sicorro, and Ogowe.

Turning to the western part of the continent a wide degree of prevalence of elephantiasis is found in many districts. It appears to be a common disorder in Senegambia. On the Gold Coast cases are not infrequently seen; and at Lagos the disease undoubtedly exists, though it is not common. It is found also on the Benin coast, the Spice coast, and at Sierra Leone, while in the Cameroons the filaria is by no means a rare parasite. In the Belgian Congo it is said that "the majority" of the negro inhabitants harbour these hæmatozoa. Elephantiasis and the other forms of filarial disease are at any rate very commonly seen in the possession. Elephantiasis is very common in Uganda[1], and filarial diseases certainly exist in Zanzibar and Mombasa. In British Central Africa, Daniels found that the *Filaria nocturna* and elephantiasis occurred with considerable frequency on the Lower Shire River, on the Zambesi, and at the north end of Lake Nyassa, while in the Shire and Angoni highlands, on the Upper Shire, at the south end of Lake Nyassa, and in Likoma neither the parasite nor the disease appeared to exist[2].

Throughout South Africa elephantiasis is apparently very rare, though not wholly unknown. One observer met with only two cases of the disease in a total of 40,000 patients in the course of seven years. Of the two patients, one came from the Kentani district, in the native territory of the Transkei, and the other from a locality not named[3].

The parasite is comparatively frequently seen in Madagascar, and at Mayotta in the Comoro Islands. In Mauritius elephantiasis, chyluria, and varicose groin glands are also common affections.

America. Filarial diseases appear to be absent from the

[1] Cook, *ibid.* June 1, 1901.

[2] *Ibid.* June 15, 1901.

[3] *South African Medical Journal.* Quoted in *Journ. Trop. Med.* March, 1899.

pathology of Canada and of the Northern States of the American Union; but in some of the Southern States they seem to exist rather extensively. They are also found with some frequency in Mexico, and in the Central American States of Nicaragua, Costa Rica, and Panama.

Filarial disorders are of frequent occurrence in many of the West Indian islands. In Cuba they are very common[1], and they are found with considerable frequency in Barbados, Martinique, Guadeloupe, Trinidad, St Vincent, St Bartholomew, St Kitts, and Montserrat. In St Lucia, on the other hand, the *Filaria nocturna* is exceedingly rare, and elephantiasis is consequently not a common disorder. A few cases are however treated yearly at the hospital and poor asylum.

In South America elephantiasis is said to be endemic on the coast and marshy plains of Guiana, in some parts of Brazil, and on the shores of Colombia, Venezuela, and Peru. In British Guiana filarial affections appear to be particularly common. Daniels found the *Filaria nocturna* in 52 out of 348 persons examined for it at Georgetown, and Ozzard found it in 28 out of 100 persons in whom it was looked for at New Amsterdam. It is also of comparatively frequent occurrence in some parts of Brazil; one observer has found it at Bahia in about one-twelfth of the persons examined for it.

Factors determining the Geographical Distribution. From the brief account already given of the group of diseases associated with the *Filaria nocturna*, it will be seen that a number of factors are primarily necessary to their production. There is firstly the specific filaria itself; there is secondly the specific mosquito (or perhaps several species of mosquito) in which it undergoes its extra-human phase; and there must obviously be a susceptible population, and a set of physical conditions favourable both to the life of the parasite, to the life of its insect host, and to its easy passage from the latter to the human subject, and again from the human subject to the mosquito.

As to the filaria itself it is very generally admitted that the parasite above described, and it alone, is the specific cause of the group of diseases here dealt with. Less is known as to the specific character of any particular mosquito or mosquitoes as the filarial

---

[1] *Commercial Cuba*, by W. J. Clark, London, 1899.

host. Bancroft, in Australia, has found *Culex ciliaris* (Linn.) an efficient host for the filaria, and has shown that two other mosquitoes, *Culex notoscriptus* (Skuse) and *Culex annulirostris* (Skuse) are not hospitable. The *Anopheles musivus* behaved in a manner which showed that it might harbour the filaria, and James has shown that *Anopheles rossi* and perhaps other species of Anopheles may act in a similar manner.

In regard to the presence of a susceptible population, it is for the present uncertain whether individuals or races do or do not differ in their susceptibility to filariasis, provided they are under exactly the same conditions of exposure to the risk of contracting it. It is true that the dark-skinned races in India, the islands off the East Coast of Africa, and the West Indies seem to be most liable to it, while white men living in countries where the filaria exists appear to largely escape it. But this may to a great extent, if not wholly, be explained by the different conditions under which the two classes live. In India, for example, the native constantly bathes in and drinks the water of tanks and streams which may well contain the eggs and embryo forms of the *Filaria nocturna*. He rarely, if ever, employs a mosquito net or a punkah at night, and is thus far more exposed than the average European to the bites of mosquitoes, so that whether the filaria finds its way to the human tissues by drinking-water, by penetrating the skin during bathing, or by means of a mosquito bite, it is certain that the native in India (and to a great extent the argument applies elsewhere) is much more exposed to its attacks than the European. Moreover when Europeans are attacked in India they are usually of the poorer classes and "country-born," and consequently their mode of life, in regard to their degree of exposure to the risk of filariasis, may be presumed more closely to resemble that of the native Indian than that of the average European. Maitland[1], indeed, after 24 years' residence in India, has stated that he never met with nor heard of a case in a well-to-do European in that country. It is certain that filariasis may occur under suitable conditions in a great variety of races, including the various Hindu, Mohammedan, and other races of the Indian peninsula, the East Indians, Malays, Annamese, Chinese, Pacific Islanders, African and American negroes, as also in persons of mixed races, as the half-castes or half-breeds in India and in South America.

[1] Brit. Med. Assoc. Annual Meeting, 1900.

It is probable therefore that race is in itself a less important factor in the geographical distribution of filariasis than certain physical conditions which affect, favourably or otherwise, the life-history of the parasite to which it is due. For a general prevalence of filariasis it is clear that the physical conditions, such as temperature, moisture etc., must be favourable not only to the life of the filaria itself, but also to that of the mosquito in which one portion of its existence is spent. A considerable degree of warmth appears to be essential to either the one or the other, or perhaps to both. It is at least certain that the principal endemic areas of filariasis both in the Old and New World are in or near the tropics. True filarial elephantiasis appears to be unknown as an indigenous disorder in regions more than 35° to the north or 30° to the south of the equator. A moist atmosphere seems also to favour its prevalence, as it is commoner on the coasts than in the interior of countries, and is most frequent in low-lying damp localities on the shores of rivers and streams, while it is absent or rare in higher and drier places. This is, no doubt, largely to be explained by the important part which water plays in the life-histories both of the filaria itself and of its mosquito-host. Both may be said to pass one stage of their existence in water, and, as already stated, it is highly probable that it is by means of filaria-containing water that the parasite usually gains access to the human being.

Finally, conditions which enable the filaria to pass with ease from the human to the insect host, and *vice versâ*, must be present if filariasis is to be a common disorder. Some of these conditions have been already named. They are, on the one hand, the exposure of filarial subjects to the bites of mosquitoes, with the risk or certainty of contaminating the mosquito ; and on the other the exposure of healthy subjects either to the bites of filaria-containing mosquitoes, or to the risk of filarial infection by drinking or bathing in contaminated water. Such risks are clearly to a great extent avoidable ; and if, in countries where filariasis exists, every individual, whether the subject of it or not, took every possible precaution to avoid being bitten by mosquitoes, and to drink and bathe in no water the purity of which was not above suspicion, it is more than probable that filariasis would cease to exist. A useful auxiliary measure would be the destruction of the mosquito, and the success recently met with in the war against the malaria-bearing Anopheles may warrant the hope that a similar campaign against the insect which carries *Filaria nocturna* would lead to equally satisfactory results.

9. FILARIA DIURNA.

The Parasite. *Filaria diurna* is the name applied by Manson to a parasite which resembles *Filaria nocturna* in all its physical characters, but differs from it in that it is found in the blood, not during the night, but during the day. His observations were made on two negro patients, one of whom came from Old Calabar and the other from the Congo ; and as one of them had some years previously had a *Filaria loa* in one of his eyes it seemed possible that the *Filaria diurna* might be the embryo form of the sexually mature *Filaria loa*.

The Geographical Distribution of this parasite appears to be limited to certain regions of Africa. Manson believes it to be extremely common among the natives of the Lower Niger, where one-fourth of the population may be affected by it. Brault states that it is found especially at Old Calabar, along the Guinea Coast, and on the Congo River. Cook has seen two cases of it in Uganda[1]. It is not known to occasion any pathological condition.

10. FILARIA LOA.

The Parasite. *Filaria loa* is a small thread-like worm, the two sexes of which occur in different individuals. The male is about an inch in length, the female rather longer. The worm is found in the connective tissues, and seems to wander about from one part of the body to another. It is most usual to find it under the conjunctiva, or in the connective tissue in the neighbourhood of the eye, or just under the skin on the bridge of the nose. It is believed to pass by wriggling movements under the skin from one eye to the other, or to other parts of the body.

The life-history of this filaria is unknown. Manson suggests that *Filaria diurna*, above described, may be the embryonic form of *Filaria loa*. No embryos have been found in the peripheral blood of persons the subjects of the loa worm, and nothing is

---

[1] *Journ. Trop. Med.* June 1, 1901.

known as to the way in which it leaves the human host, or whether (as is probable) it passes part of its life-cycle in some insect host, as is the case with *Filaria nocturna*.

Geographical Distribution. This parasite is only found in the West Coast region of Africa. It is probably rather common here and particularly in the neighbourhood of Old Calabar. It is also found on the Congo; and a case has been seen as far inland as Talagouga on the Ogowe, 120 miles from the coast[1]. The late Miss Mary Kingsley, the well-known African traveller, stated that on the Ogowe river nearly every inhabitant suffers from the worm. It is said to be endemic from a few degrees north of the equator to about ten degrees south. It seems to be absent from the Cameroon territory. Formerly it is said to have been seen from time to time among negroes in America, but since the slave-trade from Africa to the New World has ceased, this parasite is no longer found on the other side of the Atlantic.

11. FILARIA DEMARQUAII.

Filaria demarquaii resembles *Filaria nocturna* and *Filaria diurna* in shape, but, at least in dried specimens, appears to be only about one-half their size. It was first described by Manson, who found it in 10 specimens of blood (out of 152 examined) taken from natives of St Vincent in the West Indies. It shows no periodicity, and is present in the peripheral circulation day and night. Parasites apparently identical with this were also found in blood taken from natives of St Lucia in the West Indies, and from Papuans. Brault states that this filaria occurs throughout the Lower Niger district in West Africa. Manson is inclined to think that *F. demarquaii* and *F. ozzardi* of British Guiana are the same species.

12. FILARIA OZZARDI.

Filaria ozzardi is the name provisionally applied by Manson to a nematode embryo observed by Ozzard, Daniels, and himself in the blood of certain aboriginal Carib Indians in British Guiana. These people are said never to suffer from elephantiasis, and *Filaria nocturna* has never been seen in their blood,

---

[1] Yarr, *Journ. Trop. Med.* Feb. 1899.

so that the filaria now under consideration is probably of a separate species. It closely resembles *Filaria demarquaii*, but about every sixth parasite in a microscopical specimen is seen to have a pointed instead of a blunt tail. The exact relation of this filaria to both *Filaria demarquaii* and *Filaria perstans* is uncertain. Daniels found the mature form of the worm in the bodies of two Demerara Indians, whose blood during life had contained both blunt- and sharp-tailed embryos. The mature worm was about 3 inches long and very slender—about one-third the diameter of *Filaria bancrofti*.

<h3>13. FILARIA PERSTANS.</h3>

The Parasite. *Filaria perstans* is much smaller than either *F. diurna* or *nocturna*, and presents some other physical characteristics—such as the absence of a sheath, an abruptly rounded caudal end, and the power of locomotion—which markedly distinguish it from the other filariæ[1]. The mature form of the worm is unknown, and its pathological effects are uncertain. Manson has however suggested that it may possibly be the cause of negro lethargy and of some of the forms of African craw-craw. Both these diseases are treated separately elsewhere in this volume, and here it will suffice to state what is known of the geographical distribution of the filaria itself.

Geographical Distribution. This parasite appears to be confined wholly to the negro races of Africa. Manson found it in natives from Old Calabar and the Congo, both from the coast and from the interior. Firket, of Liège, also observed it in Congo natives. Brault states that it is confined to the west coast between the mouths of the Senegal and the south of Benguela; but, as just stated, it is found not only on the coast, but also in the interior. Nor is it confined wholly to the western part of the continent, for Cook records that he has found it in three instances in the Uganda Protectorate[2]. The only mention of the parasite outside Africa is a reference by Manson to its doubtful occurrence in Demerara Caribs. The latter observer failed to find it in West Indian nègroes, and in natives of the Malay Peninsula, China, and India.

[1] For a full description of this and the other filariæ the reader is referred to Dr Patrick Manson's *Tropical Diseases*.

[2] *Journ. Trop. Med.* June 1, 1901.

14. FILARIA MAGALHAESI.

Filaria magalhaesi is the name given to a filaria, of which the adult form only has been observed in a single case by Prof. Magalhães of Rio de Janeiro. It differs essentially from *F. bancrofti* (the adult form of *F. nocturna*), but it has been conjectured that it may be the mature form of *F. demarquaii* or *F. ozzardi*. Its life-history and pathological effects are quite unknown. The single case in which it was observed was that of a child who died in Rio de Janeiro.

PARASITIC ARACHNIDÆ AND INSECTS.

A. ARACHNIDÆ.

1. PENTASTOMUM CONSTRICTUM.

The Parasite. This is an elongated larval form, varying in length from half an inch to an inch and a half, and with 23 rings on the abdomen. It occurs in cysts in the liver and occasionally in the lungs. It may be the cause of serious and even fatal illness.

Geographical Distribution. The parasite was first observed in Cairo, where it has only been seen in the bodies of negroes. It is uncertain therefore whether it is truly indigenous to Egypt, or imported from the Sudan. It has also been found in a soldier who died at Bathurst, Gambia; in an African slave coming from St Helena; and in some natives of the Belgian Congo[1]. Flint mentions a case in Missouri, in which from 75 to 100 of the parasites were expectorated. Osler refers to this case and adds that in 1869 he saw a specimen which had been passed in the urine by a patient of a Toronto practitioner. This would seem to indicate that the parasite may exist in the urinary passages as well as in the liver or lungs, and that it may occur in northern latitudes.

The parasite seems to be mainly one of the negro race. Pruner, of Egypt, found it in a giraffe.

2. SARCOPTES SCABIEI.

Geographical Distribution. Itch is common among the very dirty classes of almost all countries. In *Europe* it is said to

[1] *Congrès Nat. d'Hygiène......de la Belgique et du Congo*, 1897.

be particularly frequent in Iceland and the Faröe Islands, Norway, Poland, Russia, some parts of Eastern Germany, France, and the Ionian Islands.

In *Asia* it is widely seen throughout Siberia and Kamchatka; in the Caucasus, and in Russian Central Asia; in India and China; in Persia and Arabia. In fact it is practically universal throughout the continent. In some of the *Australasian* islands, on the other hand, the disease is rare, and it is said to be unknown in New Guinea.

In *Africa* itch is apparently met with in many parts of the continent, but some confusion exists here between scabies and craw-craw, a disease which is dealt with elsewhere.

In *America* scabies is less common in Canada and the United States than in Europe. It is extremely frequent in Brazil and Peru, and is probably endemic in Central America and the West Indies.

Factors concerned in the Distribution. Scabies is found in all races and in all climates. The parasite, dirt, and neglect are apparently the only essential factors for the prevalence of the disease.

B. INSECTS.

The Flea, *Pulex irritans*, has a world-wide distribution. In itself it is harmless to man, but as a possible carrier of the infection of plague and other diseases it has a certain pathological importance. The Bed-bug, *Cimex lectularius*, has also an extremely wide distribution and is of exactly the same pathological significance as the flea.

The larvæ of various flies have occasionally been found in the human alimentary canal and given rise to some trouble and more alarm. In other instances they have penetrated the nasal passages, and even passed into the frontal sinuses and antrum. In such cases ulcerative processes and caries may be set up, leading to perforation into the cranial cavity, with resulting meningitis and death. Such troublesome larvæ have been observed mostly in hot countries, such as India, Senegambia, Mexico, Central America, Cayenne, Colombia, Brazil, and the Argentine Republic.

PULEX PENETRANS.

The Parasite. The jigger, chigo, or sand-flea (*Pulex penetrans*) is an insect closely resembling, but rather smaller than, the common domestic flea. It usually penetrates the skin of the feet and causes no little trouble from the inflammation subsequently set up. The female alone enters the human tissues in this way. Immediately after impregnation she burrows through the skin of the first warm-blooded animal that comes in her way. As ovulation progresses she becomes enormously distended, and, in consequence, much inflammation with pus formation is set up in the tissues of the host. The skin usually ulcerates, and the insect is in this way expelled. While the foot is the usual site for the chigo, it has been known to penetrate the skin of the scrotum, penis, perineum, thighs, hands, and face. The affection is not in itself a grave one, but the subject of it suffers greatly and is for the time seriously disabled; and as the resulting ulcer lends itself readily to septic absorption, or to the development of tropical phagadæna or other serious complications, the chigo disease is not one that can be lightly regarded.

Usually only one or two chigos are present in a patient, but occasionally they are very numerous; probably the largest number ever seen in a single individual was 280, removed from a boy at one time[1].

History. The chigo is said to have been known for over a century in South America and the West Indies. About the middle of the nineteenth century it seems to have been introduced to the West Coast of Africa in the sand ballast of a ship coming from Brazil. In Africa the pest at first spread slowly. Stanley's last expedition and the increase of movement brought about by advancing civilisation are said to have been the means of spreading it more rapidly over the continent. It became widely diffused over the Congo Free State; by 1891 it had crossed as far as Lake Nyassa and appeared in Uganda; in 1894 it had reached the Shire Highlands, and some time later it was carried to the East Coast of Africa. In 1899 it appeared for the first time in Madagascar, and in the same year it was imported to India.

[1] *Janus*, 1899, p. 439.

Recent Geographical Distribution. *America.* At the present day the chigo is found with great frequency in some of the West Indian Islands and in the neighbouring portion of South America. In Cuba it is particularly troublesome. "The worst of all the pests on the island," writes a recent author, "is the *nigua* or jigger, so-called, sometimes described as a cross between Satan and the wood-tick, and the torment from their attacks is said to be worse than would be possible from either of their reputed progenitors[1]." In Jamaica many cases of chigo come under treatment at the hospitals each year. It occurs in Mexico, both in the eastern and western provinces, and even at considerable heights. It has also been seen in Honduras, Guatemala, Costa Rica, San Salvador, and Panama, and has been met with further north in Florida. In French Guiana and in Brazil it is very widely prevalent, and it is known to occur in the tropical parts of Paraguay, Chile, Colombia, and Peru.

Africa. In Africa the chigo is found through the entire tropical zone from west to east. The easterly spread of the pest across this continent has been briefly described above, and is a phenomenon of very great interest. Whether it was truly imported here from America by means of infected sand from Brazil or not, it is certain that it has appeared in Africa only in quite recent times. It is now firmly established in the Cameroon district[2], and in the whole of the Congo valley. It has travelled as far south as Mashonaland, where it is widely disseminated and a cause of much trouble to the natives[3]. In Uganda it was first seen in 1891. Its appearance here was attributed indirectly to Stanley's expedition in relief of Emin Pasha. "This expedition," writes Dr Cook, on medical mission work in Uganda, "did not actually enter Uganda, but many Baganda who had been driven out of their country by a revolution met with Stanley's expedition and brought back the jiggers to Uganda. Coming up country in 1896," the same author continues, "we met the jiggers just before we got to Machakos. Following the caravan route they slowly made their way down to the coast, which they reached about 1899. Till the natives realised their nature they did great damage and caused the loss of many toes, etc.[4]" This course of the

[1] *Commercial Cuba*, by W. J. Clark, London, 1899.
[2] Plehn.   [3] Todd, *Journ. Trop. Med.* Nov. 1900.
[4] *Ibid.* June 1st, 1901.

parasite is more or less confirmed by other writers. It appears, however, that it must have reached the coast earlier than 1899.

Sir Harry Johnston states that it reached the shores of Lake Tanganyika first; and then those of Lake Nyassa, where it was first heard of at Karonga about 1891. It was in Southern Nyassa in 1892, and in 1894 was already a serious pest at Zomba and throughout the Shire highlands. It appeared at Chinde on the sea-coast in 1895. Lionel Decle states that he found it all over the shores of Lake Tanganyika, in Uha, Unyamwezi, Usukuma; on the shores of the Victoria Nyanza; and throughout Uganda and Usoga[1]. The southern part of the Masai country was alone free.

There is other proof that the pest must have been well established on the east coast before 1899. Already at the beginning of January of that year the medical officer in charge of the 4th Bombay Infantry, then stationed in East Africa, reported to Bombay that the chigo was firmly established at Mombasa[2]. It probably reached Zanzibar about the same time, and a little later it appeared in Madagascar. In a communication made to the French Académie de Médicine, on Jan. 30th, 1900, M. Blanchard stated that the chigo had reached that island some nine or ten months previously. It was first seen on the west coast (nearest the continent) and its appearance coincided with the arrival there of native troops, Senegalese and Haussas, from the mainland. Owing to the sandy nature of the coast it spread very rapidly through the island and appears now to be definitely established there.

Asia. The first known introduction of the jigger pest into India in 1899 was traced to the return from Africa to Bombay of the Indian regiment named above. That regiment left Mombasa on December 3rd, 1898, and in the course of the voyage between Aden and Bombay 26 fully developed cases of chigo were discovered among the men and followers. Many native coolies returning in the ship were also affected by it, and subsequently some indigenous cases were reported from Bombay. It is probable, however, that the insect had been imported to India before this incident without being recognised. The alarm created by its discovery led to a careful system of inspection of arrivals from Africa, and in 1899 as many as 168 cases of chigo were found on

---

[1] *Three Years in Savage Africa*, London, 1898.
[2] Kilkelly, *Indian Medical Gazette*, April, 1899.

vessels arriving in Bombay. In 1900 this figure fell to 125 and in 1901 to 46. Sir B. Franklin, to whom I am indebted for these figures, writes to me that "before 1899 cases must have occurred among the coolies returning from Africa, but so far as is known without the pest establishing a foot-hold in this country ; and it is doubtful if the damp coast districts to which the parasite would first be brought are suited to the development of the larval stage." The parasite does not seem to have spread beyond Bombay, and this is attributed to the great range of temperature and the monsoon-conditions to which the greater part of India is liable and which are probably inimical to its naturalisation.

The marked diminution in the number of imported cases is attributed to stringent measures to prevent the departure from Africa of persons the subject of chigo, and to ensure the prompt recognition and treatment of cases developing during the voyage or after arrival in India.

Factors determining the Geographical Distribution. The chigo appears to thrive best in a warm, dry, sandy soil, and has up to the present only been seen in countries with a high average temperature. In America it is found exclusively between the limits of 30° N. and 30° S. In Africa it is essentially a tropical— almost an equatorial pest. In Asia it has at present only been reported from tropical India. The future history of the insect may show that it can exist elsewhere. Its diffusion over the earth's surface seems to be principally, if not entirely, dependent upon human movements. It has been transported by infected in- dividuals over a whole continent in a few decades, in the wake of explorers and traders ; and within the last few years it has been carried by sea across the Indian Ocean. On two occasions it has been brought to Europe, though without in any way becoming endemic here. The first occasion was in 1866 at Toulon, when a sailor was found to be the subject of chigo, on board a ship which had left Cayenne six months before. The second instance occurred in Paris, in a man who had arrived from Pernambuco. Possibly the movements of animals may aid in its diffusion, for it is said that not only human beings, but dogs, cats, and pigs, and even birds and mice may harbour the insect.

It is almost certain that the habit of going about bare-footed enormously increases the chances of becoming infected in a region where the chigo exists. But the wearing of boots does not ensure absolute protection. It is possible that, like the ordinary domestic flea, the chigo may be carried about in clothing as well as buried in the tissues. In Cuba there is a general belief that it thrives and breeds in woollen garments, while a person dressed entirely in cotton or linen is rarely troubled with it.

The chigo is usually most troublesome during the period immediately after its fresh introduction into a country. It may then attack so many persons as really to deserve the name of a pest, and one of no little gravity. In Uganda, for example, for the first two years after its introduction, "its ravages were so terrible that it assumed the dimensions of a veritable plague. Many hundreds of people lost one or more toes, and I should think" (wrote the Principal Medical Officer to the Protectorate) "half the population were temporarily incapacitated for longer or shorter periods." Decle has described a village in Africa in which not a single man, woman, or child was free from chigo ulcers. As time goes on the chigo perhaps diminishes in numbers, or the people learn to protect themselves from it, and the pest becomes much less frequent and serious.

LIST OF PRINCIPAL AUTHORITIES.

GENERAL.

1. *Handbook of Geographical and Historical Pathology.* By Dr August Hirsch. (New Sydenham Society.)
2. *Geographical Pathology.* By Andrew Davidson, M.D., F.R.C.P. Ed. Edin. and Lond., 1892.
3. *Hygiene and Diseases of Warm Climates.* Edited by Andrew Davidson, M.D., F.R.C.P. Ed. Edin. and Lond., 1893.
4. *Tropical Diseases.* By Patrick Manson, M.D., LL.D. London, 1899.
5. *Traité de Climatologie Médicale.* Par le Dr H. C. Lombard. Paris, 1877–80.
6. *Traité Pratique des Maladies des Pays-Chauds.* Par le Dr Fernand Roux. Paris, 1888–89.
7. *Précis de Pathologie Exotique* (Maladies des Pays-Chauds et des Pays-Froids). Par A. le Dantec. Paris, 1900.
8. *Traité des Maladies des Pays-Chauds.* Par F. Brault. Paris, 1900.
9. *Bulletin de l'Institut International de Statistique.* Vol. x. Rome, 1897.

EUROPE.

10. *Annual Reports* (and Supplements to *Annual Reports*) of the *Registrar-General for England and Wales*.
11. *Annual Reports* (and Supplements to *Annual Reports*) of the *Registrar-General for Scotland*.
12. *Annual Reports* (and Supplements to *Annual Reports*) of the *Registrar-General for Ireland*.
13. *Annual Reports of the Medical Officer to the Local Government Board*, and Supplements to the same.

14. *Statistique Sanitaire des Villes de France et d'Algérie, pendant l'année* 1898, *et Tableaux Récapitulatifs des années* 1886–98. Melun, 1899.
15. *Annuaire Statistique de la Belgique.*
16. *Annuaire Statistique des Pays-Bas.*
17. *Statistik Aarbog.* Copenhagen, 1899.
18. *Statistik Aarbog for Kongeriget Norge.* 1897.
19. *Bidrag Till Sveriges Officiela Statistik* (för år 1896).
20. *Annual Reports* (in Russian) *of the Medical Department of the Russian Ministry of the Interior* (from 1887, the year of their commencement, to 1895). St Petersburg.
21. *The Bashkirs: An Ethnographical and Sanitary-Anthropological Investigation* (in Russian). By Dr P. Nikolski. St Petersburg, 1899.
22. *Österreichisches Statistisches Handbuch.* Wien, 1899.
23. *Ungarisches Statistisches Jahrbuch.* 1897.
24. *Statistisches Jahrbuch der Schweiz.* 1899.
25. *Annuario Statistico Italiano.* 1898.
26. *Annuaire Statistique du Royaume de Serbie.* Belgrade, 1898.
27. *The Faröe Islands.* By J. Russell Jeaffreson, F.R.G.S. London, 1898.
28. *Colonial Hospital Reports for Gibraltar.*
29. *Public Health Reports* (Annual). Malta.

ASIA.

30. *Reports* (mostly unprinted) *to the Ottoman Board of Health, Constantinople,* from Sanitary Officers in all parts of the Turkish Empire (including European Turkey, Asia Minor, Syria, Arabia, the Turco-Persian Frontier and Tripoli (Africa)).
31. *Statistiki Kavkazskoi Patologii.* I. I. Pantiukhof. Tiflis, 1898.
32. *Annual Reports of the Sanitary Commissioner with the Government of India.*
33. *Annual Reports of the Sanitary Commissioners with the separate Governments of India.*
34. *Annual Reports on Sanitary Measures in India.*
35. *Transactions of the* 1st *Indian Medical Congress.* Calcutta, December, 1894.
36. *The Western Rajputana States. A Medico-Topographical and General Account of Marwar, Sirohi, and Jaisalmir.* By Lt.-Col. A. Adams, I.M.S. London, 1899.
37. *Ceylon Administration Reports.* Reports of the Principal Civil Medical Officer and Inspector-General of Civil Hospitals.
38. The Straits Settlements. *Medical Reports* (Annual).

39. Labuan. *Colonial Hospital Report* (unprinted).
40. *Five Years in Siam.* By H. Warington Smyth, M.A., LL.B. London, 1898.
41. *Medical Reports of the Imperial Maritime Customs of China.*
42. Hongkong. *Principal Civil Medical Officer's Reports* (Annual).
43. *The Chinese; their Present and Future, Medical, Political, and Social.* By Robert Coltman, Junr., M.D. 1891.
44. *Annual Report of the Central Sanitary Bureau attached to the Home Department of the Imperial Japanese Government.* Tokio, 1897.

AUSTRALASIA.

45. *Annual Reports of British New Guinea.*
46. Queensland. *Reports of the Registrar-General.*
47. New South Wales. *Annual Reports.*
48. Victoria. *Reports of the Board of Public Health.*
49. South Australia. *Statistical Registers.*
50. Western Australia. *Statistical Registers.*
51. Tasmania. *Central Board of Health Reports.*
52. *Statistics of New Zealand.* 1898.
53. Blue Book of Fiji.
54. *The Solomon Islands and their Natives.* By H. B. Guppy, M.B., F.G.S., late Surgeon R.N. London, 1887.

AFRICA.

55. Egypt. *Reports, Statistical and Medical, of the Ministry of the Interior.* Cairo.
56. *La Tunisie. Histoire et Description.* L'Association Française pour l'avancement des Sciences. Paris et Nancy, 1896.
57. *The Arab and the African.* S. Tristram Pruen, M.D. London, 1891.
58. Sierra Leone. *Colonial Hospital Reports.*
59. Lagos. *Colonial Hospital Reports.*
60. Gold Coast Colony. *Sanitary and Medical Reports.*
61. Gambia. *Colonial Hospital Reports.*
62. Congrès National d'Hygiène et de Climatologie Médicale de la Belgique et du Congo, 1897. *Comptes Rendus.*
63. Cape of Good Hope. *Reports on the Government and State-aided Hospitals and Asylums.* (Including Reports on the Public Health, etc.)
64. *Reports of the Durban Hospital.*

65. Mauritius. *Reports of the Medical Inspector.*
66. St Helena. *Colonial Hospital Reports* (unprinted).

NORTH AMERICA.

67. *Reports of the Board of Health of the Province of Quebec.*
68.   „   „   „   „   „   „   *Nova Scotia.*
69.   „   „   „   „   „   „   *New Brunswick.*
70.   „   „   „   „   „   „   *Manitoba.*
71.   „   „   „   „   „   „   *British Columbia.*
72.   „   „   „   „   „   „   *Ontario.*
73. *Public Health Reports.* Issued by the Supervising Surgeon-General, Marine Hospital Service. Washington.
74. *Mortality Statistics in the United States for the year* 1897.
75. *Transactions of the Congress of American Physicians and Surgeons.* Washington, 1894.
76. *25th Annual Report of the Health Department of the City of Boston.*
77. *28th Annual Report of the State Board of Health of Massachusetts.* 1897.
78. *Annual Report of the State Board of Health of the State of Michigan.* 1897.
79. *The Principles and Practice of Medicine.* By W. Osler, M.D., LL.D., F.R.S. Edinburgh and London, 1901.

CENTRAL AMERICA AND THE WEST INDIES.

80. British Honduras. *Report of the Colonial Surgeon.* 1899.
81. Guatemala. *Annuario Estadistico.* 1898.
82. Jamaica. *Reports of the Island Medical Department.*
83. Grenada. *Reports of Medical Officers.*
84. Bermuda. *Reports of the Medical Officer of Health, and of the Registrar-General.*
85. Virgin Islands. *Annual Reports of the Commissioner.*
86. Antigua. *Reports of the Registrar-General.*
87. St Lucia. *Annual Reports on the Hospitals and Sanitary Institutions.*
88. Bahamas Blue Book.
89. St Vincent. *Medical Reports.*
90. Barbados. *Reports of the Poor Law Inspector.*
91. Leeward Islands. *Colonial Hospital Reports.*
92. Trinidad. *Annual Reports of the Surgeon-General.*

SOUTH AMERICA.

93. British Guiana. *Colonial Reports.*
94. British Guiana. *Annual Reports of the Surgeon-General.*
95. *Anuario Estadistico de la Provincia de Buenos Aires, año* 1897. La Plata, 1899.
96. *Anuario Estadistico de la Republica Oriental de Uruguay.* Montevideo, 1898.
97. The Falkland Islands. *Colonial Reports.*